METALLURGY OF SEMICONDUCTOR MATERIALS

SEMICONDUCTORS COMMITTEE *

H. E. Bridgers, *Chairman*
D. C. Jillson, *Past Chairman*
F. L. Vogel, *Vice-Chairman*
G. E. Brock, *Secretary*
J. B. Schroeder, *Treasurer*

INSTITUTE OF METALS DIVISION *

J. H. Jackson, *Chairman*
David Swan, *Senior Vice-Chairman*
R. L. Smith, *Vice-Chairman*
D. C. Johnston, *Secretary-Treasurer*

SOUTHERN CALIFORNIA SECTION, AIME *

J. O. McCaldin, *Chairman*
W. H. Crutchfield, *Vice-Chairman*
A. A. Lewis, *Treasurer*
W. W. Wright, Jr., *Secretary*

THE METALLURGICAL SOCIETY OF AIME *

J. S. Smart, Jr., *President*
K. L. Fetters, *Vice-President*
T. D. Jones, *Treasurer*
R. W. Sherman, *Secretary*

* OFFICERS AT TIME OF THE CONFERENCE.

METALLURGICAL SOCIETY CONFERENCES

VOLUME

15

METALLURGY OF SEMICONDUCTOR MATERIALS

Proceedings
of a Technical Conference
sponsored by the
Semiconductors Committee
of the
Institute of Metals Division,
The Metallurgical Society,
and
Southern California Section,
American Institute of Mining,
Metallurgical, and Petroleum Engineers

LOS ANGELES, CALIFORNIA, AUGUST 30—SEPTEMBER 1, 1961

Edited by JOHN B. SCHROEDER
Natick, Massachusetts

INTERSCIENCE PUBLISHERS
a division of John Wiley & Sons / New York • London

AMERICAN INSTITUTE OF MINING, METALLURGICAL, AND PETROLEUM ENGINEERS, INC.

LIBRARY OF CONGRESS CATALOG CARD NO. 62-12209

JOHN WILEY & SONS, INC.
440 PARK AVENUE SOUTH, NEW YORK 16, NEW YORK

PRINTED IN THE UNITED STATES OF AMERICA

PREFACE

This volume includes the papers and discussions which were presented at the 1961 Conference on the Metallurgy of Semiconductor Materials at the Ambassador Hotel, Los Angeles, California from August 30 to September 1, 1961. The Conference was sponsored jointly by the Semiconductor Committee* of the Institute of Metals Division, The Metallurgical Society, American Institute of Mining, Metallurgical, and Petroleum Engineers and the Southern California Section of AIME.

The 1961 Conference was the third annual meeting sponsored by the Committee and its location is significant. The decision to hold the Conference away from the Boston area reflects the Committee's awareness of its responsibility to provide a truely National forum for the presentation of timely data on materials.

Introductory remarks were made at the opening session by J. O. McCaldin on behalf of the Southern California Section, by H. E. Bridgers as Chairman of the Semiconductor Committee, and by R. McNaughton, President of AIME.

The Conference was divided into five half-day sessions; chronologically these were: one on Thermionic Energy Conversion, one on Thermoelectric Energy Conversion, two on Microelectronics, and one on Recent Technical Developments. Section I of this volume includes the papers from the Microelectronics Sessions with five of the ten Recent Technical Development papers interspersed without recognition. While many of the short papers have obvious application to fields far removed from Microelectronics, this arrangement seemed to offer the best coherence. The inversion operation was applied to the order of presentation so as not to delay publication because of some difficulties which arose in connection with what is now Section III. Section II, on Thermoelectricity, was retained in its original position. Both Sections I and II contain results of work which had been continued beyond that reported at the earlier Conferences on Semiconductor Metallurgy (volumes 5 and 12 in this series) as well as completely new work.

The session on Thermionic Converters, Section III, was the first in this country to deal exclusively with the criterion for selecting materials capable of withstanding the severe conditions encountered in thermionic converters. During the intermission of this session several attendees requested that additional phenomenological background be presented. Chairman J. J. Connelly responded with an impromptu talk which, after some editing, has been retained in this volume.

W. V. Wright acted as Program Chairman for the Conference and is to be thanked for the excellent job he did in attending to the myriad details. I would

*Because of the continually expanding scope of the Committee's activity its name was changed to the Electronic Materials Committee after the meeting.

also like to acknowledge the cooperation of the authors and discussers in the editing of the discussions, the invaluable assistance of my colleagues in reviewing several of the manuscripts, and the aid of my wife, Lynn, in preparing portions of the volume and for tolerating the drastic changes in my daily schedule as a result of this undertaking.

John B. Schroeder

Solid State Materials Corp.
East Natick, Massachusetts
March, 1962

CONTRIBUTORS

C. C. ALLEN, Texas Instruments, Inc., Dallas, Texas
J. APPEL, General Atomic, San Diego, California
H. BASSECHES, Bell Telephone Laboratories, Inc., Allentown, Pennsylvania
S. F. BEVACQUA, General Electric Company, Syracuse, New York
J. T. BROWN, Westinghouse Electric Corporation, Pittsburgh, Pennsylvania
E. G. BYLANDER, Texas Instruments, Inc., Dallas, Texas
F. D. CARPENTER, General Dynamics Corporation, San Diego, California
F. L. CARTER, Westinghouse Research Laboratories, Pittsburgh, Pennsylvania
J. J. CONNELLY, Office of Naval Research, Washington, D.C.
W. J. CORRIGAN, Motorola Semiconductor Products Division, Phoenix, Arizona
G. C. DELLA PERGOLA, Westinghouse Research Laboratories, Pittsburgh, Penn.
J. W. FAUST, Jr., Westinghouse Research Laboratories, Pittsburgh, Pennsylvania
A. E. FEIN, Westinghouse Research Laboratories, Pittsburgh, Pennsylvania
S. S. FLASCHEN, Motorola, Inc., Phoenix, Arizona
R. GLANG, IBM Command Control Center, Kingston, New York
R. J. GNAEDINGER, Jr., Motorola, Inc., Phoenix, Arizona
A. GOETZBERGER, Shockley Transistor, Palo Alto, California
J. M. GOLDEY, Bell Telephone Laboratories, Inc., Murray Hill, New Jersey
P. GOODMAN, Allied Research Associates, Inc., Boston, Massachusetts
M. GOTTLIEB, Westinghouse Research Laboratories, Pittsburgh, Pennsylvania
L. N. GROSSMAN, General Electric Company, Pleasanton, California
O. HAASE, Bell Telephone Laboratories, Inc., Murray Hill, New Jersey
L. K. HANSEN, Atomics International, Canoga Park, California
R. R. HEIKES, Westinghouse Research Laboratories, Pittsburgh, Pennsylvania
N. HOLONYAK, General Electric Company, Syracuse, New York
H. HOMONOFF, Allied Research Associates, Inc., Boston, Massachusetts
R. G. HUDSON, General Dynamics Corporation, San Diego, California
D. C. JILLSON, General Electric Company, Syracuse, New York
H. F. JOHN, Westinghouse Research Laboratories, Pittsburgh, Pennsylvania
R. G. H. JOHNSON, Westinghouse Electric Corporation, Pittsburgh, Pennsylvania
A. I. KAZNOFF, General Electric Company, Pleasanton, California
G. KEMENY, Westinghouse Research Laboratories, Pittsburgh, Pennsylvania
S. W. KURNICK, General Atomic, San Diego, California

T. B. LIGHT, Bell Telephone Laboratories, Inc., Murray Hill, New Jersey
J. P. McHUGH, Westinghouse Research Laboratories, Pittsburgh, Pennsylvania
R. C. MANZ, Bell Telephone Laboratories, Inc., Murray Hill, New Jersey
P. H. MILLER, Jr., General Atomic, San Diego, California
R. C. MILLER, Westinghouse Research Laboratories, Pittsburgh, Pennsylvania
R. C. NEWMAN, Associated Electrical Industries, Aldermaston, Berkshire, England
S. O'HARA, Westinghouse Research Laboratories, Pittsburgh, Pennsylvania
S. A. PRUSSIN, MonoSilicon, Inc., Gardena, California
N. S. RASOR, Atomics International, Canoga Park, California
F. REIZMAN, Bell Telephone Laboratories, Inc., Allentown, Pennsylvania
D. A. RESNIK, Bell & Howell Research Center, Pasadena, California
L. S. RICHARDSON, Westinghouse Research Laboratories, Pittsburgh, Pennsylvania
J. M. SCHULTZ, Westinghouse Research Laboratories, Pittsburgh, Pennsylvania
W. SHOCKLEY, Shockley Transistor, Palo Alto, California
P. STEPHAS, General Electric Company, Pleasanton, California
C. O. THOMAS, Bell Telephone Laboratories, Inc., Murray Hill, New Jersey
W. A. TILLER, Westinghouse Research Laboratories, Pittsburgh, Pennsylvania
S. K. TUNG, Bell Telephone Laboratories, Inc., Allentown, Pennsylvania
T. P. TURNBULL, Massachusetts Institute of Technology, Lexington, Massachusetts
E. S. WAJDA, IBM Command Control Center, Kingston, New York
J. WAKEFIELD, Associated Electrical Industries, Aldermaston, Berkshire, England
E. P. WAREKOIS, Massachusetts Institute of Technology, Lexington, Massachusetts
A. F. WEINBERG, General Dynamics Corporation, San Diego, California
R. K. WILLARDSON, Bell & Howell Research Center, Pasadena, California
L. YANG, General Dynamics Corporation, San Diego, California
S. M. ZALAR, Raytheon Company, Waltham, Massachusetts
S. A. ZEITMAN, Westinghouse Research Laboratories, Pittsburgh, Pennsylvania
R. ZOLLWEG, Westinghouse Research Laboratories, Pittsburgh, Pennsylvania

CONTENTS

PART I: MATERIALS AND TECHNIQUES FOR MICROELECTRONICS

PART II: MATERIALS FOR THERMOELECTRIC ENERGY CONVERSION

PART III: MATERIALS FOR THERMIONIC ENERGY CONVERSION

PART I: MATERIALS AND TECHNIQUES FOR MICROELECTRONICS

• •

Integrated Semiconductor Circuits

J. M. GOLDEY

Bell Telephone Laboratories, Inc., Murray Hill, New Jersey

Abstract

In this paper various approaches to semiconductor integrated circuits will be described and the relative merits of each in meeting the objectives of lower systems cost, improved system performance, and higher system reliability will be assessed. The assessment will be made from both the systems and the device viewpoint.

Early data on multiple diodes and multiple transistors will be discussed with emphasis on reliability and performance. Multiple diode data suggest that reliability is determined by the number of encapsulations rather than the number of devices if all devices are equally stressed. In certain critical circuits a performance improvement results when similar devices are fabricated on a common substrate. Additional improvement in performance may result by packaging components which interact closely with each other in a common encapsulation.

The results and concepts described above suggest a scheme of intergration that will lead to better performance, improved reliability, and lower costs. The implementation of this scheme for a specific logic circuit, used as a vehicle, will be described. From this it will be inferred that the method is generally applicable to electronic systems of the digital type.

INTRODUCTION

Throughout the past two years there have been a number of talks and articles on why we are, should be, and, in some cases, shouldn't be interested in integrated semiconductor circuits. The major conclusions of many authors are that the primary benefits that may be achieved from the integrated circuit approach are lower systems cost, improved systems performance, and higher systems reliability.

In this paper these "whys" of integrated circuits will be translated into "whats" and "hows." After some preliminary discussion, a specific proposal will be made as to what, in the opinion of the author, is the best way to proceed toward attaining these goals with the present state of the art.

DISCUSSION

One approach to integrated circuits that has been suggested is to use devices which perform functions more complex than those that can be achieved in individual present-day devices such as diodes, transistors, etc. When devices of this type become available, they will most certainly find wide application, but since, with few exceptions, they are not available today, they do not provide the means by which the expected benefits can be realized.

It has also been suggested that the right approach is integration of a small number of standard Boolean circuits from which a variety of systems could and would be built. This approach is different from that mentioned above in that it can be done today. This fact in itself, however, does not justify its use. Such an approach has a number of disadvantages. If the building blocks are designed to render the highest performance in terms of gain and speed needed in the system then the use of these blocks in other parts of the system may be wasteful. If lower performance building blocks are chosen, on the otherhand, then it may be necessary to use several of these to perform a function that could have been done with a single high performance circuit. An additional argument against the use of standard building blocks is based on the fact that many subsystem functions can be built with far fewer components if they are designed directly.

In a recent paper Rice (1) has given an example of the difficulties involved when one attempts to build an electronic computer from a few basic building blocks. The original objective was to design the bulk of the logic circuits for the computer using eight standard Boolean circuit packages. When the drawings were trurned over to the manufacturing area, there were however 486 instead of eight circuit packages. Although the system could easilty have been built using the eight packages, it had better performance at lower cost by using the 486 circuits than it would have had with eight. We may conclude that, at least for many systems, the use of a small number of standard Boolean circuits as building blocks, integrated or not, is not the correct approach.

Before progressing further, let us consider what is really needed in order to meet the objectives of lower systems cost, improved systems performance, and higher systems reliability.

From the systems point of view flexibility must be maintained to permit optimum design, to a given performance, if cost reduction is to result. This is almost a truism, yet still not recognized by many, for if standardization on a relatively small number of specific building blocks does anything, it takes away the flexibility of design so vital to the systems designer.

System performance is, of course, determined to a large degree by device performance and, in addition, for optimization requires clever circuit design as well. In most systems there are a number of special circuits which require pairs or larger multiples of like devices with closely matched characteristics over wide temperature ranges. Thus schemes of integration which provide devices with such characteristics will lead to better system performance at lower cost. Several examples are described below.

Systems reliability is determined by device reliability and by the margins and redundancy incorporated into the design. Any integration technique, to be useful, must provide reliability at least equal to and preferably better than that which can be obtained by the use of conventional individual components. Of nearly equal importance is the development of reliable evaluation techniques such as stress aging which provide information on failure laws and on the correlation of failures of different devices with usage. Such data will lead to improved systems reliability because it will provide the system designer information which will permit him to optimize his design from margin and redundancy considerations.

From the viewpoint of the device designer, flexibility must also be maintained in order to enable him to take advantage rapidly of important advances in technology. If device manufacturing costs are to be reduced, the use of processes which are not excessively difficult to control is an important factor.

A few examples of integrated circuits which are with us today are described below. They may not be particularly elegant, but they are, nevertheless, useful, relatively inexpensive, and highly reliable.

Fig. 1. Multiple diode. This figure illustrates a multiple of six silicon computer diodes fabricated on a common substrate. Lead wires are thermocompression bonded to the mesas and to the external leads of the eight pin header. The larger mesa in the center is used only in bonding the wafer to the header.

As a first example consider a parallel combination of silicon computer diodes. This multiple diode is illustrated in Figure 1. A group of six diodes with one side common is fabricated on a common substrate and bonded to a header. Lead wires are attached, by thermocompression bonding in this case, to the individual diodes and the external lead wires. The fabrication procedure for these diodes is essentially indentical to that for single diodes of the same type. The only differences are that the die includes six diodes rather than one and that a large central mesa is included so that header bonding may be accomplished without contacting any of the electrically active regions. A major advantage lies in the use of a single header which one expects to lead to lower costs. In order to realize this benefit, however, a satisfactorily high yield must be achieved. Results at Bell Telephone Laboratories show that this can be the case and, more specifically, that the yields are much higher than would be expected if drop-outs were statistically independent. For statistical independence of drop-outs, the yield of a multiple of six diodes would be equal to the yield of the diodes tested individually raised to the sixth power. Actual yields on the above multiple diodes (6 out of 6) have run about Y^3. The reasons for this improvement are clear. The diodes in the multiple are immediately adjacent to one another through all stages of fabrication and thus receive common processing throughout manufacture. Furthermore, some of the major factors controlling yield are functions of the encapsulation process and not of the individual diode.

Results of stress aging of these diodes have been recently described by Howard and Hare (2). The important results of that report were the following:

1. The reliability of multiple diodes is approximately equal to that of single diodes of the same type.

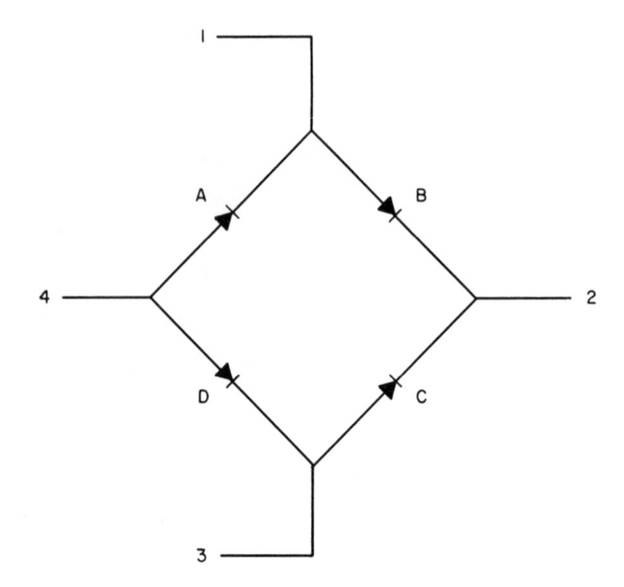

Fig. 2. Diode bridge circuit. When used as a modulator the performance of this circuit is determined by the match of the forward characteristics of diodes A and D and of B and C. Values of match in forward voltage required vary from 0.1 to 100 mv depending on the application.

2. All diodes within a common encapsulation fail at essentially the same stress level.

3. Failure, when it does occur, is most likely caused by contamination arising from the can itself.

These results imply that no reliability penalty is incurred and that the opportunity for a systems reliability improvement is present.

Another diode example is in order. Figure 2 shows a bridge circuit which is commonly used as a modulator. The performance of this circuit is determined by the closeness of the match between the forward characteristics of the diode pairs. It has been common practice for several years now for device manufacturers to make individual diodes and then to search for matching pairs by the very laborious method of measuring and cataloging large numbers of diodes. Once found, matching pairs are packaged together as a unit. Currently, laboratory-made multiple diodes are being evaluated for suitability as elements in bridge circuits. Preliminary results, based on a sample of 100 cans, six diodes per can are as follows. For a random choice of two diodes per can, the median difference in forward voltage is 4.3 mv and 88% of the pairs have a difference of less than 10 mv. For the best matched pair in each can, the median difference is 0.35 mv in forward voltage; 100% of the pairs are matched to 10 mv and 82% to 1 mv. In addition to the close match at room temperature, a good match over a wide temperature range is expected because of the intimate thermal contact between the diodes of the pair.

The usefulness of multiple devices is by no means restricted to single junction diodes. As another example consider transistors, which are already being packaged as multiples in a variety of configurations. Several examples are shown in Figure 3 including an "or" gate where both emitters and collectors are tied together, a chopper where collectors are tied together, and a differential amplifier where all leads are brought out independently. In most of these configurations, the closer the characteristics of the two transistors

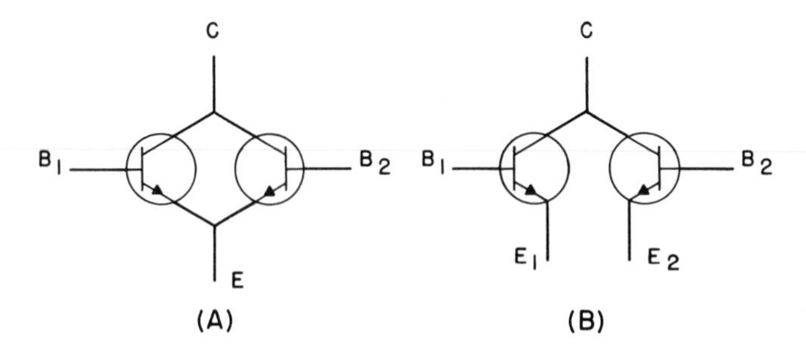

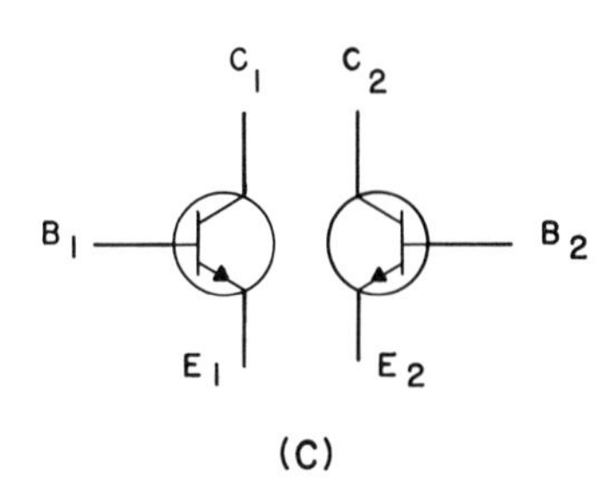

Fig. 3. Transistor multiples. Figure 3A illustrates an "or" gate common to many logic circuits. The emitters and collectors of the two transistors are connected together while the base leads are separate. Figure 3B is a chopper used to convert dc to ac. In this case the collectors are common while both emitters and bases are brought out separately. Figure 3C illustrates the use of transistors in a differential amplifier where all leads are brought out independently.

are matched, the higher the performance of the circuit. Improved performance at lower cost results by using two transistors fabricated together and mounted in the same package.

The p-n-p-n diode, one of the few devices currently available, incidentally, which performs a complex function directly, will serve as a final illustration of the multiple device. In one type of electronic switching system under development at Bell Telephone Laboratories, switching is carried out on a time division basis. In such a system, high-speed low-loss switching elements are needed as gates. An attractive solution consists of a pair of p-n-p-n diodes connected as shown in Figure 4. Since the diode pair always appears in the same configuration, they can be fabricated in a common encapsulation to reduce the number of cans, thus hopefully reducing the cost, without impairing system performance.

In some of the examples of multiple devices that have been described, there was at least one common connection between different devices, thereby facilitating use of a single substrate. By using this type of multiple, the advantage of elimination of some interconnections is achieved. In some of the transistor multiples and in the p-n-p-n diode, on the other hand, fabrication on a common substrate is difficult and therefore not done.

It is important to realize that different elements need not be on the same semiconductor substrate. Although fabrication on the same slice of material leads, in certain instances, to the elimination of interconnections, the multiple use of single material is advantageous only when the complexity of the structure and the performance of the circuit are not degraded. The major advantages of integrated circuits come from the use of a common package not a common substrate. This suggests that in some cases it may make sense to

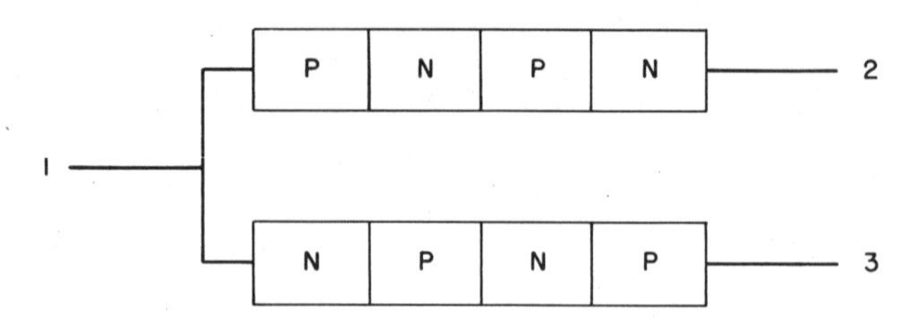

Fig. 4. The p-n-p-n diode gate. The p-n-p-n diodes connected as shown provide high-speed low-loss switching elements for time division gates. The occurrence of diodes in pairs as shown suggests a common, three-leaded encapsulation for the two devices.

encapsulate different types of devices in a common package without regard to whether or not they are on a common substrate. Indeed, such a scheme is highly desirable in many instances.

Now let us examine a particular circuit and see how these concepts can be applied to it. The circuit is the diode transistor logic gate known also as LLL or low level logic. The basic circuit configuration is shown in Figure 5. The basic features of this logic circuit are as follows. The input diodes are all connected to collectors of transistors of the previous stage as indicated in the figure. When all transistors connected to these diodes are in their off, or high

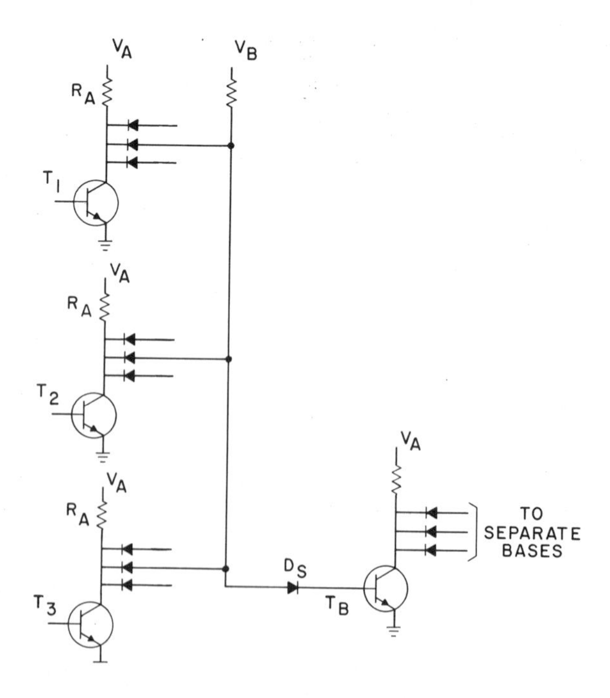

Fig. 5. The LLL gate. This logic gate performs the "and not" Boolean function and in addition provides pulse amplification. If all of the transistors T_1-T_N are off, then transistor T_B will be on. If any of the transistors T_1-T_N are on, transistor T_B will be off. The "and" function is performed by the diodes and the inversion or "not" function and amplification by the transistors. The shifter diode D_s provides circuit margin against noise, etc.

impedance state, then current will flow from supply B through the shifter diode D_S (which is incorporated to provide margins and in many cases actually consists of three series diodes) into the base of the next transistor turning it on and thus forward biasing all the diodes connected to its collector. This in turn robs the base current of the next set of transistors turning them off, etc. The function performed by this circuit is referred to as an "and not" function, since when transistor 1 and transistor 2, etc., are off, then transistor B will not be off. In this circuit the diodes are the logical elements and the transistors serve as pulse amplifiers and inverters.

Let us now consider several approaches which might be employed to integrate this circuit. To begin, it is desirable to have at hand some additional information from the circuit designer concerning the specific use of the gate and its many variations. When one examines the use of this logic scheme in any real system, one finds that the number of diodes feeding into any pulse amplifier, called the "fan-in" in circuit parlance, and the number of output diodes, called the "fan-out," varies widely. In typical computers, the fan-in and fan-out may vary from one to ten throughout the system. Therefore any integration scheme that will leave the system designer flexibility must be capable of providing from one to ten diodes per gate. Further examination of the LLL circuits shows that the shifter diode never appears in the circuit except in the input lead to the transistor, the base in this case.

Now for the possible methods of integration of this circuit. There are, of course, a considerable number of ways of doing it, but we shall consider only four here. One method consists of complete integration, and two ways of doing this are illustrated. First, one could integrate fully on a single substrate of silicon. One method of doing this, chosen for the purpose of illustrating disadvantages as well as advantages and not because it is the simplest or best way, is shown in Figure 6. The lines on the figure indicate leads either going to external connections or providing internal interconnections. Planar technology permits many of these leads to be made by evaporation of metal films over an insulating layer so that internal interconnections need not be made by thermocompression bonding. In order to provide electrical isolation between the p-side of the input diodes and the transistor base, an n-type region is interposed. The junctions formed between the n-side of the shifter diodes

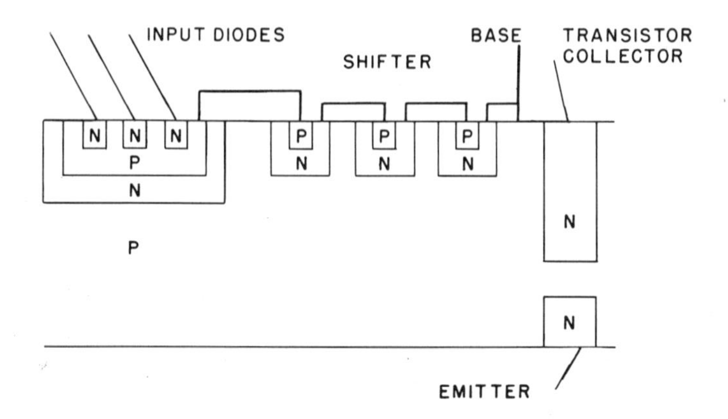

Fig. 6. Integration of LLL gate on a single substrate. This figure illustrates how the active components of the LLL gate can be fabricated on a single substrate. Seven diffusion and oxide masking operations plus fifteen lead attachments are required. Performance of such a gate is inferior to one made of individual components, and fabrication difficulties are greatly increased.

and the transistor base are back biased in circuit use and therefore dc isolated, but their presence does give rise to extra capacity. This scheme provides, through the use of a common substrate, the advantage of either elimination of or easier methods of fabrication of interconnections. However, the yield picture on complex structures is not yet clear and it is not obvious that the advantages to be gained will not be outweighed by the fact that any defective component requires the replacement of the complete structure. The loss of the ability to pretest and select components for performance is severe and may lead to difficulties when high performance circuits are fabricated in this manner.

Before the second method of fabricating the complete circuit is described, consider another scheme where part of the circuit is integrated on a common substrate. This is illustrated in Figure 7 which shows the transistor and the output diodes fabricated on a single piece of silicon. This scheme appears to offer advantages because the n-collector and the n-side of the diodes are connected together in the circuit, and therefore, fabrication on a common substrate eliminates interconnection without introducing difficulties. In this scheme the gate could be completed by adding the shifter as a separate wafer

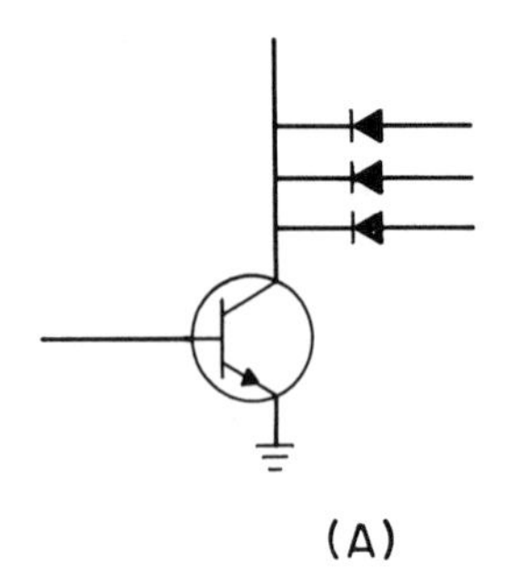

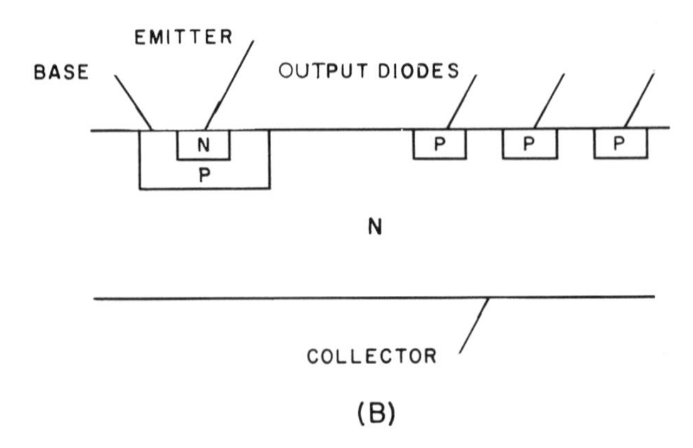

Fig. 7. Partial integration of LLL gate on a common substrate. The top of the figure illustrates the transistor and the output diodes of the LLL gate. The bottom part shows how the transistor and diodes may be fabricated on a common substrate. Use of a common substrate here introduces negligible fabrication difficulties and provides the advantage of elimination of interconnections between the transistor collector and the diodes. Circuit considerations, however, dictate against the general use of this scheme.

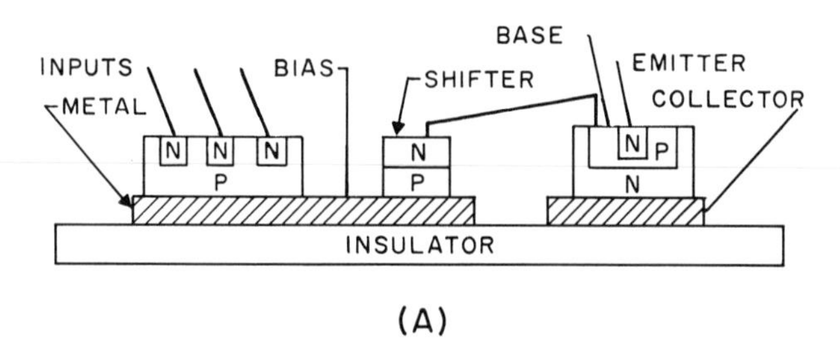

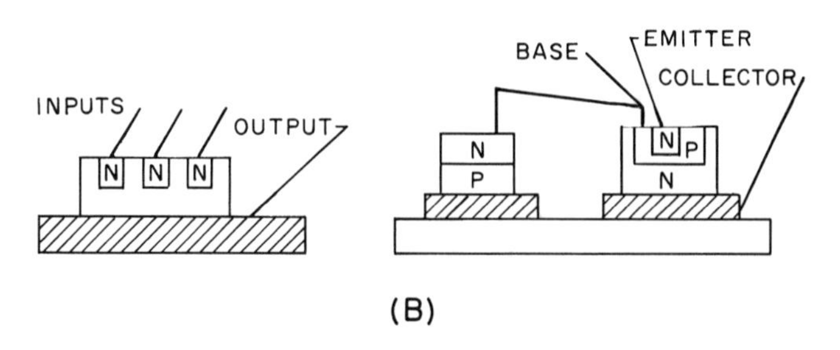

Fig. 8. Integration of LLL gate with different devices on different semiconductor wafers. In the top part of the figure, integration of the complete LLL gate is shown. The computer diodes are fabricated on a common substrate, the shifter diode and the transistor separately. All components are packaged together. The lower portion shows separate packages for the computer diodes and the shifter-transistor combinations.

and encapsulating all in one package. However, one disadvantage of this scheme, not immediately apparent, does come to light when it is regarded from the systems point of view. The external lead wires connecting the different packages will be from the computer diodes to the shifter diodes. This is the worst place for long leads from noise considerations. Thus, though this is indeed a useful scheme, some price may be paid in system performance and reliability because of decreased circuit margins.

The next scheme illustrates the second method of complete integration, but no longer on a common substrate. Figure 8 shows the diodes fabricated on a common substrate, a second wafer for the shifter and a third for the transistor. Again, common packaging is used so that the advantages of potential cost reduction, increased performance, and higher reliability are gained and this is accomplished without forcing different types of components onto a common substrate. The fourth scheme also shown in the figure differs only in that the diodes are in one package and the shifter and transistor are packaged together in another.

Which of these integration methods is preferred when cost, performance, reliability, and circuit flexibility (many different values of fan-in in this case) are taken into account? The first two schemes make use of a single semiconductor wafer for either the complete gate or at least a substantial part of it. While offering many advantages, the complexity of the fabrication procedure in the first case and the degradation of performance in the second tends to rule these out at present. Lack of sufficient data precludes a definitive choice between the third and fourth method at this time. However, some of the factors which will influence this choice can be named.

Spare parts are always needed and are usually carried in inventory by both the systems users and the device manufacturer. One such factor is the cost differential between an inventory of complete gates and one consisting of multiple diodes and shifter-transistor combinations. If complete gates are packaged, the inventory cost per package will be higher. On the other hand, if inventories of diodes in single and multiple up to, say, ten are carried separately from the shifter and transistor package, then more packages will be required, though they will be cheaper. In addition to inventory, other considerations of importance are cost of manufacture and the mechanics and cost of interconnection at the next level. An additional factor relates to the circuit design itself. Should the circuit design be modified in such a way that several stages of diode logic occur without pulse amplifiers and inverters then the separate package method would be desirable. Other considerations will most certainly enter, but these appear to be important ones.

The arguments presented and examples cited above have been primarily concerned with the methods to be used in fabricating the semiconductor components of an integrated circuit. The same line of thought applies equally well, with appropriate modifications, to passive components. It follows that the fabrication of resistors and capacitors of silicon, or other semiconductor material should be done only if a performance, reliability, or cost gain is achieved. Although semiconductor resistors and capacitors may find application in integrated circuits, in many cases their inclusion leads to a degradation in overall circuit performance.

SUMMARY AND CONCLUSIONS

In this paper various approaches to integrated circuits have been discussed and evaluated with regard to their capability of providing lower systems cost, improved systems performance, and higher systems reliability. In order to meet these objectives, it was stated that flexibility must be maintained for both the systems designer and the device designer and that devices which are larger runners and are not excessively difficult to fabricate are required to meet lower costs.

Consideration of several basic approaches leads to the conclusions that devices which perform complex functions directly do not satisfy today's needs primarily because they do not exist; that integrating basic Boolean functions has its shortcomings, because it takes away a great deal of the systems designers flexibility; that multiple devices and common encapsulations of different device types go a long way toward meeting the objectives for a wide variety of digital systems. The fabrication of different device types on a common substrate for its own sake however can lead to increased cost with no apparent gains in performance or reliability in many instances. The major advantages to be realized from integrated circuits come from the use of a common package not a common substrate.

ACKNOWLEDGMENTS

The author would like to thank a number of colleagues for helpful discussions and for supplying data. He is particularly indebted to Messrs. I. M. Ross, R. M. Ryder, M. M. Atalla, B. T. Howard, E. G. Walsh, and R. W. MacDonald.

References

1. Rice, R., "A System Designer Views 'Micro-Integrated Electronics,' " paper presented at the Solid State Device Research Conference, Palo Alto, California, June 1961.

2. Howard, B. T., and W. F. J. Hare, "Accelerated Aging of Multiple Diodes in a Single Encapsulation," paper presented at the Solid State Device Research Conference, Palo Alto, California, June 1961.

DISCUSSION

F. H. BOWER (Sylvania Electric Products): In those multiple diodes in which all six diode elements did not meet the specification, what was the parameter they failed to meet? Have you analyzed the cause of the failure?

J. M. GOLDEY: In most cases it was breakdown voltage. A detailed analysis of the specific cause has not been made.

Surface Passivation Techniques for Microelectronics

STEWARD S. FLASCHEN and ROBERT J. GNAEDINGER, JR.

Motorola Inc., Phoenix, Arizona

Abstract

The influence of the semiconductor surface on the device performance is reviewed with particular emphasis upon the integrated devices. Methods of isolating the surfaces are described and discussed.

INTRODUCTION

The primary purpose and promise of microelectronics or integrated circuits is to bring about a major improvement in electronic equipment reliability at a cost that compares favorably with present equipment. The improvement in reliability and reduction in cost ultimately to be expected from microelectronics will depend upon the furthering of our understanding of the physics and chemistry of the semiconductor and upon the degree of control achieved over its materials and processes. The purpose of this paper is to discuss one of these major areas of work, the understanding and control of the properties of the active surfaces of the semiconductor.

It is not unusual in transistor device design that the most carefully synthesized device will exhibit electrical properties that are both unexpected and indeed out of line with those predicted from solid state theory. This lack of agreement between design and performance occurs primarily because of the uncontrolled properties of the semiconductor surface. The lack of control of semiconductor surfaces and interfaces has been among the most pressing and challenging problems of solid state technology. Beyond these problems which are concerned with semiconductor device design and fabrication, there arise the further and critical problems of device parameter changes and failure during operation. It has been established that many of these also are due to surface changes.

Up to the present time the problem of understanding and controlling the properties of surfaces has been associated with the single semiconductor device. The integrated circuit concept, however, has further extended the problem by requiring solutions to be compatible with different semiconductor and associated elements, operating at high packing densities and under high field strength conditions. For integrated circuitry to achieve its primary purpose, it is essential that both of these surface problems be solved.

THE PROBLEM OF THE SEMICONDUCTOR SURFACE

The two major factors determining the cost of manufacture of semiconductor or microelectronic devices are (1) the cost associated with the necessity for hermetic sealing and (2) the yield to a given specification.

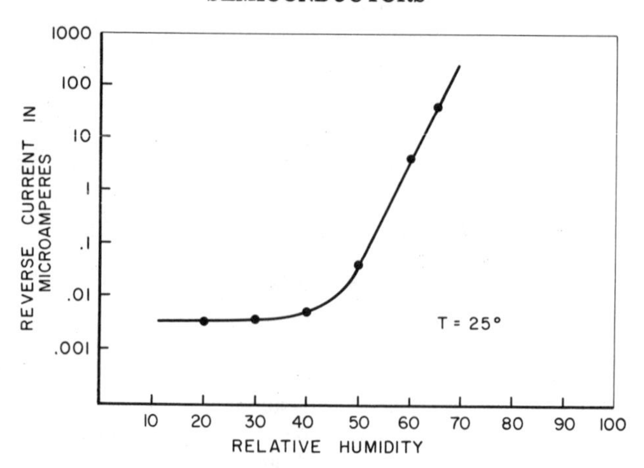

Fig. 1. Effect of relative humidity on the leakage current of silicon diodes.

The first factor arises because of the need for surface protection of the semiconductor; the second arises often because of a lack of control over the electronic properties of the surface.

One of the primary causes of deterioration or change in semiconductor device properties is due to the effect of moisture on p-n junctions. This sensi-

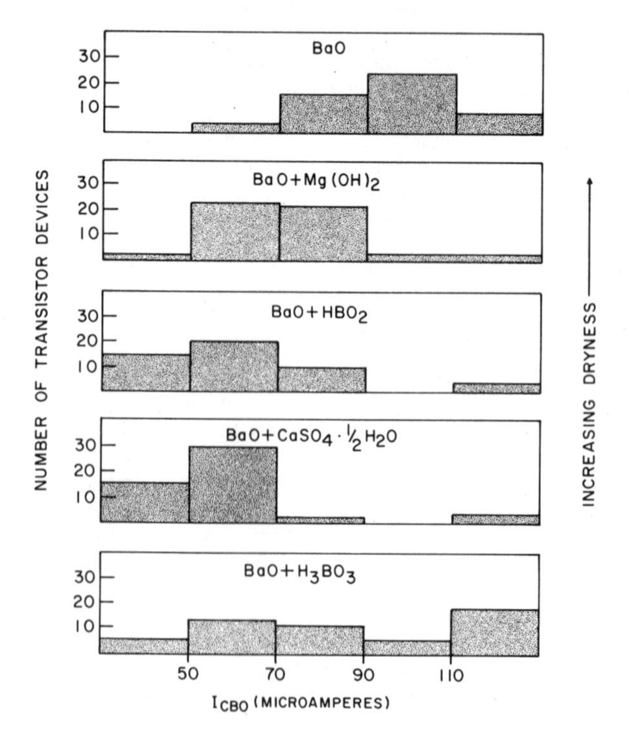

Fig. 2. Effect of moisture level on low-voltage saturation current of germanium p-n-p power transistors.

tivity to moisture was observed soon after the invention of the transistor itself in 1948. A typical effect of moisture on p-n junction properties is shown in Figure 1. The catastrophic deterioration of reverse currents at humidities above 40% led to the requirement of hermetic sealing of all devices.

The role that water vapor and other surface impurities play in determining device parameters is not a simple one (1). It is clear that moisture condensation causes gross shorting across a p-n junction. When water has been adsorbed to a thickness of approximately four molecular layers, it takes on the properties of an ionized solvent. At this level, obtained at approximately 40% relative humidity, large changes and instabilities in device parameters are observed. However, it has also been found that device parameters continue to be sensitive to moisture even down to very low levels. This sensitivity has been associated with the amount of water adsorbed onto the semiconductor surface down to fractions of a molecular layer (2).

Figure 2 shows, at relative humidities well below 1%, the sensitivity of the saturation current, I_{CO}, to moisture of a p-n-p germanium alloy transistor. Figures 3 and 4 show the effect of varying the moisture level on the power gain of this same device. This strong relationship between adsorbed moisture and surface recombination velocity (which affect the device parameters β, I_{CO}, and their derivatives) has been found to be common to a variety of devices. The establishment of a controlled environmental condition at the surface for maintaining a fixed amount of adsorbed water is therefore important in realizing the goal of a reproducible and stable device parameter. Similar effects are known to be present with other gases such as oxygen and nitrogen.

A second cause of change in device properties during operation is due to the diffusion, or drift, of ion species in the fringing field of the semiconductor

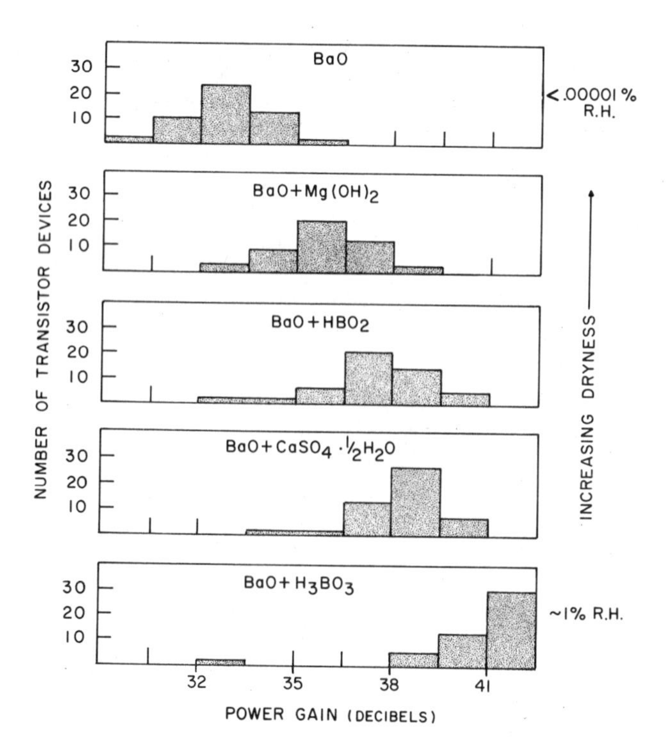

Fig. 3. Effect of moisture level on power gain of germanium p-n-p power transistors.

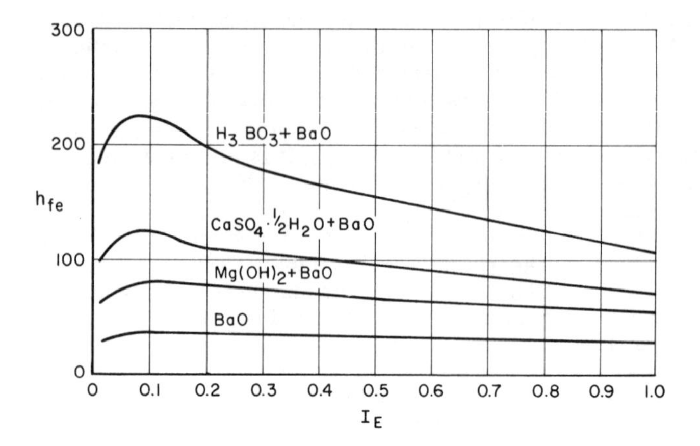

Fig. 4. Effect of moisture level on h_{fe} vs I_E for germanium p-n-p power transistors.

p-n junction. Figure 5 is a schematic representation of the surface of such a device. The surface oxide formed by exposure of silicon to moisture and/or oxygen at room temperature is typically 30 to 60 A thick. Included within this oxide, even after stringent chemical cleaning, is a concentration of ionic species of approximately 10^{14} ions cm^{-2} (3). Ionic compounds of sodium, potassium, fluorine, water, and copper are among the dominant impurities. Because of the unusually high field strength conditions encountered at the surface of a normal p-n junction during rectifier or transistor operation, these ionic compounds find themselves under the influence of fields ranging up to 10^5 v/cm. (It is interesting to note that such fields approach, within two to three orders of magnitude, those characteristic of the ionic chemical bond itself.) There is considerable evidence that at high temperatures solid state diffusion of these ions occurs in the fringing field of the junction with corresponding changes in the electrical properties of the junction.

APPROACHES TO SOLUTION OF THE PROBLEM

The problem of surface control and stabilization (passivation) of semiconductor devices and integrated circuits containing exposed p-n junctions con-

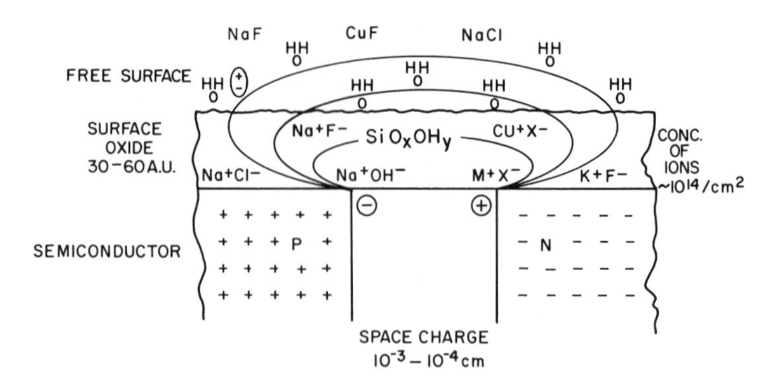

Fig. 5. Surface of semiconductor p-n junction (idealized.)

sists of: (a) the control of the effect of adsorptive gases and moisture on the electrically active areas of the junction, and (b) the isolation of the junction from ionic contaminants. The experience of industry over the past 10 years with a large variety of individual devices has provided methods of approach to the control problem. The most common of these has been the hermetically sealed package. The ideal hermetic seal isolates the device from a changing external environment, but by itself may leave the device subject to a varying internal environment. Such variations can be the result of contamination transfer to the active surface, desorption of moisture from internal surfaces, and hydrate-dehydrate chemical changes at the semiconductor surface itself (4). To minimize the problem of internal moisture buildup, desiccants have been introduced into the hermetic package. A wide variety of moisture sorbing materials have been used by the industry, producing sizable increases in the stability and reliability of devices. The nature of some of these materials and their results in operation have been discussed by Gnaedinger et al. (2).

The use of a hermetically sealed package with a controlled internal environment for integrated circuits will obviously be desirable in many cases. However, it has some potential disadvantages. The first is the large package volume and cost associated with the hermetic seal. The second is the requirement that all circuit components operate satisfactorily within a common ambient.

Another method of surface control is the formation of protective coatings on the active regions of the device. The primary function of such coatings is to act as a barrier to prevent the access to the sensitive semiconductor surface of deleterious environmental gases and nongaseous contaminants. Additional requirements for such a coating is that it should produce a desirable distribution of electronic surface states, be atomically bonded and tightly adherent to the semiconductor (or its natural surface oxide), and not be adversely affected by stresses imposed by environmental changes. An additional desired property of such coatings is that they be capable of resisting damage from impact and from erosion or abrasion, to minimize the extend of damage occurring during the processing of the device and the fabrication of the circuit.

A potential disadvantage in the general application of coatings to single chip or monolithic integrated circuits is that the various device types must be simultaneously electrically and thermally compatible with a given coating and its process of application. However, the diversity of coatings available makes this approach very promising for multiple chip construction.

Coatings in advanced stages of development, and in some cases application, are of two general types, organic and inorganic, and include the following: (1) Organic polymer coatings—epoxies, silicones, others (diallylphthalates, fluorocarbons, etc.). (2) Inorganic oxides and glasses—high temperature oxidation, hydrothermal oxidation, accelerated thermal oxidation, low melting glasses, pyrolytically formed glasses, and evaporated and sputtered films.

The protective coatings in the organic polymer group show a relatively high permeability to moisture, because of their generally low molecular packing densities. (The diffusion coefficients of moisture in these materials is characteristically in the range of 10^{-7} to 10^{-9} cm^2/sec.) As a consequence, they are generally unsatisfactory for extended direct exposure to high humidity. They have been found to be of value, however, when used within a hermetically sealed enclosure by providing an additional degree of protection of the junction from the effects of gross moisture.

For long-term protection or for effective operation outside of a hermetic seal the use of an inorganic oxide or glass coating is required. These coatings satisfy the requirement of low moisture permeability (diffusion coefficients of water are less than 10^{-12} cm^2/sec).

High Temperature Oxidation

The formation of a high temperature oxide on the surface of silicon has produced a significant degree of passivation of semiconductor diodes and transistors (5). This approach, and its modifications that follow, is the most widely used method in the manufacture of "passivated" devices.

This method depends upon an oxidation reaction of silicon $Si + O_2$ (or H_2O)$\rightarrow$ SiO_2 forming a protective and electrically insulating oxide coating. The rate of the thermal oxidation of silicon is inversely proportional to the oxide thickness, the reaction rate being controlled by the rate of self-diffusion of silicon and oxygen. Temperatures of around 1000°C and times of several hours are required for the formation of protective thicknesses of the oxide. The thermal oxide, if properly formed, has provided a high degree of protection of the active semiconductor substrate.

The problems that have been encountered in application of thermal oxidation to devices have been of two types: First, at the high temperature required for passivation the surface reaction can cause a redistribution of the doping impurities at the semiconductor-oxide interface; this effect can produce in some devices undesirable parameter changes. Second, it has been found difficult to produce a film that is completely free from pinholes, cracks, and other nonuniformities. Also a problem exists in maintaining film protection through thermal quenching and the chemical etching processes required to form electrical contacts.

Hydrothermal Oxidation

Hydrothermal (high pressure steam) oxidation (6) has been developed in order to avoid the problems associated with high temperature oxidation. This significantly lowers the temperature required for the formation of a thermal oxide on the surface of silicon.

Under the influence of moisture at high pressure, the rate of oxidation of silicon is significantly accelerated and is independent of thickness, resulting in formation of protective thicknesses of oxide films at 600—700°C in a few hours. In a typical oxidation reaction, an amorphous SiO_2 coating 3000 A thick is formed in 2 hr at 650°C and 750 psi H_2O. Though this work is still in the stages of advanced development, the experimental results suggest that the undesirable fractionation of impurities at the interface is considerably less than in high temperature oxidation. Because of the lower temperatures required, electrical contacts can normally be made prior to passivation. Thus, the masking and etching step, a source of film degradation, can be eliminated.

Accelerated Thermal Oxidation

Accelerated thermal oxidation is another recently developed method for lowering the temperature of oxide film formation. This temperature reduction is possible because of the addition of rate accelerating additives (7). The accelerating additives found most effective have been the lead oxides and halides. The addition of the accelerating agent is through the vapor phase. The amount added to the oxidation reaction is controlled as a function of the vapor pressure-temperature dependency of the compound. By this means, using PbO, it is possible to form a thick and continuous glass coating at 550°C that is integrally bonded to the silicon surface. Similarly, using PbBr2, germanium oxide glasses have been formed on germanium at 350°C (8).

The degree of passivation that can be achieved by this and similar oxidation processes is shown in Figure 6. Typical aging behavior of passivated silicon rectifiers is shown under combination high temperature, high humid-

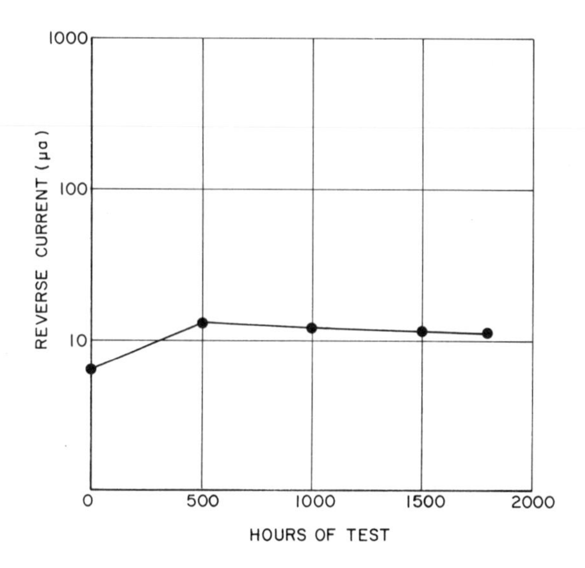

Fig. 6. Aging behavior of surface passivated diodes at high humidity.

ity, and power stress. Reverse currents observed after 1500 hr continuous aging are well within the tolerance level established for the hermetically sealed device.

The hydrothermal and accelerated thermal oxidation processes, because of their lower formation temperatures, do not exhibit the deleterious effects associated with body or surface impurity redistribution. An additional attribute of these modified processes is that they can be used to form passivating coatings on completed devices with electrodes and contacts in place.

Low Melting Glass

The recently discovered low melting (150—300℃) arsenic-sulfur-iodine glasses offer a fourth method for surface passivation. These glasses are applied to the semiconductor surface rather than formed thereon as in the previous methods.

Since these glasses are more fluid at lower temperatures than the thermal oxide glasses, it can be expected that the degree of protection they offer against the transfer of ionic contamination to the device surface will be less. Yet, for certain devices, such as those of the indium alloy type, where electrode melting occurs at 156℃, these glasses may offer valuable protection properties (9). They are alone among known glasses in their ability to be directly deposited by evaporation. Other methods of application include dip coating and potting.

Pyrolytically Formed Glasses

Gaseous reactions of certain chemical compounds can be used to form a coating on the surfaces of semiconductor devices. These reactions occur at, or near, the device surface and may consist of the decomposition of single chemical species or the reaction between several chemical species. Under certain conditions masking techniques can be used to restrict the deposition of the deposited films to particular regions of the device rather than to the entire surface. This can be of particular value where the deposited coating is ideal for one component of an integrated circuit, but not for another.

One of the simpler of such deposition techniques is the high temperature dissociation (pyrolysis) and oxidation of organo-silicon compounds to deposit silica-like coatings (10). Other oxides that have been deposited by this technique include TiO_2, SnO_2, Al_2O_3, ZnO, Fe_2O_3, and a number of mixed oxide phases (11). This process is well suited to the deposition of dielectrically isolating coatings over large surface areas. The recent development of electron beam dissociation makes it possible to selectively deposit inorganic dielectric films at spacings as low as a micron (12).

Much work remains to be done with these types of coatings, however, to assess their passivating value on given devices, electronic effect on surface states, and long-term stability. These coatings can be applied at lower temperatures than required for thermal oxidation and as in the previous method do not use up substrate material.

Evaporated and Sputtered Coatings

Coatings on semiconductor substrates may also be deposited by evaporation or sputtering. Materials having appreciable vapor pressures at temperatures in the range of a few hundred degrees to approximately 2000°C may be evaporated in high vacuum systems onto devices or integrated circuits. Typical of such processes is the evaporation of silicon monoxide (as well as the arsenic sulfide glasses previously discussed). In this process as in most others the perfection and properties of the resulting coating are strongly influenced by the conditions of the process such as pressure, deposition rate, filament temperature, and substrate temperature.

Many materials which cannot be satisfactorily evaporated can be deposited by sputtering. Sputtering allows more complex phases as well as the highly refractory and insulating oxides to be deposited as coatings.

CONCLUSION AND SUMMARY

The degradation of unprotected semiconductor surfaces is frequently so rapid as to be termed catastrophic; this has necessitated the use of hermetic sealing methods in current device manufacture and has led to extensive exploration and development of passivation coatings of devices.

The hermetic seal with ambient controlling getters has the advantage of being a well-known and familiar packaging method. Its disadvantages in addition to size and cost are that it requires all components of the integrated circuit to operate within a fixed internal ambient, an ambient that may be optimum for only a portion of the components.

The surface passivation coatings are newer and less well understood but offer the potential advantages of low cost, small size, and improved stability of device parameters. Of the two major approaches to surface passivation, organic coatings have been extensively tried to form long-term moisture protection but have been found wanting. The inorganic surface protective coatings have been recently and rather extensively explored. As a result, passivation coatings to reliably protect silicon and germanium surfaces appear attainable for a number of device and integrated circuit applications. The same considerations apply to the passive components of integrated circuits but to a lesser degree since their surface is less sensitive to impurities and environmental changes.

In order to fully evaluate a passivating coating, answers must be obtained to the following questions: (a) Is the coating compatible with the device throughout operational and environmental uses? (b) How long and how well is protection provided? (c) Is the process applicable to more than one device design (a necessity for application to monolithic-type integrated circuits)?

(d) How practical and economical is the coating process? (e) Is reprocessing of a damaged or imperfect coating possible? In this respect it will be desirable to develop standard procedures for evaluation of passivating coatings.

References

1. Bardeen, J., and W. H. Brattain, Bell System Tech. J., 32, 1 (1953). R. H. Kingston, J. Appl. Phys., 27, 101 (1956). H. Statz, G. A. deMars, L. Davis, and A. Adams, Phys. Rev., 101, 1272 (1956). J. T. Wallmark, RCA Rev., 18, 255 (1957). N. B. Hannay and J. T. Law, Semiconductors, Reinhold, New York, 1959, p. 676. J. J. A. Ploos van Amstel, Philips Tech. Rev., 22, 181 (1961).

2. Gnaedinger, R. J., Jr., S. S. Flaschen, M. A. Hall, and E. J. Richez, J. Electrochem. Soc., in press.

3. Law, J.T., Proc. Inst. Radio Engrs., 42, 1367 (1954).

4. Wallmark, J. T., and R. R. Johnson, RCA Rev., 18, 512 (1957).

5. Atalla, M. M., E. Tannenbaum, and E. J. Scheiber, Bell System Tech. J., 38, 749 (1959).

6. Spitzer, W.G., and J. R. Ligenza, J. Phys. Chem. Solids, 17, 196 (1961).

7. Kallander, D. A., S. S. Flaschen, R. J. Gnaedinger, Jr., and C. M. Lutty, J. Electrochem. Soc., Electronic Div. Abstr., 10, 1, 129, (1961).

8. Flaschen, S. S., D. A. Kallander, and C. M. Lutfy, J. Electrochem. Soc., Electronic Div. Abstr., 10, 1, 128 (1961).

9. Flaschen, S. S., A. D. Pearson, and I. L. Kalnins, J. Appl. Phys., 31, 431 (1960).

10. Klerer, J., J. Electrochem. Soc., 108, 1070 (1961).

11. Holland, L., Vacuum Deposition of Thin Films, Wiley, New York, 1956.

12. Communication from Stanford Research Institute.

DISCUSSION

J. WATSON (Texas Instruments): You state that the lead oxide additive to the grown oxide has a lower diffusion coefficient than pure grown oxide. Is it possible that the lead oxide actually opens the structure for oxygen diffusion, and therefore would it not also open the structure for diffusion of water vapor?

S. S. FLASCHEN: The question is I believe, what effect does the additive in accelerated thermal oxidation have on the rate of diffusion of the water at normal temperatures. There is not any correlation here. The acceleration of thermal oxidation is due to bond weakening in the silicon-oxygen-silicon network by means of the partial substitution of a divalent additive (lead) in place of the tetravalent silicon. The self-diffusion coefficients for silicon and oxygen in the SiO_2 lattice are primarily functions of bond strength and the stretching modes of the bonds as a function of temperature. With the addition of lead, the rate of oxidation changes from a parabolic oxidation rate characteristic of a rigid SiO_2 network to the faster more nearly linear oxidation rate which is a characteristic of the lowered viscosity or quasi-liquid lead glass at the temperatures 500—600°C.

The question of the permeability of the solid glasses to water is a question of the comparative densities of molecular packing of both glasses. At room temperature both are characterized by a highly rigid and fixed molecular network. I think we can visualize the process of moisture diffusion in both glasses as the diffusion of the water molecule through a molecular-sized sieve with fixed pore sizes.

The atomic packing of SiO_2 consists of Si-O tetrahedra joined in pseudo-hexagonal symmetry with a fairly large hole in the plane of the hexagon. Additions to SiO_2 glass, such as lead, act to close up or block these atomic holes.

As a result these glasses normally show a much lower permeability to water vapor, helium, and other gases than does SiO_2. Actually pure SiO_2 glass has the highest permeability to gases, including water, of all of the silicate glasses.

J. WATSON: In Fig. 6, which showed the reverse current as a function of time, were the devices biased before the relative humidity was increased or was the humidity increased and then the bias applied across the units.

S. S. FLASCHEN: The process was a random one with no deliberate effort to introduce the specimen at any given part of the cycle.

A. J. ROSENBERG (Tyco, Inc.): It was my understanding that the oxidation of silicon is controlled by the interstitial diffusion of oxygen through an oxide film. I do not quite see how the inclusion of lead would promote such an interstitial diffusion by weakening the film or creating a somewhat weakened structure.

S. S. FLASCHEN: The question of the controlling mechanism in solid state diffusion is a very complicated one. It is difficult to put relative ratings on the different mechanisms by which oxidation of a given metal, or reaction of any compound with a gas phase, takes place. It depends upon a number of variables including temperature, the gas dynamics of the reaction, the condition of the interface, and the structural perfection of the solid among others. Under these conditions several different controlling mechanisms are possible. One of these as you mention is interstitial diffusion. Another mechanism is grain-boundary diffusion. Another is vacancy diffusion or Schottky-type diffusion.

Our hypothesis was that the latter was one of the possibly important mechanisms of the silicon oxidation reaction. The additives chosen were on the basis of lowering the bond strength of the SiO_2 network to lower the activation energy for Schottky-type diffusion. I must say that the actual kinetics of this reaction are not as yet well fixed. The overall results, however, are not at all at variance from what would have been expected from glass theory. The addition of a network breaker or network modifier, such as lead, to silica glass is well known, for example, for effectively reducing the melt viscosity.

P. J. SCHLICHTA (Jet Propulsion Lab.): The catastrophic oxidation of stainless steels by fuel ash appears to be due to complex vanadates, which carry oxygen through the corrosion barrier by means of oxidation reduction of the vanadium [cf. Foster, Leipold, and Shelvin, Corrosion, 12, 539 (1956)]. Is it not possible that lead oxide, which like vanadium is capable of several oxidation states, accelerates the oxidation of silicon by a similar mechanism?

S. S. FLASCHEN: We were not aware of that study. Thank you very much for calling it to our attention. We would like to look further into this.

P. S. FLINT (Fairchild Semiconductor): In Fig. 6 which showed the leakage current for the lead oxide case, could you give us some idea of the area of the junction involved?

S. S. FLASCHEN: This was a fairly large power device. The diameter of the unit was approximately 90 mils.

P. S. FLINT: Also, I did not quite follow your explanation of why you do not observe channels with the low temperature oxidation.

S. S. FLASCHEN: There are two reasons. First, oxidation under the conditions of the accelerated reaction appears to be fairly uniform. Preferential oxidation between silicon and the common doping impurities such as aluminum, boron, and phosphorus does not appear. This is in contrast to the standard thermal oxidation reaction. Second, at the lower temperatures of the accelerated reaction the diffusion rate of doping impurities both in the body and at the surface is negligible. As a consequence we do not find redistribution of impurities occurring.

P. S. FLINT: Suppose you attribute this channel formation to the ability of the doping in the bulk to diffuse away from the moving boundary when you are oxidizing. At low temperatures, then, perhaps you have the case where the bulk dopant is not going to be able to diffuse away as readily, i.e., pile-up will occur, and in this case you might think that a channel would form less rapidly.

S. S. FLASCHEN: In practice we have looked for impurity pile-up in both p + n and n + p structures. We have not as yet been able to find any evidence for it.

Status of Vapor Growth in Semiconductor Technology

R. GLANG and E. S. WAJDA

IBM Command Control Center, Federal Systems Division,
Kingston, New York

Abstract

The deposition of single crystalline layers from the gas phase has been added to the field of semiconductor processing technology. Although the vapor growth technique is applicable to a variety of semiconductor elements and compounds, it has been developed most successfully for silicon.

There are several chemical reactions and deposition systems which have been used to grow silicon epitaxially. Some of these processes present particular technical difficulties or have a limited practical value. The two processes most widely used employ the reduction of silicon tetrachloride or trichlorosilane with hydrogen.

Results concerning crystal perfection, surface quality, and uniformity of silicon deposits vary widely depending on a variety of process parameters. The presence of impurity halides in the gas phase has also been studied in order to obtain deposits with varying resistivities. It has been found that the impurity concentration in the deposit is related to the amount of added impurity halides in a complex way. Close process control is necessary to yield reproducible results and make the vapor growth technique a versatile and reliable process.

INTRODUCTION

Within a few months of the first announcements of successfully developed semiconductor vapor growth processes, in the middle of 1960 (1-5) single crystal deposition from the gas phase has become one of the standard methods in industrial transistor manufacturing (6-10). The term "vapor growth" is used to describe processes by which a semiconductor is deposited from one of its gaseous compounds onto a seed crystal under conditions such that single crystalline growth occurs. Special process requirements, such as controllability of crystal perfection, impurity concentration, and thickness of the deposited layer are also imposed.

It is interesting to review briefly the major technical areas which contributed important background knowledge for the successful development of the first vapor growth processes for semiconductors (Table I). A substantial part of this background was acquired during the last decade from work concerning the preparation, analysis, and study of impurity behavior of semiconductor crystals. These efforts have been directed toward the same objective as vapor growth processes but differed insofar as the material to be added

TABLE I

Vapor Growth of Semiconductor Single Crystals

	Vapor deposition of elements	Semiconductor crystal fabrication	Vacuum evaporation of epitaxial films
Objective	High purity materials	Highly perfect single crystals with controlled impurity content	Crystal surface study Nucleation Crystal growth
Contributions	Chemical processes Thermodynamics	Crystal preparation and analysis	Substrate surface preparation Conditions for controlled nucletion and growth

to the seed crystal was supplied from a liquid phase such as in the Czochralski (11) or Bridgman (12) techniques and later by the zone melt (13) and floating zone methods (14).

The idea that solids can be crystallized from a gaseous as well as from a liquid phase is not new. Systematic exploitation of this concept began in 1925 when Van Arkel published his work on the deposition of tungsten from the gaseous hexachloride (15). Since then, polycrystalline metals like molybdenum, tantalum, zirconium, and titanium have been obtained from similar reactions, and iron and nickel from the decomposition of their carbonyls (16). In the mid-fifties, the same principle was applied to semiconductors. Many investigations have been concerned with the preparation of high-purity polycrystalline silicon either by pyrolytic decomposition (17, 18), hydrogen reduction (19-22), or disproportionation of various silicon halides (23,24). Some of these reactions have been used to deposit silicon in a single crystalline form and the knowledge of their chemistry and thermodynamics helped considerably in the development of vapor growth processes.

Finally, a third area to be considered in this connection is the vacuum evaporation of epitaxial films. Since 1935, it has been shown in many instances that completely oriented metal films can be vacuum-deposited on certain single crystalline substrates (25,26). Some of these studies concerning surface preparation and conditions necessary for controlled nucleation and growth on crystalline substrates added substantially to the recent advance in semiconductor technology.

VAPOR GROWTH PROCESSES FOR ELEMENTAL AND COMPOUND SEMICONDUCTORS

At the present time, various vapor growth processes for semiconductor elements and compounds are under investigation (Table II). These data are neither complete nor final because of the limited information available and the continued advancement in the field. Although the less common compounds like sulfides or selenides have been omitted, the list of epitaxial processes for semiconductors is already impressive. Various types of reactions have been employed using either halides, elements, suboxides, or intermetallic compounds as source materials. Reactions like hydrogen reduction or gaseous cracking are usually conducted as open-tube processes with a constant flow

TABLE II. Vapor Growth Processes for Semiconductor Elements and Compounds

Semi-conductor	Type of process	Source Material	Ambient or carrier gas	Deposition temp., °C	Has epitaxial growth been achieved?	Is process being used for commercial devices?	Ref. No.
Si	H_2 reduction	$SiHCl_3$	H_2	1150-1300	Yes	Yes	27,10
Si	H_2 reduction	$SiCl_4$	H_2	1200-1300	Yes	Yes	3,6,10
Si	H_2 reduction	$SiBr_4$	H_2	1100-1300	Yes	No	20
Si	H_2 reduction	SiI_4	H_2	1000-1200	Yes	No	21,28
Si	Prol. decomposition	SiI_4	Vacuum	~1000	?	No	18
Si	Disproportionation of SiI_2	Si	I_2	~950	Yes	No	4,5
Ge	Disproportionation of GeI_2	Ge	I_2	~400	Yes	No	4
Ge	H_2 reduction	$GeCl_4$	H_2	~830	Yes	Yes	6
SiC	Sublimation	SiC	Vacuum	>2000	Yes	No	29
SiC	Gaseous cracking	$SiCl_4$-toluene	H_2	~2000	?	No	30
GaAs*	Evapor.-disproport.	GaAs	I_2	620/690	Yes	No	31,32
GaAs*	Evapor.-disproport.	GaAs	HCl	900	Yes	No	33
GaP*	Reduction and compound	$P/Ga/Ga_2O_3$	Vacuum	~1000	?	No	34
GaP*	Evapor.-disproport.	GaP	I_2		Yes	No	32
InSb	Complex formation and decomposition	$In(CH_3)_3/SbH_3$	Vacuum	>150	?	No	35

* See N. Holonyak, Jr., D. C. Jillson, and S. F. Bevacqua, this volume, p. 49.

of carrier gas and the substrate in the hottest portion of the system. Disproportionating or sublimating processes, on the other hand, are more commonly carried out in closed systems and the transport of material occurs from a hotter source region to a cooler substrate zone. The deposition temperatures vary for different materials, increasing with the melting points of the semiconductors. Epitaxial growth has been accomplished in most of the cases listed.

Attempts to grow epitaxial germanium started simultaneously with silicon but with less emphasis. It is surprising to note that the preparation of an atomically clean germanium surface proved to be almost as difficult as it is in the case of silicon. Of the two processes listed in Table II, the tetrachloride reduction has been developed to almost the same level of perfection as the silicon chloride reduction.

In silicon carbide, both processes known thus far suffer from the necessity of operating at unusually high deposition temperatures, which inherently leads into equipment and contamination problems. Although epitaxial growth has been reported (29), the state of the art is still far away from a controllable process.

Because of its high potential as a semiconductor material, gallium arsenide has been studied extensively in connection with vapor growth. The preparation of single crystals of most of the III-V compounds from the melt requires complicated experimental techniques due to the high vapor pressures of some of the elements involved. Therefore, vapor growth medthods may eventually assume a dominant position in the preparation of these compounds. At the present time, two different transport gases, iodine and hydrochloric acid, have been tried successfully. The transfer mechanisms are different for the two elements. The gallium forms lower halides which disproportionate on the substrate surface, whereas the arsenic simply evaporates and recondenses. The details of the epitaxial growth of the III-V compounds is given by Holonyak et al. (this volume, p. 49).

An interesting new approach has been reported recently for indium antimonide. Indium trimethyl and stibine are reacted in the vapor phase to form a complex containing both elements in a 1:1 ratio. This complex deposits on surfaces at room temperature or below and decomposes above 150°C to yield a film of indium antimonide. Although epitaxial growth has not as yet been reported, the process would probably lend itself to this purpose.

The most advanced state of the art is represented by silicon. Two processes, the hydrogen reduction of trichlorosilane and of silicon tetrachloride, are used commercially to fabricate transistor structures. Other systems, as shown in Table II have been investigated with varying degrees of success. A comparison of these processes will be made in the following section. Silicon deposition is chosen as an illustrative example of the technical difficulties encountered in semiconductor vapor growth because of the large amount of available data and commercial interest.

COMPARISON OF VAPOR GROWTH SYSTEMS FOR SILICON

Experience has shown that some of the reactions in Table II do lend themselves to the requisite process control for the reproducible fabrication of multiple p-n structures with specified physical and electrical properties.

The silicon di-iodide disproportionating reaction reported by Wajda and Glang (5), illustrates the importance of tight process control. This process employed a sealed quartz tube, in which silicon was transported from a source region into a substrate region of lower temperature. Iodine vapor at a pressure of about 5 atm served as the transporting medium. Early results obtained with this method were quite encouraging because impurities like boron, phosphorus, and arsenic could be transferred together with the source silicon so that controlled doping was conveniently achieved. However, the purity of the gas atmosphere and the temperature distribution in the system could not be controlled to a degree necessary for reproducing deposition rates and good crystal perfection. The lack of a purging capability caused compensation effects which restricted the type of junctions formed when multiple doping was attempted. The latter consideration leads to the conclusion that an open-tube system with a constant flow of reaction gases is to be preferred if a deposition process of maximum control and flexibility is desired.

Another important factor is the vapor pressure of the silicon compound. To obtain deposition rates of about 1 μ /min, the silicon compound pressures are usually adjusted between 1 and 20 mm of mercury. This is not difficult with silicon tetrachloride and trichlorosilane, which have considerably higher vapor pressures at room temperature. The pressure adjustment proved to be a problem, however, in the development of a deposition process based on the hydrogen reduction of silicon tetra-iodide. To maintain the necessary compound pressure, the tetra-iodide source and all the feed lines have to be kept at temperatures between 100 and 200°C. This, in turn, is not compatible with vacuum-tight valves or stopcocks needed to manipulate the gas flow.

These considerations show the advantage of using one of the two silicon chlorides for an open-tube deposition process. Neither of the two processes has shown an advantage over the other and they are very similar as far as their mechanisms, operating conditions, and results are concerned. Theurer (6) found that trichlorosilane is a major by-product of the tetrachloride reaction, which seems to suggest that a common equilibrium composition of the reaction gases is being approached regardless of the initial silicon compound employed.

TRICHLOROSILANE SYSTEM

Although the choice of an open-tube system in combination with one of the silicon chlorides eliminates some difficulties, the development of a refined deposition process is still problematical. To further illustrate the situation, the results of our own investigation of the trichlorosilane process will be discussed.

Among the many vapor growth systems being used, there are probably no two exactly alike. Details concerning apparatus design are often company secrets, more carefully guarded than operational data. The choice of construction materials, the geometry of the reaction chamber, and the mode of operation have a very significant effect on the properties of the deposited silicon. Therefore, numerical process data and quantitative relationships among process parameters and deposit properties are meaningful only in conjunction with the specific system they are derived from. The basic trends and the complex interdependency of process parameters and deposit properties are a common feature of all the different systems.

The data presented in the following discussion were obtained from the deposition apparatus shown in Figure 1. The trichlorosilane is picked up in one of the three flasks by bubbling a small flow of hydrogen through the liquid. The saturated gas is then injected into the main hydrogen stream. The silicon substrate is resting on a graphite block heated by rf energy. A different reaction chamber with a resistive heated graphite strip is shown in Figure 2. Although this method of heating gives a more uniform temperature distribution than rf heating, the various metal parts in the reaction tube offer additional chances for contamination. Therefore, the rf heated version has been preferred.

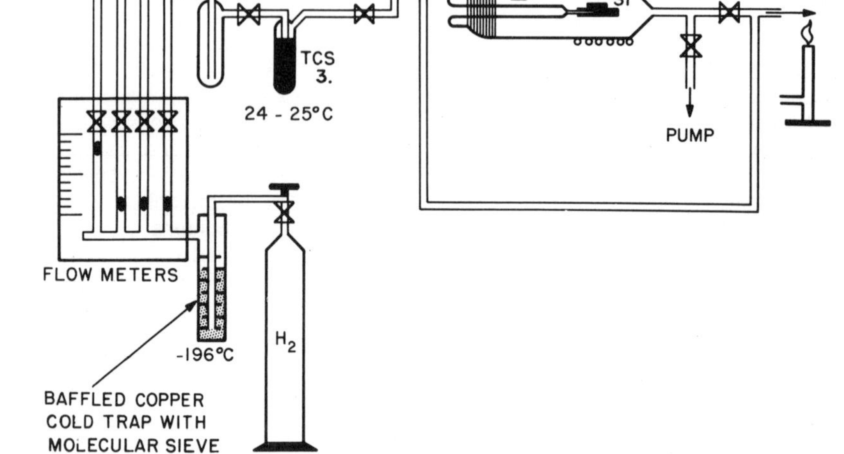

Fig. 1. Radio frequency heated silicon deposition system.

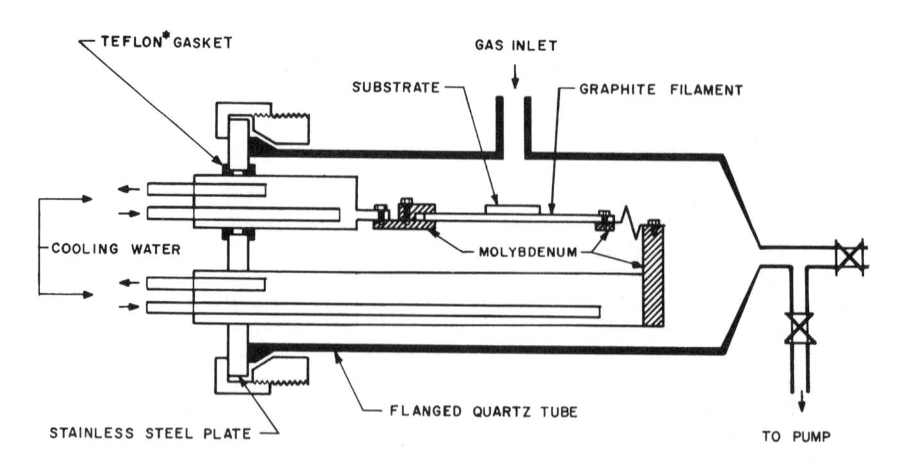

Fig. 2. Resistance heated silicon deposition chamber.

All deposition temperature values reported are uncorrected pyrometer readings. The true temperatures are estimated to be approximately 75°C higher.

The results obtained from about 200 deposition experiments with widely varying conditions do not fit easily into a pattern. Table III shows the complex dependency of deposit properties on some of the process parameters. The properties listed at the top determine the value and applicability of the process. An "X" indicates the process parameters by which these properties are affected. The multiplicity of parameters makes it necessary to begin the process development by standardizing as many of the parameters as possible and to maintain the established optimum conditions within narrow limits.

Deposit Crystal Perfection

The deposit crystal perfection was characterized by the dislocation density as revealed through chemical etching and by the surface topography. All deposits were made on {111} oriented wafers of 18 mm diam. Substrate preparation and deposition conditions severely affect the crystal perfection, and their influences are difficult to identify and control. With the system shown in Figure 1, optimum results were obtained at a deposition temperature of 1100°C. The substrates had dislocation densities of about 3×10^4 cm^{-2} and these were also the lowest densities observed in the deposits. Mechanically polished substrate surfaces, which had not been etched, gave deposits with dislocation densities at least twice as large. All deposits had imperfection densities of 10^6 cm^{-2} or higher near the periphery of the wafers. This is attributed to temperature gradients at the edge of the wafer.

A serious problem is the valves manipulating the flow of reaction gases. Ordinary glass or Teflon stopcocks require grease to be gas tight. The grease, however, is slowly attacked by the halide vapors and after several days of exposure, the system no longer holds a vacuum below 50 μ. Deposits grown under these conditions had dislocation densities higher than 10^6 cm^{-2}, and their surfaces showed pyramids with twinned or polycrystalline tips. The same effects occurred when insufficiently dried hydrogen was used or when the graphite heater had not been baked out properly. Consequently, the exclusion of oxidizing gases appears to be essential for the preparation of highly perfect deposits.

There seems to be a correlation between etch-pit density and the texture of the deposit surfaces. In Figure 3 various surface types of vapor grown deposits are reproduced. The needle or chevron patterns are usually associated with dislocation densities higher than 10^5 cm^{-2}. Figures 3e and 3f are the smoothest surfaces which could be achieved, and these deposits had dislocation densities below 10^{-5} cm^{-2}.

Deposit Thickness and Uniformity

The thickness and cross-sectional uniformity of the deposit is another important criterion with respect to practical applications. Figure 4 shows some of the reaction gas flow patterns that have been investigated. Curved deposit profiles are formed when the reaction gas enters the tube perpendicular to the substrate surface (Figs. 4a and 4b) whereas a wedge-shaped profile is grown from gas streams parallel to the surface. The situation pictured in Figure 4c has been preferred because the overall variation in thickness could be reduced by increasing the total hydrogen flow rate. This is shown in Figure 5 where thickness profiles are plotted and numerical data of variations are listed for several flow rates. The figures apply for round substrate wafers of about 18 mm diam. but neglect the pronounced accumulation of deposit at the

TABLE III

Dependency of Deposit Properties on Various Process Parameters

Deposit properties / Process parameters	Crystal perfection: (a) mirror surface (b) dislocations	Deposit thickness: (a) controllable (b) uniform	Deposition rate: (a) controllable (b) uniform	Deposit resistivity: (a) variable (b) reproducible
Apparatus and Experimental Techniques				
Flow pattern, tube, and heater geometry		X	X	
Purity of system: (a) trichlorosilane (b) heater material (c) hydrogen carrier gas (d) stopcocks and lines	X			X
Substrate preparation	X			
Temperature gradient in substrate	X			
Process Variables				
Deposition temperature	X		X	X
Deposition time		X		
Trichlorosilane feed rate	X		X	X
Hydrogen flow rate		X	X	
Impurity halide concentration in trichlorosilane reservoir	?			X

leading edge indicated on the graph. On the basis of these experiments, a standard flow rate of 4 standard liters per minute has been adopted.

If all deposition conditions are kept constant, the overall thickness of the deposit is strictly a function of time as shown in Figure 6. The scatter of experimental points is typical and indicates that the deposition system needs further refinement to allow a better degree of control.

Deposition Rate

The deposition rate was determined mostly by the trichlorosilane feed rate. According to Figure 7, the growth rate increases linearly with the amount of

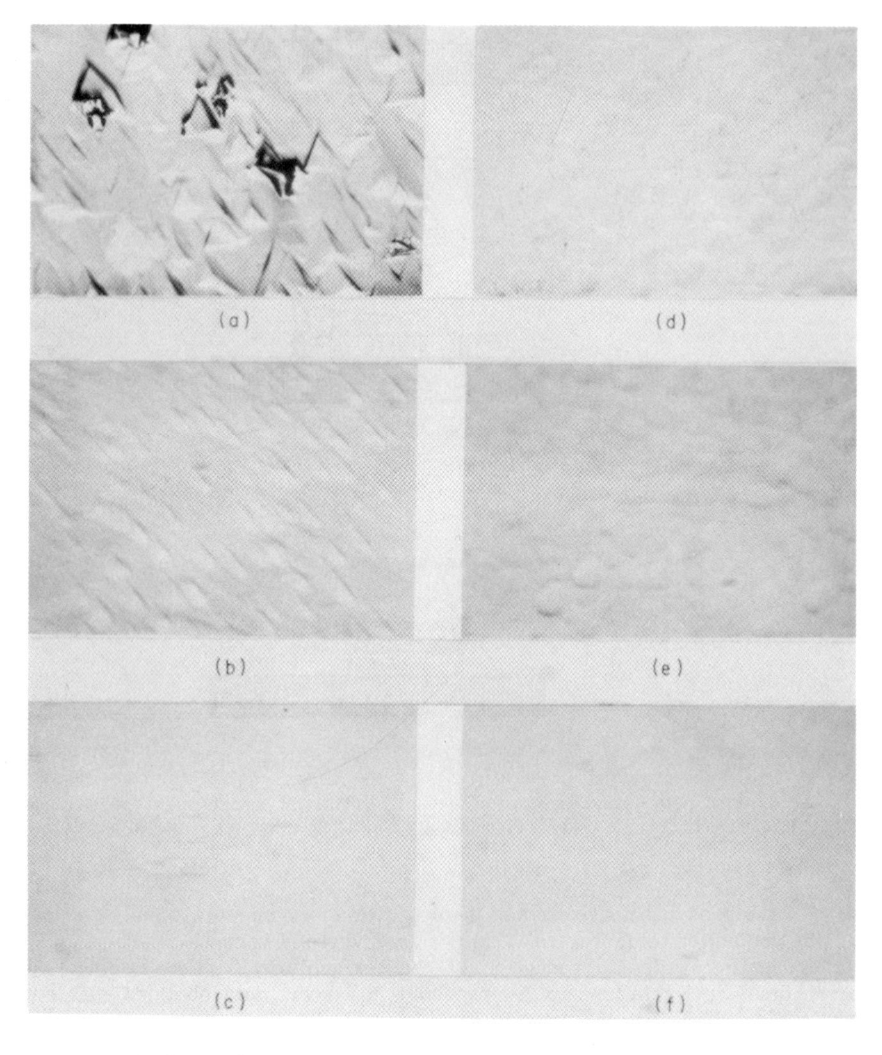

Fig. 3. Surfaces of vapor grown silicon deposits: (a) polycrystalline growth, (b) chevron pattern, (c) needle pattern, (d) speckled surface, (e) micro-orange-peel pattern, and (f) smooth surface.

silicon compound in the hydrogen gas. Beyond 1 mole-% of trichlorosilane the curve seems to level off and will probably reach a maximum, but this range has not been investigated. A variation of the hydrogen flow rate, with the trichlorosilane feed rate being constant, has little influence on the deposition rate and shifts the curve only slightly up or down, depending on how the residence time of the gas is affected.

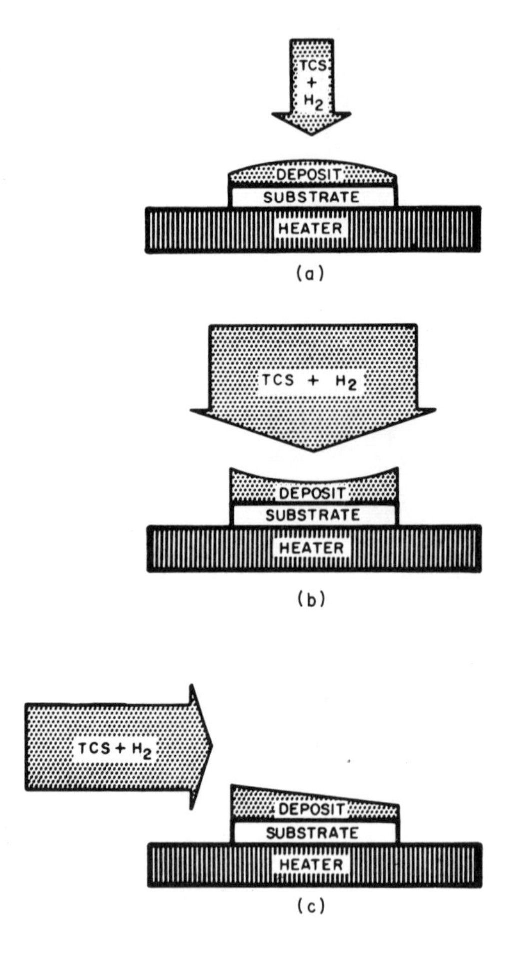

Fig. 4. Flow pattern and deposit thickness distribution: (a) narrow jet perpendicular to surface, (b) very wide jet perpendicular to surface, and (c) wide jet parallel to surface.

For thermodynamical reasons, the deposition rate should also be a function of the deposition temperature. This has been verified recently by Theurer (6) in connection with the reduction of the tetrachloride, where he found the Arrhenius-type relationship as expected. However, data obtained with our system do not follow the same pattern. In Table IV, the results of two series of experiments are listed, and there is no significant variation of growth rates within the narrow temperature range investigated. This is attributed to

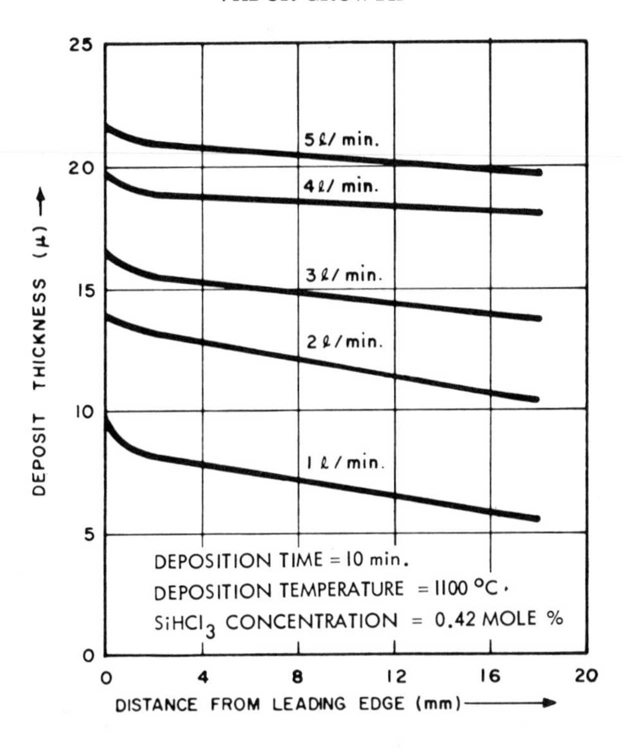

Fig. 5. Thickness profiles and thickness variation of silicon deposits for various hydrogen flow rates.

H_2 flow rate, l/min	Avg thickness, μ	Thickness variation, %
1	6.8	± 19
2	11.7	± 12
3	14.6	± 6
4	18.4	± 2
5	20.3	± 3

the particular design of the substrate heater and gas flow pattern in the reaction tube.

Controlled Impurity Introduction

The various factors influencing the resistivity of the deposit will now be considered. Before controlled doping can be attempted, it is necessary to determine the intrinsic impurity level of the system. The commercial semiconductor grade trichlorosilane is pure enough to yield n-type layers of 50 ohm cm resistivity or high without additional purification. If this is not achieved, the deposition apparatus is contaminated. As as example, a molybdenum block used as rf adapter and substrate heater was found to be responsible for producing 1 ohm cm n-type deposits with great consistency.

In order to grow deposits with specified resistivities, varying amounts of either phosphorus or boron tribromide were added to the trichlorosilane. If

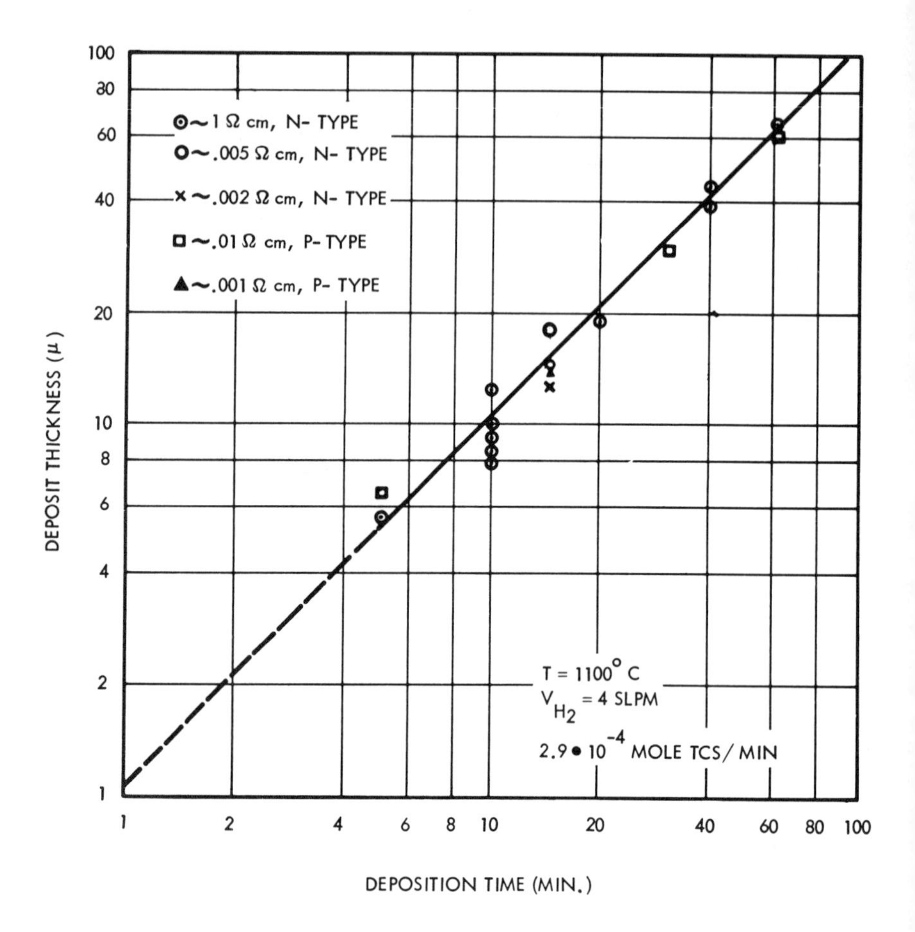

Fig. 6. Deposit thickness as function of deposition time.

these doping agents are mixed with the liquid trichlorosilane, they do not evaporate at the same rate as their solvent; therefore, during evaporation, a trichlorosilane solution prepared to a certain impurity concentration will constantly change. Since both impurity halides have vapor pressures smaller than trichlorosilane, they tend to accumulate in the liquid. As a consequence, deposits grown from the same solution show decreasing resistivities according to the sequence of growth. The impurity accumulation effect can be counteracted by bubbling wet hydrogen through the trichlorosilane. The water vapor hydrolyses the impurity halides preferentially and deposit resistivities are increased after the treatment. The trend of resistivities obtained with a phosphorus trichloride doped solution and the effect of the wet hydrogen treatment are shown in Figure 8.

Phosphorus is easily incorporated into silicon deposits, and a wide range of n-type resistivities can be produced as shown in Figure 9. To achieve the same p-type impurity levels, higher boron concentrations are required. The same rule applies also for the silicon tetrachloride process as has been shown by Theurer (6).

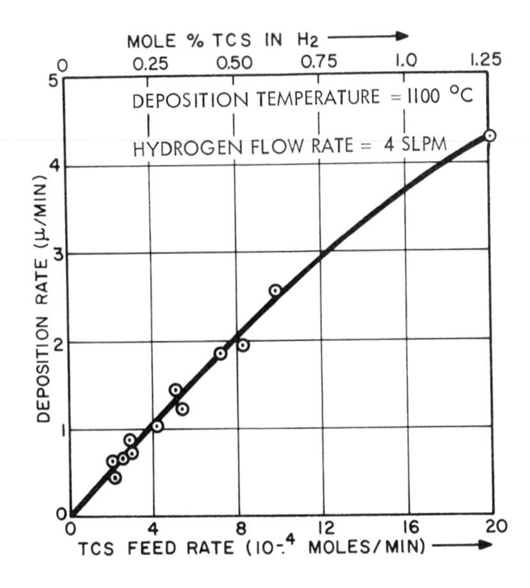

Fig. 7. Deposition rate as function of trichlorosilane feed rate.

The quantitative relationship between resistivities and impurity halide concentrations as represented in Figure 9 is only valid for the special depositon conditions listed above the curves. A change in deposition temperature, for example, slightly alters the deposit resistivity. Table V shows two consecutive series of experiments made with the same solution under identical conditions but at different deposition temperatures. More phosphorus is incorporated at the higher temperatures. It has not been investigated, however, if the same effect occurs with more strongly doped solutions.

The resistivity of a deposit is not yet fully determined when deposition temperature and impurity concentration of the trichlorosilane solution have

TABLE IV

Effect of Deposition Temperature on Deposition Rate

Deposition temp., °C	Deposition rate, μ/min
Experiment 1	
1100	0.7
1125	0.65
1150	0.8
1175	0.75
1200	0.8
Experiment 2	
1100	0.7
1125	0.7
1150	0.85
1175	0.8
1200	0.8

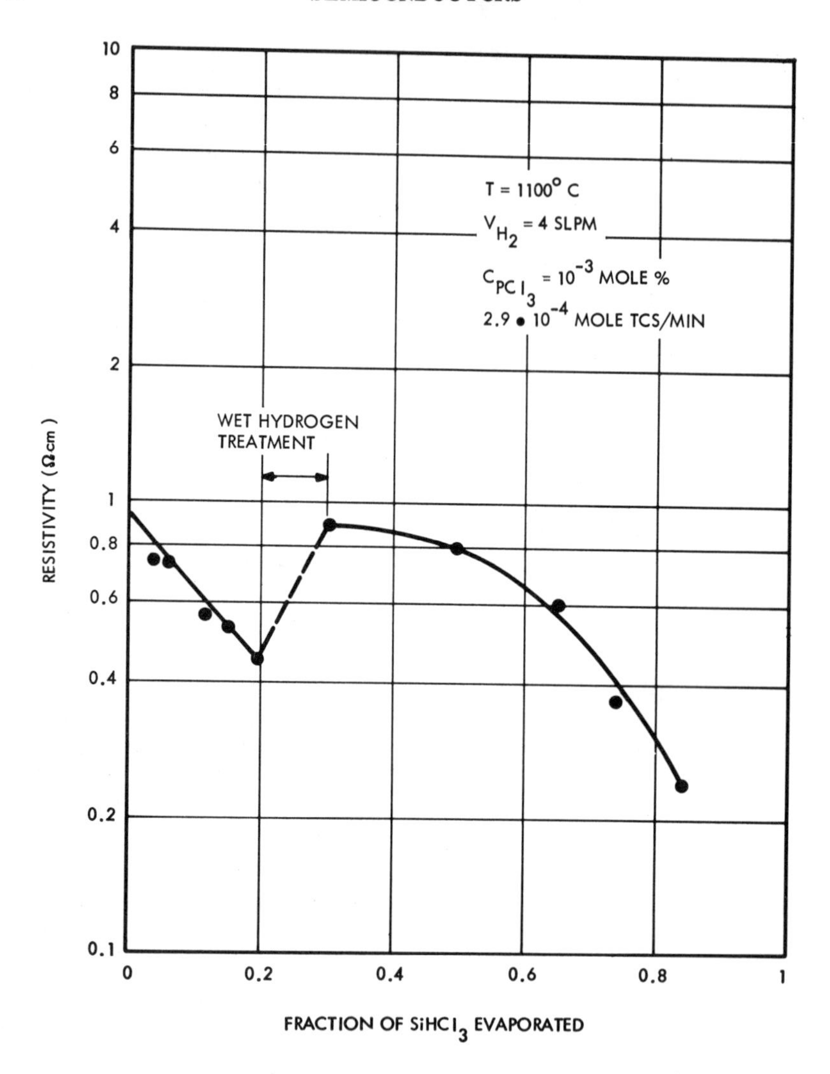

Fig. 8. Resistivities of silicon deposits vs fraction of evaporated trichlorosilane solution.

been chosen. The same solution may produce resistivities varying by an order of magnitude depending on the rate of evaporation. As plotted in Figure 10, resistivities decrease sharply if the amount of combined trichlorosilane and impurity halide vapors injected into a constant stream of hydrogen is increased. The resistivity changes are most drastic for halide concentrations below 1 mole-%. At higher concentrations the curves seem to level off, and the effect may not be observed. For the operating conditions chosen, control of the trichlorosilane evaporation rate was most critical in order to reproduce resistitivities.

A limitation of the system shown in Figure 1 is the residual contamination left in the feed lines after one type of dopant has been used. Subsequent deposits are partially or totally compensated depending on the solution concen-

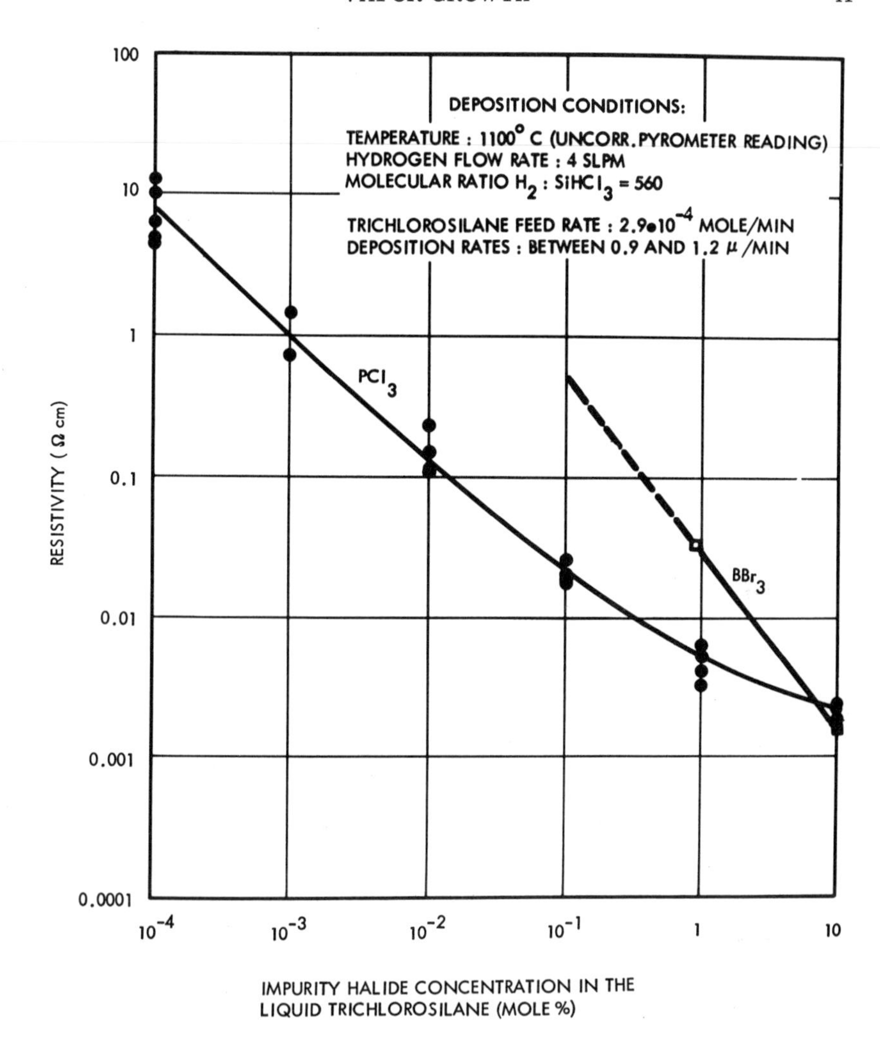

Fig. 9. Resistivity of silicon deposits vs impurity halide concentration in the trichlorosilane solution.

tration used. This effect restricts the possibilities of growing electrical junctions and multilayered structures with this system.

CONCLUSIONS

Summarizing the results, it is apparent that many of the problems associated with vapor growth of semiconductors are not as yet solved satisfactorily. Even for silicon, deposition processes need further refinement in process control, modes of operation, and materials used in building deposition systems.

To fully exploit the potential of vapor growth methods for semiconductor devices, controlled growth through masking will have to be developed.

TABLE V

Effect of Deposition Temperature on Resistivities Obtained from One and the Same Trichlorosilane Solution

Deposition temp., °C	Resistivity (n), ohm cm
Experiment 1	
1100	4.8
1125	4.3
1150	3.6
1175	3.3
1200	2.9
Experiment 2	
1100	4.9
1125	5.3
1150	3.9
1175	3.1
1200	2.9

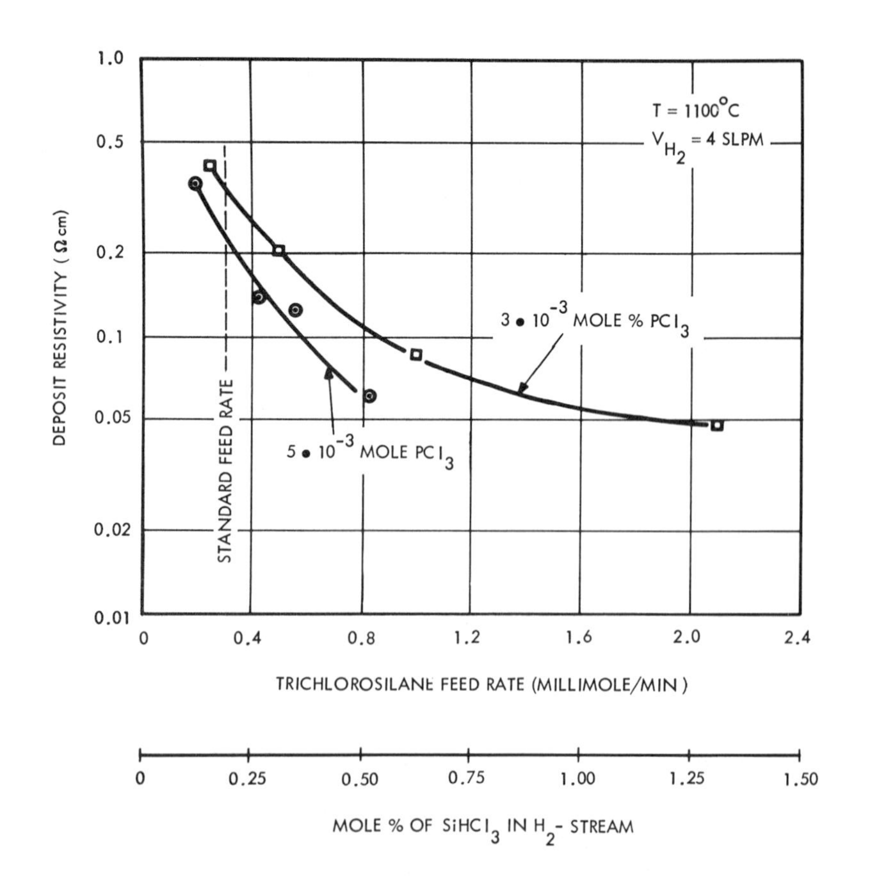

Fig. 10. Deposit resistivity vs amount of trichlorosilane solution injected into the hydrogen stream.

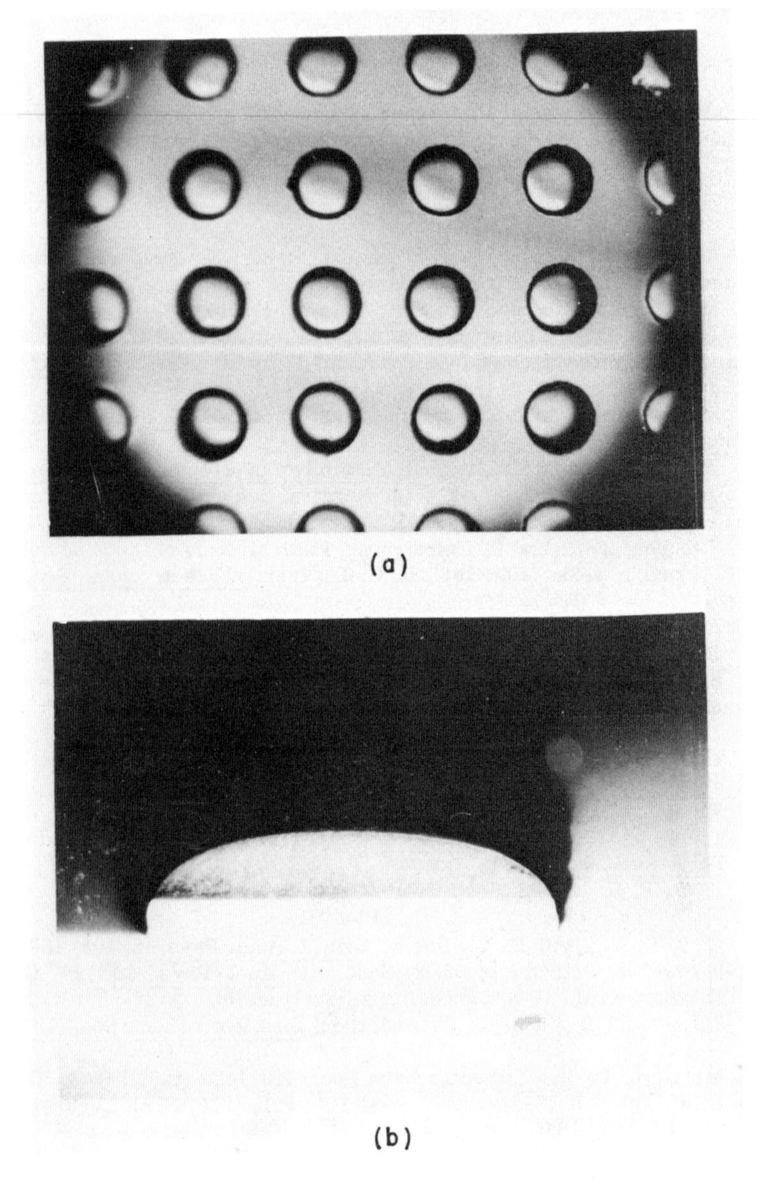

Fig. 11. Array of n-type silicon dots (1.5 mm diam) deposited on p-type substrate through silicon mask. (a) Top view. 6.7 ×. (b) Cross section. 40 ×.

Figure 11 shows an array of n-type dots which have been deposited through a silicon mask resting on a p-type substrate wafer.

Considering the multitude of new and complex structures, which are conceivable as a result of vapor growth methods, it is felt that deposition processes will lead to the development of a variety of solid state devices far beyond the scope of today's conventional epitaxial transistor.

ACKNOWLEDGMENTS

The authors wish to acknowledge the valuable assistance of Messrs. John G. Kren, who performed the deposition experiments, and William J. Patrick, who analyzed the samples. The paper is based on an experimental program which has been supported by the U. S. Army Signal Research and Development Laboratory, Fort Monmouth, New Jersey.

References

1. Allegretti, J. E., and D. J. Shombert, Recent Newspaper, 117th Meeting of the Electrochem. Society, Chicago, Ill., May 1960.

2. Mark, A., J. Electrochem. Soc., 107, 568 (1960).

3. Loar, H. H., H. Christensen, and J. J. Kleimac, Joint Conference of the AIEE and IRE on Research of Semiconductor Devices, Pittsburgh, Pa., June 1960.

4. Eleven papers on vapor growth of Ge and Si, IBM J. Research and Development, 4, 248 (1960).

5. Wajda, E. S. and R. Glang, in Metallurgy of Elemental and Compound Semiconductors, R. O. Grubel, ed. (Metallurgical Society Conference, Vol. 12), Interscience, New York-London, 1961, p. 229.

6. Theurer, H. C., and H. Christensen, 118th Meeting of the Electrochem. Society, Houston, Texas, Oct. 1960; H. C. Theurer, J. Electrochem. Soc., 108, 649 (1961).

7. Early, J. M., Electron Device Meeting, Washington, D. C., Oct. 1960.

8. Russel, G., Electron Device Meeting, Washington, D. C., Oct. 1960.

9. Sandler, N. P., S. A. Prussin, and A. Stevenson, Electron Device Meeting, Washington, D. C., Oct. 1960.

10. LaFond, Ch. D., Missiles and Rockets, p. 24 (Dec. 1960).

11. Czochralski, J., Moderne Metalkunde, Springer Verlag, Berline, 1924, and Z. Phys. Chem., 92, 219 (1918).

12. Bridgman, P. W., Proc. Am. Acad. Arts Sci., 60, 305 (1925).

13. Pfann, W. G., Trans. AIME, 194, 747 (1952); W. G. Pfann and K. M. Olsen, Phys. Rev., 89, 322 (1953).

14. Keck, P. H., and M. S. E. Golay, Phys. Rev., 89, 1297 (1953).

15. VanArkel, A. E., Physica, 61, 316 (1925).

16. Gow, K. W., and B. Chalmers, Brit. J. Appl. Phys., 2, 103, 106, 300, 675 (1951); A. A. Petranskas and F. Gandry, J. Appl. Phys., 20, 1257 (1949).

17. Hirshon, J. M., U. S. Signal Corps., Contract No. DA-36-039-SC-72768.

18. Litton, F. B., and H. C. Anderson, J. Electrochem. Soc., 101, 287 (1954).

19. McCarty, L. V., J. Electrochem. Soc., 106, 1036 (1959); C. S. Herrick and J. G. Krieble, J. Electrochem. Soc., 107, 111 (1960); H. C. Theurer, Bell Labs. Rec., 33, 327 (1955).

20. Sangster, R. C., E. F. Maverick, and M. L. Croutch, J. Electrochem. Soc., 104, 317 (1957).

21. Szekely, G., J. Electrochem. Soc., 104, 663 (1957).

22. vanDerLinden, P. C., and J. deJonge, Rec. trav. chim., T78, 962 (1959).

23. Kempter, Ch. P., and C. Alvarex-Tostado, U. S. Pat. 2,840,489 (1958).

24. Schaefer, H., and B. Morcher, Z. Anorg. u. Allgem. Chem., 290, 279, (1957).

25. Lassen, H., and L. Brueck, Ann. Physik, 22, 65 (1935).

26. Mayer, H., Physik duenner Schichten, Vol. II, Wissenschaftliche Verlagsgesellschaft m.b.H., Stuttgart, 1955, p. 105.

27. Allegretti, J. E., D. J. Shombert, E. Schaarschmidt, and J. Waldman, "Vapor Deposited Silicon Single Crystal Layers" Electronic Chemicals Division, Merck & Co., Rahway, N. J.

28. Beatty, H. J., R. Glang, E. S. Wajda, and W. H. White, U. S. Signal Corps, Contract No. DA-36-039-SC-87395, First Quarterly Progr. Rept., p. 10.

29. Hergenrother, K. M., S. E. Mayer, and A. I. Mlavsky, "Silicon Carbide," Proceedings of the Conference on Silicon Carbide, Boston, Mass., April 1959, p. 60.

30. Five papers by various authors in "Silicon Carbide," Proceedings of the Conference on Silicon Carbide, Boston, Mass., April 1959, p. 67-114.

31. Hagenlocher, A., Recent Newspaper No. 2, 119th Meeting of the Electrochem. Society, Indianapolis, Ind., May 1961.

32. Antell, G. R. and D. Effer, J. Electrochem. Soc., 106, 509 (1959).

33. Newman, R. L., and N. Goldsmith, Recent Newspaper No. 3, 119th Meeting of the Electrochem. Society, Indianapolis, Ind., May 1961.

34. Gershenzon, M., and R. M. Mikulyak, J. Electrochem. Soc., 108, 548 (1961).

35. Keith, J. N., and E. H. Thompkins, Armour Research Foundation, Final Report 3819-4, Project C819.

DISCUSSION

C. O. THOMAS (Bell Telephone Labs.): On the data showing the effect of growth temperature upon resistivity, what was the nature of the substrate wafer that you were using and how was the resistivity measured?

R. GLANG: All n-type deposits were made on p-type substrate wafers with resistivities of at least 1 ohm cm, usually they were higher than that. Resistivities were measured with a four-point probe apparatus.

C. O. THOMAS: Do you find the same relationship if the substrate is n^+?

R. GLANG: We have not investigated n^+- substrates.

T. B. LIGHT (Bell Telephone Labs.): What etch did you use to determine dislocation densities in silicon?

R. GLANG: We used an etch consisting of 30 ml HF, 15 ml HNO_3, 0.1 ml Br_2, 1.1 g $Cu(NO_3)_2$, and 450 ml of water.

D. M. MATTOX (Sandia Corp.): It has been reported by some Russians that they are having a little trouble with a deposition such as this on a quartz substrate. They found that the type, whether it is n- or p-type, is very dependent upon the grain size. Apparently this is a result of oxidation although they do not mention it. Do you think the oxidation and backstreaming in your open tube process is affecting the layer somewhat?

R. GLANG: I do not think that we have any significant oxidation in our system; at least not if the hydrogen is sufficiently dry and free of oxygen.

D. M. MATTOX: Prevention of backstreaming is quite often a very difficult thing to do. Bruce and some co-workers at Westinghouse have recently reported some Hall measurements on epitaxial films in which heat treating changed the conduction type. As I recall they did not give an explanation, but this could be caused by oxidation and I was wondering if you took any special precautions.

R. GLANG: I have no observations in this direction and, considering the relatively high flow rate of 4 standard liters/min of hydrogen being squeezed through the 2 mm opening at the outlet, I do not think that back diffusion of air has occurred.

J. H. SUNG (Hoffman Semiconductors): You showed six types of wafer surface in your figures. One is polycrystalline; are the rest single crystals?

R. GLANG: Yes, they are.

J. H. SUNG: On your dislocation counts, I wonder if they were made on the surface of the epitaxial layer or on a cross section.

R. GLANG: The count was made on the surface of the deposit. Since our substrate was of the opposite conductivity type, we checked with a hot probe after etching to verify that the layer had not been removed. In certain instances we determined that the amount of silicon removed did not exceed 1 or 2 μ.

S. K. TUNG (Bell Telephone Labs.): As I recall from your Figures you used a deposition rate of between 0.7 and 0.8 μ/min with a temperature of 1100 and 1200°C, respectively. Do you consider this a real difference in deposition rate or do you consider the rate constant within this temperature range?

R. GLANG: I do not think that the variation of the deposition rate with temperature is systematic. I think the variation is random.

S. K. TUNG: That means you do not consider the deposition temperature as a process parameter. Is this statement correct?

R. GLANG: In respect to deposition rate, yes. The system which we have been using has this particular feature. But the temperature has an influence on the resistivity and crystal perfection and therefore is one of the process parameters.

S. K. TUNG: Do you have any deposition rate data outside of this temperature range?

R. GLANG: No, I do not.

V. R. ERDELYI (Knapic Electro-Physics): What etchants did you used prior to deposition?

R. GLANG: We used a 10:1 mixture of $HN0_3$ and HF.

V. R. ERDELYI: Have you tried any electrolytic etching?

R. GLANG: No, we wanted to but did not quite make it.

V. R. ERDELYI: You said that the etching is followed simply by mechanical polishing. From this one can conclude that you have not cleaned up the surface after mechanical polishing. Is this conclusion right?

R. GLANG: No, we have used two types of wafers; those which have been mechanically polished and cleaned in organic solvents, and others that had been polished and then chemically etched. The etched substrates give superior results.

V. R. ERDELYI: Do you think that the two-color optical pyrometer is to be preferred over a brightness instrument?

R. GLANG: I do not know. Our temperature control was achieved by means of a thermocouple inside the carbon block. The pyrometer was only used to determine surface temperature.

H. BASSECHES (Bell Telephone Labs): What thickness were your deposited layers? I do not recall seeing any figures.

R. GLANG: We have deposited layers between 5 and 60 μ thick. Most of the layers included in this work were between 10 and 20 μ.

H. BASSECHES: Have you noticed any relationship between thickness and resistivity of these layers?

R. GLANG: No, we did not investigate this particular point.

J. J. McMULLEN (Rheem Semiconductor Corp.): Most of my questions seem to have been answered with the exception of one relating to the pretreatment of the substrates: I do not believe you mentioned whether they were Czochralski or floating-zone crystals. What relation did you find between dislocation count on the substrate and that on the layer?

R. GLANG: All wafers were cut from a floating-zone rod with dislocation densities of 2 to 3×10^4 cm^{-2}. The lowest dislocation densities obtained in the deposits were of the same order. If we had deteriorating effects during the deposition, higher dislocation densities would have been observed.

J. J. McMULLEN: You remarked that on your good wafers you had a dislocation count on the deposit of 2×10^4 cm^{-2}. I presume, therefore, that the substrate must have had that or larger.

R. GLANG: We did not investigate every single substrate wafer, but only representative samples from each crystals. These samples always had dislocation densities from 2 to 3×10^4 cm^{-2}.

J. J. McMULLEN: Again on the pretreatment, you did not mention your thermal treatment.

R. GLANG: Normally we did not use a thermal treatment.

J. J. McMULLEN: You put the wafers immediately into the system?

R. GLANG: Yes, our usual preparation consisted of heating the wafer up to the deposition temperature, leaving it there for about 10 min, and then starting the deposition. We did not pretreat at 1275°C as is recommended by some investigators.

J. J. McMULLEN: To re-emphasize your point that the data collected are quite dependent on the particular system used, the three configurations you showed (humps, depressions, and wedge shapes) have all been observed in our laboratory in the same system.

R. GLANG: I would expect the basic behavior to be the same regardless of the system, but the numerical data may vary considerably depending on the mode of operation and the dimensions of the apparatus.

P. WANG: (Sylvania Electric Products): In your results dealing with the layer thickness and resistivity, have you any profiles of these properties, and if so, is there any variation in these parameters?

R. GLANG: No, we have not investigated resistivity variations within the layer.

C. O. THOMAS: In Table II you show a normal gross temperature range of 1200 to 1300°C for the hydrogen reduction process, but most of the samples that you discussed were grown between 1100 and 1200°C. Can you say anything about the electrical properties of the films that you grew in the higher range, in addition to the resistivity data that you have already presented?

R. GLANG: The temperatures given in the disucssion of our own experiments are pyrometer readings. In the table which you mentioned the temperatures are corrected. When corrected, our experimental temperatures fall into the range given in the table.

J. WATSON (Texas Instruments): Regarding the electrical characteristics, were the junctions actually usable?

R. GLANG: We did not investigate the junction properties.

H. M. MANASEVIT (Autonetics): Have you studied the effects of Pyrex or quartz containers on the properties of the trichlorosilane or tetrachlorosilane?

R. GLANG: In our experiments we did not find an indication that the Pyrex glass, which was used as a container for the liquid trichlorosilane, had an effect on the resistivity or the crystalline perfection of the deposit.

H. M. MANASEVIT: You mentioned the use of stopcocks. Do you use any particular stopcock grease?

R. GLANG: We have tired Dow-Corning silicone grease and Kel-F grease and are unsatisfied with both.

Halogen Vapor Transport and Growth of Epitaxial Layers of Intermetallic Compounds and Compound Mixtures

N. HOLONYAK, JR., D. C. JILLSON, and S. F. BEVACQUA

General Electric Company, Syracuse, New York

Abstract

Methods of synthesizing or regrowing and doping a number of intermetallic compounds and mixtures from the vapor state are described and some results are presented.

The general procedure is to place the constituent elements or source intermetallic compound, together with suitable quantities of halogen or halide and a seed crystal or substrate, if desired, in different parts of a quartz tube, seal under high vacuum or with halogen gas back-fill, and heat in a manner such that the source elements or crystals are at high temperatures relative to the seed end of the tube. Halogen vapor attacks nonvolatile material, transports it to a cooler part of the tube, deposits it on a seed crystal (or on the tube itself), and repeats the cycle, thus providing continuous epitaxial deposition and growth of the compound until the source is depleted.

Gallium arsenide, gallium phosphide, gallium antimonide, and indium phosphide have been transported in this manner and deposited, with controlled doping, either as large crystallites when seeded on the walls of the tube, or as epitaxial layers when deposited on seed crystals. Material has been transported simultaneously from different compounds and deposited as a mixed compound, e.g., Ga(As,P). Compounds have been synthesized from the elements and deposited, e.g., GaP. The p-n junctions, including low current density tunnel junctions, have been prepared by the deposition of epitaxial layers. Heterojunctions have been formed by deposition of compounds on dissimilar substrates, e.g., GaAs on Ge, $GaAs_xP_{1-x}$, on $GaAs_yP_{1-y}$, and GaP on GaAs.

INTRODUCTION

Recently, Marinace (1) has described a system in which iodine reacts in a closed quartz tube with a germanium source crystal, transports the germanium as an iodide vapor, and deposits it epitaxially, via disproportionation reactions, upon a germanium or gallium arsenide seed crystal maintained at a lower temperature. If the source crystal is doped properly or if a doping impurity is included in the closed tube, it is possible to grow an epitaxially deposited layer of conductivity type opposite to that of the seed crystal and thus form a junction layer, or indeed a layer of arbitrary conductivity type and resistivity. The purpose of this paper is to describe a simple closed-tube process, which is basically an extension of that of Marinace, in which a free halogen or a halogen supplied by a metal halide is used to transport and grow epitaxial layers of intermetallic compounds and mixed intermetallic compounds.

From Marinace's experiments with germanium it is not obvious that similar results could be expected with compounds. Experimentally, however, we have established that a number of intermetallic compounds can be transported by means of a halogen and that separate compounds such as gallium arsenide and gallium phosphide can be simultaneously transported and combined into mixed compounds, $GaAs_xP_{1-x}$.

Intermetallic compounds such gallium arsenide and phosphide, and indium phosphide are very readily transported and deposited either heterogeneously upon the walls of the reaction vessel or epitaxially upon seed crystals. These results and the fact that gallium antimonide is transported slowly and with great difficulty suggest that only those compounds with at least one relatively volatile constituent are good candidates for transport and epitaxial growth in the closed-cycle system. Although this paper is not concerned with the detailed chemical processes which govern transport and epitaxial growth of compounds, it should be mentioned that the statements above and results presented later for compounds suggest that as the source crystal is heated some of the more volatile constituent (arsenic or phosphorus) is driven off until the source is in equilibrium with its more volatile component; the halogen sealed in the system attacks and transports the less volatile constituent (gallium), and by disproportionation reactions at an appropriate lower temperature the less volatile constituent is released and combines with some of the gaseous volatile component and reforms the compound. The coolest part of the reaction tube is held at a temperature such that the more volatile constituent of the intermetallic compound is at one or more atmospheres pressure. This apparently insures that the transported elements will combine and deposit as the compound and aids in suppressing thermal decomposition of the newly grown compound.

In the following paragraphs we shall describe the experimental conditions and give typical results obtained in the transport and epitaxial growth of gallium arsenide, phosphide, and mixed $GaAs_xP_{1-x}$. Various junction structures and heterojunctions which have been fabricated will be described.

EXPERIMENTAL PROCEDURE

All experiments were carried out in 1 cm diam closed quartz tubes ranging in length from 15 to 37 cm. A preselected quantity of gallium arsenide or phosphide, gallium arsenide and phosphide, or gallium was placed near one end of the tube, and 5 to 20 mg of excess arsenic or phosphorus and usually 5 to 50 mg of either $ZnCl_2$, $CdCl_2$, $SnCl_2$, $MgCl_2$, $HgCl_2$, $AlCl_3$, or $SbCl_3$ were also placed into the tube. Seed crystals, if used, were placed in the tube at suitable locations as discussed later in specific examples. The tube and its contents were very carefully heated and outgassed to insure removal of water before evacuation and sealing. If the halogen for transport purposes was not introduced by means of a metal halide, the quartz tube was backfilled with approximately 1/4 atm of chlorine, or alternatively approximately 10 mg of iodine was sealed into the tube. The end of the tube containing the relatively nonvolatile materials or the source crystals was heated from 900° to 1100°C and the cooler end from 600° to 900°C. When seed crystals were used for the deposition of epitaxial layers, lower temperatures were frequently employed than when deposition was allowed to proceed on the walls of the reaction vessel, and a given run was sometimes as short as an hour. When deposition is sought on the vessel walls, a run may be allowed to proceed for as long as two weeks to produce more massive crystals.

Gallium Arsenide

Following the methods outlined above, gallium arsenide has been transported and large polycrystals have been grown directly on one end of quartz reaction tubes. Iodine, $ZnCl_2$, $HgCl_2$, $CaCl_2$, $SnCl_2$, $AlCl_3$, and $MgCl_2$ have been used to transport GaAs. When one of the metal chlorides is employed, it is obvious that a doping impurity is introduced along with the halogen necessary for transport purposes. The $ZnCl_2$ and $CdCl_2$ have been used to transport GaAs and dope it degenerately p-type. Tunnel diodes have been alloyed on the materials so prepared and have yielded peak-to-valley current ratios as high as 30:1. Crystals transported and doped with $HgCl_2$ or $MgCl_2$ were p-type but not degenerate. Crystals transported and doped with $SnCl_2$ were n-type with carrier concentrations of 3.5×10^{18} cm^{-3} and were easily made degenerate p-type when free zinc was sealed in the reaction vessel with the $SnCl_2$.

By the methods described above both n-type and p-type gallium arsenide have been grown epitaxially upon, respectively, p-type and n-type seeds. X-ray determinations indicate that the deposited material has the orientation of the seed crystal, and such structures exhibit the usual junction properties. Since one of the goals of this activity was to grow tunnel junctions, it was necessary to attempt the crystal growth at low temperature in order to obtain "step" junctions. With seed crystals held at approximately 600°C and source crystals in the range from 800° to 1000°C it is possible to get good epitaxial growth and to get junctions which exhibit tunneling with low current density.

Figure 1 shows a metallurgical cross section of a gallium arsenide p-n junction obtained by gowing p-type gallium arsenide on an n-type seed doped approximately 10^{19} atoms cm^{-3}. The $ZnCl_2$ (10 to 20 mg) was sealed into the reaction vessel to effect the vapor transport of the source gallium arsenide. The source crystal was held at approximately 900°C and the seed crystal at ap-

Fig. 1. Cross section of the gallium arsenide epitaxially grown p-n junction. P-type gallium arsenide grown on n-type seed. $ZnCl_2$ transport and doping agent. Cross-diffusion effects or surface damage evident in seed.

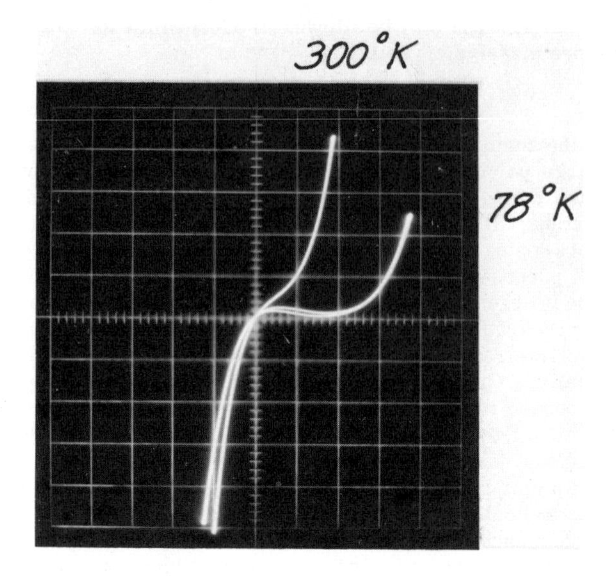

Fig. 2. Room temperature and liquid nitrogen current-voltage characteristics of epitaxially grown gallium arsenide p-n junction which exhibits tunneling. P-type gallium arsenide grown on n-type seed. $ZnCl_2$ transport and doping agent. Vertical scale: 0.01 ma/division; horizontal scale: 0.1 v/division.

proximately 600°C in an overnight run. At low magnification there appear to be two junctions in the crystal. However, as is evident from Figure 1 only one junction is present. Below the etched line indicating the junction there appears a band that may result from surface damage on the n-type seed or impurity cross diffusion from the epitaxially grown p-type layer (top) into the seed or vice versa. The I-V characteristic of the junction is shown in Figure 2. The room temperature (300°K) characteristic indicates low current density tunneling. This is clearly established by the negative resistance that is evident at liquid nitrogen temperature (78°K).

Perhaps a more interesting type of structure which has been fabricated in GaAs by vapor transport and epitaxial growth is an $n^+ \pi p^+$ diode. Crystals doped n^+ with approximately 10^{19} atoms cm^{-3} were used as seeds. The center π-type regions were grown roughly 0.025 mm thick by utilizing $CuCl_2$ as the transport and doping agent. A copper-doped GaAs layer grown by using approximately 20 mg $CuCl_2$ in the closed-tube process appears to be sufficiently compensated to give the high resistivity gallium arsenide which Gooch and coworkers (2) call semi-insulating (SI). A layer of p^+ GaAs was grown on the π or SI regions in overnight runs employing approximately 20 mg $ZnCl_2$.

The $n^+ \pi p^+$ diodes fabricated in our laboratory have the expected capability of blocking current in the reverse direction while in the forward direction they conduct by means of space-charge-limited emission into the π or SI region. Gooch and co-workers (2) have also observed space-charge-limited emission in their experiments with SI GaAs. Depending upon the particular experimental run, forward currents have been observed over a range of four decades, which increase as V^n, where n ranges from 2.5 to 3.6. These diodes have the very interesting property of being more photosensitive in the forward than in the reverse direction.

Other interesting applications exist for gallium arsenide grown epitaxially on various substrates. Utilizing the methods described above and $ZnCl_2$ as a

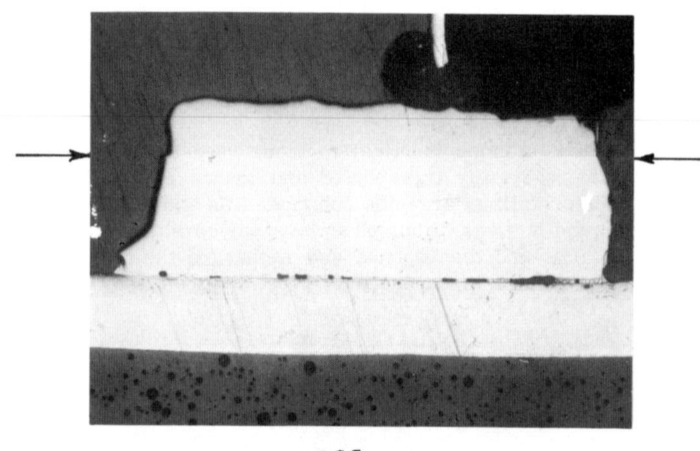

Fig. 3. Cross section of GaAs-Ge heterojunction. P-type GaAs grown on n-type Ge. $ZnCl_2$ transport and doping agent.

transport and doping agent, p^+ gallium arsenide has been grown on semi-insulating gallium arsenide and Hilsum et al.'s (3) recently announced low-inductance tunnel diode configuration has been duplicated. Germanium substrates are perhaps more interesting for deposition and epitaxial growth of gallium arsenide because of the possibility of fabricating wide band gap emitters on germanium transistors. Figures 3 and 4 show cross sections of a p-n junction prepared by growing p-type gallium arsenide (top) on n-type germanium (bottom). The $ZnCl_2$ was used as the transport and doping agent; the seed germanium crystals were held at 550 to 600°C and the source gallium

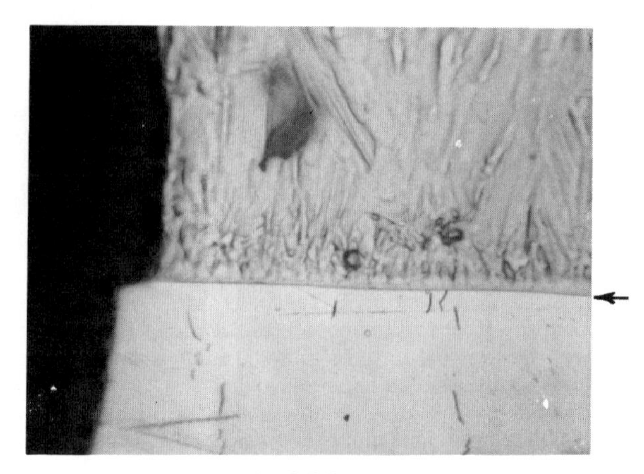

Fig. 4. Cross section of GaAs-Ge heterojunction. P-type GaAs grown on n-type Ge. $ZnCl_2$ transport and doping agent. Step at left produced by electrolytic etching of junction.

arsenide at 900°C in an overnight run. Electrically, the junction of Figures 3 and 4 displayed a forward characteristic intermediate between those of germanium and gallium arsenide p-n junctions.

Gallium Phosphide

Using prodedures almost identical to those used for gallium arsenide, gallium phosphide is readily transported and grown epitaxially, on either gallium phosphide or gallium arsenide substrates. In some early experiments we transported and grew gallium phosphide polycrystals into one end of a quartz reaction tube and then opened and recharged the tube and continued the growth with a change in conductivity type. Figure 5 shows an I-V char-

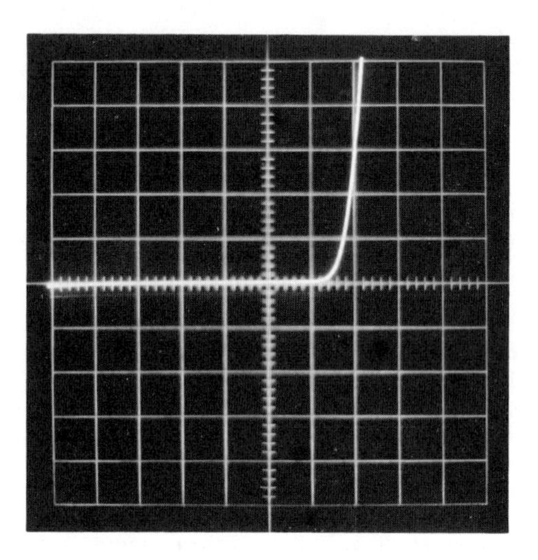

Fig. 5. Current-voltage characteristic of gallium phosphide junction. P-type gallium phosphide transported with $ZnCl_2$ and grown on n-type GaP transported with $SnCl_2$. Vertical scale: 1.0 ma/division; horizontal scale: 1.0 v/division.

acteristic of a gallium phosphide junction prepared in this manner. The polycrystalline seed (initial growth) was grown in a three-day run with a gallium phosphide source, 5 mg $SnCl_2$, 5 mg P, and 2 mg Te. The source end of the tube was held at 1120°C and the cool end of the tube at 920°C. After the tube was recharged with more GaP, 5 mg $ZnCl_2$, and 5 mg P, the run was repeated with the same temperature conditions for one day. A longitudinal section through the polycrystalline growth showed clearly that individual crystallites had continued their growth, with a change in conductivity type, from the first to the second run. Some of the junctions obtained in this experiment were cooled to 78°K and in the forward direction of bias gave visible red light emission as expected.

A rather striking demonstration of the propensity for epitaxial growth with gallium phosphide is shown in Figure 6. Here a polycrystalline seed (shown below line A-B) was employed. With the source at 1050°C and the seed at 750°C, the epitaxial layer shown above A-B was grown. The $ZnCl_2$ was used as the transport and doping agent. The crystallite indicated by the arrow was perpetuated perfectly from the seed into the epitaxial layer.

With a slight modification of the usual transport procedure, it is readily possible to synthesize gallium phosphide from the elements and under certain

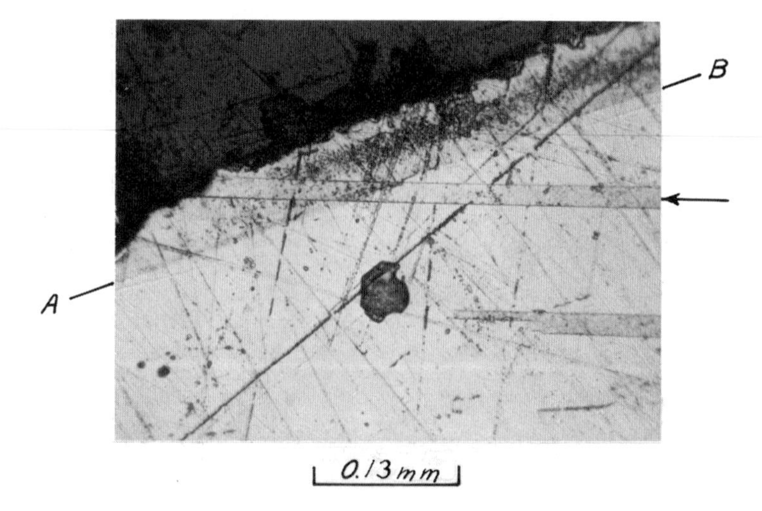

Fig. 6. Cross section of gallium phosphide epitaxial layer, above line A-B, grown by means of $ZnCl_2$ transport on a polycrystalline seed. Crystallite designated with the arrow perpetuated from seed into epitaxially grown region.

circumstances cause deposition to occur epitaxially upon a seed crystal. In experiments for this purpose the source materials for epitaxial deposition are gallium and red phosphorus. The gallium is sealed at one end of the tube, to be held at 1000 to 1100°C, and an excess of phosphorus at the opposite end, to be held between 450 and 500°C or high enough to insure at least 1 atm of phosphorus pressure. One of the metal halides, for example 20 to 50 mg of $CuCl_2$, is also sealed into the closed tube along with a seed crystal which is located in the region from 700 to 800°C. With this procedure, but without the seed crystal, as much as 4 g of gallium has been combined with 2 g of phosphorus

Fig. 7. Surface of gallium phosphide grown by means of $SnCl_2$ transport on a gallium arsenide seed with a $(\bar{1}\bar{1}\bar{1})$ or arsenic surface.

to form GaP which was deposited on the vessel walls between 700 and 800°C. Since this system, when used to grow epitaxial layers, has given the same results as the system employing a gallium phosphide crystal source, it seems likely that the transport and epitaxial growth of compounds proceed essentially in the manner outlined in the early part of this paper.

The various metal halides found satisfactory for transport and epitaxial growth of GaAs are also satisfactory for gallium phosphide, and transported and doped with $CuCl_2$ behaves like SI GaAs, again because copper doping results in compensation. Just as with gallium arsenide, this has been the basis on which to build a hybrid p+in+ diode, described below.

Even though the gallium phosphide lattice does not match that of gallium arsenide, it is possible to seed and grow gallium phosphide on gallium arsenide epitaxially. An x-ray analysis of a sample crushed from such a specimen indicated the presence of $GaAs_{0.24}P_{0.76}$. This may mean that a graded transition occurs between the gallium arsenide seed and the gallium phosphide epitaxial layer. Such a layer, like a graded glass seal, might help to accommodate lattice mismatch. Figure 7 shows the surface of GaP grown on a GaAs seed with a $(\bar{1}\bar{1}\bar{1})$ or arsenic surface (4-6). The GaP was transported in a three-hour run, with source at 1050°C, and seed at 750°C, and with 20 mg $SnCl_2$ and 2 mg phosphorus in the closed tube. X-ray analysis of the growth shown in Figure 7 indicates single crystalline growth with the same orientation as the substrate. Under transmitted illumination, cross sections of the growth taken normal to the wafer surface reveal reddish-orange gallium phosphide grown on opaque gallium arsenide.

Figure 8 shows a cross section of a hybrid $p^+\pi n^+$ (or p^+in^+) structure consisting of a gallium arsenide seed, 0.0075 mm of GaP π or i region, and a thick n^+ GaP region. The π or i region was grown with 20 mg $CuCl_2$ and 5 mg phosphorus in an hour. The n^+ region was grown with 22 mg $SnCl_2$ and 5 mg of phosphorus in three hours. During the growth of each layer the source end of the tube was held at 1050°C and the seed end at 750°C. As in the previous example, the epitaxial π and n^+ regions were grown on the $(\bar{1}\bar{1}\bar{1})$ or arsenic face of the gallium arsenide seed. X-ray analysis of each epitaxial

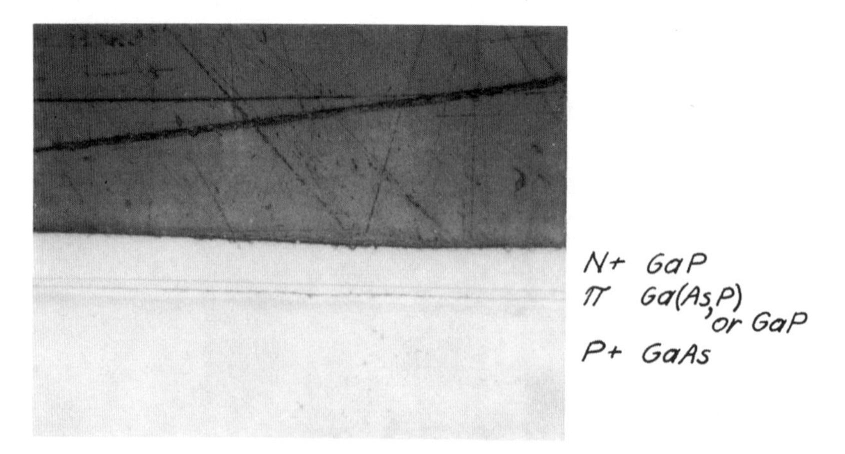

Fig. 8. Cross section of hybrid $p^+\pi n^+$ (or $p^+ in^+$) structure consisting of p^+ GaAs seed, GaP or Ga(As,P) π or i region grown with $CuCl_2$, and GaP n^+ region grown with $SnCl_2$.

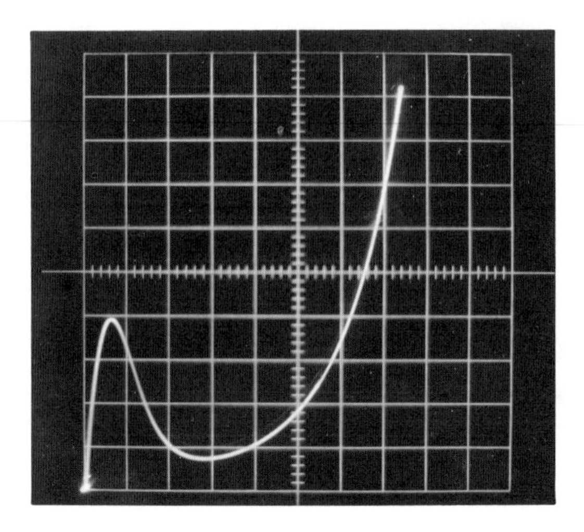

Fig. 9. Current-voltage characteristic of $GaAs_xP_{1-x}$ tunnel diode at room temperature. Vertical scale: 0.5 ma/division; horizontal scale: 0.1 v/division.

layer indicated single crystalline growth with the orientation of the seed. Although it is not evident from Figure 8, under transmitted illumination the middle region (π or i) appears red, as would expected. Electrically the structure conducts in the forward direction and blocks in the reverse. An unetched square die with edges over 1 mm in length, mounted as a diode, switched from 10 ma in the forward direction to essentially no reverse current in 2 nanoseconds. The capacitance of the diode was 24 pf, a value which could easily be reduced below 1 pf by merely decreasing the area.

Gallium Arsenide-Phosphide

Following the basic procedure outlined above, one can readily synthesize gallium arsenide and phoshide into $GaAs_xP_{1-x}$. Now, instead of using only gallium arsenide or phosphide, preselected quantities of both are introduced simultaneously as source material for transport and co-deposition as the mixed compound. Generally the source end of the tube is heated to 1000 to 1100°C while the cool end of the tube is held at 850 to 950:C. These conditions, with no seed, will grow massive polycrystalline $GaAs_xP_{1-x}$ with crystallite diameters from 2 to 3 mm.

In most experimental runs the ratio of arsenic to phosphorus in the source crystals (and in the $GaAs_xP_{1-x}$) was either ~ 10:1 or ~ 1:1. Relatively pure polycrystals have been grown using chlorine gas (back-fill) as the transport agent. In one specific run 677 mg of gallium arsenide and 576 mg of gallium phosphide yielded a mixed polycrystal (2 to 3 mm crystallite dimensions) which transmits red light. Preliminary measurements indicated a band gap near 2 ev. Other similar crystals have been grown with band gaps from 1.8 to 1.9 ev. The measured band gaps in general have been close to those predicted on the basis of the known amounts of starting material. In addition to runs in which uniformly doped material has been prepared, one run has been performed in which $GaAs_xP_{1-x}$ of one conductivity type and band gap has been grown on $GaAs_yP_{1-y}$ of the opposite conductivity type and of different band

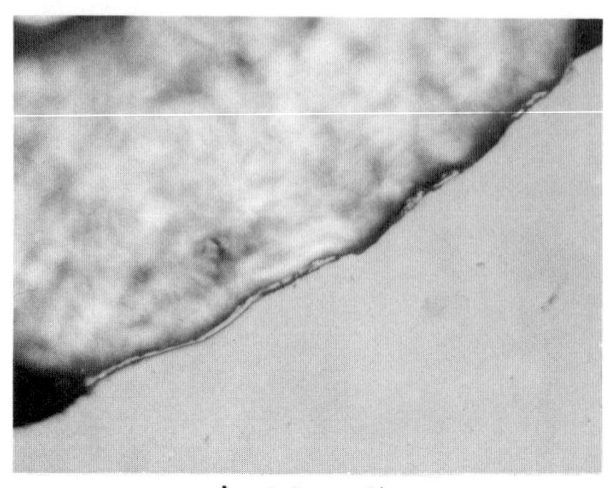

Fig. 10. Cross section of tellurium-doped regrown region (p-n junction) on zinc-doped $GaAs_xP_{1-x}$ grown by means of $ZnCl_2$ transport from 519 mg of GaAs and 378 mg of GaP.

gap and thus yielded another form of heterojunction. In this particular run individual crystallites continued their growth across the junction boundary.

Most of the $GaAs_xP_{1-x}$ crystals which have been synthesized have been deliberately doped degenerately p-type. This is readily accomplished by using $ZnCl_2$ as the transport and doping agent. Tunneling and negative resistances have been observed regularly in junctions formed by alloying tellurium-doped dots on this material. To date the largest number of tunnel junctions have been fabricated upon a crystal grown from 519 mg of gallium arsenide and 378 mg of gallium phosphide. Ten mg of $ZnCl_2$ was used to effect the transport and synthesis of the crystal and to obtain p-type doping. Tunnel diodes alloyed upon this crystal, and upon other similar crystals, have yielded peak-to-valley current ratios as high as 8:1 and voltage swings from tunnel region to thermal region as high as 0.8. Figure 9 shows a typical I-V characteristic of a tunnel junction alloyed on the crystal described. Figure 10 shows a highly magnified cross section of a tunnel junction alloyed upon the same crystal. It will be observed that an adequate regrown region, tellurium-doped, exists directly under the alloy at the left. The regrown region at the right has been damaged in sectioning.

DISCUSSION AND CONCLUSIONS

In addition to the work described above on GaAs, GaP, and $GaAs_xP_{1-x}$, exploratory runs have been completed with InP and GaSb. Indium phosphide has been transported and grown with $SnCl_2$ and gallium antimonide with either iodine or $SbCl_3$. In the case of gallium antimoide the transport is exceedingly slow as mentioned earlier. If this is because antimony has a substantially lower vapor pressure than either arsenic or phosphorus, we can look for similar effects with compounds consisting of relatively nonvolatile elements.

From a materials and device point of view it is obvious that vapor transport and epitaxial growth of intermetallic compounds are of considerable importance. Some techniques have been illustrated that can be employed to regrow and dope compounds for use in tunnel diodes, to grow junctions and heterojunctions of many configurations including types which exhibit space-charge-limited emission, and to prepare mixed intermetallic compounds. Other potential applications include wide band gap emitters for transistors, hybrid devices involving coupling of optical effects in gallium phosphide to electrical effects in gallium arsenide, and solar heterojunctions with one side of the junctions consisting of a wide band gap material of considerable transparency.

ACKNOWLEDGMENTS

We should like to thank C. V. Bielan, F. A. Carranti, R. E. Hysell, B. G. Hess, R. C. Thomas, and R. Stermer for measurements and other assistance in this work. Also, we are indebted to A. E. Blakeslee, R. N. Hall, and F. H. Horn for helpful discussions. This work supported by the Electronics Research Directorate, AFCRC, under Contract No. AF19(604)-6623.

References

1. Marinace, J. C., I.B.M. J. Research Development, 4, 248 (1960).
2. Gooch, C. H., C. Hilsum, and B. R. Holeman, J. Appl. Phys., to be published.
3. Hilsum, C., N. J. Coupland, and R. J. Sherwell, I.R.E.-A.I.E.E. Solid State Development Research Conference, Stanford University, California, June 1961.
4. Gatos, H. C., and M. C. Lavine, J. Electrochem Soc., 107, 427 (1960).
5. White, J. G., and W. C. Roth, J. Appl. Phys., 30, (1960).
6. Richards, J. L., and A. J. Crocker, J. Appl. Phys., 31, 611 (1960).

Texture and Orientation of Evaporated Bismuth Films

T. P. TURNBULL and E. P. WAREKOIS

Lincoln Laboratory,* Massachusetts Institute of Technology,
Lexington, Massachusetts

Abstract

X-ray and electron diffraction studies on evaporated bismuth films show that $(00\cdot\ell)$-oriented single crystals up to 10 μ thick can be prepared by deposition on substrates heated to about 200°C or by annealing at the same temperature. If bismuth is deposited on room temperature substrates, films less than 4 to 5 μ thick are also $(00\cdot\ell)$-oriented single crystals, but in forming thicker films an outer polycrystalline layer with $(11\cdot0)$ and $(01\cdot4)$ planes preferentially oriented parallel to the surface of the film is deposited on top of the $(00\cdot\ell)$ layer.

INTRODUCTION

In the search for new materials and methods of fabricating them into micro-electronic devices, considerable interest is now being focused on the semimetals (1). This group of materials is characterized by very small electron effective masses and small negative energy gaps. Bismuth and bismuth alloys are at the present the most promising of these new systems. Since metallurgical grinding and polishing of such soft materials are usually undesirable because they produce residual strains and distortions, it was decided to investigate the feasibility of producing oriented thin films of bismuth by evaporation techniques.

An annealed evaporated bismuth film usually has a strong basal $(00\cdot\ell)$ texture. (Hexagonal coordinates are used throughout this paper.) This texture imposes serious limitations on the devices which can be produced, since the desired effects are very much dependent on the orientation of an external magnetic field relative to the crystallographic axis of the specimen. It would be highly desirable to be able to produce thin films with different orientations.

EXPERIMENTAL METHODS

In the present investigation thin layers of bismuth were deposited on various crystalline and amorphous substrates by conventional vacuum evaporation techniques. Tungsten baskets and tantalum boats were used as crucibles for zone-refined bismuth. The distance between crucible and substrate was approximately 10 cm. Evaporations were carried out at pressures less than 10^{-5} mm Hg, and pressures during annealing were less than 5×10^{-4} mm Hg. In order to prepare layers with flat surfaces, during each evaporation the bismuth was initially deposited on a shutter. After evaporation began to

*Operated with support from the U. S. Army, Navy, and Air Force.

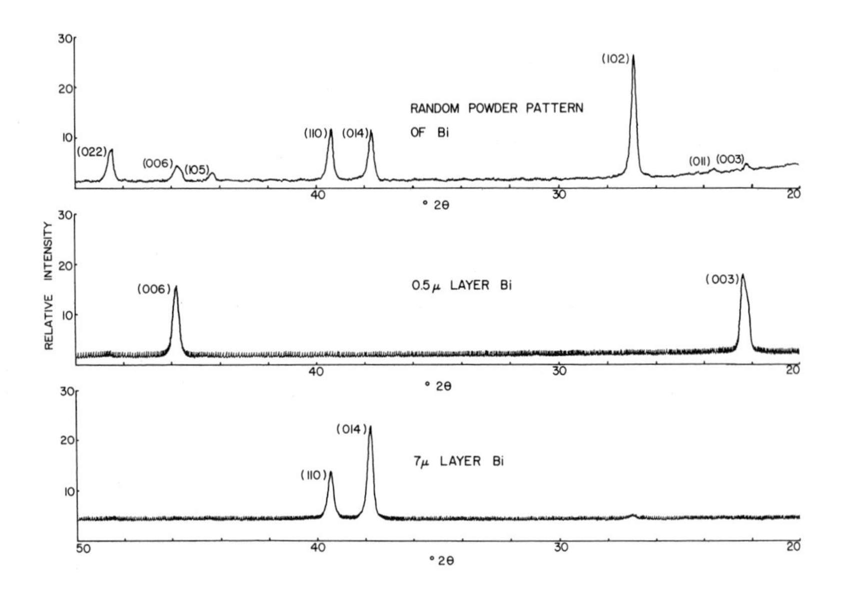

Fig. 1. X-ray diffractometer data for powdered bismuth and evaporated bismuth films with (00·ℓ) and (11·0) (01·4) textures.

occur at a uniform rate, the shutter was removed to permit deposition on the cleaned substrate. When the temperature of the crucible was held constant to minimize thermal agitation, mirror-like surfaces were produced for layers less than 5 μ thick. It was found that films with cleaner surfaces were obtained when deposition on the substrate was stopped by replacing the shutter before the melt was completely evaporated. Large area films free from pits and voids were obtained by rotating the substrate during evaporation in much the same manner as in rotary shadowing of replicas for electron microscopy.

RESULTS

Orientation of the films as a function of film thickness between 0.05 and 10 μ , substrate temperature, substrate composition, and annealing conditions was determined by x-ray diffraction. Regardless of thickness, films deposited on substrates at -75°C were randomly oriented, while those deposited at 200°C had a strong (00·ℓ) texture. A (00·ℓ) texture was obtained when randomly oriented films were annealed at 230°C for an hour with a heat lamp or when a hot wire was passed over them. When the substrate was at room temperature, the texture depended on film thickness. Films less than 4 μ thick had a (00·ℓ) texture, while for films thicker than 5 μ only (11·0) and (01·4) reflections were observed. For films between 4 and 5 μ thick, a number of

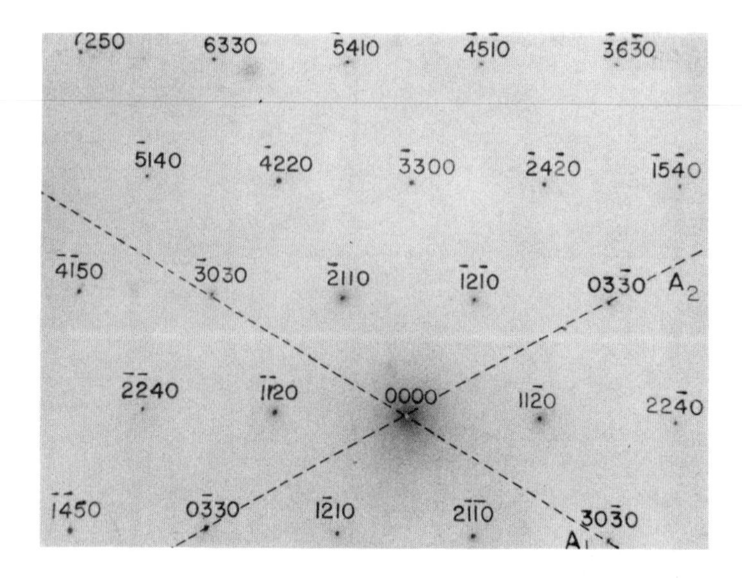

Fig. 2. Electron diffraction pattern for single crystal bismuth film with $(00\cdot\ell)$ orientation.

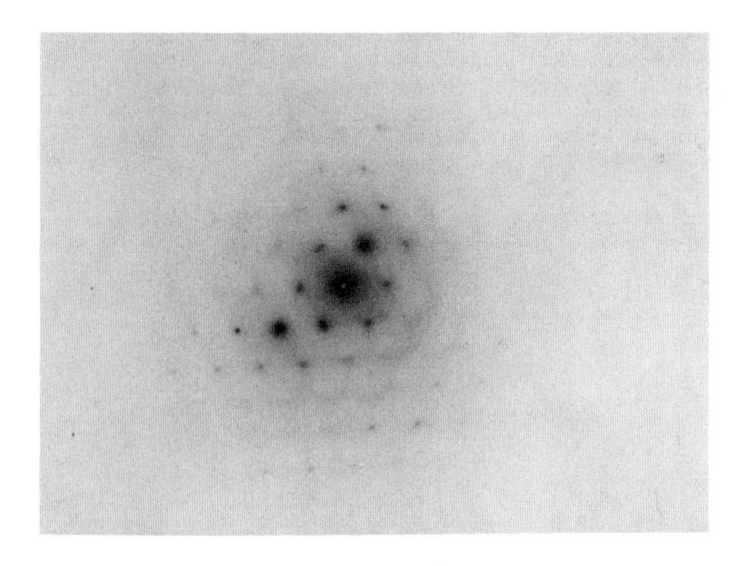

Fig. 3. Electron diffraction pattern for bismuth film with $(11\cdot0)(01\cdot4)$ texture, etched on upper surface only.

diffractometer patterns containing both $(00\cdot\ell)$ and $(11\cdot0)$ $(01\cdot4)$ lines were obtained: in other cases, only $(00\cdot\ell)$ lines or only $(11\cdot0)$ $(01\cdot4)$ were observed. The results for room temperature substrates are illustrated in Figure 1, which shows the diffractometer patterns for a $(00\cdot\ell)$-textured film 0.5 thick and for a $(11\cdot0)$ $(01\cdot4)$ film 7 μ thick. (A random pattern obtained from

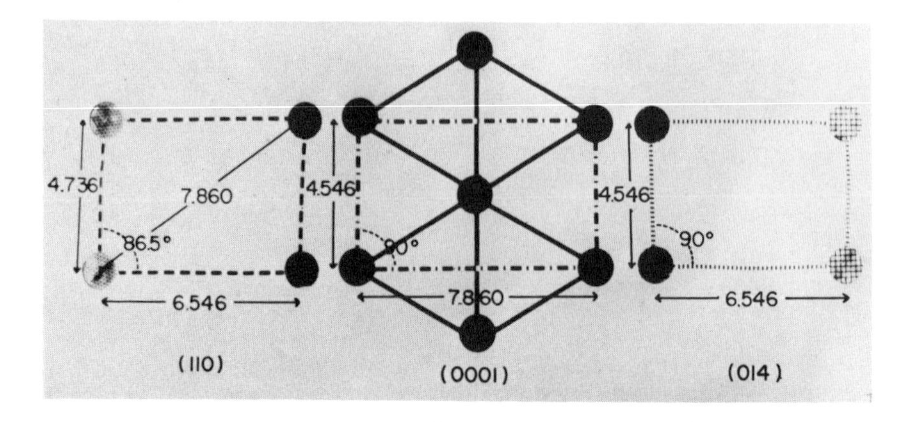

Fig. 4. Arrangement of atoms in (00·1), (11·0), and (01·4) bismuth planes.

powdered bismuth is also included for reference.) The same results were obtained with all the substrates used, whether crystalline or amorphous. The crystalline substrates included single crystal bismuth and antimony oriented in the (00·ℓ) direction, LiF, and Al_2O_3, while the amorphous substrates included Pyrex and polyethylene.

From the x-ray data we conclude that when bismuth is deposited on room temperature substrates, films more than 4 to 5 μ thick consist of a (00·ℓ)-textured layer adjacent to the substrate and an outer layer with (11·0)(01·4) texture. There is some evidence that increasing the deposition rate favors the (11·0) (01·4) texture. The change in the relative growth rates of the two textures with thickness does not appear to result from a change in their relative stabilities, since (00·ℓ)-textured films up to 10 μ thick are produced by deposition at 220°C or by annealing (11·0)(01·4) or even randomly oriented films at 230°C.

A number of films which had been found by x-ray diffraction to have either (00·ℓ) or (11·0) (01·4) textures were examined by electron diffraction to

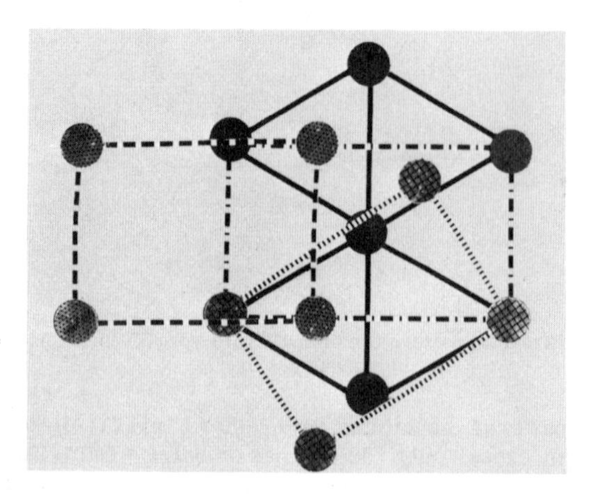

Fig. 5. Possible transitions from growth on (00·1) plane to growth on (11·0) or (01·4) plane.

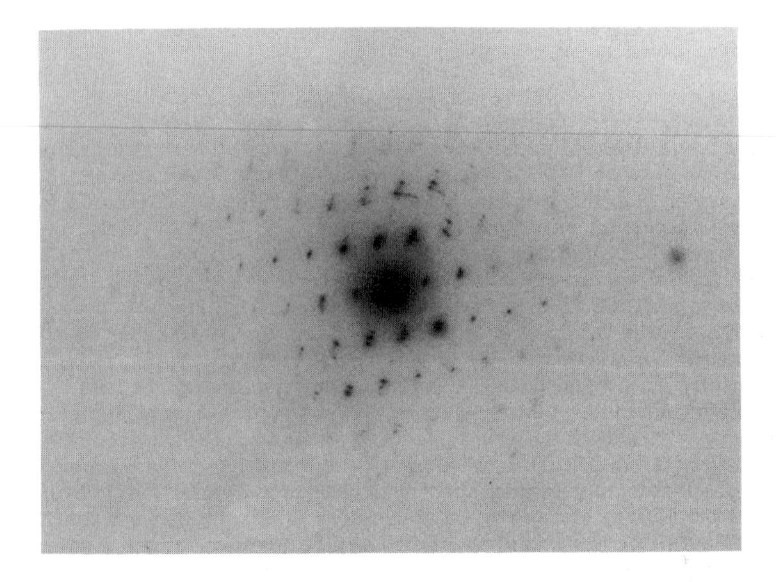

Fig. 6. Electron diffraction pattern for bismuth film with (11·0) (01·4) texture, etched on both upper and lower surfaces.

determine whether they were single crystal or polycrystalline. A diffraction pattern typical of the (00·ℓ) films is shown in Figure 2. The well-defined spots show that the area covered by the electron beam is single crystal. The lattice parameters calculated from the pattern are in good agreement with the published values obtained by x-ray diffraction (2). By scanning the films with the electron beam, it was found that in many cases the single crystal

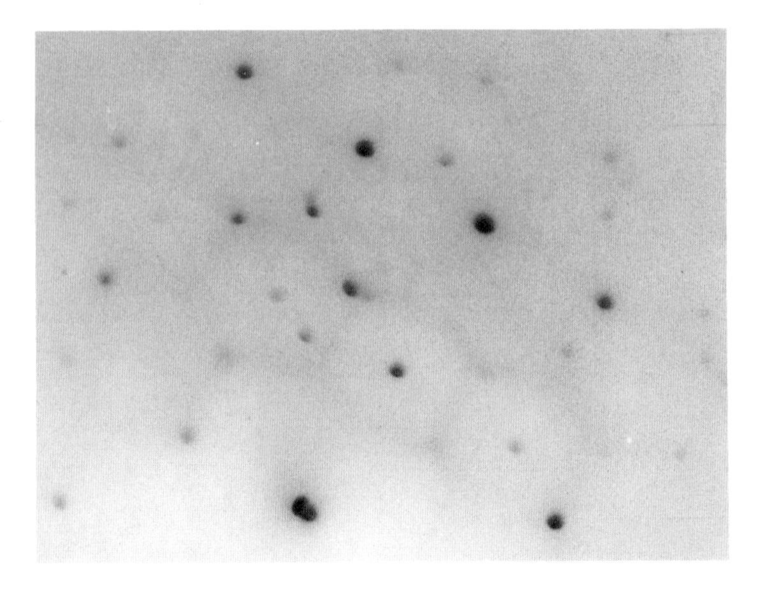

Fig. 7. Electron diffraction pattern in reflection for bismuth film with (11·0)(01·4) texture.

region extended over 10 mm^2, the maximum area which could be examined with the diffraction camera.

DISCUSSION

Electron diffraction patterns for (11·0)(01·4) films etched on the outer surface include not only well-defined spots but also a network of streaks, as illustrated in Figure 3. Similar patterns have been observed by Cochrane (3) for films of nickel deposited on copper single crystals. Cochrane found that the nickel films consisted of an inner layer with the same orientation as the copper substrate and an outer layer with different orientation, and he attributed the streaks to multiple twinning caused by the mismatch in orientation at the interface between the two layers. In the present case the streaks may be attributed by analogy to multiple twinning at the interface between the (00·ℓ) layer which is initially deposited and the (11·0)(01·4) layer formed after the film is 4 to 5 μ thick. The mismatch between these layers is apparent from Figure 4, which shows the characteristic arrangements of atoms in the (00·1), (11·0), and (01·4) planes, and from Figure 5, which gives two examples of the manner in which the transition from growth on the (00·1) plane to growth on the (11·0) and (01·4) planes might take place. In view of the close similarity between the (11·0) and (01·4) planes, as illustrated in Figure 4, it is not surprising that layers containing both these orientations can be formed.

Electron diffraction patters were also obtained for (11·0)(01·4) films which had been etched on both the outer surface and the surface adjacent to the substrate. A typical pattern is shown in Figure 6. No streaks are present, presumably because the twinned interface region has been etched away. The pattern consists of short intense arcs similar to those observed by Yearian (4) in polycrystalline films with a high degree of orientation. Such a pattern indicates

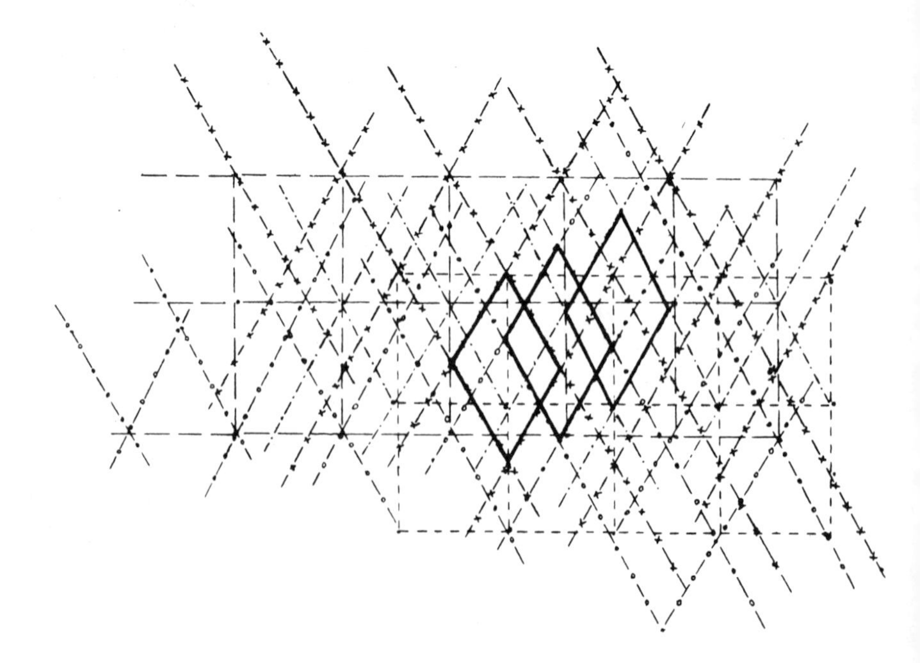

Fig. 8. Relationship between (11·0) (01·4) texture and (00·ℓ) texture, for electron diffraction pattern of Fig. 7.

that the film is composed of crystallites which have either their (11·0) or (01·4) planes oriented parallel to the surface of the film but which are randomly oreiented with respect to the axis normal to the surface.

Electron diffraction patterns were also obtained in reflection for a number of films. The pattern for one film with (11·0)(01·4) texture is shown in Figure 7. The relationship of this texture to the (00·ℓ) basal plane orientation is shown in Figure 8, where rhombi (indicated by the heavy lines) similar to those characteristic of the basal plane have been constructed on the basis of the (11·0) (01·4) spots.

In summary, x-ray and electron diffraction studies on evaporated bismuth films show that (00·ℓ)-oriented single crystals up to 10 μ thick can be prepared by deposition on substrates heated to about 200°C or by annealing at the same temperature. If bismuth is deposited on room temperature substrates, films less than 4 to 5 μ thick are also (00·ℓ)-oriented single crystals, but in forming thicker films an outer polycrystalline layer with (11·0) and (01·4) planes preferentially oriented parallel to the surface of the film is deposited on top of the (00·ℓ) layer.

ACKNOWLEDGMENTS

The authors are indebted to R. E. Hanneman, T. B. Reed, and A. J. Strauss for valuable discussions.

References

1. Lax, B., Bull. Am. Phys. Soc., 6, 58 (1961).
2. Swenson, H. E., R. K. Fuyat, and G. M. Ugrinic, Natl. Bur. Standards Circ. 539, 21 (1954).
3. Cochrane, W., Proc. Phys. Soc. (London), 48, 723 (1936).
4. Yearian, H. J. in L. Marton, ed., Methods of Experimental Physics, Vol. 6, Academic Press, New York, 1959, p. 277.

Factors Affecting the Resistivity of Epitaxial Silicon Layers

H. BASSECHES and S. K. TUNG

Bell Telephone Laboratories, Inc., Allentown, Pennsylvania

and

R. C. MANZ and C. O. THOMAS

Bell Telephone Laboratories, Inc., Murray Hill, New Jersey

Abstract

From studies in both a hot-wall tube furnace and a cool-wall, rf heated system, an understanding has been obtained of the factors which influence the resistivity of epitaxial layers produced by the reaction of silicon tetrachloride and hydrogen on single crystal silicon substrates. In order to explain how the substrate and the pedestal influence the resistivity of the layer, a model is proposed which involves a gas phase transfer of dopant atoms from the back side of the substrate and the pedestal. When using a molybdenum pedestal in an rf heated system, it is possible to minimize the interaction of the substrate with the layer by resting the substrate on a high resistivity silicon spacer. By a mass transfer process, lightly doped silicon is transferred to the back of the substrate sealing up and reducing its influence. The chemical reactions in the mass transfer process may be in part similar to those responsible for releasing dopant atoms when a silicon spacer is not present. This transfer occurs normally when a silicon pedestal is used for deposition. Experiments which show that the dopant atom transport processes can be controlled suggest that it is feasible to control conductivity type and level in epitaxial layers by solid state doping.

INTRODUCTION

Recently there has arisen a widespread and rapidly growing interest in the use of epitaxial semiconductor films for a variety of new and improved devices. The term epitaxy ("arrangement on") was introduced by Royer (1) to denote the phenomenon of the oriented growth of one crystal upon another. Most previous work (2) has pertained to the oriented relationship of different chemical species. In semiconductor technology, considerable attention is now being focused on the growth of thin films or layers of semiconductors on substrates of the same parent material and having the same crystal orientation as the substrate. The impurity content and the defect density of this layer may be considerably different (in terms of semiconductor usage) from that of the substrate. It is on such a basis that it is reasonable to refer to films or layers as epitaxial films.

The device implications of epitaxial semiconductor films have been discussed elsewhere (3-5). In addition, there are a number of publications which discuss methods for preparing epitaxial semiconductor films (6-9).

There are three general properties of epitaxial silicon which must be controlled if material suitable for device applications is to be prepared. These are the impurities (type, concentration, and distribution), thickness, and crystal perfection. The first of these properties is the subject of this paper, and it will be discussed in terms of the various factors which affect the impurities and hence the resistivity of epitaxial silicon layers.

From studies of the epitaxial growth of silicon by the hydrogen reduction of silicon tetrachloride in both a hot-wall tube furnace and in a cool-wall rf heated system several factors have been shown to influence the film resistivity. These are (a) the background contamination of the system (which includes impurities in the hydrogen carrier gas), (b) the purity of the silicon tetrachloride, and (c) the resistivity of the substrate upon which the epitaxial layer is deposited. If a silicon pedestal is used to support the substrate wafer, its resistivity is considered as a fourth variable. The contribution of other pedestal structures is considered a part of the system background contamination.

It is convenient to summarize the impurities in the layer by the equation:

$$N_{total} = N_{SiCl_4} + N_{system} + N_{substrate} + N_{pedestal}$$

where N_{total} is the total number of impurities in the layer, N_{SiCl_4} is the impurities present in the silicon tetrachloride, N_{system} the impurities coming from the fixtures and reaction vessel, $N_{substrate}$ impurities from the substrate, and $N_{pedestal}$ from the pedestal. Methods for determining the contributions separately are discussed. In the case of the last two factors, where the impurity content is relatively well known, only a fraction of the impurities becomes available for doping. This fraction is not known at present. The resultant conductivity type and level are governed by the balance of these factors. In order to explain how the substrate and the pedestal influence the resistivity of the layer, a model is proposed for doping. This fraction is not known at present. The resultant conductivity type and level are governed by the balance of the factors on the right side of the equation.

A model involving a gas-phase transfer process is proposed which appears to explain how the substrate and pedestal interact with the growing epitaxial layer. Methods are discussed by which this substrate and pedestal effect may be minimized. The model suggests that doping can be controlled from the solid state and experiments illustrating this are discussed.

EXPERIMENTAL

All of the epitaxial silicon films discussed in this paper were grown by the hydrogen reduction of silicon tetrachloride in a flow system. The rf apparatus is similar to that described by Tung (11) and Theurer (5,7). Some of the films were grown in a horizontal hot-wall quartz tube furnace, with temperature measured by means of a thermocouple inserted into the furnace. Slice temperatures in the rf system were monitored by an optical pyrometer. The pyrometer correction was determined by observing the color temperature of solid silicon in equilibrium with liquid silicon in the apparatus being used.

SYSTEM BACKGROUND CONTAMINATION

The system background contamination level can be estimated by making several runs in which high resistivity wafers are heated in the system with hydrogen flowing but with no subsequent deposition of silicon. For example, consider a 500 ohm cm p-type substrate for which the doping level is somewhat less than 10^{13} cm^{-3}. If the resistivity of the substrate measured after a blank run drops in value but remains p-type then the system contribution is identified as p-type. If the resistivity increases or converts to n-type, an n-type background is indicated. Results from several systems have indicated that the background contamination is n-type and provides doping of between 10^{13} cm^{-3} and 10^{14} cm^{-3}.

The hydrogen gas stream is an obvious potential source of contamination. This was checked in a system in which the hydrogen is purified by means of a Deoxo unit followed by a large column of molecular sieves at room temperature. The hydrogen gas then passes through a liquid-nitrogen trap. Because of the possibility of doping contaminants (such as phosphine) in the hydrogen, a series of runs was made with and without molecular sieves in the cold trap. When no sieves were present the trap was filled loosely with clean quartz wool. The substrates used were 0.004—0.007 ohm cm n-type (arsenic doped) and were heated upon a 110—120 ohm cm n-type (arsenic doped) pedestal. The films were grown from silicon tetrachloride purified as described in the next section.

With sieves in the cold trap film resistivities of 6—22 ohm cm, n-type were obtained. Without sieves 9—20 ohm cm, n-type films were obtained. It is apparent that with our present hydrogen supply molecular sieves in the cold trap do not contribute to higher film resistivities. Since the molecular sieves may produce dust (and film pitting), and also necessitate extra precautions to avoid nitrogen carryover from the purge cycle, they were discontined in this apparatus. The room-temperature molecular sieve column (followed by a Pyrex frit) is retained to remove moisture. These conclusions should be verified for each apparatus used since variations in the hydrogen supply may produce different results.

The effect of the wall temperature and miscellaneous side reactions at the wall upon the film resistivity was checked. In a hot-wall tube furnace there is an accumulation of polycrystalline silicon as well as other products on the inner wall. When the reaction chamber wall is air cooled (rf system) a nondescript brown material accumulates which occasionally may flake off onto a wafer and cause pitting. When the reaction chamber wall is water cooled (rf system) a viscous polymeric material, pyrophoric with water, accumulates on the inside of the wall.

In the rf system the two cooling conditions do not appear to have much effect upon film resistivity. For films grown as described above, resistivities of 6—15 ohm cm n-type were obtained with an air-cooled wall and 8—10 ohm cm n-type were obtained with a water-cooled wall.

The absence of a wall-temperature effect upon system background contamination, and consequently film resistivity, is further illustrated by the fact that 10-15 ohm cm n-type films can be grown on heavily doped n-type substrates in a hot-wall tube furnace provided the backside of the substrate wafer is "sealed." This point is discussed in more detail in a later section (see Table III).

These experiments provide useful information regarding system design, but they do not pinpoint the source of the system background contamination. However, with present contamination levels a wide range of film resistivities can be grown. Therefore, the background contamination does not appear to be a major problem.

SILICON TETRACHLORIDE PURITY LEVEL

Having determined the contribution of the system background contamination, the contribution of the silicon tetrachloride may be evaluated by depositing epitaxial silicon on high resistivity substrates. It has been demonstrated that 100 ohm cm n-type bulk silicon may be produced by the hydrogen reduction of silicon tetrachloride in apparatus similar to that used for epitaxial silicon. The fact that the films generally are n-type is therefore not surprising.

Silicon tetrachloride is obtained from a number of commercial sources. Generally no assay information is provided with the material. Films with resistivities of 1–10 ohm cm, n-type, have been grown on high resistivity substrates using silicon tetrachloride as received. However, the results are not uniform from batch to batch, and film resistivities less than 1 ohm cm, n-type, have been produced by several batches. When the silicon tetrachloride is purified, [saturated with chlorine and aluminum trichloride, distilled into a chromatographic column, and passed through a silica gel column packing (10)] film resistivities 30–50 ohm cm, n-type, are obtainable. The same trend is observed when the films are grown on degenerate n-type substrates. Therefore, in the interest of both uniformity and high film resistivity it appears to be well worthwhile to purify the silicon tetrachloride.

SUBSTRATE EFFECTS

The influence on the layer of heavily doped n-type substrates by diffusion into the growing film is considered negligible for most of the thicker layers studied here. Calculations, based on the assumption that diffusion in epitaxial material is the same as in bulk silicon, indicate that for the conditions used during deposition on arsenic-doped substrates, only the first 1–2 μ of the epitaxial layer would be affected.

However, the substrate can influence the resistivity of thicker epitaxial layers if it contains a suitable number of impurities and deposition occurs under certain conditions. This interaction has been found in both the tube furnace and the rf system. The interaction was first observed when the substrate resistivity was lowered from 0.1 ohm cm n-type to 0.004 ohm cm n-type. The layer resistivity decreased from values around 2–4 ohm cm to 0.2–0.4 ohm cm. At that stage of developing of epitaxial silicon transistors, it was desired to produce a high resistivity n-type layer on a highly doped n-type substrate. It was, therefore, of concern when the low resistivity layers were obtained. The experiments made in both types of systems are described below.

Tube Furnace

The important features of the tube furnace from the standpoint of the present resistivity considerations are as follows. The substrates which are to be deposited are laid on a flat quartz plate which rests in a horizontal position in the bottom of the tube furnace. Usually more than one substrate is deposited in a run and the substrates and test probe pieces are arranged in a row along the direction of gas flow. The temperature is held uniform (±10°C) in the deposition region.

Table I shows representative runs (each line represents one run) which indicate that on low resistivity, arsenic-doped substrates, the layer resistivity is considerably lower than on higher resistivity substrates. Table II shows results obtained on deposits on boron-doped substrates of different resistivity. On very low resistivity substrates (0.002 ohm cm) the resultant layer is p-type; on high resistivity p-type substrates the layer resistivity is n-type and high. Figure 1 shows the values of resistivity obtained on different resistivity n-type

TABLE I
Layer Resistivities on n-Type Substrates Obtained in a Tube Furnace

Substrate properties	Film properties		
Resistivity, ohm-cm	Type	Resistivity, ohm-cm	
		4 pt[a]	C-V[b]
0.004	n	.66	.3
0.004	n	0.48	
0.004	n		.26
0.004	n		0.2
0.004	n		0.3
0.005	n		0.2-0.3
0.1	n		8-17
0.1	n		10-31
0.1	n	17	
0.1	n	5	
0.1	n	22	19.5
0.1	n	16-40	5-16.2
0.1	n	13	
0.1	n	31	1.6
0.1	n	45	

[a]From p-type probe piece placed near sample.
[b]From measurements made on alloyed-aluminum diodes.

TABLE II
Layer Resistivities on p-Type Substrates Obtained in a Tube Furnace

Substrate properties	Film properties	
Resistivity, ohm-cm	Type	Resistivity, ohm-cm 4 pt. probe
0.002	p	
0.002	p	
0.002	p	
0.002	p	
0.7	n	12
0.7	n	30

substrates which were deposited at different temperatures. All the values are based on 4-point probe measurements on a 7 ohm cm p-type test probe piece that was placed in between the n-type samples. Although the measurement is not made on the n-type substrate itself and there is some scatter, the influence of the resistivity of the substrate on that of the epitaxial layer is indicated.

A series of tests were carried out in which various sandwich combinations were prepared where one silicon wafer was placed on top of another. The conductivity types and level of resistivity were varied and the area relationships were different. The values and the geometrical pattern are indicated in Table III. Runs were made with and without deposition of silicon. The arrow in the table indicates the direction of gas flow. In runs BDS113 and BDS115 the procedure was normal except that no deposition of silicon took place. Any changes in resistivity were thus a function primarily of the interaction of spe-

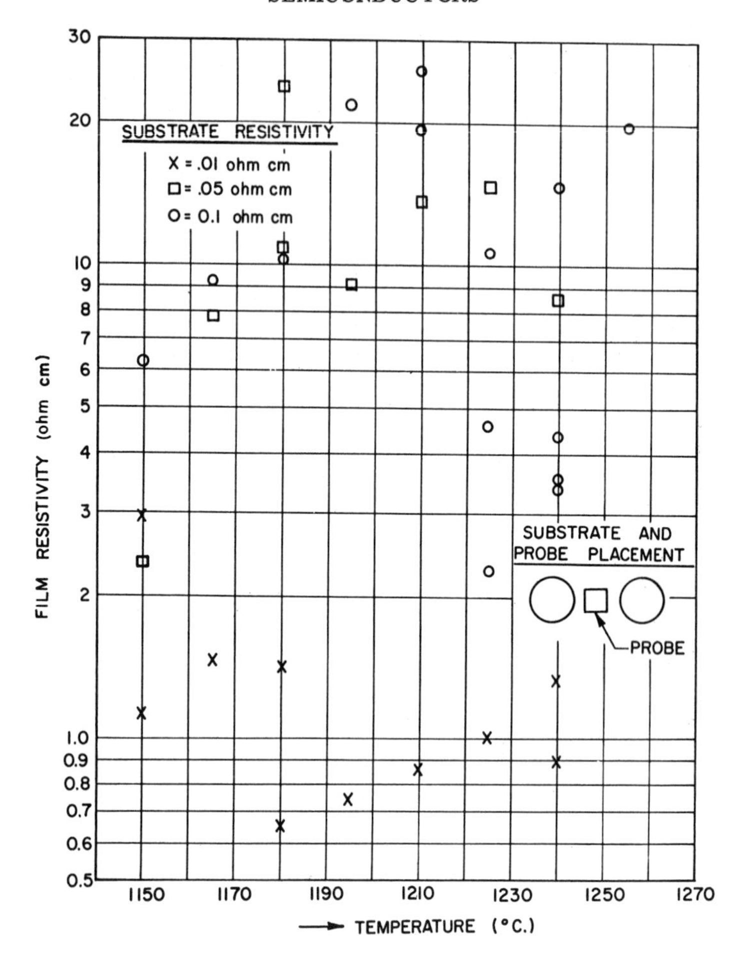

Fig. 1. Resistivity of epitaxial layers on silicon substrates grown in a tube furnace.

cies coming from the two members of the sandwich. In BDS113 where the two pieces initially were heavily doped, the conductivity types remained the same. In BDS115, however, the heavily doped n-type sample caused an n-type layer to form on the lightly doped p-type sample. In BDS114, 126 and 127 heavily doped n-type wafers of small area were placed on larger diameter high resistivity p-type wafers and epitaxial layers were deposited. The layers were n-type and, as measured on the p-type wafer, had quite high values. In BDS126 the layer resistivities, as measured on the first two wafers following the arrow, were the same, while on BDS127 they were different. In BDS114, 126, and 127 the top wafer tended to stick to the bottom wafer and had to be pried gently to remove it. It appeared to be sealed around the edges. In BDS113 and 115, the top wafer did not stick.

Evidence from rf System

In one of the rf heated systems studied (11) a molybdenum pedestal is used to support the substrate and act as a susceptor. In the initial studies with this

TABLE III
Substrate and Flow Effects for Films Grown in a Tube Furnace

	Substrate properties			Film properties		
Run No.	Type	Resistivity, ohm-cm	Dopant	Type	Resistivity, ohm cm 4 pt.	
BDS 113	n	.005	As		Unchanged	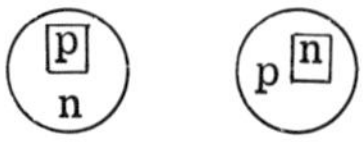
	p	.002	B			
BDS 115	n	.005	As	n	Skin all over	
	p	7	B	p	wafer	
BDS 114	n	.005	As	n	13	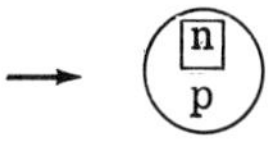
	p	7	B			
BDS 126	n	.005	As	n	5.4	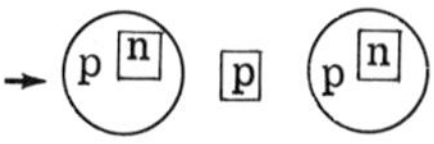
	p	80	B			
BDS 127	n	.005	As	n	3.9, 7.3	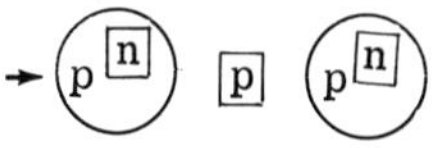
	p	80	B			
BDS 111	n	.005	As	n	5.7, 0.28	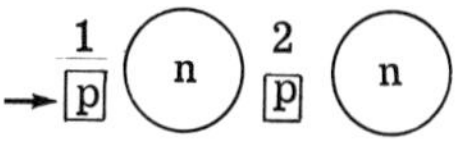
BDS 129	p	.002	B	p and n	n Type	
	n	30	As		41 on p'	
BDS 180	p	.002	B	p		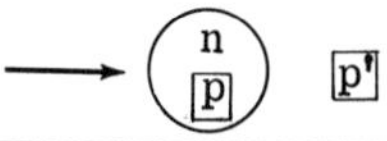
	n	.05	As			

system, a quartz spacer about 0.2 mm thick was first placed on the pedestal and the substrate was placed on it.

The results of runs made on different resistivity substrates are presented in Table IV. Figure 2 summarizes results over a range of resistivities. For simplicity, we can divide the curve into three regions. In region I, the n-type layer resistivity decreases as the n-type substrate resistivity decreases. In region II, the n-type layer resistivity is more or less independent of the substrate where the substrate resistivity is high; i.e., the impurity concentration is low. In this region, the impurities are picked up from the system and the silicon tetrachloride. In region III, the n-type layer resistivity increases as the p-type substrate resistivity decreases. This indicates compensation is occurring within the layer. Finally when the n-type impurity is overcompensated, the resultant layer is p-type. The data confirm the results obtained in the tube furnace as to the influence that the substrate has on the resistivity of the epitaxial layer. The dotted lines separate the various regions.

Since the substrate can affect the layer resistivity, p-type substrates (high or low resistivity) have limited value as a means of measuring epitaxial resistivity in a system normally used to grow n-type films on n^+ substrates (or vice versa). High resistivity p-type substrates may be satisfactory to monitor stability and reproducibility in the system, but they should not be used to infer the resistivity of n-type films unless a dependable correlation has been established.

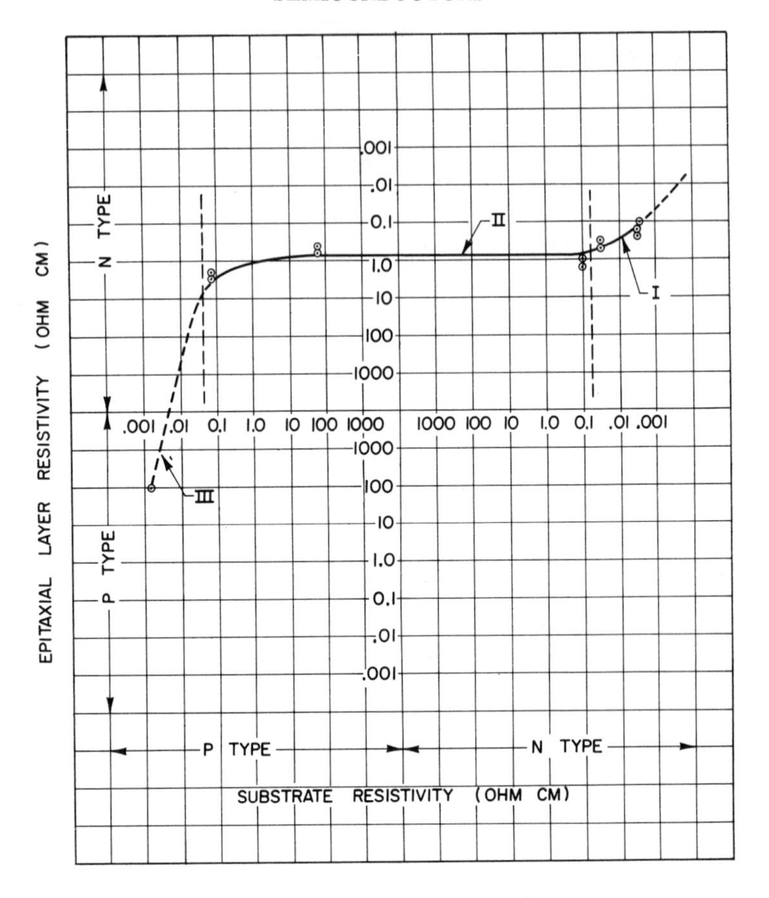

Fig. 2. Resistivity of epitaxial layers on silicon substrates (quartz spacer–molybdenum pedestal).

INFLUENCE OF SILICON PEDESTAL ON LAYER RESISTIVITY

Experiments with a silicon pedestal show that impurities from it can also affect the resistivity of the layer. The pedestal described by Theuerer (7) consisted of a low resistivity p-type insert that was fused into a high resistivity pedestal. This p-type insert was used to facilitate rf coupling into the pedestal. High resistivity n-type films on heavily doped n^+ substrate films have been produced and it is believed that these result in part from a compensation phenomenon. For films of greater than 25 ohm cm resistivity this is probably the dominant factor. Several types of evidence support this. With one pedestal the low resistivity insert was not adequately melted in, so it was unsymmetrically located in the pedestal. (Normally, after melting, a uniform toroidal band of p-type material exists.) The layer was nonuniform in resistivity with the high values (3.5–14 ohm cm) located on the same side of the substrate as the insert was located and with the lower values observed at the other side.

In the normal procedure the pedestal is coated with a layer of silicon. This is normally n-type material. During one series of tests this step was omitted. The resultant films came out p-type. This suggests that the thin n-type layer

TABLE IV
Layer Resistivities Using Molybdenum Pedestal with Quartz Spacer

Substrate Properties			Film Properties		
Type	Resistivity ohm-cm	Dopant	Type	Resistivity, ohm cm 4 pt.	Resistivity, ohm cm C-V[a]
n	0.1	As	n		$\rho > 1$ ohm cm[b]
p	0.08	B	n	3.8	
p	0.08	B	n	2.7	
n	0.004	As	n		0.07-0.12
n	0.056	P	n		0.5
n	0.056	P	n		0.3
n	0.005	As	n		0.1-0.3
n	0.005	As	n		0.32-0.37
p	0.002	B	p		—
n	0.005	As	n		0.3
n	0.005	As	n		0.3
p	0.002	B	p		—
p	0.002	B	p		—
n	0.1	As	n		1.5
n	0.1	As	n		1.6

[a]From measurements made on alloyed aluminum diodes.
[b]Deduced from data on transistors, no direct resistivity measurement.

TABLE V
Silicon Pedestal vs Epitaxial Film Resistivity

Pedestal resistivity, ohm cm		Film resistivity, ohm cm	
0.03-0.13	n-type	0.4	n-type
0.03-0.13	n-type	1.2	n-type
0.03-0.13	n-type	1.6	n-type
23-42	n-type	1.4-3.2	n-type
23-42	n-type	5-6	n-type
23-42	n-type	3-4.6	n-type
23-42	n-type	1.5-2.7	n-type
23-42	n-type	3.5	n-type
110-120	n-type	11	n-type
110-120	n-type	6	n-type
110-120	n-type	5.6	n-type
110-120	n-type	18	n-type
3000-5000	p-type	13-20	n-type
3000-5000	p-type	60-70	n-type
2900-4000	p-type	21	n-type
2900-4000	p-type	42-70	n-type
1.2-1.5	p-type		mixed n and p-type
1.2-1.5	p-type		p-type

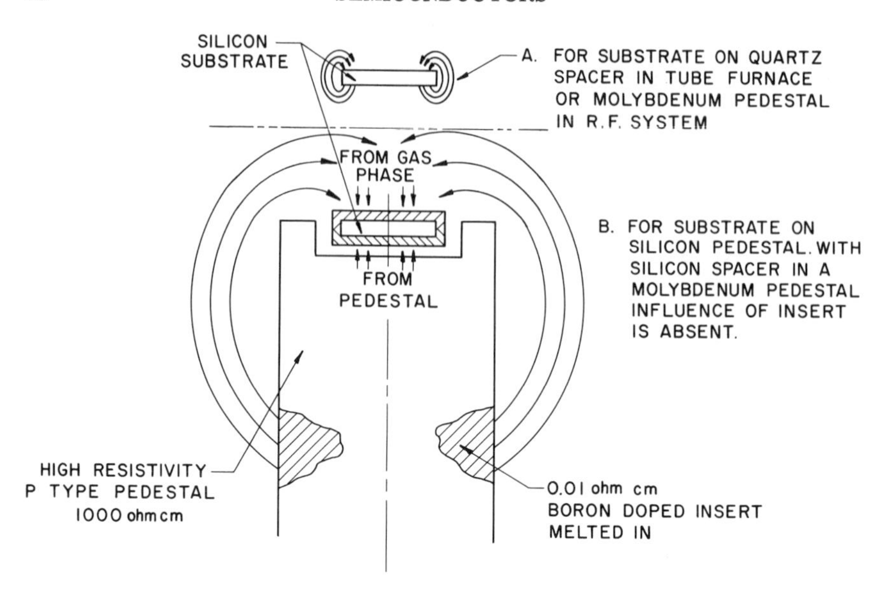

Fig. 3. Model for interaction of substrate and pedestal with epitaxial layer.

which normally deposits acts as a barrier, letting just enough p-type atoms through to help provide the proper amount of compensation. When the pedestal is sufficiently coated, high resistivity n-type layers can be obtained.

Silicon pedestals of various resistivities (without doped starting inserts) have also been used. If the resistivity of the pedestal is low, it is relatively easy to couple to it and to heat it by rf. The higher resistivity pedestals were "started" with heat lamps.

Table V shows the average maximum epitaxial film resistivity grown upon n^+ substrates (arsenic doped 0.003—0.007 ohm cm) using the various pedestals. All other conditions were identical from run to run. It can be seen that the film resistivity increases as one increases the resistivity of the n-type pedestals, but that it will not reach a maximum unless a p-type pedestal is used. This fact indicates that dopant from the pedestal gets into the epitaxial film by a gas phase transfer (in spite of the fact that the pedestal is downstream from the wafer), and that the higher film resistivities involve compensation. The view is further supported by the fact that the film resistivity increases and eventually becomes p-type.

In a number of runs both p- and n-type layers have been obtained. This suggests some delicate unbalance in the system which shifts the conductivity type.

MODEL FOR INTERACTION OF SUBSTRATE AND PEDESTAL WITH FILMS

The evidence from the experiments described support the following model for the interaction of the substrate with the film. The epitaxial layer is influenced by the substrate because impurities come out of the backside of the substrate during the deposition and are carried via the gas phase to the region of the growing epitaxial layer. They are there incorporated into the layer. The diagram in Figure 3, part A, illustrates qualitatively the path proposed. A similar gas phase transfer, shown in part B, is proposed for the contribution of the pedestal. At the present time, it is not certain whether it is by some chemical reaction (discussed later) or by evaporation, or by a combination of both that the impurity is removed from the doping source. Data from the tube furnace have shown that when a hydrogen stream passes over a wafer at the temperatures used in deposition, the wafer will lose weight.

In the case of the experiments made in the tube furnace, it is quite clear that the transfer has to be through the gas phase since it is difficult to find any other way to explain the influence of one sample on another located several centimeters away. The directional effect of the flow is illustrated particularly well in experiments in the tube furnace shown in Table III. In BDS111 the resistivity of the n-type layer on wafer 1 is considerably high than on 2 which is subject to the flow passing over the intermediate n wafer. In BDS129 the two p-type wafers which rested on the n-type wafers did not stick so that the effect of the backside predominated, doping the layers p-type. Upstream on p', however, there was just the right balance to result in a high resistivity n-type layer. In BDS180 the single p' wafer remained p-type even though it was 7 ohm cm in resistivity. However, since the highly doped p-type wafer upstream did not seal to the n-type wafer the dopant atoms produced a p-type layer on the n-type wafer as well as on the downstream p' wafer.

The influence of the backside of the substrate is indicated by the sandwich experiments in the tube furnace in Table III. When the top wafer stuck to the bottom wafer, the influence of the substrate was materially decreased. When it did not stick, (BDS113, 115, 129, and 180) the substrate influenced the layer resistivity. It is shown later in this paper that when the back can be effectively covered up, its influence is substantially eliminated.

The proposed gas-phase transfer model is further supported by the fact that the pedestal effect is flow sensitive in experiments with the silicon pedestal. High resistivity n-type films are obtained at higher flow rates (1–2 l/min) and p-type films are obtained at lower flow rates (400–600 cm^3/minute), if a low resistivity p-type silicon pedestal is used.

Figure 4 illustrates the results of some of the effects which have been discussed. The epitaxial film was grown on an n^+ (0.009 ohm cm) substrate. The gas flow rate was 1 liter/min and the silicon pedestal resistivity was 1.2–1.5 ohm cm, p-type. These conditions are near those required to produce a mixed n- and p-film as discussed earlier in this paper (see Table V).
The initial portion of the film, where the effect of the n^+ substrate is still strong, is n-type. As the top of the substrate becomes "sealed" by the growing film and the bottom by transfer from the pedestal to the backside of the wafer, the gas-phase transfer contribution of the p-type pedestal becomes im-

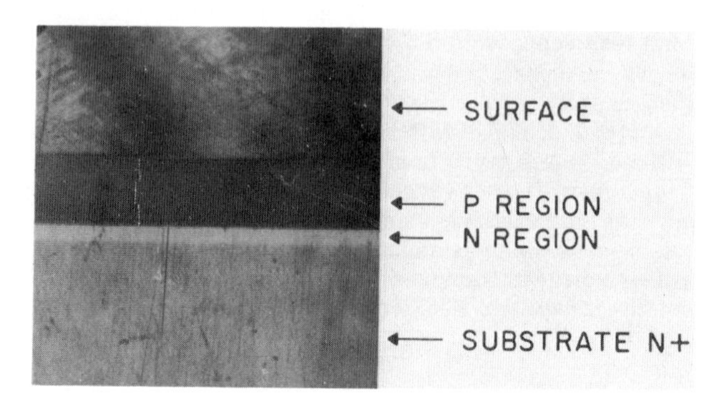

Fig. 4. The p–n–n+ epitaxial silicon structure.

portant, and the film becomes p-type. The balance necessary to obtain this effect is quite difficult to control and to reproduce.

MECHANISM FOR RELEASE OF DOPANT ATOMS

It is of interest to consider mechanisms which might be responsible for the release of dopant atoms. It is possible that sublimation of the silicon would carry along dopant atoms or that sublimation of the dopant itself might have the same effect. This may not be enough, however, to account for the amounts found. Here the nature of the dopant in the substrate could play an important part.

There are various chemical reactions which can be proposed to account for the release of dopant atoms from the silicon substrate when it rests on a quartz surface in the tube furnace and on a quartz spacer in a molybdenum pedestal. There must be a competition among several reactions. Since the backside of the wafers appear uncoated, the deposition reaction that occurs on the frontside must be inhibited on the back. The more restricted reaction region might well alter the balance of the reactions since the diffusion path length before a surface (either the quartz plate or the underside of the wafer) is met is different on the exposed side compared to the bottom of the substrate. Due to different lifetimes, some species might be favored over others. One reaction that might occur at the underside of the substrate is

$$SiCl_4(g) + Si\,(s) \rightarrow 2SiCl_2\,(g)$$

As the silicon is removed, dopant atoms are carried off in the gas phase to the region of the growing film. Possibily the dopant atoms react to form gaseous chlorides according to a reaction such as

$$4B\begin{pmatrix} s\,? \\ g \end{pmatrix} + 3SiCl_4(g) \rightarrow 4BCl_3(g) + 3Si\begin{pmatrix} s\,? \\ g \end{pmatrix}$$

The chlorides would then react at the top side of the substrate. This might be the type of reaction which occurs at the doped insert in a pedestal.

The doping level and the area of the substrate can influence the availability of dopant atoms. Temperature can also have an effect since the reactions are temperature sensitive.

METHODS FOR MINIMIZING INFLUENCE OF THE SUBSTRATE

Assuming the validity of the above model, it is proposed that if the influence of the back of the substrate could be reduced or eliminated the layer resistivity would become less dependent on the substrate resistivity.

Theuerer (7) had observed that silicon was transferred to the backside of a wafer when deposition took place on a silicon wafer supported on a silicon pedestal. By virtue of this mass transfer the backside of a substrate is covered with silicon. If this layer is of high resistivity, it effectively "seals off" the back of the substrate and prevents or inhibits the transfer of dopant atoms.

The nature of the reactions which may be responsible for this effect is as follows. It is necessary to postulate a vertical temperature gradient in the pedestal with the pedestal surface being at the higher temperature than the underside of the substrate. Schafer et al (12,13) have studied the reactions

$$SiCl_4 + Si \rightleftharpoons 2SiCl_2$$

and shown that the forward reaction is favored at high temperature and the reverse reaction is favored at low temperature. This suggests that high resistivity material can be picked up from the pedestal and deposited at the

TABLE VI
Layer Resistivities Using Molybdenum Pedestal with Silicon Spacer, 1000 ohm cm, p-Type

Substrate properties		Film properties	
Resistivity, ohm cm	Dopant	Type	Resistivity, ohm cm C-V[a]
0.056	P	n	2.0
0.056	P	n	2.6
0.056	P	n	2.4
0.005	As	n	6.0
0.005	As	n	2.3
0.005	As	n	4.5
0.056	P	n	8.7
0.056	P	n	4.4
0.056	P	n	9.0

[a]Alloyed aluminum diode used.

underside of the substrate. The use of a similar halide disproportionation for growing epitaxial silicon and germanium in sealed tubes also has been reported (8). An alternative mechanism for silicon transfer involves a series of reactions between hydrogen chloride and silicon.

The influence of the substrate can be eliminated in the case of a molybdenum pedestal, on the basis of the above reaction, by replacing the quartz spacer upon which the substrate rested by a high resistivity silicon spacer. When this was done it became possible to get resistivity values above 2 ohm cm n-type using highly doped n-type substrates. This has been confirmed by hundreds of runs, some of which are shown in Table VI.

Some further insight into the various reactions which are occurring is obtained when the sealing action is not good over the entire underside surface. This occurs when etchant action during chemical polishing attacks the backside of the substrate giving rise to a tapered section at the edge. A warped substrate may also give rise to the same problem since dopant atoms continue to be available.

Apparently the reaction which gives rise to mass transfer, and hence sealing, does not occur or is minimized and those responsible for erosion dominate. Resistivity profiles for silicon layers grown on molybdenum pedestals with intrinsic silicon spacers indicate that regions close to the edge of a slice have lower resistivities than at the center. Near the tapered edge of the backside of the substrate, the conditions apparently are not favorable for deposition of high resistivity silicon from the hydrogen-silicon tetrachloride reaction. The gas mixture is probably depleted of silicon tetrachloride due to reaction at the neighboring hot walls of the pedestal. Thus, the edge remains uncovered and, consequently, remains available as a doping source.

Some support for this picture comes from the following experiment by G. H. Edmunds. A low resistivity n-type substrate (0.01 ohm cm) received a deposit of epitaxial material. This side was then placed downward so that it faced the molybdenum pedestal without a silicon spacer. The top surface was then coated (after being properly prepared). The layer formed was n-type and had a resistivity of 2 ohm cm. Normally, a layer deposited on a low resistivity substrate without a spacer would have resulted in a low resistivity layer. The coating acted as a sealant, particularly at the edges where it would be difficult to get such intimate contact by other materials or means.

The attempts to get sealing action in a tube furnace by use of a spacer were not reproducible, as is indicated by results in Table III. The requisite conditions for the mass transfer do not seem to exist in the tube furnace used since they do not favor a vertical temperature gradient as does the rf heated pedestal. Consequently, although the forward reaction may be occurring,

$$SiCl_4 + Si \rightarrow 2SiCl_2$$

the reverse reaction may not be. However, the forward reaction can release the dopant atoms as proposed before.

Other techniques have been tried in the tube furnace to attempt to seal off the back but they have not met with particular success. These included the use of slices which were oxidized on the bottom side. Oxide layers (3,000—10,000 A) did not act as a sealant. The oxide appeared to be removed. Silicon nitride might prove to be a more effective coating.

SOLID STATE DOPING

The phenomena described above can be used to advantage to dope from the solid state. It would appear that by a combination of such factors as area of substrate, temperature of substrate, position in flow stream, level of doping in substrate or pedestal, nature of dopant in substrate or pedestal, and placement of separate doping slugs the conductivity level and type of epitaxial layers can be controlled and junctions produced.

Table VII is illustrative of some of the results obtained. Normally it has been difficult to obtain n-type layers on heavily doped p-type substrates. However, by adjusting the resistivity of the space, it is possible to obtain layers of varying resistivity. With appropriately doped silicon pedestals, as indicated in Table V, doped layers can be obtained.

TABLE VII
Film Resistivities Obtained Using Different Pedestals and Spacers

Substrate properties		Pedestal		Film properties		
Type	Resistivity, ohm cm	Material	Spacer	Type	Resistivity, ohm cm 4 pt	C-V[a]
p	0.002	Si	0.003-n	n	0.053	
p	0.002	Si	0.003-n	n	0.051	
p	0.002	Si	0.001-n	n	0.03	
p	0.002	Si	0.001-n	n	0.027	
p	0.002	Si	1-n	n	29.9	
p	0.002	Si	1-n	n	29.5	
n	0.001	Si	7-p	n		22
p	0.002	Mo	0.004-n	n	0.03	16
p	0.002	Mo	0.004-n	n	0.045	

[a]Deep diffused gallium diodes used here.

SUMMARY AND CONCLUSIONS

This study has elucidated the factors that are important in controlling the resistivity of epitaxial layers that have been grown in several types of experimental systems. These are summarized by the equation.

$$N_{total} = N_{SiCl_4} + N_{system} + N_{substrate} + N_{pedestal}$$

where the source of the impurity is indicated by the subscript.

Doping of the film by diffusion out from the substrate is limited to the first 1–2 μ. However, for thicker films the substrate and the pedestal both contribute to film doping by means of a gas-phase transfer mechanism. When a molybdenum pedestal is used, the substrate may be minimized by resting the substrate on a high resistivity spacer. Lightly doped silicon is transferred to the backside of the substrate which tends to seal it and reduces its effect. If a high resistivity silicon pedestal is used, the pedestal provides the sealing material. The gas phase transfer is flow sensitive as would be expected.

The desired goal in growing the epitaxial layer is to produce a layer of any resistivity and type in a controlled and reproducible manner. If any of the factors vary uncontrollably, this goal cannot be achieved. If the effect of the system, substrate, and pedestal is minimized the control of resistivity can be obtained by suitable doping of the silicon tetrachloride. Alternatively highly purified silicon tetrachloride could be used and the control of the doping could be obtained by judicious manipulation of the substrate, spacer, or pedestal. In any case, it is necessary to reduce contamination to levels where it has no unwanted effect.

By use of an appropriately doped spacer the doping of the film can be achieved. In order to use the solid state doping in a more quantitative manner, it will be necessary to know more about the mechanism and the rates of removal of impurities from the solid. Some data are available from molten silicon (14). Most of the evidence for the movement of dopant atoms is indirect in nature. The use of radioactive dopants would provide direct proof of the transfer processes.

ACKNOWLEDGMENTS

The authors wish to thank A. E. Blakeslee, T. D. Jones, and P. P. Martinko for experimental results obtained in the tube furnace.

References

1. Royer, L., Bull soc. franc. mineral., 51, 7 (1932).
2. Pashley, D. W., Advances in Phys., 5, 173 (1956).
3. Loar, H. H., H. Christenson, J. J. Kleimack, and H. C. Theurer, IRE-AIEE Solid State Device Research Conference, Pittsburgh, June 1960.
4. Theurer, H. C., J. J. Kleimack, H. H. Loar, and H. Christenson, Proc. I.R.E., 1642 (1960).
5. Theurer, R. C., Bell Labs. Rec., 273 (1960).
6. Christenson, H., and G. K. Teal, U. S. Pat. 2, 692, 831 (Oct. 26, 1954).
7. Theurer, H. C., J. Electrochem. Soc., 108, 649 (1961).
8. IBM J. Research, (July 1960).
9. Reizman, F., and H. Baseches, this volume, p. 169.
10. Theurer, H. C., J. Electrochem. Soc., 107, 29 (1960).
11. Tung, S. K., this volume, p. 87.
12. Schafer, H., and J. Nickl, Z. Anorg. u. Allgem. Chem., 274, 250(1953).
13. Schafer, H., Z. Anorg. u. Allgem. Chem., 274, 265 (1953).
14. Ziegler, K., Z. Metallk., 49, 491 (1958).

DISCUSSION

J. J. McMULLEN (Rheem Semiconductor Corp.): I believe you pointed out, and rightly so, that diffusion from the substrate up into the epitaxial deposit is only possible for about 1 or 2 μ. During deposition the pedestal is at a higher temperature than the wafer, which is on top of the pedestal, and silicon tetrachloride, or trichlorosilane, and hydrogen are passing downward so some deposition is occurring on the pedestal; therefore, after 1 or 2 μ have been deposited the pedestal is completely covered. How can you account for contamination in the film after the pedestal is completely covered?

H. BASSECHES: In the case of the pedestal you must realize that our heavily doped inserts are perhaps 1 to 1-1/2 in. below the very hot region. The temperature may only be 800 or 900°C so there is much less deposition there. Although there is some coverage, the impurities simply get out.

J. J. McMULLEN: Do you feel that you have a great deal of dependence on the length of run? In other words, the longer you run the thicker the layer on the pedestal, and therefore the lower the contamination of the layer on the wafer.

H. BASSECHES: Yes, I would say that run sequence is an important factor. With the geometries we employed, however, we have been able to run for a considerable period with this p-type pedestal for example, and still have enough p-type dopant come out to produce high resistivity films.

J. J. McMULLEN: On the reproducibility of doping, do you etch the pedestal back to the original resistivity after each run?

C. O. THOMAS: No, we do not. We have obtained figures for over 15 runs in a sequence, and the average deviation from the mean film resistivity is about plus or minus 50%.

T. A. LONGO (Sylvania Electric Products): I tend to agree with McMullen, I do not believe there is as much dependence on the resistivity of the substrate as you have indicated. We have certainly looked for this and have grown layers in the same system with molybdenum susceptors, silicon spacers, both n- and p-type degenerate substrates; and we have grown subsequent layers with the same resistivity in the layer, regardless of the substrate resistivity, as determined by the doping in the silicon tetrachloride. We have made junctions with aluminum alloy junctions on the n-type layer. When we grew n-type layers on the p-type substrate, we obtained just about the same breakdowns in the full-grown junctions, or slightly higher as would be expected.

H. BASSECHES: We have had hundreds of runs which all fit into this pattern. May I ask, were you using a silicon pedestal?

T. A. LONGO: We were using a moly pedestal with a silicon spacer, and a spacer with a high resistivity.

H. BASSECHES: Well, that is fine. That effectively eliminates the effect of the substrate by this model. I do not see any contradiction here. I commend you on your practice. The spacer is apparently sealing the back of the substrate, both adequately and reproducibly.

T. A. LONGO: I would say that if this were so the results would be much, much more random than is observed since the thickness of the layer grown and the amount of sealing that would be accomplished during the growth run would be a very important factor in determining how much impurity could get out of the substrate and into the growth layer. We have not seen this.

H. BASSECHES: Well, I would have to sit down and look into the details of the particular system and experiments. There can be a balance according to our equation which would make the layer independent of the substrate resistivity.

J. J. McMULLEN: I feel it is my duty to say that we have purchased silicon tetrachloride from Sylvania which was doped to a certain level and gave a specified resistivity when deposited in their system. I am happy to say that we must have achieved the same degree of contamination in our system because we measured the same resistivity.

N. P. SANDLER (Pacific Semiconductor): We just finished a brief study on this problem, and have found the same thing that Dr. Basseches has reported, with one important difference. We used a high-resistivity n-type susceptor, and deposited silicon upon low-resistivity p-type substrates. The deposit was converted to p-type. This indicated that the transfer of impurities during the vapor deposition must have come only from the substrate, since if they had come from the susceptor, they would have tended to maintain an n-type deposit.

H. BASSECHES: There are geometrical factors, involving the registration of the substrate and the spacer which can cause a variability in the effectiveness of the sealing of the back of the substrate. To illustrate this we made a number of experiments with a molybdenum pedestal in which the substrate overhung the spacer at one edge. Diodes were made on the layer near this overhang and at the far edge where sealing should have been effective. The resistivity, as deduced from measurements on the diodes, was of the order of 1 ohm cm or less near the overhang and between 5 and 15 ohm cm far from the overhang. It appears that all of the observations qualitatively are explained by our model.

N. P. SANDLER: Such a scheme would indicate that the impurities are not coming from the susceptor in as great a quantity as they are from the substrate.

H. BASSECHES: Apparently very few impurities were coming from your high resistivity n-type susceptor.

N. P. SANDLER: Figure 3 indicated that most of the impurities were coming from the susceptor.

H. BASSECHES: No, this is not the case. Remember, this is only schematic and does not represent the number of impurities coming from any particular region. The resistivity of the insert also enters the picture.

N. P. SANDLER: The schematic for impurity transport as proposed seems to depend almost entirely upon the resistivity of the susceptor and/or the insert. If this were true, the deposit would not vary with deposition time, i.e., the deposit resistivity would remain essentially constant regardless of the length of the deposition process. We made several runs varying the time of deposition and found that the deposit resistivity decreased with increasing time. In our process, the impurities are p-type while the deposit is normally n-type. This indicates an occlusion of the substrate and a subsequent decrease in the apparent deposit resistivity. From these data one could deduce that the resistivity of the substrate does indeed play an extremely significant role in impurity transport during vapor deposition and also that gaseous interaction at the substrate surface may be a more direct mechanism for explaining this process.

H. BASSECHES: And remember, it depends on the relationship of your rf coil to the insert in the pedestal. After all, the hotter the insert the more important it will tend to be.

C. O. THOMAS: I think that I can clarify something on this pedestal business. The effect of the pedestal is going to depend somewhat critically upon the spacing between the pedestal and the wall because, as the gap is decreased, the linear velocity of the gas through that annular space is going to increase. The gap will also influence the temperature gradient which is essential for the back diffusion. As a result there may be large variations in the magnitude of this effect between systems, depending upon the relative size of the pedestal and the chamber and the doping level may be flow sensitive under certain conditions. At rather high doping levels (0.5 to 1.0 ohm cm) where a heavily doped substrate is employed, the effect of the pedestal may be masked by other factors. As the resistivity is increased to 10 or 15 ohm cm the effect of the pedestal becomes distinguishable.

The Influence of Process Parameters on the Growth of Epitaxial Silicon

S. K. TUNG

Bell Telephone Laboratories, Inc., Allentown, Pennsylvania

Abstract

The factors affecting the growth of epitaxial silicon layers on {111} silicon substrates by the hydrogen reduction of silicon tetrachloride using rf heating have been studied. These factors include deposition temperature, time, deposition rate, and $SiCl_4$ concentration. The effect of these variables on such layer properties as thickness, resistivity, and surface perfection is discussed. From the temperature study, an activation energy of 26 kcal/mole was determined for the overall reaction in the range 1100—1300°C. A number of types of surface defects were observed. The occurrence of some of these can be controlled by temperature. Others seem associated with the deposition rate.

INTRODUCTION

Early results (1) utilizing epitaxially grown, high resistivity silicon layers indicated that marked improvements could be realized in transistor performance. This paper describes the results for depositing layers by using the rf heating method in the hydrogen reduction of silicon tetrachloride. It is similar to that employed by Theuerer (2). The effect of such growth variables as deposition temperature, time, and concentration on such layer properties as thickness, resistivity, and surface perfection is discussed.

EXPERIMENTAL APPARATUS

The deposition of a thin single crystal layer was accomplished by the reaction of a hydrogen-silicon tetrachloride mixture at the surface of a hot single crystal substrate.

A commercially available floating-zone machine, originally designed for silicon single crystal growing, was modified for the growth of epitaxial silicon films. A train utilizing a Deoxo purifier, a molecular sieve at room temperature, and a hot titanium trap (800°C) was used for purifying the hydrogen. The silicon tetrachloride supply was kept in a Pyrex flask. A two-stage condenser operated at -10°C above this supply was used to insure that the hydrogen carrier gas was saturated with silicon tetrachloride vapor. The stream of hydrogen containing the silicon tetrachloride was diluted with a stream of pure hydrogen to achieve the desired concentration. The system could be evacuated by means of a mechanical pump to test for leaks. Suitable bypasses allowed flushing of inert gases through the system, as well as the establishment of equilibrium of the reactant mixture. Figure 1 shows a schematic diagram of the apparatus utilized for epitaxial growth.

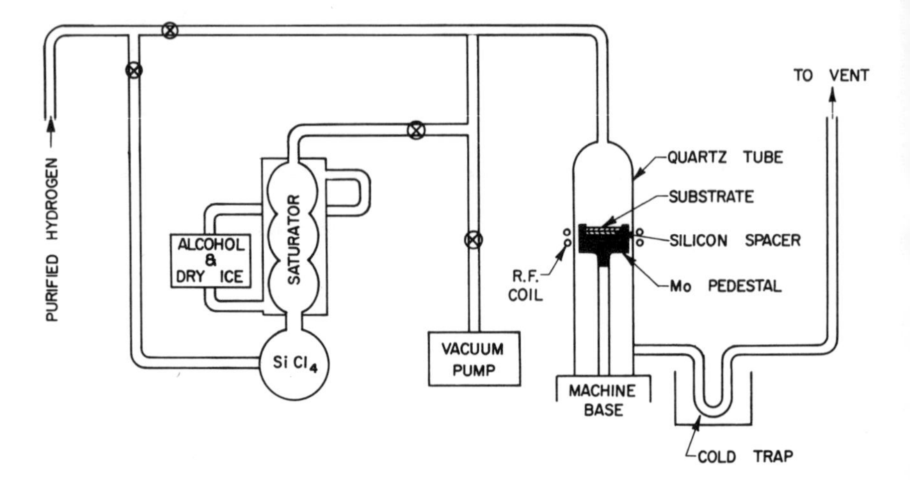

Fig. 1. Schematic drawing of vapor decomposition by rf heating method.

A high purity molybdenum pedestal was utilized to hold the substrate in position inside a quartz tube. Some depositions were also carried out using a silicon pedestal. A two-turn rf coil was used to restrict the heated zone. This was done to minimize the effect on resistivity of the layer that might come from impurity atoms vaporized from other parts of the system.

The system possessed the following features.

1. Low thermal inertia with a restricted hot region comprising the substrate and molybdenum pedestal.
2. A reproducible and constant rf source of heat.
3. A rather uniform temperature can be obtained by means of rotation of a molybdenum pedestal which has a relatively high thermal conductivity.
4. A gas flow directed normal to the substrate surface, minimizing the concentration depletion of silicon tetrachloride encountered in other systems, e.g., a horizontal tube furnace.

EXPERIMENTAL PROCEDURE

A chemically polished substrate was placed into the molybdenum pedestal. The substrate rested on either a quartz or silicon spacer. The purpose of these will be described later.

The system was first evacuated to a pressure of less than 5 μ and then helium gas was admitted. The pedestal was heated in this gas. The hydrogen gas was introduced as soon as the desired deposition temperature was reached. Hydrogen was then passed through the system displacing the helium.

Hydrogen was passed over the hot pedestal containing the substrate for 15 min. During this period the surface of the substrate was cleaned so that satisfactory layers could be obtained. (Other factors affecting film quality will be discussed later.) After this cleanup period, the reactant mixture was passed into the heated reaction zone. The deposition time (starting from the time when silicon tetrachloride vapor and hydrogen were admitted) varies depending on the film thickness desired. After the prescribed deposition time, the hydrogen gas flow is bypassed and replaced by helium gas flow. The temperature is

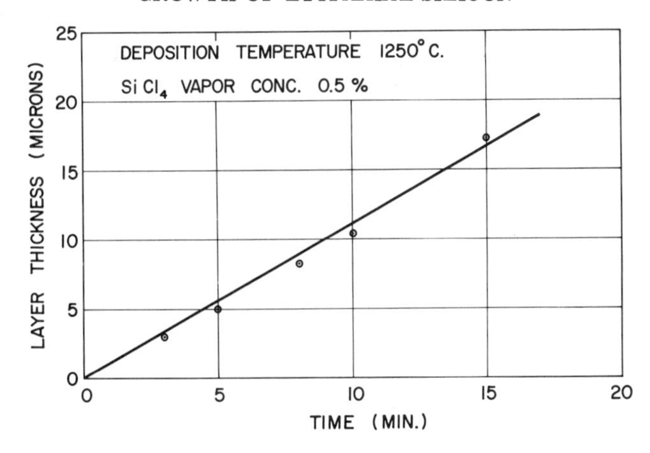

Fig. 2. Typical deposition curve obtained at 1250°C.

brought quickly back to room temperature by cutting the power off and a "deposition run" is then considered to be over.

As a means of checking background contamination, runs are occasionally made in which a high resistivity p-type substrate is placed in the system. The operating conditions are similar to a regular deposition run, except that no silicon tetrachloride is admitted to the system. Analysis of resistivity changes of the p-type wafer helps to indicate the nature of the background. The background has been found to contribute between 10^{13} and 10^{14} n-type impurities cm^{-3}. Runs are also made to check the purity of the silicon tetrachloride each time a new lot of material is used.

EXPERIMENTAL RESULTS AND DISCUSSION

Factors Affecting Rate of Film Deposition

The silicon substrate used for epitaxial growth is normally 7 mils thick by 1 in. in diameter with a nominally $\{111\}$ orientation. The wafers are lapped both sides with A.O. 305 grit (5 μ particle size) and then chemically polished on one side to give a bright and smooth surface. The deposition temperature used is usually between 1200 and 1250°C for device work. However, single crystal films can be grown anywhere from 950 to 1380°C. Qualitatively, a higher deposition temperature produces a better film surface. As the temperature is decreased other growth defects appear, as discussed later. At temperatures below 950°C, polycrystalline deposits are observed.

If the temperature, flow rate, and concentration of silicon tetrachloride vapor is held constant, the deposited film thickness is a linear function of time. Figure 2 shows a typical curve obtained at 1250°C. Figure 3 shows the dependence of deposition rate on temperature. A plot of the logarithm of the deposition rate versus $1/T$, where T is absolute temperature results in a straight line as shown in Figure 4. This yields an activation energy for the overall reaction of 26 kcal/mole.

As shown in Figure 5 the deposition rate increases as the vapor concentration of silicon tetrachloride increases. In the present experimental apparatus, both the diluent gas flow and the saturator temperature affect the vapor concentration. Increasing the diluent flow decreases the silicon tetrachloride vapor concentration, and a net decrease in deposition rate results. Figure 6

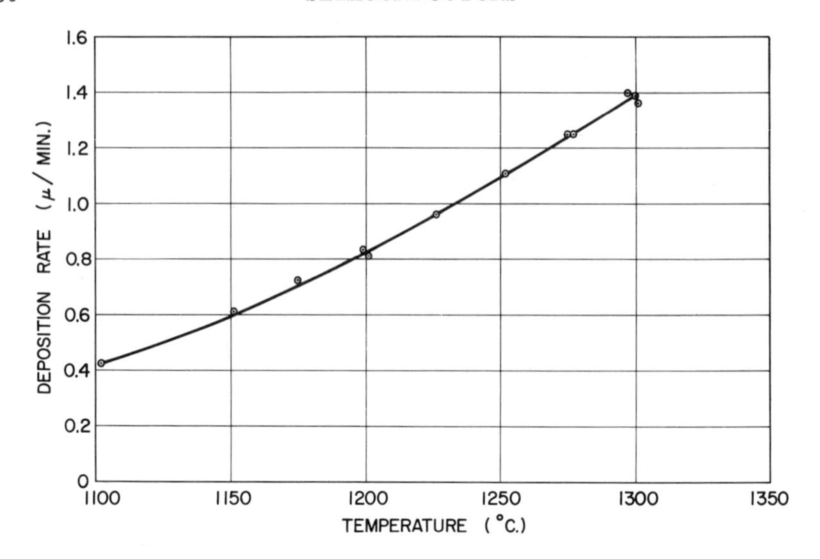

Fig. 3. Dependence of deposition rate on temperature.

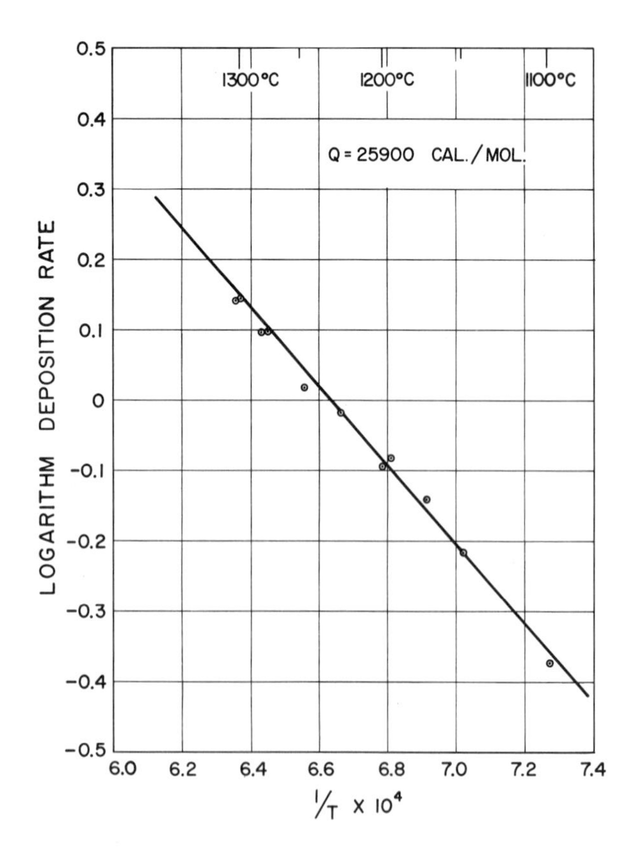

Fig. 4. Plot of the logarithm of the deposition rate versus 1/T.

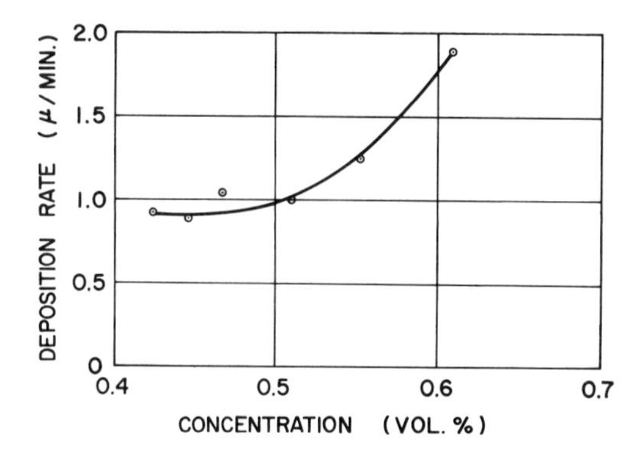

Fig. 5. Plot of deposition rate versus concentration of silicon tetrachloride. The concentration was controlled by varying the diluent gas flow.

shows the deposition rate as a function of saturator temperature at a constant diluent gas flow of 1710 cm^3/min and deposition temperature of 1250 °C. The hydrogen flow through the saturator was 110 cm^3/min.

The deposition rate is very sensitive to both deposition temperature and saturation temperature, as the latter determines the $SiCl_4$ vapor concentration. In general, these two factors are equally important to the deposition rate control. In order to hold the deposition rate to ±10% at 1200°C, the temperature should be held to ±15°C and the saturation temperature should be held to less than ±0.5°C.

Reproducibility and Evaluation of Film Thickness

As discussed in the preceding section, if the temperature and the $SiCl_4$ vapor concentration (or saturator temperature) are held constant, the deposition rate should be constant, and film thickness should vary linearly with time. Figure 7 shows the film thickness produced at a deposition temperature of 1230°C and a $SiCl_4$ vapor concentration of 0.3%. These data indicate a standard deviation (σ) of 0.12 μ/min, corresponding to a fractional standard deviation of 13.5%. All the thickness data reported above were based on weight gain meas-

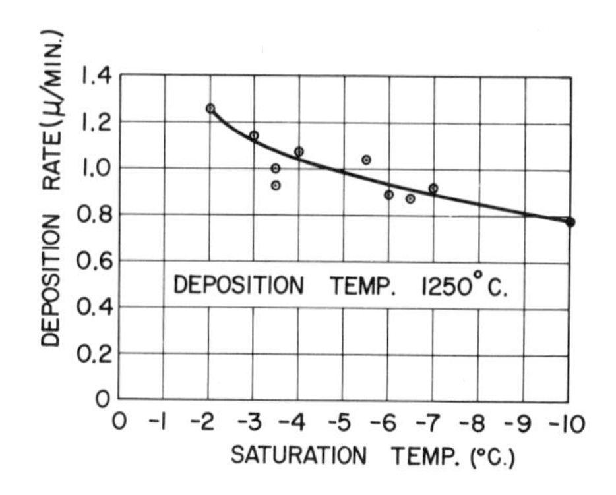

Fig. 6. Deposition rate as a function of saturator temperature at a constant diluent gas flow of 1710 cm^3/min and deposition temperature of 1250°C.

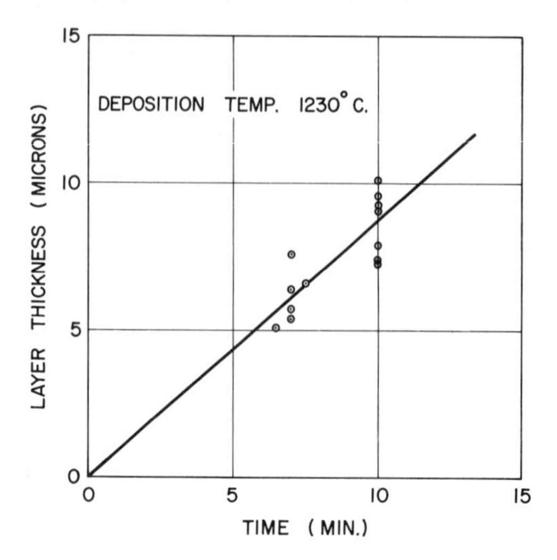

Fig. 7. Film thickness produced at a deposition temperature of 1230°C and a $SiCl_4$ vapor concentration of 0.3%.

urement. The formula used for calculation is

$$\Delta y = \frac{w_2 - w_1}{w_1} y_1$$

where Δy is the film thickness, w_2 the total weight after deposition, w_1 the total weight before deposition, and y_1 the thickness of substrate before deposition. The basic assumptions for use for the above formula are: (a) a substrate with uniform thickness, (b) a deposition layer that is uniform across the substrate area, and (c) the weight gain is from the deposited layer due to thermal decomposition and is all on the top surface.

Thus, regardless of the irregularity of the surface area, the only requirement is to measure substrate thickness precisely. This eliminates the need to make a complicated surface area calculation. It provides a measure of average layer thickness.

As shown in Table I, the thickness obtained from weight gain when a quartz spacer is used agrees quite well with values obtained using the angle lapping technique.

TABLE I
Comparison of Thickness Determined by Angle Lap and Weight Measurements

Slice No.	Thickness	
	Weight gain, μ	Angle lap, μ
TDS-226	6.88	6.5
TDS-232A	7.99	7.4
TDS-232B	7.1	5.0
TDS-703	8.4	8.6
TDS-704	12.6	11.5
TDS-705	6.7	8.9
TDS-706	17.2	16.8

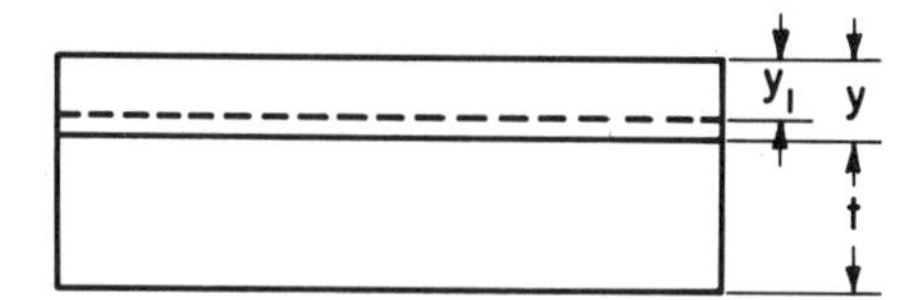

Fig. 8. Measured layers: t is the substrate thickness, y the theoretical epitaxial layer calculated by weight gain, y_1 the epitaxial layer revealed by staining technique, and $y - y_1$ the distance of junction migration by diffusion.

There is a small difference between the actual deposited film layer and the thickness layer revealed by angle lapping staining techniques. This is due to the diffusion that occurs at high temperature during deposition. The stain is sensitive to differences in resistivity, and diffusion from the substrate effectively dopes a thin section of the epitaxial layer. Figure 8 illustrates the different layers measured by the two methods. The distance of migration, i.e., $y-y_1$ as the figures show, is a function of deposition temperature, deposition time, and substrate dopant. In normal operation, with a deposition temperature of 1200°C, a 20-min deposition time, and an arsenic-doped substrate, $y-y_1$ is on the order of 1 μ.

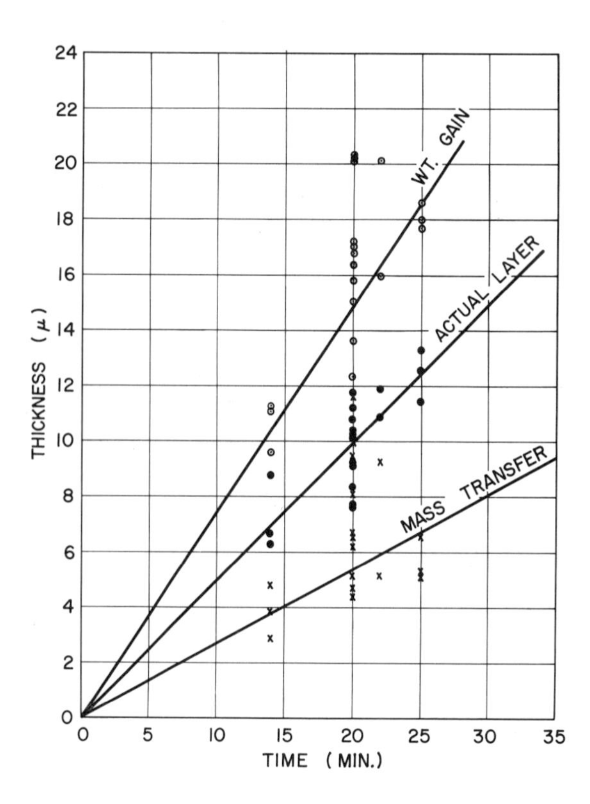

Fig. 9. A comparison of layer thickness obtained from weight gain data, the actual deposited layer, and the mass transfer layer.

Mass Transfer Phenomena and Film Thickness Uniformity

A layer is produced at the underside of the substrate by a mass transfer process if a silicon spacer is introduced between the substrate and the molybdenum pedestal. The same phenomena was observed by Theuerer (2) with a silicon pedestal. The main purpose of introducing such a spacer is to improve film resistivity, i.e., to minimize the interaction of the low resistivity substrate with the deposited layer. The mechanism of this interaction is discussed elsewhere (3).

The extent of the mass transfer is obtained by comparing the thickness deduced from weight gain with that from angle lapping performed on the same specimen. Figure 9 shows the layer thickness obtained from weight gain data (top curve). The middle curve is the true actual layer obtained by adding a diffusion correction term to the value obtained from angle lapping and staining. The diffusion correction was based on the assumption that ordinary solid state diffusion would occur during the deposition process. The bottom curve obtained from the difference of the other curves is a measure of the mass transfer. The spread of thickness data where mass transfer occurs is greater than when it is absent (see Fig. 7). The true layer thickness shown in Figure 9 has the smallest spread. The variables affecting mass transfer are temperature, time, partial pressure of silicon tetrachloride, and possibly the geometrical shape of the substrate. This latter affects the spacing between the substrate and spacer over the area of their contact. It is the variability in this factor which is probably responsible for the variations in mass transfer.

The weight gain method is based on an assumption that a uniformly thick layer has been deposited. The same assumption must be made if a single

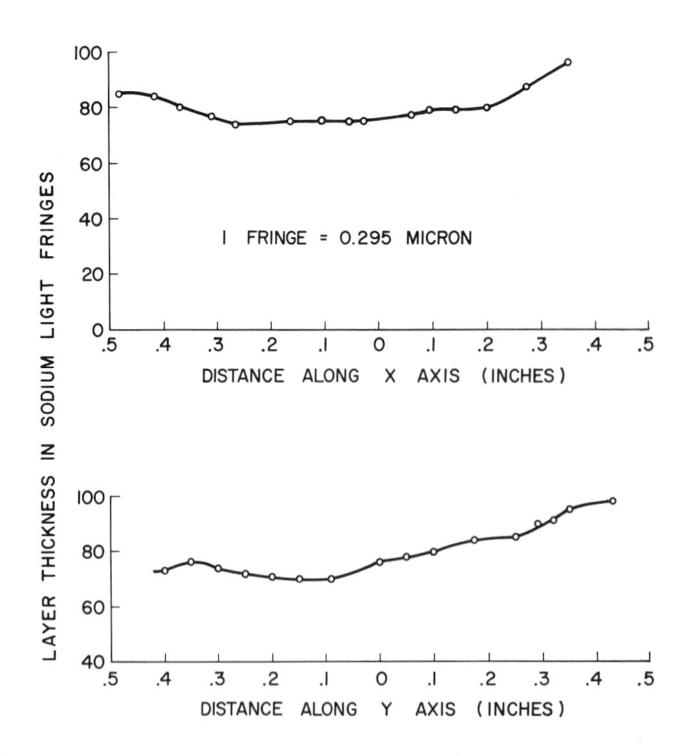

Fig. 10. Thickness variation on DCL-287.

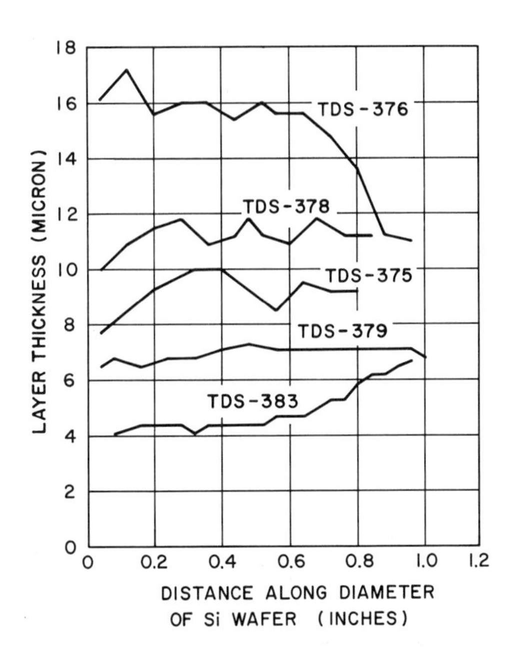

Fig. 11. Thickness profiles.

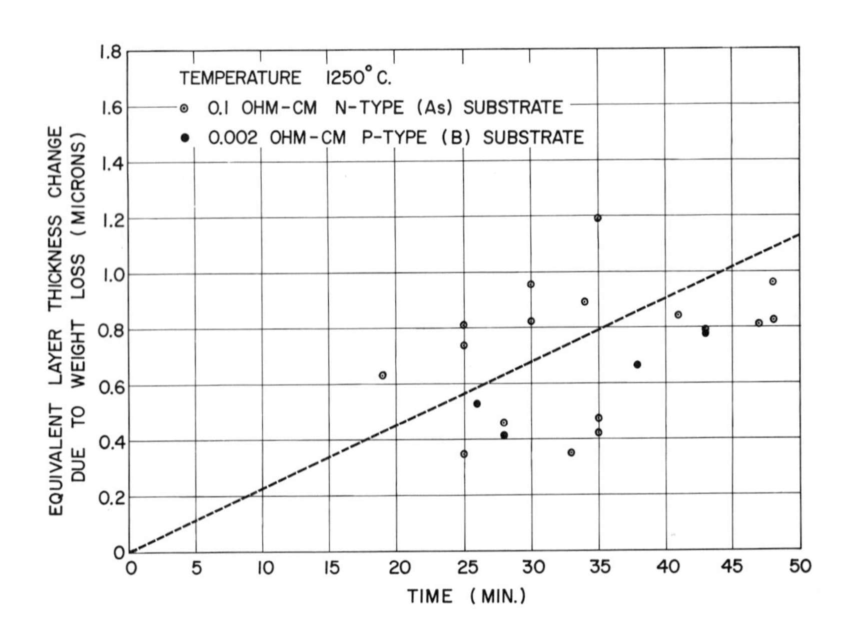

Fig. 12. Random weight loss data.

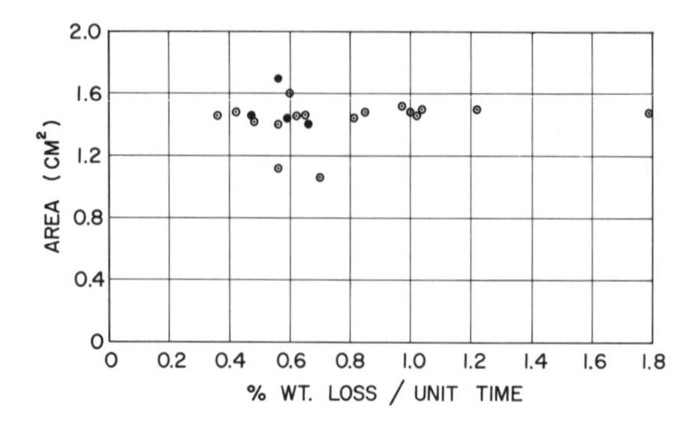

Fig. 13. Replot of Fig. 12 as the percentage of weight loss per unit time.

measurement by angle lap and staining technique is to give a true picture of layer thickness. The data indicate that this assumption may not be true. Figure 10 shows the thickness profile across both x and y axis of DCL-287. Other single profiles are shown in Figure 11. It is seen from the figures that there is a gradient in thickness around the edge portion, but it is fairly uniform at the central portion.

Several factors could produce the thickness profiles observed. A small nonuniformity in temperature would cause local variations in the deposition rate. The crystallographic growth plane is different on the tapered edges than in the center of the wafer. This suggests that the deposition rate may be a function of orientation. The flow pattern over the slice also affects the uniformity of the deposit. The surface finish might also have an effect. Thus, only if a uniform layer is produced can a single point determination of the thickness provide a valid picture of the entire layer. The weight gain method is attractive because it is simple, quick, and nondestructive.

Weight Loss

It was found in the rf system that chemically polished substrates lost weight when heated in a stream of hydrogen with no silicon tetrachloride present. The mechanism responsible for this weight loss is not yet known. The erosion may be due to some chemical reaction or to evaporation or to a combination of the two processes.

As shown in Figure 12, the weight loss data are quite random, i.e., at any particular run time there is a spread in the equivalent layer thickness removed. This run-to-run variation is believed due to the fact that the erosion over the surface is nonuniform. This is confirmed by microscopic observations of the surface which show some areas to be unaffected, while others are seriously affected. Figure 13 shows a replot of the data of Figure 12 as the percentage weight loss per unit time. There is a spread of a factor nearly 3 to 4 for the same slice area.

Fortunately, the weight loss observed in rf heating is comparatively small. If weight gain proved to be a usable method for thickness determination, the weight loss would not play a major role for thick films. However, for thinner films more data would be necessary to quantitatively determine the loss at various operating conditions, and methods for minimizing it would have to be sought.

The weight loss has significance from other standpoints. The material lost from the substrate is distributed to other places in the system. It thus repre-

TABLE II
Typical Layer Resistivity Produced by RF Heating Method

Run No.	Type	Substrate resistivity, ohm cm	Layer resistivity, ohm cm	Spacer
TDS-55	n	0.056	0.5	Quartz
TDS-56	n	0.056	0.3	Quartz
TDS-60	n	0.005	0.1-0.3	Quartz
TDS-73	n	0.002	0.74	Quartz
TDS-102	n	0.056	0.4-0.9	Quartz
TDS-124	n	0.005	6	Silicon
TDS-135	n	0.005	2.4-10	Silicon
TDS-138	n	0.056	8.7	Silicon
TDS-139	n	0.056	4.4	Silicon
TDS-142	n	0.056	9	Silicon
TDS-143	n	0.005	1.6-6	Silicon

sents, at least in the case of low resistivity substrates, a source of contamination, and hence can affect the resistivity.

Film Resistivity

The film resistivity is an important property of epitaxial layers since it directly influences the device parameters. A detailed discussion of the factors influencing the resistivity is presented elsewhere (3). The data indicate that a film resistivity level on the order of 0.3 ohm cm can be produced on a n^+ substrate without a silicon spacer and greater than 2 ohm cm utilizing a silicon spacer. Table II shows typical data. Without doping, n-type layers with resistivities as high as 30 ohm cm have been prepared.

Resistivity Control and Uniformity

Figure 14 shows resistivity profiles across 1-in. diam slices. In general, the data indicate that low resistivity regions are obtained around the periphery

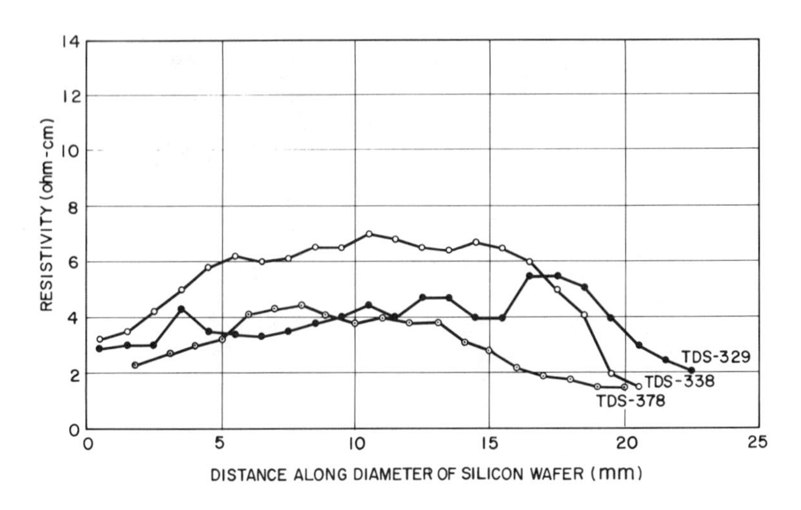

Fig. 14. Resistivity profile of epitaxial silicon film.

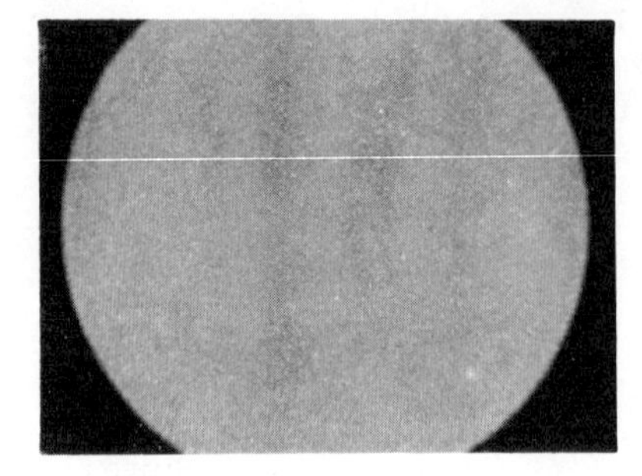

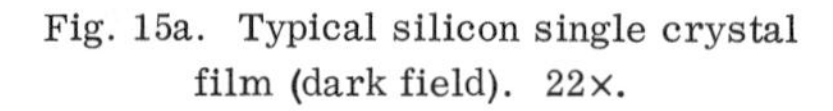

Fig. 15a. Typical silicon single crystal film (dark field). 22×.

Fig. 15b. Typical fringe pattern of silicon single crystal film. 22×.

of the substrate and high resistivity regions at the central portion. A possible explanation of nonuniform resistivity might be due to the "back doping" phenomena (3). This is associated with the geometrical shape of the back of the substrate, i.e., the flatness of the underside of substrate. If it is bowed the sealing action due to mass transfer may be ineffective near the edge, and hence this region acts as a doping source. In addition, problems of proper registration of the spacer and the substrate also contributed to asymmetry in the profile. The latter problem can be overcome by careful attention to the geometrical relationship of spacer and slice (4).

Film Surface

The film surface in general is a duplication of the substrate surface. It will not be superior, at least in appearance, to the initial substrate surface. In addition, some defects occasionally are introduced in the course of deposition. A typical silicon single crystal film is shown in Figures 15a and 15b. Another typical texture revealed from the electron microscope is shown in Figure 16.

The surface defects, often observed with the rf heating process, are shown in Figure 17. The conditions causing such defects (triangular pits) are not un-

Fig. 16. Silicon single crystal film texture by electron microscope. 8400×.

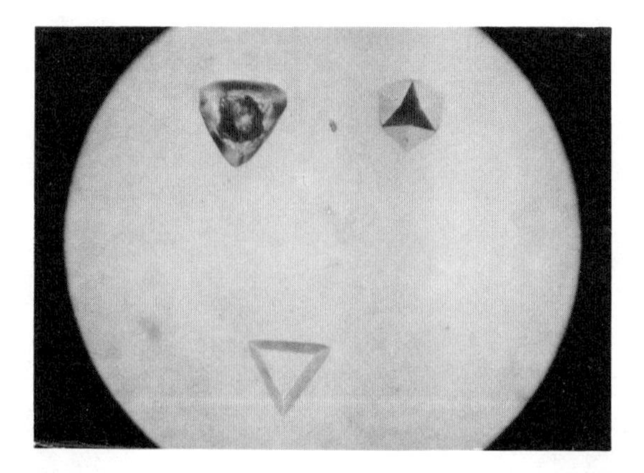

Fig. 17. Surface defect normally observed with rf heating process. 320×.

Fig. 18. Other surface defects. 70×.

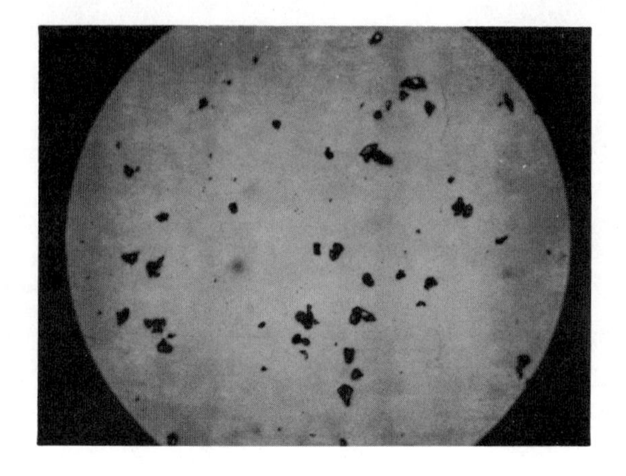

Fig. 19. Other surface defects. 70×.

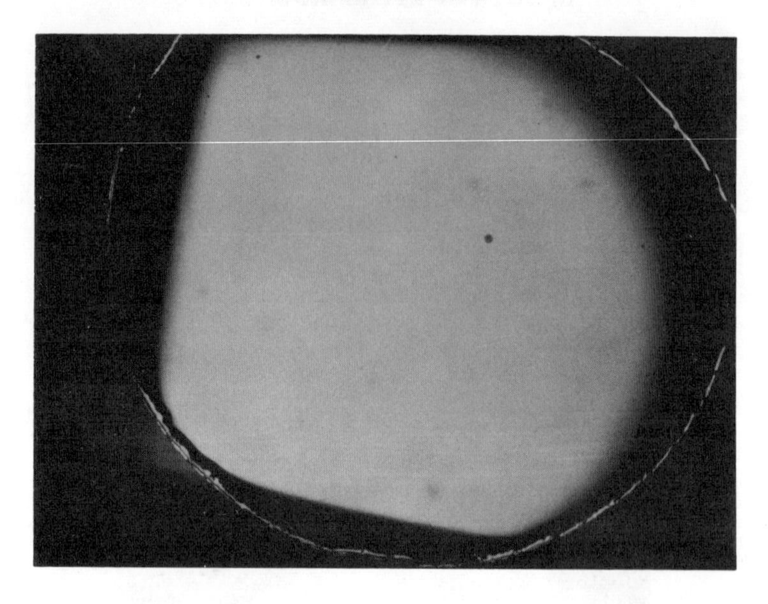

Fig. 20. Surface appearance with low deposition rate. 3×.

derstood at present. Experimental results indicate that fewer triangular pits are produced if the film is grown at a high deposition temperature and if extreme caution is used in cleaning the surface. Other surface defects, such as shown in Figures 18 and 19, are occasionally observed in the rf system. Figure 18 is perhaps due to the growth at low temperature and Figure 19 is from the uncleaned substrate surface. Substrate surface preparation is undoubtedly an

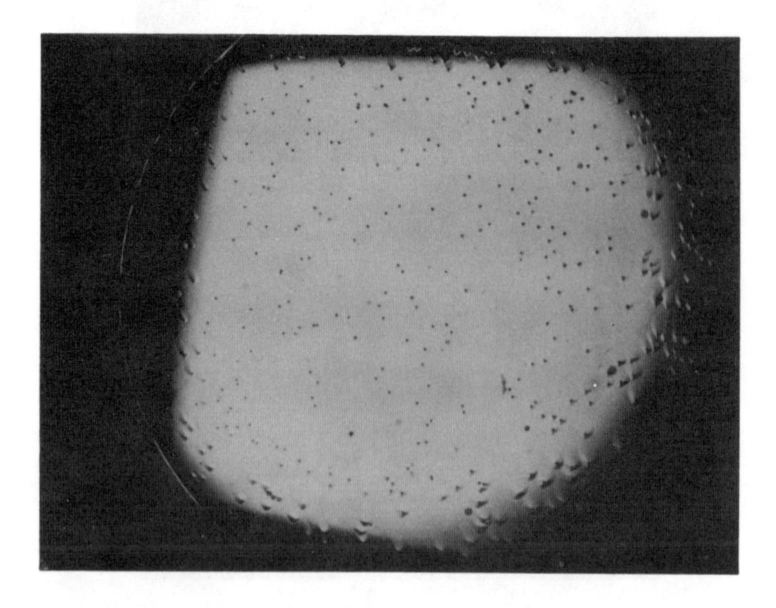

Fig. 21. Surface appearance with high deposition rate. 3×.

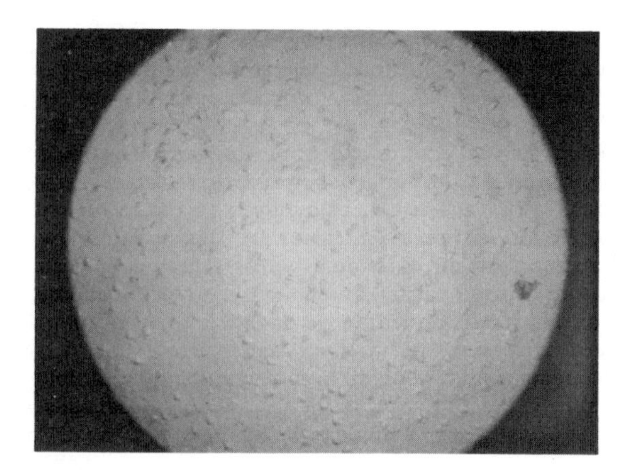

Fig. 22. Epitaxial film surface appearance after chemical etching. 128×.

important factor in producing epitaxial surfaces that are free of defects. Predeposition cleaning *in situ* is also important.

There are indications that surface perfection is related to deposition rate. Figure 20 shows a surface obtained with a $SiCl_4$ vapor concentration of 0.2% and Figure 21 shows one obtained at the same temperature, but with a concentration of 2%. The deposition rates were 0.6 and 4.5 μ/min, respectively.

The surface of an epitaxial film was etched with 1 part of HF, 3 parts of HNO_3, and 4 parts of acetic acid. Figures 22 and 23 show two types of etching characteristics of eptiaxial layers. Figure 22 is comparable with the usual etch figures obtained on a highly perfect bulk silicon crystal. Figure 23 is similar to the features ordinarily obtained from thermal etching of bulk material. This suggests that a thermal etching process may be going on simultaneously with deposition.

SUMMARY OF RESULTS

The factors affecting the growth of epitaxial silicon layers on $\{111\}$ silicon substrates using an rf heating method have been studied.

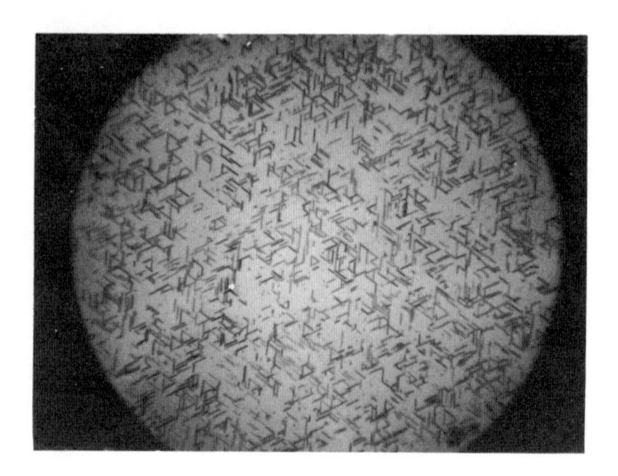

Fig. 23. Epitaxial film surface appearance after chemical etching. 128×.

The thickness, resistivity, and surface perfection of the epitaxial layers have been determined as a function of the growth variables. The activation energy of 26 kcal/mole was determined for the overall reaction over the temperature range 1100—1300°C.

The run-to-run variation of thickness as determined by a weight gain method was ±14%. This evaluation method, however, gives an average value. Thickness variations of ±25% are indicated by angle lap technique where measurements are taken at a given point. Thickness profiles by angle lapping on a slice indicate thickness variation can occur across the layer. This suggests that the greater run-to-run variation as indicated by angle lapping was a reflection of the thickness variation within a slice.

The resistivity produced on n-layers ranged from 2 to 30 ohm cm by capacitance voltage measurements on diodes. A resistivity profile is found to exist across the slice with the lowest values being predominantly at the edge. A possible explanation of this feature might be attributed to the back doping phenomena.

The film surface has also been studied. Certain surface defects seem associated with deposition rate.

ACKNOWLEDGMENTS

The author is indebted to Dr. H. Basseches for his helpful discussions and valuable suggestions.

References

1. Theuerer, H. C., H. H. Loar, J. J. Kleimack, and H. Christensen, Proc. I.R.E., 48, 1642 (1960).

2. Theuerer, H. C. J. Electrochem Soc., 108, 649 (1961).

3. Basseches, H. R. Manz, C. O. Thomas, and S. K. Tung, this volume, p. 69.

4. Edmunds, G. H., unpublished data.

Doping of Silicon Epitaxial Layers

WILFRED J. CORRIGAN

Motorola Semiconductor Products Division, Phoenix, Arizona

Abstract

The epitaxial deposition of silicon was investigated over a wide range of deposition rates and substrate temperatures. The growth temperature was varied from 930°C to the silicon melting point, and growth rates of 0.01 to 5.0 μ/min were employed. Controlled quantities of boron and phosphorus halides were injected during the growth process enabling the resistivity of the epitaxial layer to be controlled within 10%. The resistivity range investigated was from 1.0 to 0.0005 ohm cm both n- and p- type. Many multilayer structures have been produced with excellent electrical properties. Interdiffusion between layers can be a limiting factor on the design of a given structure.

INTRODUCTION

The epitaxial deposition of silicon, using the hydrogen reduction of silicon tetrachloride, has been described previously by several workers (1,2). Addition of controlled amounts of dope during the growth cycle makes this a convenient method for producing device structures.

Horizontal flow, inductively heated, epitaxial furnaces were used. The temperature range used in this work was from 900°C to the melting point of silicon. Growth rates of from 0.01 to 5 μ/min were employed. The dependence of the type of growth on both temperature and growth rate was established. The inherent resistivity of the distilled silicon tetrachloride used was above 100 ohm cm. Controlled amounts of boron and phosphorus halides were injected into the gas stream. The resistivity of the growing layer was shown to have a direct dependence on the gas-phase impurity/solidus silicon ratio. The resistivities covered in this work were between 1 and 0.0005 ohm cm both n- and p- type. Some limitations are imposed on the epitaxial structure by interdiffusion between layers during growth.

Epitaxial Furnace Construction

The furnace layout is shown in Figure 1. The furnace tube and the tetrachloride pickup system are constructed of quartz. The furnace tube is water cooled by using a water curtain. The water cooling inhibits growth on the tube walls which may cause spurious nucleation by small particles falling on the surface. The heating element is a high purity graphite strip, inductively heated by a 10-kw rf generator. The wafers lie on a thin quartz plate in intimate contact with the graphite susceptor.

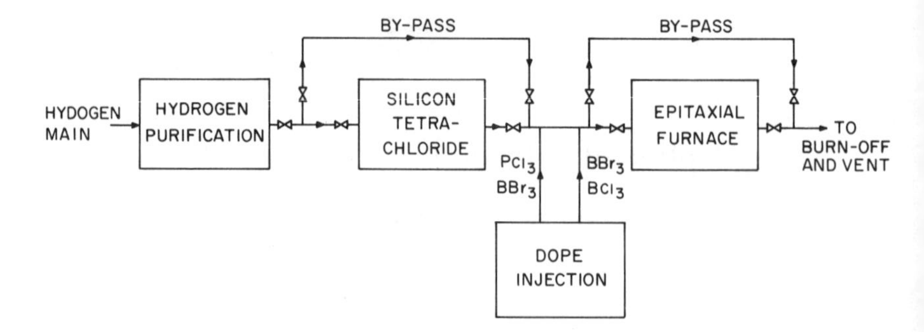

Fig. 1. Furnace layout.

WAFER PREPARATION

Oriented {111} wafers 6 mil thick were used. The wafers, after lapping, are given a final polish with $1/4\,\mu$ of diamond paste. Several cleaning steps are employed finishing with organic solvents. The wafers are not etched before growth.

INFLUENCE OF PROCESS VARIABLES ON GROWTH

The important variables affecting the growth process are: gas phase silicon tetrachloride concentration, substrate temperature, gas temperature, furnace geometry, and overall flow rate.

The furnaces were deliberately designed to be highly inefficient from the reaction yield standpoint. This is necessary, using a horizontal flow system, in order to get the same growth on every wafter. Provided that the reaction products from the upstream wafers represent a small fraction of the overall gas stream, the downstream wafers will be unaffected and each wafer will see essentially the same gas phase composition above it. Variations in thickness of less than 10% can be obtained with more than 20 wafers in line; adjacent wafers across the tube show the same thickness. Variation in thickness across the wafer can be eliminated by careful control of flow rate, temperature, and furnace geometry. Small variations in any of these parameters can give rise to either troughed or mounded epitaxial layers. These defects normally show symmetry about the longitudinal axis of the tube as shown in Figure 2. The maximum variation in thickness is normally about $0.2\,\mu$ in a $10\,\mu$ film.

Temperature is measured using an optical pyrometer with the appropriate corrections applied. Estimated reproducibility is ±5°C. Temperature has two

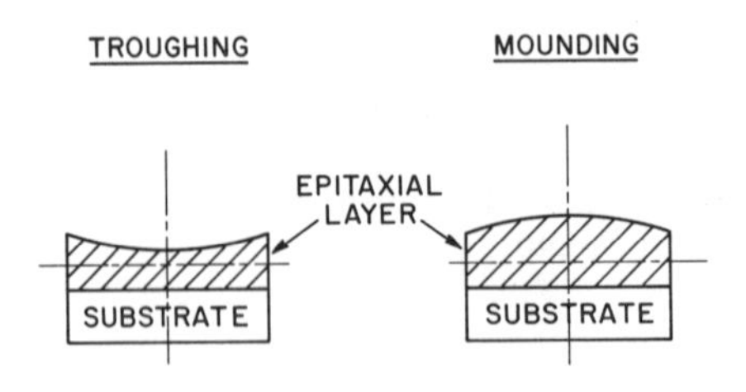

Fig. 2. Troughed and mounded epitaxial layers.

important effects. First, it affects the growth rate and, secondly, it determines whether the growth will be single crystal, terraced, or polycrystalline. At each temperature in the range of 900°C to the melting point, there is a definite maximum growth rate above which polycrystalline growth is initiated. As the growth rate is increased at constant temperature the growing film progresses from a single crystal plane surface, through a terraced structure, to a completely polycrystalline growth. A similar sequence is followed, as the temperature is dropped, with a constant gas phase composition. It is necessary to preclean the wafers at 1200°C in hydrogen for 5 to 15 min before growth. This removes surface oxides, probably by the reduction of silicon dioxide by silicon to silicon monoxide which then evaporates.

The silicon tetrachloride content of the gas phase has a fairly direct relationship to the growth rate. The relationship is not linear as may be anticipated for a first-order gas reaction. The numerical data obtained follow a systematic pattern when considering either growth rate as a function of temperature, gas phase composition, or overall flow rate. However, these data have little theoretical importance, since other variables such as gas inlet temperature, tube-wall temperature, and furnace geometry are involved. The activation energies for the growth process reported by other workers must be considered inaccurate for this very reason. For example, does the reaction rate increase with temperature by virtue of the reaction kinetics at the hot surface, or by the change in gas temperature and flow patterns due to the increased heat from the relatively massive susceptor? The changes in growth rate that we have observed from small changes in furnace geometry, or wafer position, would argue that gas phase mass transfer effects are at least a partially controlling factor in the reaction mechanism. The flow rates were from 2—8 liters min, the composition ranges used were from 0.3—5.0 mole-%.

Contamination of the surface of almost any sort can initiate polycrystalline growth. Surface defects such as scratches are washed out to some extent, but in general the defect is propagated through the film, retaining its initial shape with some rounding of the edges.

Surface defects do not appear to act as nucleation sites for polycrystalline growth until the limiting conditions for single crystal growth are approached.

PURIFICATION OF TETRACHLORIDE

A variety of purification techniques including adsorption, chemical reagents, and distillation have been evaluated. The best results have been obtained from distillation in an all-quartz still (3), operating at a high reflux ratio in an argon atmosphere. The height of the column was such as to correspond to a minimum of 10 theoretical plates, the reflux condenser was water cooled. The minimum batch size was two liters.

The inherent resistivity of this material, as evaluated by growing on high resistivity p-type substrates is of the order of 400 ohm cm n-type. It is likely that silicon tetrachloride is capable of higher resistivity using more refined purification techniques, although it is possible that an upper limit may be placed on the film resistivity by the system itself.

DOPE INJECTION SYSTEM

Several techniques for the addition of dope may be envisaged; the three most obvious are: (1) addition of dope to the liquid silicon tetrachloride source, (2) direct injection into the gas stream immediately before the reaction chamber, and (3) use of a doped gas source.

Method (1) gives excellent results from the point of view of reproducibility. For most of the halides used, the change in concentration of the halide due to the differences in vapor pressure between the halide and silicon tetrachloride does not become important until a large number of runs have been made.

However, this system is rather inflexible for experimental work. Even though a variable resistivity system could be easily devised it would be unnecessarily complex.

A reasonably flexible system requires a silicon tetrachloride source capable of producing high resistivity material, with a separately controlled dope injection system between the silicon tetrachloride source and the furnace. This can be either direct injection or doped gas. We have been using direct injection. Direct injection is accomplished by a gas phase diffusion-controlled process.

The vapor pressure of the halide source is controlled by temperature. Provided the diameter of the injection orifice is small, transfer by convection can be neglected, and transfer of the halide through the orifice will be by diffusion alone. Application of Fock's law to this system shows that the mass transfer rate will depend directly on the square of the orifice diameter, and the vapor pressure of the halide, and inversely with the length of the orifice; i.e., $N_a \propto P_a d^2/l$, where P_a is the halide vapor pressure, d the orifice diameter, and l orifice length.

This was found to hold true over a wide range of orfice dimensions and vapor pressures.

Once the dope is injected it is necessary to adequately mix the gas. If this is not done correctly, large resistivity variations are obtained both from wafer to wafer and across individual wafers.

The dope injection does not appear to have a significant effect on the silicon deposition rate. We have doped down to 0.0005 ohm cm both n- and p-type without seeing any difference in silicon growth rate provided all other conditions remain constant.

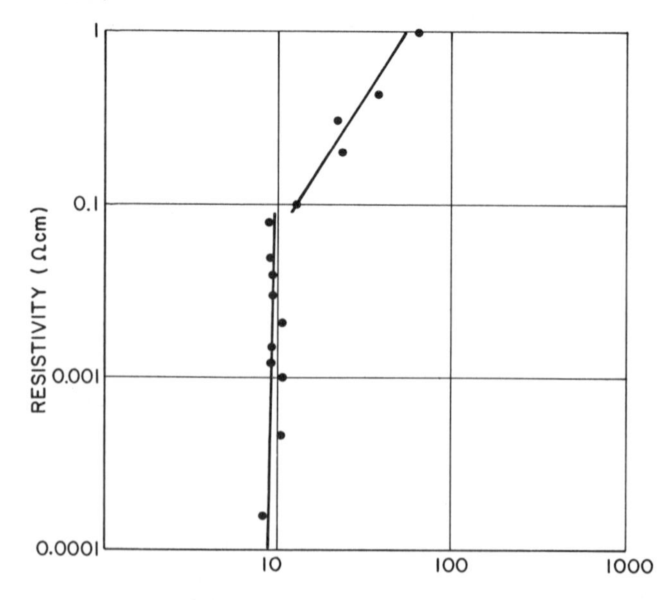

Fig. 3. Graph showing $\left(\frac{[B]}{[Si]}\right)_{gas} / \left(\frac{[B]}{[Si]}\right)_{solid}$ against resistivity of the epitaxial layer.

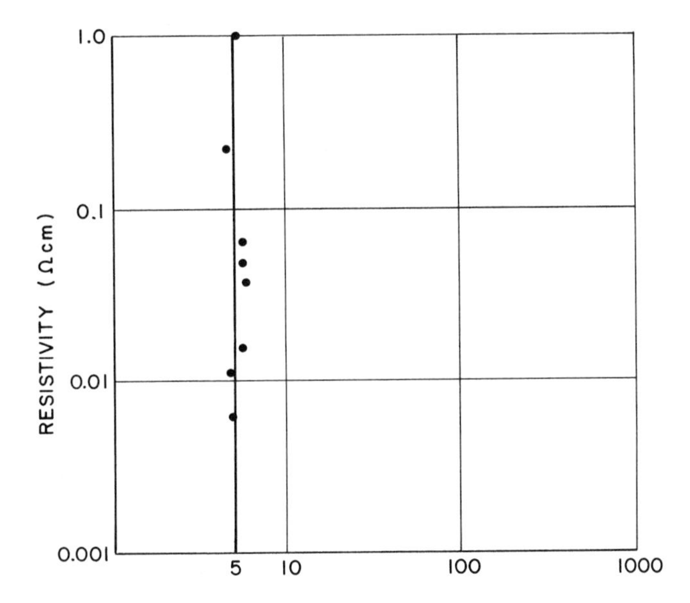

Fig. 4. Graph showing $\left(\frac{[P]}{[Si]}\right)_{gas} \Big/ \left(\frac{[P]}{[Si]}\right)_{solid}$ against epitaxial layer resistivity.

P-TYPE DOPING

Boron tribromide and boron trichloride have been used. The deposition rate of boron is not dependent upon whether the boron is present as the trichloride or the tribromide, provided the same boron/silicon ratio is used.

Figure 3 is a graph of

$$\left(\frac{[B]}{[Si]}\right)_{gas} \Big/ \left(\frac{[B]}{[Si]}\right)_{solid}$$

against resistivity of the epitaxial layer. It can be seen that this ratio is approximately 10:1 over a wide range. In the higher resistivity range 0.1 ohm cm and up

$$\left(\frac{[B]}{[Si]}\right)_{gas} \Big/ \left(\frac{[B]}{[Si]}\right)_{solid}$$

begins to rapidly increase being 50:1 at 1 ohm cm. This is probably due to the system rather than a change in the reaction kinetics.

N-TYPE DOPING

Phosphorous trichloride and phosphorous tribromide have been used. Again, the important parameter is the phosphorous/silicon ratio. Figure 4 is a graph of $(P/Si)_{gas}/(P/Si)_{solid}$ against epitaxial layer resistivity. No significant change is seen at the higher resistivity end of the curve. The ratio remains fairly constant at about 6:1 over the whole range.

VARIATION IN DOPING LEVEL WITHIN THE FURNACE

For boron doping a maximum variation of 20% over 8 wafers in line is obtained. The lead wafer is the lowest resistivity, which then increases gradually to the eighth wafer. The variation is expressed as the percentage increase of the last wafer relative to the first.

The maximum spread with phosphorus doping is less than 10%. Resistivity variation across the individual wafer is small, generally much less than 5%.

Resistivity measurements for the curves shown in Figures 1 and 2 were by a simple 4-point probe measurement. The 10-ohm cm substrates were used for evaluation on both n- and p-type. A standard 5-μ thick layer was used in each run at a standard growth rate of 0.7 μ/min at 1130°C. Under these conditions, considering the accuracy of the 4-point probe method, and the thickness measurement, the effect of interdiffusion was neglected. Recent work by McNamara and Robertson (4) has shown this assumption to be justified.

The reaction temperatures suitable for epitaxial growth are also temperatures at which boron and phosphorous diffuse rapidly. This necessitates the use of a fast growth rate and as low a growth temperature as possible. It is necessary to cut the high temperature time down to a bare minimum. Multilayer structures must be designed with the fact in mind that the resistivity will change across the layer even though dope is being injected at a constant level.

Structures requiring a very thick final layer, of the order of 1 mil, are difficult to produce if this thick layer covers several thin layers. The time taken to grow the final thick layer is often sufficient for a complete layer to be eliminated by interdiffusion from both sides. We are limited in the growth rate that can be used by the earlier considerations with respect to terracing and polycrystalline growth.

Interdiffusion is magnified with adjacent layers of widely differing impurity concentration. If the impurity concentration in adjacent layers is similar, the junction itself will remain in almost the same position, although there will still be higher resistivity regions in the vicinity of the junction.

EVALUATION TECHNIQUES

Epitaxial layer thickness is measured by weight change and bevel and stain methods (1). Provided the wafer lies on an accurately flat, inert surface during the growth process, little undergrowth is observed. Generally the undergrowth is of the order of 0.3 μ during the growth of 10 μ on the upper surface, and even this small thickness only occurs at the edge. Weight change is an accurate measure of the amount of silicon deposited when a 6-place microbalance is used. However, due to the effects of interdiffusion during growth the mechanical junction will not coincide with the electrical junction in normal circumstances. For this reason bevel and stain techniques must be used to delineate the electrical junctions and to measure the thickness of the various regions of differing resistivity type. Once a time and temperature cycle is established for a given structure, a simple relationship between the relative position of the mechanical and electrical junctions can be obtained.

The bevel and stain method involves angle lapping the desired section, staining the lapped surface, then measuring the various thicknesses using an optical interferometer.

This method is superior to weight measurement in that the thickness profile across the wafer in any direction may be established, whereas weight measurement simply gives a mean thickness.

Resistivity is measured using a 4-point probe with 25 mil spacing between points. This method is only applicable when the sublayer, or substrate is of opposite conductivity type. With n on n^+ or p on p^+ structures, alloy junctions are made to the upper surface and the resistivity estimated from breakdown voltage and junction capacitance measurements.

STRUCTURES PRODUCED EPITAXIALLY

The two structures to be described are both structures which have been produced many times over.

Figure 5 shows a simple transistor structure. The n^+ substrate is about 0.02 ohm cm. On this we have grown an n-region of 5 ohm cm, 12 μ thick. Some spreading of the n-type dope can be seen from the substrate into this layer. A 4-μ p-type layer of 0.1 ohm cm has been grown on top of the n-type layer. Note the high resistivity n- and p-type regions in the vicinity of the junction caused by interdiffusion. The final layer is an n^+-region of 0.002 ohm cm. Again, high resistivity n- and p-regions are seen at the junction.

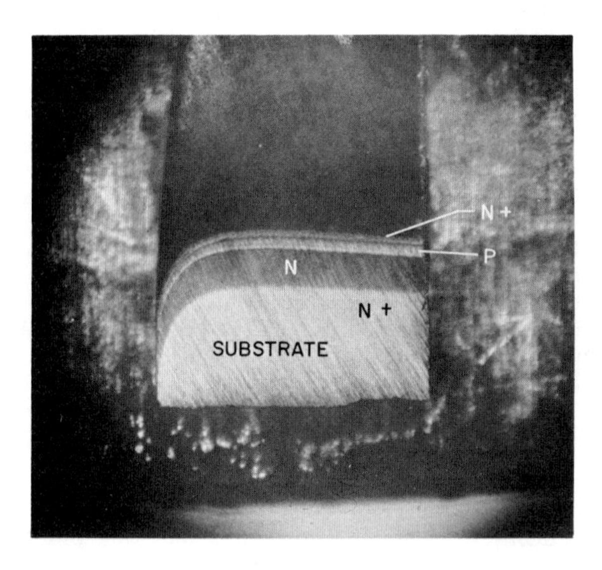

Fig. 5. Transistor structure.

Devices produced by these techniques show similar properties to comparable diffused structures. Leakage current is slightly lower, but as this is a surface sensitive property, it would be unwise to draw premature conclusions concerning the nature of the grown p-n junction.

The well-known epitaxial collector 2N834 has been used as a standard of comparison. Grown base and completely epitaxial devices have been constructed which show almost identical properties to the standard production double-diffused 2N834. Many other structures have been made, such as Zener diodes, 4 layer diodes, power transistors, and solar cells. Some yield difficulties are experienced with very large area power transistors. Minute imperfections or inclusions in the growth can cause very low breakdown voltages. A wafer showing greater than 90% yields on a small area device can often give only one or two good devices out of six on a large area device. By "large" 1/8 by 1/4 in. or bigger is implied.

References

1. Russell, G. V., Epitaxial Growth of Germanium and Silicon from the Vapor Phase," P.G.E.D. Meeting, Washington, Oct. 1960.

2. Theurer, H. C., Proc. I.R.E., (1960).

3. Fenske, Larowski, and Tonberg, Ind. Eng. Chem., 30, 297 (1938).

4. McNamara and Robertson, "Delineation of Epitaxial p-n Junctions," Electrochem. Soc. Meeting, Indianapolis, May 1961; also, private communication.

DISCUSSION

C. O. THOMAS (Bell Telephone Labs.): Was the thickness control figure 1%?

W. J. CORRIGAN: Yes. This is on our development trends.

C. O. THOMAS: How do you measure 1%? One fringe is a third of a micron, and out of 10 μ that is 3%.

W. J. CORRIGAN: It is well within the fringe line.

C. O. THOMAS: You are estimating within the fringe line?

W. J. CORRIGAN: Yes.

C. O. THOMAS: You mentioned that you had grown films as low as 900°C and they were still smooth.

W. J. CORRIGAN: I did not say "as low as." I said "approaching 900." About 930 is the lowest we have gone.

C. O. THOMAS: Are these single crystal, and if so, do you have any device characteristics in them?

W. J. CORRIGAN: We have not examined them. We just grew about 10 μ on n^+, at temperatures near 930°C, at very slow growth rates—around 0.01 μ/min. Perfect single crystal surfaces were obtained.

C. FA (Hughes Semiconductor): Could you elaborate more about the electrical characteristics of these junction evaluations?

W. J. CORRIGAN: The three-layer structure is as far as we have gone, but Dr. Trevor Law has gone to more layers than this in his experiments.

H. BASSECHES (Bell Telephone Labs.): What type of resistivity uniformity do you have in your layers?

W. J. CORRIGAN: I would say within 5%, in general.

H. BASSECHES: Across a 1-in. slice?

W. J. CORRIGAN: We have not done a great deal of work with 1-in., but generally with 0.7 in. wafers it is well within 5%. This is over the whole range, up to 1 ohm cm. Above that, large resistivity variations are seen.

T. A. LONGO (Sylvania Electronics Systems): You showed a curve or graph for doping with phosphorus and boron impurities with a constant ratio of doping in the gas phase to doping in the grown layer. Do you have a similar result for arsenic doping?

W. J. CORRIGAN: We have only used arsenic at around 0.1 ohm cm and we found it to be quite good. With arsenic you get even less variation across the wafers than you do with phosphorus doping.

T. A. LONGO: You mentioned that you used the alloy junction to evaluate p-type layers also. What did you use for the alloy?

W. J. CORRIGAN: We would generally grow an n-type layer.

T. A. LONGO: So you only used aluminum junctions for n-type layer, but not alloy junctions for p-type layer?

W. J. CORRIGAN: Yes.

J. J. McMULLEN (Rheem Semiconductor Corp.): You mentioned that the silicon tetrachloride that you used was capable of producing deposits with resistivities up to 400 ohm cm and yet covering the range only from 0.05 up to 1 ohm cm. Am I to assume that your multiwafer system in long horizontal tubes is incapable of producing the 400 ohm cm, and that 400 ohm cm was sustained only in a one-wafer system?

W. J. CORRIGAN: No. I pointed out that in a reasonably flexible system you must have a source capable of producing almost intrinsic material, which, to all practical purposes, is 400 ohm cm. We can produce 400 ohm cm material quite easily.

J. J. McMULLEN: On this multiwafer system, then, all the way down the horizontal tube you obtain 400 ohm cm with no variation?

W. J. CORRIGAN: No, in the high resistivity range large fluctuations are observed.

J. J. McMULLEN: Do you consider anything above 1 ohm cm to be high resistivity?

W. J. CORRIGAN: No. I would say that anything above 10 ohm cm would be high resistivity. In a run resistivities between 200 and 500 ohm cm might be obtained. This is measured by four point probe techniques, I would not like to estimate the accuracy.

J. J. McMULLEN: Most of the work is usually done in a low resistivity range?

W. J. CORRIGAN: Yes, but using the 400 ohm cm material as a source.

J. J. McMULLEN: You discuss gas phase doping, which I also believe is the best way of injecting a dopant into the gas stream. Did you find large resistivity variations when liquid boron tribromide was used as the source?

W. J. CORRIGAN: No.

J. J. McMULLEN: This has been our experience, and I wondered whether anyone else had corroborated it.

W. J. CORRIGAN: Our data were consistent as long as the boron tribromide was used continuously. However, if it stands a brown precipitate forms, after which strange results are obtained. You have to use it within a few days after you have prepared the dope.

J. J. McMULLEN: You mentioned that in pretreatment of the wafers you were removing the silicon dioxide by the reaction of silicon with silicon dioxide to get silicon monoxide, which was evaporating. You pointed out that a hydrogen reduction as such did not work. I wondered what evidence you had to say that the silicon dioxide was being removed by this particular reaction.

W. J. CORRIGAN: We use a lot of quartz in our system and cannot observe any reaction with the quartz which is basically silicon dioxide. Reaction of silicon dioxide with hydrogen does not commence until 1350—1400°C. [Wartenburg, Anorg. Chem., 79, 71 (1913); G. Flusin, Chim. & ind. (Paris), 3, 740 (1920)].

J. J. McMULLEN: Have you compared the silicon weight loss in an inert atmosphere at those temperatures with those in an atmosphere of hydrogen?

W. J. CORRIGAN: Yes, and they are very similar, using helium.

J. J. McMULLEN: We found quite the contrary, actually; that the rates were four times different.

W. J. CORRIGAN: Also, we have found that you do get a continuous weight loss. If you continue to heat the substrates in a hydrogen stream for a long time you continue to get a weight loss, apparently indefinitely, which probably indicates that direct evaporation of silicon is taking place as well.

Evaluation Techniques for and Electrical Properties of Silicon Epitaxial Films

C. C. ALLEN and E. G. BYLANDER

Texas Instruments Inc., Dallas, Texas

Abstract

Techniques for thickness and electrical measurements of epitaxial silicon films are described. The application of these techniques to device manufacture is discussed.

INTRODUCTION

A complete evaluation of epitaxial films would involve measurement of resistivity, thickness, carrier mobility, lifetime, surface quality, heat treatment effects, and crystalline perfection in layers that are usually a few tenths of a mil thick. Production control requires that at least resistivity, thickness, and surface quality be measured routinely.

FILM THICKNESS MEASUREMENTS

Layer thickness may be measured by the conventional angle lap and staining technique (1). Unfortunately, the method is destructive and time-consuming, and uneven surfaces will tend to make it inaccurate. Further, high resistivity material is difficult to stain reliably. A nondestructive method, which has proved more adaptable to production testing, involves the use of an electronic thickness gage (Fig. 1). The layer thickness is determined by taking the difference between the slice thickness readings before and after deposition.

The equipment used is a Cleveland Instrument Co. Model PA 615 PAR-AC production gage. A special anvil was constructed with a hemispherical pin raised a few thousandths of an inch above the otherwise flat anvil surface for making contact to the backside of the slice. This small contact area reduces the chance of obtaining an erroneous reading due to dust, small surface imperfections, or film deposition on the under surface. There is generally a deposit on the back of the slice which is normally restricted to the edge so that correct measurements may be made near the center. The pin extends down through the anvil plate and rests on one of a set of gage blocks which may be changed for range selection. Gage blocks with 1 mil increments provide suitable range coverage. The gage readings are accurate to better than 0.05 mil. Comparison of the thickness gage with the lap and stain technique gives good agreement. For example, the difference in the means of thickness gage and angle lap and stain measurements on 25 slices was only 0.017 mil.

Lap and stain samplings of one out of 10 production slices have shown that the major difficulties arise from gross errors in arithmetic or mixed sub-

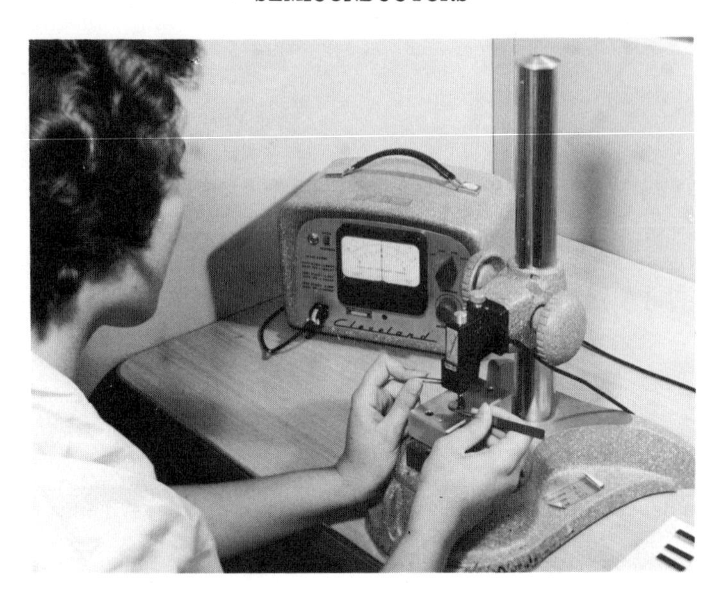

Fig. 1. The operator is measuring a slice with the electronic thickness gage. Incremental range changes are made with the gage blocks shown on the right.

strate measurements. Gross errors are easily checked by lap and stain. The gage has proved to be difficult to use by production personnel. The use of a skilled operator may however be justified by the savings in material and labor. Other errors arise from substrate thickness variations and failure to reposition the slice accurately. In order to reduce this source of error to a minimum, normal care must be used to insure flat and parallel substrate surfaces. In addition, a jig may be used to accurately position the slice before and after deposition. An oxide mask of approximately 20,000 A (0.8 mil) may be used to provide an undeposited spot for difference measurements. The amorphous silicon overdeposit is removed by under-cutting the oxide with HF, then scrubbing the film with a cotton swab.

FILM RESISTIVITY MEASUREMENTS

Layer resistivity is measured on a test piece of opposite conductivity type. At least one test piece is used per reactor run. The four-point probe resistivity meter is of special design (2) in order to insure proper results from unlapped surfaces (Fig. 2). It uses ac measuring with a superimposed dc bias on the probes in order to decrease the probe contact resistance. The bias supply also provides a reverse bias across the junction in order to electrically isolate the layer from the substrate. An auxiliary circuit is used to obtain bulk resistivity readings directly by setting in a correction for the film thickness on a helipot dial.

Another technique which is under development is based on the use of a thermoelectric probe. This method has been used for measuring the resistivity of diffused layers in germanium (3). It uses a hot probe with a point diameter less than the measurement depth in order to avoid heating the substrate. The

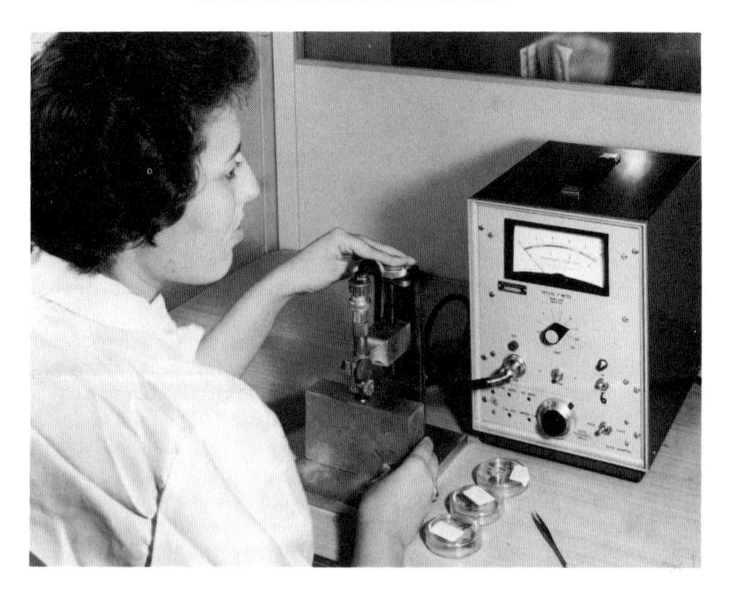

Fig. 2. The ac four-point probe resistivity meter is used to measure resistivity of films on opposite type substrates. A fifth point is used for dc injection to reduce contact resistance. The film thickness correction is made automatically by means of the 20-turn potentiometer.

Seebeck voltage is measured with a high impedance electronic voltmeter such as an H.P. 425A. The set is calibrated for a given heater power by using slices of known resistivity which have been electropolished. In preliminary runs, using approximately 200 mw input to the heater, the curve in Figure 3 was obtained. It is not necessary to electropolish the epitaxial slices if the

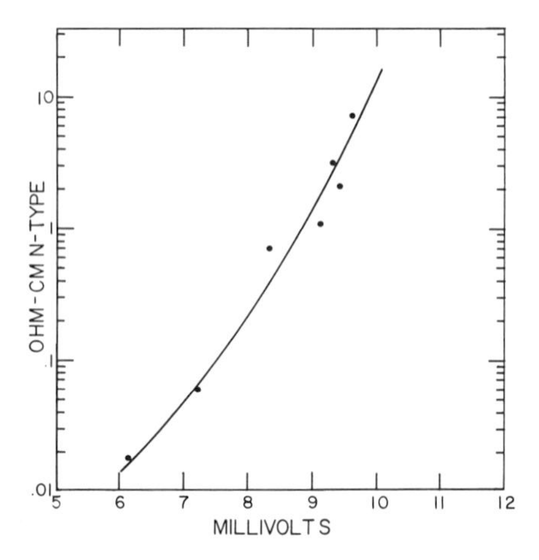

Fig. 3. Plot of Seebeck voltage vs n-type resistivity. The voltage is seen to be an insensitive function of resistivity.

surfaces are kept clean so that reliable electrical and thermal contact can be made.

The best evaluation procedure probably uses one or a combination of the above techniques to cover the spectrum of thickness, resistivity, and type. The ideal production procedure would be to improve the deposition control of thickness and resistivity to such a degree that a statistical sampling plan could be used for evaluation. This would enable one to use a more tedious, but more accurate, evaluation technique.

The following electrical data were taken on films deposited from hydrogen atmosphere containing approximately 1 mole-% halide. The deposition was carried out at 1250°C.

The diode breakdown voltage may be used as a measure of the film resistivity provided the film is thicker than the depletion layer width. The depletion layer may expand through the high resistivity film to the substrate upon the application of a reverse voltage less than that required for avalanche breakdown in the film. This factor along with the film resistivity fixes the breakdown voltage.

Figure 4 is a plot of breakdown voltage versus n-type resistivity for various film thicknesses. The portion of the curve which relates breakdown voltage to resistivity is a replot of McKay's (4) and Miller's (5) results.

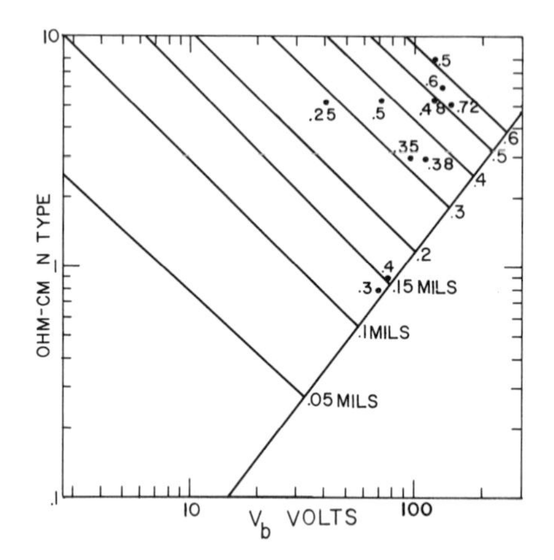

Fig. 4. Plot of abrupt junction breakdown voltage vs n-type resistivity with depletion layer thickness as a parameter.

The other portion shows the depletion layer thickness at a given voltage and resistivity. For abrupt junctions and a given film thickness, the breakdown voltage goes up as resistivity goes up, until a point is reached where the depletion layer reaches through to the substrate. The breakdown voltage is then set by the film thickness, assuming a low resistivity substrate.

The layer thickness, of course, must be corrected for the diffusion depth, the substrate diffusion distance (typically 0.05 mil), and the thickness of material removed during diffusion. In order for Figure 4 to be valid, the diffused layer concentration must be singificantly higher than the film concentration. For example in some devices, where a low surface concentration is used, it has been found that the breakdown voltage has been set by punch-through to the surface. This and other surface effects usually produce soft breakdowns.

There was some doubt as to whether these films gave the same breakdown voltage as grown material of the same resistivity, therefore a number of diodes were fabricated and analyzed to compare the two materials. The diodes were boron-diffused at 1080°C for 50 min with a boron surface concentration of 6×10^{20} atoms/cm^3. Metal contacts were applied, to assure accurate results, and 30 mil diam mesas were etched. Each of the experimental points plotted in Figure 4 represents the average values from approximately 20 diodes. The number of each point is the measured film thickness in mils remaining after diffusion. Some variability is apparent and is attributed to the diffusion process. Diodes made by a deposited junction technique have been shown to exhibit less spread in the breakdown voltage across a slice (6). As a result it may be better to use deposited junctions rather than diffused junctions for evaluation purposes.

Soft breakdown or high leakage due to surface layers has proved more of a problem with boron-diffused n on n^+-layers because of the close approximation of the low resistivity, n-type and subsequent bridging by inversion layers. It is especially important to remove all HF and metal fluorides following mesa etching.

High leakage, due to bulk properties, arises from the mechanisms described by Goetzberger and Schockley (7), Plummer (8), and Batdorf et al. (9). The

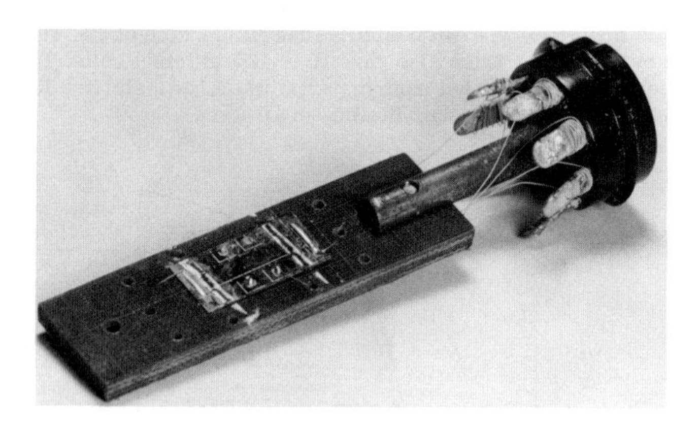

Fig. 5. Hall bar and mount. The bar was prepared by a photomasking technique.

usual procedures of surface cleaning before diffusion, oxide gettering, and slow cooling following diffusion vitiates most of these problems. These treatments are important since dislocation densities are high in the films (typically 10^4 cm^{-3} on a 10-mil slice) and there appear to be some quantities of unwanted metallic impurities incorporated there. The above procedure was closely followed for the diode construction proveiously described.

HALL MOBILITY

Hall mobilities have been measured on n-type films. An isolation of greater than 10^5 ohms was obtained with n-type films on p-type substrates which is sufficient for good isolation. The bars (Fig. 5) are prepared by a photoresist technique which allows 5 mil wide sidearm contacts. Room temperature mobilities for a 6 ohm-cm and a 7 ohm-cm film were 1.5×10^3 cm^2/v-sec and

1.3×10^3 cm^2/v-sec, respectively. These results are close to the values reported for bulk silicon (10). The p-type samples are still under study at this time.

CARRIER LIFETIME

Lifetime has been investigated by three techniques: the diode recovery technique (11), the calculation of lifetime by an analysis of the current voltage curves (12), and the use of a solar cell technique (13). The recovery time diodes were made by fusing an aluminum wire approximately 1/4 mil into an n-type surface. Films approximately 10 mils thick were used in order to assure a film thickness greater than the diffusion length. The recovery times of diodes with 2 ohm cm n-type films were close to 1/2 μ sec. The short-circuit current of a solar cell is easily shown to be a measure of the minority carrier lifetime in the bulk. When a standard process is used, the efficiency is a function of short-circuit current and therefore lifetime. Smythe, of this laboratory, has shown by subjecting lifetime bars to a standard solar cell diffusion treatment that a 6% efficient p on n cell has a hole lifetime of a few tenths of a microsecond, and an 8% cell has a lifetime of a few microseconds. Other workers report similar results (14). Calculations for p-type base material indicates that similar electron lifetimes result in corresponding efficiencies.

Solar cells made in p-type epitaxial films had efficiencies of 6%. These cells were made by a process which also gave 10 to 12% efficient cells. The higher efficiency cells were fabricated from material which had a starting lifetime of 30 μ sec and an estimated finished lifetime of 10 μ sec.

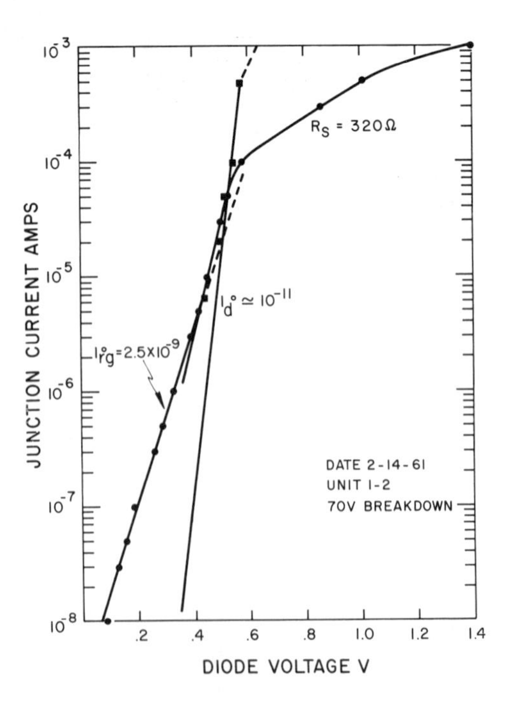

Fig. 6. Current-voltage curve of a boron-diffused diode made in a 5 ohm-cm n-type film.

It is well known that the analysis of current-voltage curves of p-n junctions can give carrier lifetimes. Such an analysis carried out on boron-diffused diodes yielded a lifetime of about 1/2 μ sec for both the diffusion region and the space-charge region. The current-voltage curve of one of these diodes is shown in Figure 6.

Trap depths were determined by plotting the reverse current, at fixed bias, divided by $T^{5/2}$ versus 1/T (14). Trap depths for the boron-diffused diodes were about 0.35 ev and trap depths for phosphorous-diffused p-type films were determined to be about midband (15).

The origin of traps was initially supposed to be either diffusion of gold from the substrate or the presence of iron in the film. However Walters and Estle (16), by using an electron paramagnetic resonance technique, have found a density of 2.5×10^{16} cm^{-3} of centers which have the same q-value as found for mechanically damaged surfaces. They have shown that these centers are electron traps and are tentatively identified as point defect clusters. These centers are, in analogy with the effects of gold, believed to cause the low lifetimes found in this material.

FILM QUALITY

Epitaxial film quality is greatly affected by the quality of the substrate surface. Microscopic examination has shown that surface imperfections such as scratches and pits are propagated into the film. In addition surface twinning, which causes pyramids or dendritic formations, results from a strain relief mechanism. The electrical properties of these blemishes are so far unknown. The primary objection to these formations is that they hold the photomasters off the surface; therefore, finished slices possessing more than a few of them must be rejected.

The surfaces of the slices are inspected before and after depositon for scratches, pits, hillocks and stains under a microscope spotlight. For quality control inspection purposes standard slices are used for reference.

The parameters normally measured for production control are resistivity, film thickness, and surface quality. No routine techniques for lifetime measurement or dislocation density measurement have been devised. Evaluation diodes may be fabricated at intervals to serve as an additional process check. Some work remains to be done on resistivity measurement on configurations with the same conductivity type in both film and substrate. The problem of evaluating epitaxially grown multiple junctions has not been considered here.

ACKNOWLEDGMENTS

The author wishes to acknowledge the help of Messrs. B. Queen, B. Smythe, L. Larsen, V. Paulos, G. Pearson, and S. Watelski of Texas Instruments.

References

1. Fuller, C. S., and J. A. Ditzenberger, J. Appl. Phys., 27, 544 (1956).
2. Allen, C. C., and W. R. Runyan, Rev. Sci. Instr., 36, 824 (1961).
3. Batifol, E., and G. Duraffourg, J. phys. radium, 24, Suppl. to No. 11, 207A (1960).
4. McKay, K. G., Phys. Rev., 94, (1954).
5. Miller, S. L., Phys. Rev., 105, 1246 (1957).
6. Watson, J., private communication.
7. Goetzberger, A., and W. Schockley, J. Appl. Phys., 31, 1821(1960); also, this volume, p. 121.
8. Plummer, A. R., J. Elect. and Control, 5, 405, (1958).

9. Batorf, R. L., et al., J. Appl. Phys., 31, 1153 (1960).

10. Morin, F. J., and J. P. Maita, Phys. Rev., 96, 28 (1954).

11. Van Roosbroeck, W., and T. M. Buck, in Transistor Technology, Vol. III, F. J. Biondi, ed., Van Nostrand, Princeton, N. J. 1958, p. 315.

12. Sah, C. T., et al., Proc. Inst. Radio Engrs., 95, 1228 (1957).

13. Smythe, B., private communication.

14. Terman, L. M., Stanford Univ. Technical Report No. 1605-1, 1959. Similar values may be inferred from the measured lifetimes and maximum efficiencies reported by M. Wolf, Proc., Inst. Radio Engrs., 48, 1246 (1960).

15. Watelski, S., private communication.

16. Walters, G. K., and T. L. Estle, J. Appl. Phys., to be published.

The Role of Imperfections in Semiconductor Devices

W. SHOCKLEY and A. GOETZBERGER

Shockley Transistor, Unit of Clevite Transistor, Palo Alto, California

Abstract

Recent work covering the role of imperfections in semiconductor devices is reviewed in this paper. The discussion includes the effect of metal precipitates and their influence on junction characteristics, avalanche effect, and microplasmas occurring at irregularities in the diffusion pattern caused by dislocations. Pipes through diffused layers are shown to be caused by a diffusion mechanism. The statistical limitations of junction uniformity and thinness of base layers are compared with the existing technology.

INTRODUCTION

The immediate need for improved yield and reliability spurred early investigations on the role of imperfections in semiconductor devices. The effects of imperfections have become of even greater importance with the advent of microelectronics and with the introduction of high power devices which require a much larger area of structurally perfect silicon.

The purpose of this paper is to review a number of imperfections occurring in diffused silicon devices, whose importance has been recognized more clearly during the past two years. These include precipitates of metals and dielectrics, irregularities in the diffusion pattern, and the effect of statistical fluctuations of donors and acceptors.

PRECIPITATES

Metals Precipitates

Metal precipitates in the space-charge layer have been found (1) to cause a "soft" appearance of the reverse characteristic of p-n junctions. Normally junctions have a hard breakdown indicating avalanche effect. The reverse characteristics of hard and soft diodes are given in Figure 1. The -70°C curve for the soft diode shows a current that varies as $V^{5.5}$ where V is the reverse voltage. Since this soft current obeys the V^n law to voltages as low as 0.8 v it cannot be due to avalanche, but could possibly be due to Zener tunneling at localized high-field points caused by precipitates.

The mechanism by which metal precipitates can be introduced into silicon becomes obvious from the solubility versus temperatures curves for various elements in silicon. A number of metals such as copper, gold, and iron have a strongly temperature-dependent solid solubility in silicon which causes super-

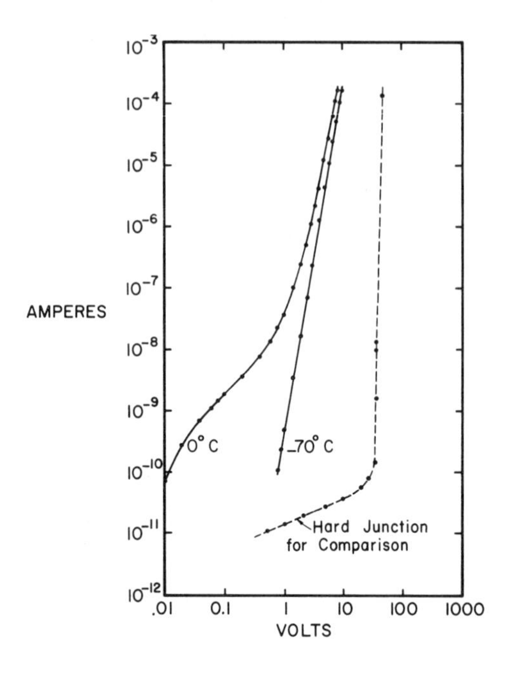

Fig. 1. The V-I characteristic of a soft junction.

saturation and formation of precipitates during the cooling cycle.* All the precipitate-forming metals have a very high diffusivity which enables them to diffuse to the precipitates.

Evidence that the reverse current in a soft diode flows at a localized spot was obtained by using a potential plotting method applied to one diode that had been treated by in-diffusion of iron (Fig. 2). A constant reverse current was fed into the diode through a small wire contact. A second probe at fixed location was used as a voltage reference, and the voltage between it and a third movable probe was measured for an array of points spaced typically 0.05 mm apart across the surface.

Copper, iron, manganese, and gold were all found to be effective in producing junction softness (Fig. 3). Since the solid solubility of all these metals in silicon is extremely small, minute amounts of metal can cause saturation and precipitation. In most cases a fraction of an atomic layer on the surface can be harmful. The small amount of contamination necessary to cause defective junction behavior can be introduced in many ways—from the furnace area during diffusion, as surface residues from chemical processing, as an impurity in the original silicon crystal, etc.

One of the results of the investigations of metal precipitates was the discovery of a method of removing or preventing these precipitates by means of gettering. Two mechanisms are theoretically possible for gettering. One is application of a liquid alloy phase on the surface. All the impurities which have a high distribution coefficient between alloy and silicon will be enriched in the surface alloy which can be subsequently removed.

---

* Precipitation of copper was first observed by Dash (2) and utilized for the study of dislocations by infrared transmission.

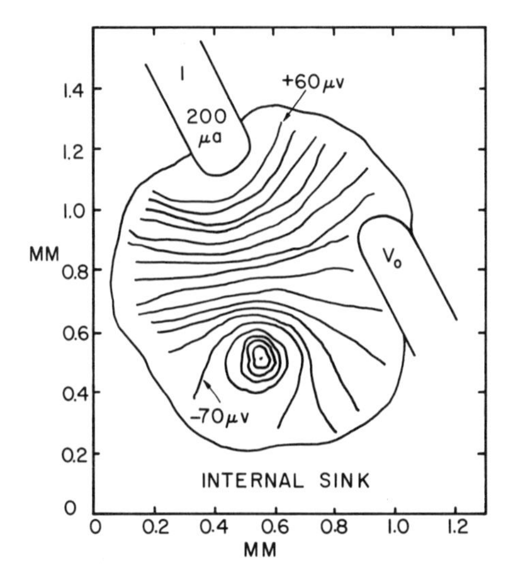

Fig. 2. Equipotential plot of iron-diffused junction.

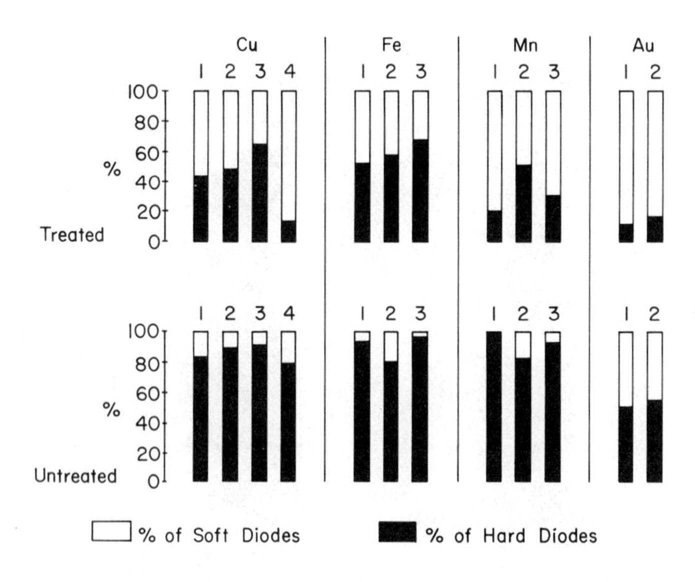

Fig. 3. Percentage of hard and soft diodes after treatment with different metals.

Another way of gettering consists of applying a substance to the surface of the silicon which forms a chemical compound with softness-causing elements. This second method was found to be the more effective one in most cases. Phosphorus pentoxide and boron trioxide can be utilized as getters. They form glassy mixtures with the surface oxide of the silicon which are liquid at the temperature of treatment.

Since the oxide getters are also used for doping of silicon, the techniques for applying them are well developed. They can be applied either from the gas

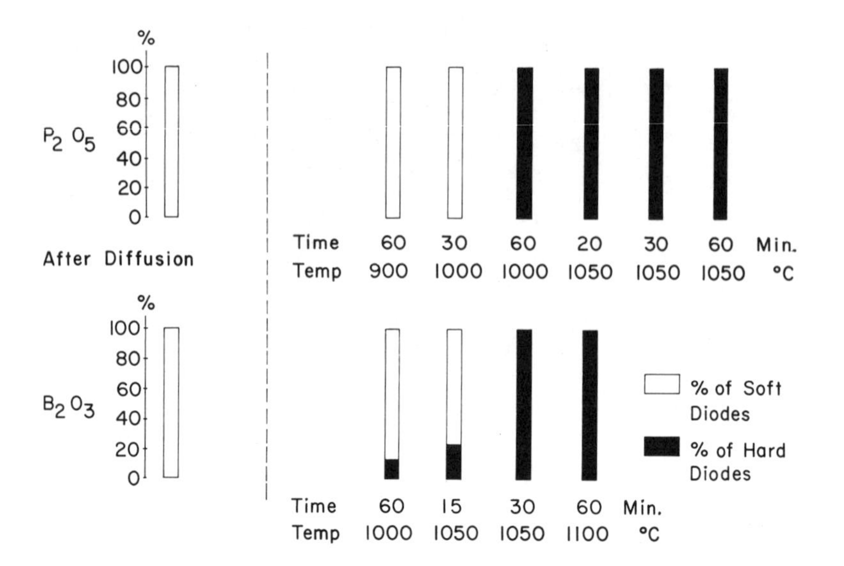

Fig. 4. Time and temperature dependence of gettering with P_2O_5 and B_2O_3 predeposit.

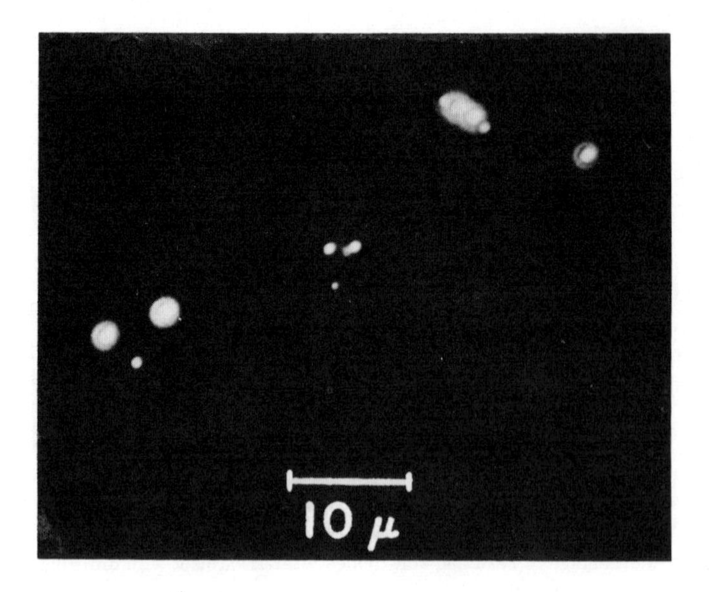

Fig. 5. Photograph of microplasmas.

phase (3) or by the paint-on technique (4). An example showing the temperature and time dependence of gettering with gaseous oxides is given in Figure 4.

Precipitates of Dielectrics

Much less is known about the effects of dielectric precipitates on junctions. Application of field theory shows that the electric field near a spherical precipitate can be increased up to 1.5 times the value found in uniform material. Shockley (5) suggested that silica precipitates might be the cause of reverse breakdown in many junctions by means of this localized field enhancement. If this assumption is true, then the so-called "microplasmas," the little light-emitting spots similar to those seen in Figure 5, might be nothing more than points where oxide precipitated within the junction. A similar mechanism was suggested by Kikuchi and Tachikawa (6, 7) who showed experimentally that microplasmas are introduced into germanium by pulling the crystal out of a quartz crucible thus saturating it with silica. Germanium grown out of a graphite crucible does not show microplasmas.

The two most important experimental observations that have to be explained by a microplasma model are (1) the fact that no decrease in voltage is observed after breakdown has set in, and (2) the "on-off" noise. The precipitate model can account for both effects. Precipitates are sufficiently small so that the field disturbance extends only over a region corresponding to a drop of about one volt. The on-off noise can be accounted for by assuming surface traps (Fig. 6) around the precipitates which provide a "lock-on" mechanism.

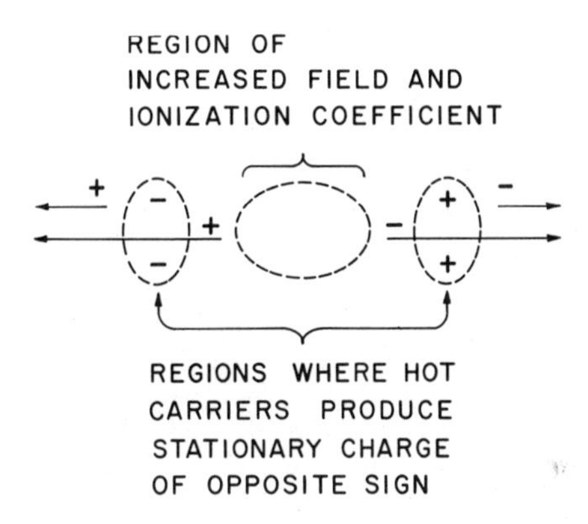

Fig. 6. Schematic representation of the way in which traps may enhance the field.

That microplasmas are not essential for breakdown in silicon has been shown by Batdorf et al. (8) who succeeded in preparing a junction that exhibited comparatively uniform light emission and no microplasma noise. This indication that microplasmas are caused by defects is augmented by a recent measurement of the distribution of breakdown voltages of all the microplasmas within one diode (9). A rather broad distribution (Fig. 7) is obtained showing a depletion of possible sites as voltage is increased. The total number of microplasmas is in good agreement with the number of oxide precipitates expected in the space-charge layer of the junction investigated. At the present time, however, there is insufficient experimental evidence to prove or disprove the oxide precipitate hypothesis.

The same can be said of the influence of dislocations which have been shown by Chynoweth and Pearson (10) to be a preferential site of microplasmas. Goetzberger and Stephens (9) investigated a junction containing a low-angle grain boundary. A line of microplasmas could be seen along the grain boundary (Fig. 8). The density of microplasmas, however, was less than 1% of the density of dislocations along the grain boundary indicating that only a small percentage of the dislocations participate in breakdown. It appears likely that disloca-

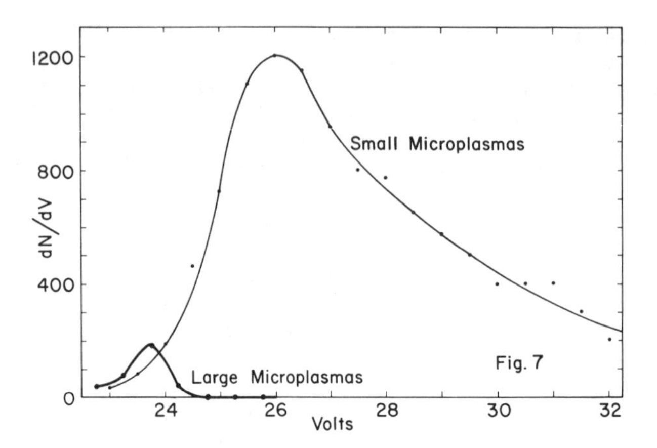

Fig. 7. Microplasma density vs voltage.

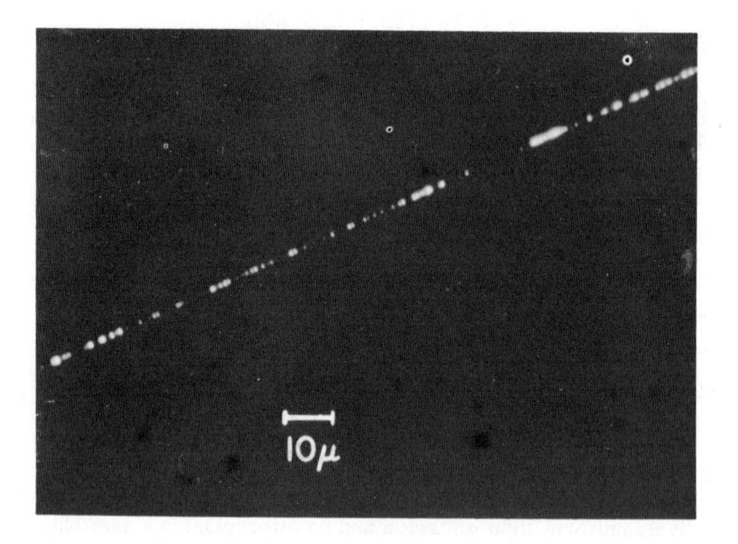

Fig. 8. Microplasmas in grain boundary.

tions provide preferential sites for precipitations, thus being only indirectly responsible for the microplasmas.

IRREGULARITIES IN THE DIFFUSION PATTERN

Irregularities in the diffusion pattern frequently cause faulty junctions as well as shorts between emitter and collector layers of transistor-like structures. Only a few causes for such irregularities, which have been studied most thoroughly, will be treated here.

Dislocations

Dislocations have a considerable influence on the diffusion of impurities. Particularly useful for the study of such diffusion effects are low-angle grain

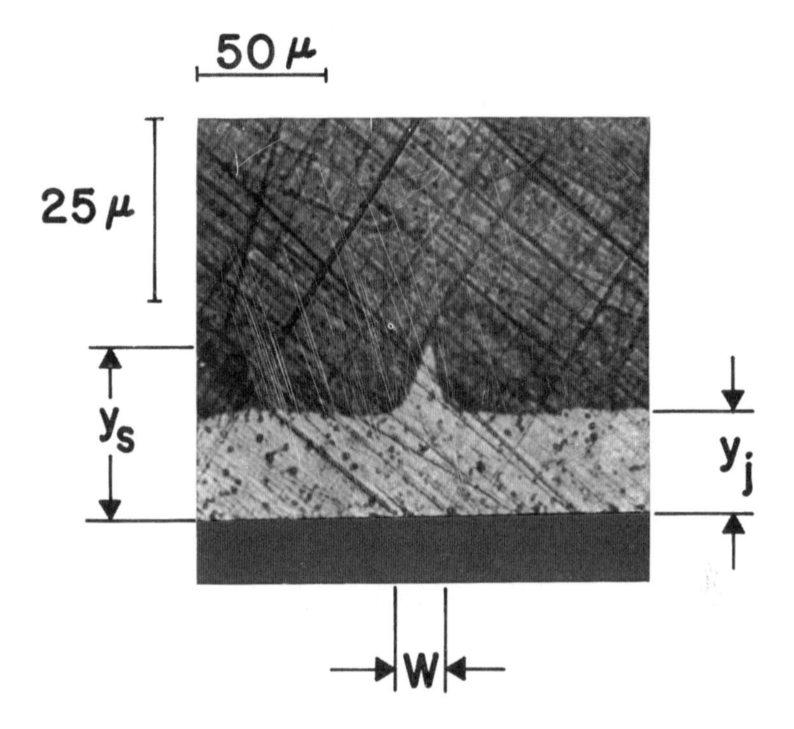

Fig. 9. Diffusion spike on grain boundary.

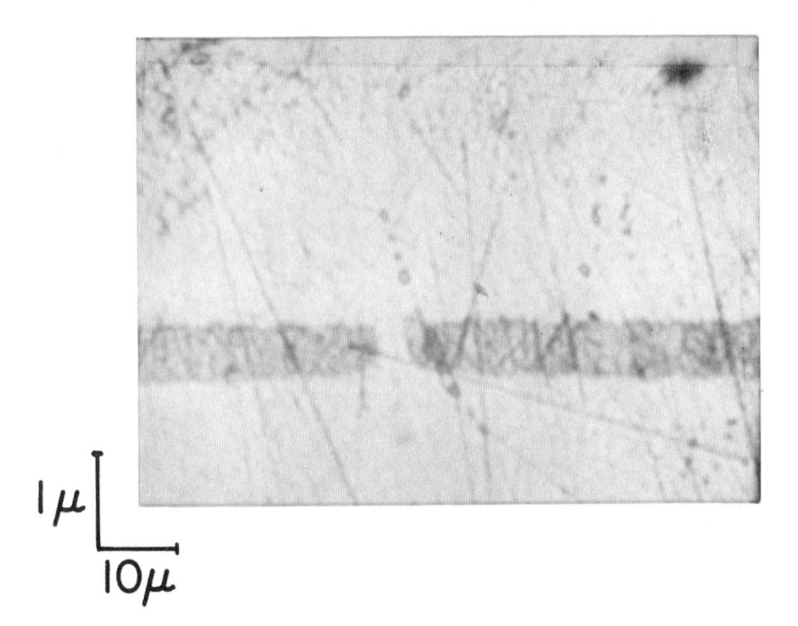

Fig. 10. Phosphorus diffusion breaking through boron diffusions.

boundaries which are made up of a linear array of dislocations. Diffusion along such grain boundaries has been studied by several investigators (11) who found an enhancement of diffusion which resulted in so-called "diffusion spikes" (Fig. 9). Theoretical evaluation showed that each dislocation can carry a chemical diffusion current density about 300,000 times larger than the surrounding bulk. Diffusion is thought to be enhanced by three mechanisms: (1) an increased concentration of impurities near the dislocations caused by the Cottrell potential, (2) higher concentration of vacancies, and (3) a decrease in the activation energy required for jump due to the stain field associated with the dislocation.

Although boron and phosphorus have approximately the same diffusion coefficients in bulk silicon, they react quite differently along grain boundaries. The diffusion of phosphorus is enhanced to a much greater extent than that of boron. Figure 10 shows a phosphorus diffusion breaking through a boron-diffused layer along a grain boundary.

A single, isolated dislocation will carry this enhanced diffusion current, but the resulting penetration spike is much smaller than for a grain boundary. This is due to the one-dimensional diffusion away from the grain boundary being less effective than the radial diffusion away from the dislocation in dispersing the spike. Single dislocations are therefore expected to cause difficulties only in devices having extremely thin base layers. The probability of shorts is much greater for clusters of dislocations which are able to cause pipelike diffusion patterns extending far beyond the junctions into the surrounding normal material. This effect is probably the explanation for the pipe effect observed by Miller (12).

Another effect involving the influence of dislocations was found recently by Queisser (13) and Prussin (14). Diffusion of impurities into silicon was shown to give dislocations which can be detected by their slip patterns. The undersized impurity atoms cause a strain in the lattice which is relieved by formation of dislocations parallel to the surface. Development of strain is dependent on the concentration of impurity atoms in the diffused layer and the misfit in size. For boron a minimum of 3×10^{15} atoms cm^{-3} is required to deformation in silicon. This was derived theoretically and confirmed by the experiment. Slip patterns made visible by etching a $\{100\}$ surface containing a grain boundary are shown in Figure 11. The influence of such slip planes on device properties is still largely subject to speculation. Effects will be most pronounced

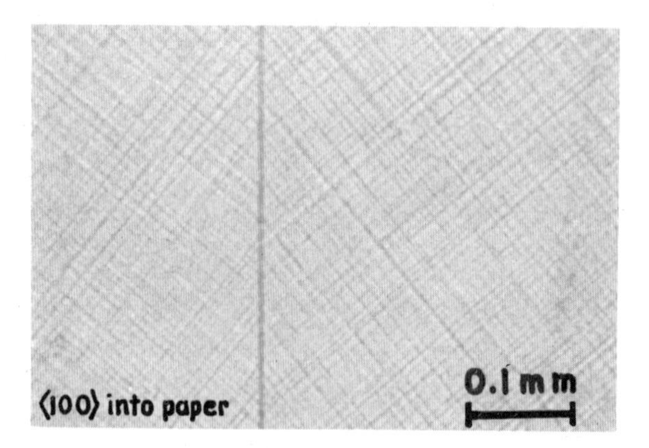

Fig. 11. Slip pattern on boron-diffused $\{100\}$ surface containing grain boundaries.

when highly doped layers are present, for instance in solar cells and emitters of diffused transistors.

Pipes Induced by Contamination

Another type of pipe that can cause serious trouble stems from tiny specks of surface contamination containing doping impurities. It has been found that boron-diffused p-layers of a low surface concentration contained pipelike, highly n-doped regions connecting the surface with the bulk (15, 16). Due to the high doping in these regions, the breakdown voltage is considerably lowered around the periphery of the pipes and ring-shaped light emission emerges from a pipe when reverse bias is applied (Fig. 12). It was shown recently (17)

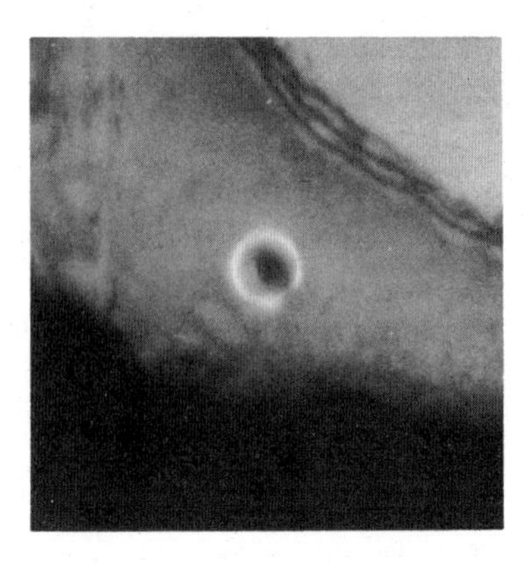

Fig. 12. Light emission from periphery of pipe.

that these light emission patterns can be produced by a short exposure of the silicon surface to phosphorus vapors at room temperature prior to diffusion. The same effect was achieved by processing in an ambient containing phosphorus contaminated dust.

It was further shown that these pipes originate from a diffusion effect by subjecting a planar diode containing a number of pipes to heat treatment. Figure 13 shows the same diode four times. Between each photograph there was a heat treatment of 15 min at 1200 °C. The diameters of the pipes can be observed to grow linearly with the square root of time, a behavior that had been predicted by application of the diffusion laws. Figures 14 and 15 show five pipe shapes that are theoretically possible. Curves of the type shown in Figure 14a are obtained when the surface concentration within the pipe is higher than in the surrounding area. The curve in Figure 15b is called a "hidden pipe" because it does not reach the surface. This type develops when the amount of impurities in the pipe is very small. It cannot be observed with the light emission technique, but it can cause shorts through base layers like the other types of pipes.

Evaluation of experimental data shows that only 10^{-14} g of phosphorus are sufficient to create one pipe; a fact that shows the importance of extreme cleanliness in the handling of devices.

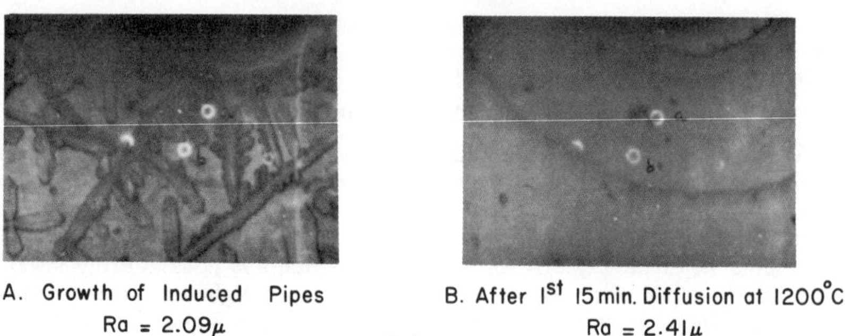

A. Growth of Induced Pipes
$R_a = 2.09\mu$
$R_b = 2.09\mu$

B. After 1st 15 min. Diffusion at 1200°C
$R_a = 2.41\mu$
$R_b = 2.37\mu$

10μ

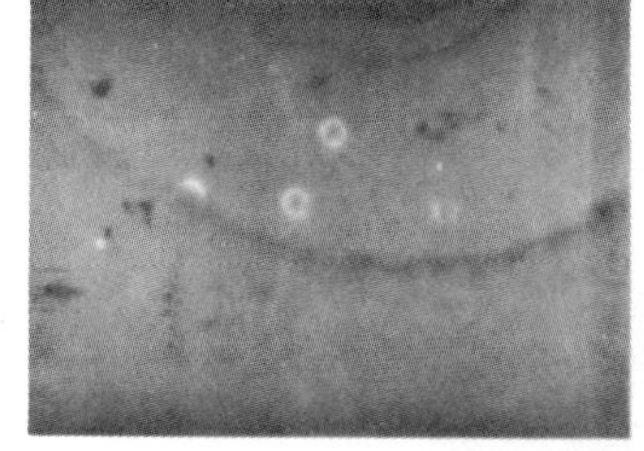

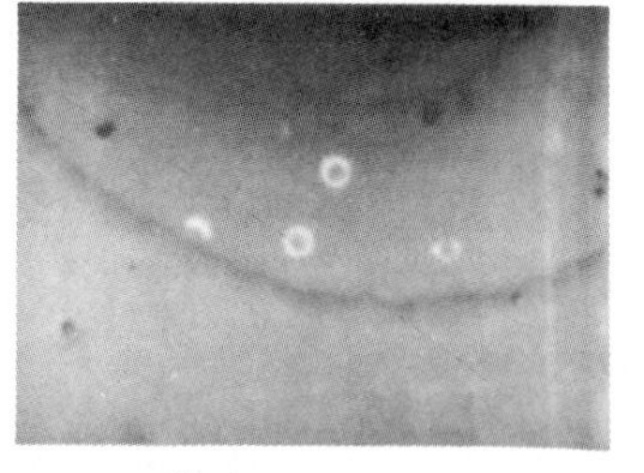

C. After 2nd 15min. Diffusion at 1200°C
$R_a = 2.50\mu$
$R_b = 2.56\mu$

D. After 3rd 15min. Diffusion at 1200°C
$R_a = 2.74\mu$
$R_b = 2.74\mu$

Fig. 13. Growth of pipes upon heat treatment.

$K > 0$ identical with $M(D_1t_1)^{1/2} > 4\pi Q(D_2t_2)^{3/2}$ or $C_{02} > C_{01}$

(a)	(b)	(c)
$D_1t_1 > D_2t_2$	$D_1t_1 = D_2t_2$	$D_1t_1 < D_2t_2$
Ellipsoid	Cylinder	Hyperbola of Rotation
Semi-axes	Radius	Semi-axes
$y_1^2 = 4D_2t_2K$	$y_1^2 = 4D_2t_2K$	$y_1^2 = 4D_2t_2K$
$x_1^2 = \frac{4D_1t_1D_2t_2}{D_1t_1 - D_2t_2}K$		$x_1^2 = \frac{4D_1t_1D_2t_2}{D_2t_2 - D_1t_1}K$

Fig. 14. Theoretical pipe shapes $\left(C_{01} < C_{02}\right)$.

RANDOM FLUCTUATIONS OF CHEMICAL CHARGE DENSITIES

The limitation of junction uniformity that is to be described here is of a more basic nature than the previously treated ones. Even if we succeed in preparing a junction that is free from any type of gross defect, there will still be a limitation due to statistical variations in donor and acceptor concentration,

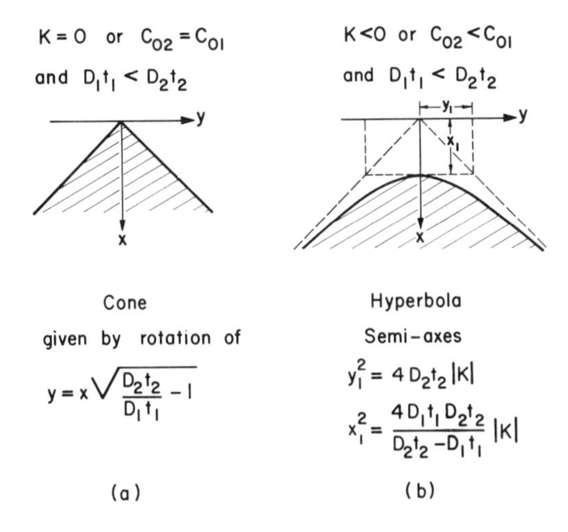

Fig. 15. Theoretical pipe shapes $\left(C_{01} \geqq C_{02}\right)$.

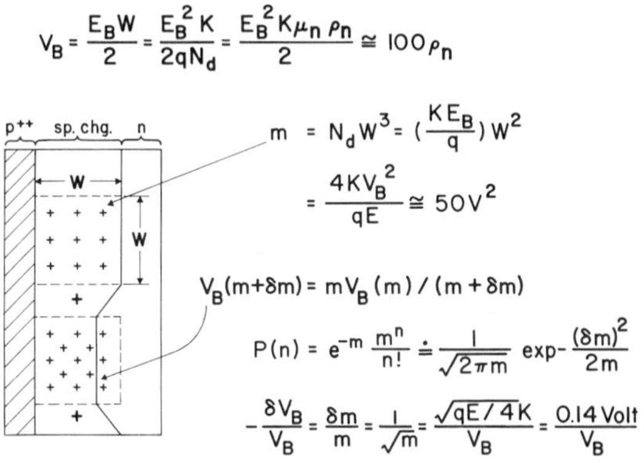

$$V_B = \frac{E_B W}{2} = \frac{E_B^2 K}{2qN_d} = \frac{E_B^2 K\mu_n \rho_n}{2} \cong 100\rho_n$$

p++ sp. chg. n

W

$$m = N_d W^3 = \left(\frac{KE_B}{q}\right) W^2$$

$$= \frac{4KV_B^2}{qE} \cong 50V^2$$

$$V_B(m+\delta m) = mV_B(m)/(m+\delta m)$$

$$P(n) = e^{-m}\frac{m^n}{n!} \doteq \frac{1}{\sqrt{2\pi m}} \exp - \frac{(\delta m)^2}{2m}$$

$$-\frac{\delta V_B}{V_B} = \frac{\delta m}{m} = \frac{1}{\sqrt{m}} = \frac{\sqrt{qE/4K}}{V_B} = \frac{0.14 \text{ Volt}}{V_B}$$

Fig. 16. Method of estimating the fluctuation of breakdown voltage.

which follow a Poisson distribution (5). These fluctuations have an influence on avalance breakdown and punch-through effect in devices.

The condition for avalanche breakdown can be approximated by the assumption that the peak field reaches a breakdown value F_B at avalanche. The peak field will occur at the junction. Fluctuations of the peak field at any point will be largely determined by ion density fluctuations in the immediate neighborhood. By dividing the space-charge layer into cubes (Fig. 16) and supposing that in each cube the breakdown is determined by the average ion density in that cube, we arrive at the characteristic mean-squared fluctuation of voltage—which is 0.08 v for silicon. We further find the cube with the lowest breakdown voltage to be 0.7 v below the average breakdown voltage for a junction without compensation of impurities. This value will be higher for a junction with compensation of donors and acceptors, as is always the case for diffused junctions. This predicted statistical behavior has not been verified experimentally to date because of large-scale imperfections.

The same type of consideration can be used to determine the limitation of the thickness of a base layer due to statistical fluctuations causing punch-through effect in one cube with low impurity density (18). If one assumes a base layer of uniform doping which can withstand the avalanche field before punch-through is reached, and then asks for the condition that will produce punch-through in one cube at one half of the avalanche voltage. The result is that in this case the base layer has to be less than 0.1 μ thick.

Since most of the presently produced devices have base-layer thicknesses far greater than this, we can conclude that at the present state the technology is still far away from its absolute limit.

ACKNOWLEDGMENT

Research in part supported by Air Force Cambridge Research Center.

References

1. Goetzberger, A., and W. Shockley, J. Appl. Phys., 31, 1821 (1960).
2. Dash, W. C., J. Appl. Phys., 27, 1193 (1956).
3. Frosch, C. J., and L. Derick, J. Electrochem. Soc., 104, 547 (1957).
4. Hughes, H. E., J. H. Wiley, and P. Zuk, IRE Convention Rec., Pt. 3, 80 (1957).
5. Shockley, W., Solid State Electronics, 2, 35 (1961).
6. Kikuchi, M., and K. Tachikawa, J. Phys. Soc. Japan, 14, 1830 (1959).
7. Kikuchi, M., and K. Tachikawa, J. Phys. Soc., Japan, 15, 837 (1960).
8. Batdorf, R. L., A. G. Chynoweth, G. C. Dacey, and P.W. Foy, J. Appl. Phys., 31, 1153 (1960).
9. Goetzberger, A., and J. Stephens, J. Appl. Phys., to be published.
10. Chynoweth, A. G., and G. L. Pearson, J. Appl. Phys., 29, 1103 (1958).
11. See for instance, Queisser, H. J., K. Hubner, and W. Shockley, Phys. Rev., in press. This paper contains references to other papers on this subject.
12. Miller, L. E., in Properties of Elemental and Compound Semiconductors, H. C. Gatos, ed. (Metallurgical Society Conferences, Vol. 5) Interscience New York-London, 1960, p. 303.
13. Queisser, H. J., J. Appl. Phys., in press.
14. Prussin, S., J. Appl. Phys., to be published.
15. James, B. O., and P. S. Flint, paper given at the spring meeting of the Electrochemical Society in Chicago, 1960.
16. McNamara, M. C., paper given at the spring meeting of the Electrochemical Society, Indianapolis, 1961.
17. Goetzberger, A., paper given at the IRE Solid State Device Research Conference, Stanford, 1961, to be published.
18. Shockley, W., Bull. Am. Phys. Soc., 5, 161 (1960).

DISCUSSION

S. A. PRUSSIN (MonoSilicon): The effect of dislocations on the properties of silicon and germanium has been thoroughly described in a recent review article (C. W. Bardsley, in Progress in Semiconductors, Vol. 4, A. F. Gibson, ed., Wiley, New York, 1960, p. 155). The preceding paper goes a step further in reviewing the effects of dislocations on device parameters. Continuing in the same vein, I would like to describe the results of an experimental study which correlated a specific device property, the reverse characteristics, with dislocation arrays.

The technique chosen was developed specifically to enable us to compare a number of simple junctions which were all identically prepared and differed from one another only on the basis of the dislocation density. Silicon crystals were chosen which contained dislocation arrays extending along the length of the crystal. One such array has been identified by the term "slippage," another by the term "lineage," characterized by small angle grain boundaries.

In practice the crystals were sliced, and all odd numbered slices were etched, using the technique described by W. C. Dash [J. Appl. Phys., 27, 1193 (1956)], and examined. Where a series of successive odd numbered slices exhibited identical dislocation arrays, it was assumed that this array would be found in the even numbered slices sandwiched between them.

The silicon used was n-type, with a resistivity of approximately 10 ohm cm. Even numbered slices were lapped, cleaned, and sealed in a quartz capsule with a fixed amount of potassium borohydride. The capsule was diffused at 1200°C, the resulting junction being developed at a depth of approximately 2.5 mils under the surface. To this point, the treatment of the silicon was identical to an actual diode production technique. The next production step would have been to lap off one of the junctions and dice the remaining wafer. Instead the diffused wafer was coated with a layer of acid-resistant wax. Grooves were scribed through this wax 25 mils apart, and in two directions, 90° apart.

Etching the wafer in a refrigerated solution consisting of equal parts of hydrofluoric, nitric, and acetic acids resulted in the production of a network of mesas. The single junction devices were physically connected but electrically separate. Over 700 individual devices were obtained in this manner from a single 7/8-in. diam wafer. It was now possible to measure the reverse charactérisitcs of each individual device. This was done and the results were tabulated.

The peak inverse voltage (PIV) was strongly affected by the presence of lineage. In areas where there were lineage dislocation arrays, the mesa devices experienced a decrease in PIV to a fraction of what it was in the unaffected areas.

The PIV did not appear to be affected by arrays of dislocations lying in slip planes. Here the voltage breakdown values were determined solely by the relative positions of the mesa junctions on the slice. The further a mesa was from the center of the slice the higher the breakdown voltage. This corresponded to the resistivity profile of the crystal which had higher resistivity values along the edge than in the center of the wafer. The correlation between resistivity and PIV indicated that the breakdown here was a bulk rather than a surface phenomenon.

In addition to the breakdown voltage, the nature of the breakdown was noted. The value of the leakage current at breakdown was used to arbitrarily separate the breakdown into four categories. These were termed hard, slightly soft, soft, and very soft. It was possible to plot the distribution of soft junctions and compare their distribution with that of the dislocation arrays.

Discussion Figure 1 illustrates oscilloscope traces typical of three of the categories mentioned above. Discussion Figure 2 compares the distribution of soft breakdowns on one silicon wafer with the array of dislocations on an adjacent wafter. The dislocation array produced by slippage has a fourfold symmetry characteristic of a {100} crystal orientation. The distribution of soft junctions can be seen to have a direct correlation with dislocation density.

P. S. FLINT (Fairchild Semiconductor): You stated that a number of metals, when they precipitate on the junction, will cause soft junctions. Some of these same metals are well known to be getters also. For example, among the ones that you named would be gold, copper, or iron. How do you attribute, then, that

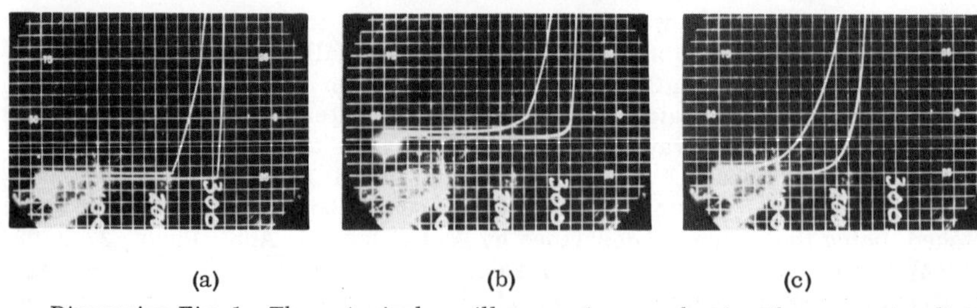

(a) (b) (c)

Discussion Fig. 1. Three typical oscilloscope traces showing the current-voltage relation at breakdown. The ordinate represents 50 ma per division: (a) represents a hard breakdown, (b) a soft breakdown, and (c) a very soft breakdown.

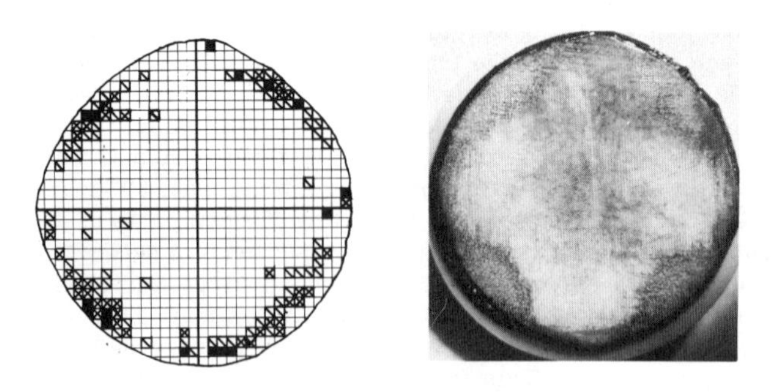

Discussion Fig. 2. Comparison of the slippage etch pit pattern on one wafer with the distribution of soft breakdowns on an adjacent wafer. Unmarked mesas are hard breakdowns, singly hatched mesas are slightly soft breakdowns, cross-hatched mesas are soft, and darkened mesas are very soft breakdowns.

they can also be producing soft junctions? Is this a matter of the rate of annealing? What annealing conditions were used?

A. GOETZBERGER: Yes. This depends on the rate of annealing to a large degree. I have only experiences with gold, but here you can get either hard junctions by quenching or you can get soft junctions by slow cooling.

P. S. FLINT: It would seem that if you want to prove this metal precipitate hypothesis, you should examine the specimen with infrared radiation, which would be certainly one method to determine if you actually do have precipitates. Do you have any estimates of the size of these precipitates? Would they be observable to this means?

A. GOETZBERGER: This was tried on some samples that were sent to Dash. Unfortunately it was necessary to remove a silicon layer thicker than our diffused layers in order to get the proper surface preparation for infrared observation. Therefore we could not see precipitates in the samples that were electrically investigated, but since our conditions were exactly identical with those that had been previously shown to cause precipitates, we are quite sure that we had precipitates.

P. S. FLINT: It is possible to decorate dislocation by copper decoration and observe them by infrared. It would seem to me you should be able to do this with precipitates also.

A. GOETZBERGER: This is theoretically possible.

J. SPANOS (Hoffman Electronics): How deep did the junctions have to be to achieve sufficient gettering to harden up those characteristics?

A. GOETZBERGER: Our junctions had a depth of 12 μ, but one can harden up much deeper junctions.

J. SPANOS: As your junction depth increases, do you have any feel for what the relationship of the actual gettering time required would be as a function of junction depth? In other words, would it be exponential, linear, square root, or what?

A. GOETZBERGER: It does not seem to be any type of a diffusion law, because the temperatures required for gettering are much higher than would be required for the metals to diffuse out to the surface. All those softness-causing metals have very high diffusion coefficients. So we assume that it has something to do with the chemical reactions taking place on the surface of the slice. We have been able to apply the getter on one side of a slice and harden up junctions on the front side, which required thickness of between 100 to 200 μ to diffuse.

J. SPANOS: You attribute this, then, to chemical reaction rather than to perhaps the increased differential diffusion rate of the metal?

A. GOETZBERGER: I do not know. I only know it is not diffusion.

G. H. SCHWUTTKE (General Telephone and Electronics Lab.): Dr. Flint stated that it is quite easy to decorate dislocation in silicon by copper decoration. We investigated the concentration of copper necessary to achieve decoration of dislocations so that they are visible in the infrared light, and we found that a minimum concentration of 10^{15} atoms of copper is necessary to achieve decoration.

A. GOETZBERGER: Yes, I think this is an important point. Precipitates, in order to be visible in the infrared light, have to have a certain size, whereas much smaller precipitates, of course, cause soft currents in junctions.

P. J. SCHLICHTA (Jet Propulsion Labs.): Have you found, or do you suspect, the presence of any electronic effects due to large vacancy clusters in the crystal?

A. GOETZBERGER: We have not investigated this field, and I have reasons to assume that vacancy clusters are very difficult to study or even to obtain because there would be a nonequilibrium condition at room temperature if you want to have a large concentration of vacancies, and I do not know if anyone has shown that these large vacancy concentrations really exist.

P. J. SCHLICHTA: Tweet has found them in unannealed dislocation-free germanium [J. Appl. Phys., 29, 1520, (1958)] and I think they have also been found in some silicon crystals.

A. GOETZBERGER: I do not have any answer to this.

P. J. SCHLICHTA: I understood Dr. Prussin to state that an effect was observed for lineage structure but not for slippage of dislocation clusters. Could this be related to an impurity segregation effect?

S. A. PRUSSIN: I certainly think it could be. Our idea was that within lineage you had the possibility of interaction of dislocations, and it seemed to us from the appearance of the pits that we were getting a different kind of pitting along the lineage structure than we were getting along the slippage plane.

W. L. TOWLE (Mallinckrodt Chemical Works): I would like to ask Dr. Prussin for his definition of slippage and lineage, and specifically how he distinguishes one from the other.

S. A. PRUSSIN: I would define slippage as the presence of pits along a $\{111\}$ plane, and when they are being examined as in this case, in a $\{100\}$ cross section, they have a very particular appearance. The techniques that we use to identify lineage would be simply by taking a Laue diffraction pattern having the focus across a lineage line. You will find that you will get a twin pattern, and from the fact that your points are twins you can determine quite easily what the angle difference is on both sides of the lineage plane.

Imperfections in Germanium and Silicon Epitaxial Films

T. B. LIGHT

Bell Telephone Laboratories, Inc., Murray Hill, New Jersey

Abstract

The continuation of a dislocation array forming a small angle boundary in a germanium substrate into an epitaxial germanium film grown on this substrate has been observed. Defects that may have been formed from vacancies have been observed in germanium epitaxial films in surface densities up to 3×10^8 cm^{-2}. The probable origin of these defects and the effect of contamination at the substrate-film interface are discussed.

Defects in germanium substrates nucleated by surface damage give rise to an "open-triangle" type defect in germanium epitaxial films. The size of this defect is a function of film thickness and a vacancy growth mechanism is proposed. These open triangle defects are also observed in silicon epitaxial films.

INTRODUCTION

There has been a recent renewal of interest in defects and dislocations in semiconductors arising from the realization that they affect the performance of semiconductor devices (1-7). It is of interest then to study the occurrence and causes of defects and dislocations in silicon and germanium epitaxial films.

Etching techniques have been used in this study and in one instance the results of transmission electron microscopy have been used in conjunction with the etching. The defects in the epitaxial films have their origins in three different regions of the substrate-film system. For convenience the defects will be classified according to the region or origin: (1) substrate, (2) interface, (3) epitaxial film.

METHOD FOR STUDYING DEFECTS

The Westinghouse silver etch (8) (WAg) was used for both silicon and germanium. It has been used for studying dislocations in germanium by Wagner (9). The WAg etch has a 50-fold advantage over CP-4 type etches in area resolution. This advantage is shown in Figure 1 where the WAg etch pit is the small triangle in the middle of the CP-4 etch pit. The WAg etch also shows dislocations and defects not shown by CP-4 type etches. Figure 2 shows an as-grown germanium sample which was given a 20:1 HNO_3-HF etch before the WAg etch. Small indistinct triangular pits [background such as at (A)] of

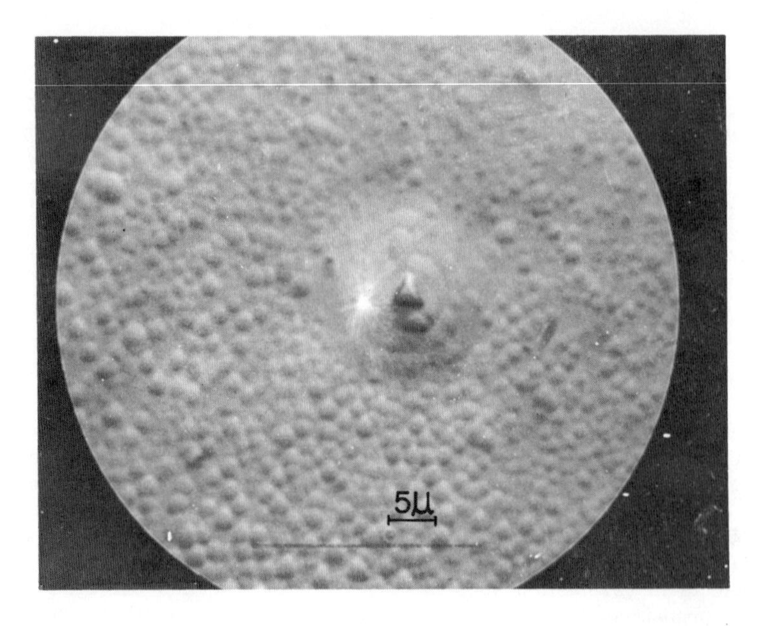

Fig. 1. Edge dislocation in germanium, CP-4 etched followed by WAg etch (small triangular pit).

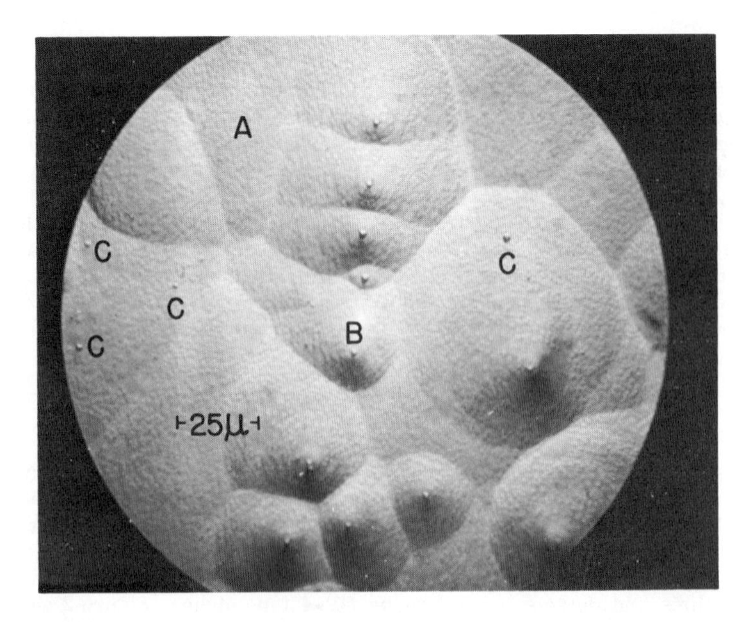

Fig. 2. Germanium $20HNO_3:1HF$ etched (large conical pits) followed by WAg etch (small triangular pits).

surface density 3×10^7 cm^{-2} are seen in addition to the distinct triangular dislocation etch pits corresponding to the large conical etch pits (B). Also there are distinct triangular pits that are unrelated (C) to the large conical etch pits.

Further experiments pin down the nature of these defects that occur in such large numbers in germanium. Germanium samples (1 ohm cm p-type) were heated to 700°C in 10^{-6} mm Hg vacuum and slow cooled with the surface under compression (1°C per minute), quenched (130°C per minute), and quenched with the surface of the sample stressed radially in tension. The WAg etch showed no defects in the slow-cooled sample (Fig. 3), circular and

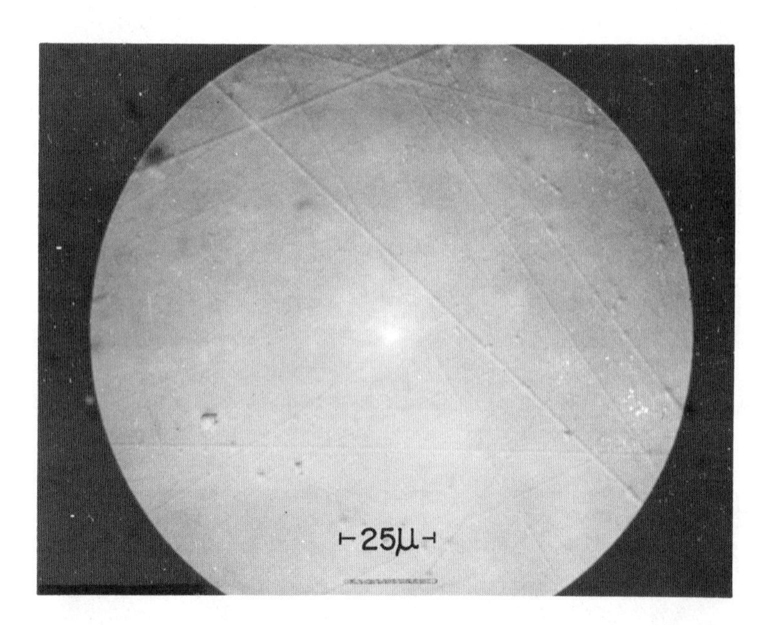

Fig. 3. Germanium, Linde B polished surface heated to 700°C in vacuum and cooled at 1°C per minute—followed by WAg etch.

semitriangular pits on the surface of samples quenched with no stress (Fig. 4), and distinct triangular etch pits in the quenched and stressed samples (Fig. 5). This is believed to be a vacancy supersaturation and precipitation phenomena in which vacancy clusters are formed under stress-free conditions and stress causes the formation of small stacking fault type defects. These results indicate the sensitivity of the WAg etch to micro disorder in germanium and this sensitivity holds in silicon as well.

DEFECTS IN EPITAXIAL FILMS

Defects Originating in Substrate

Those defects originating in the body of the substrate tend to occur in the film with no decrease in number. Those originating in the surface layers tend to become atteuated in number with increasing thickness of film. The former category includes low angle boundaries, slipped regions, and impurity segregation; the latter includes surface damage. Those dislocations with radius

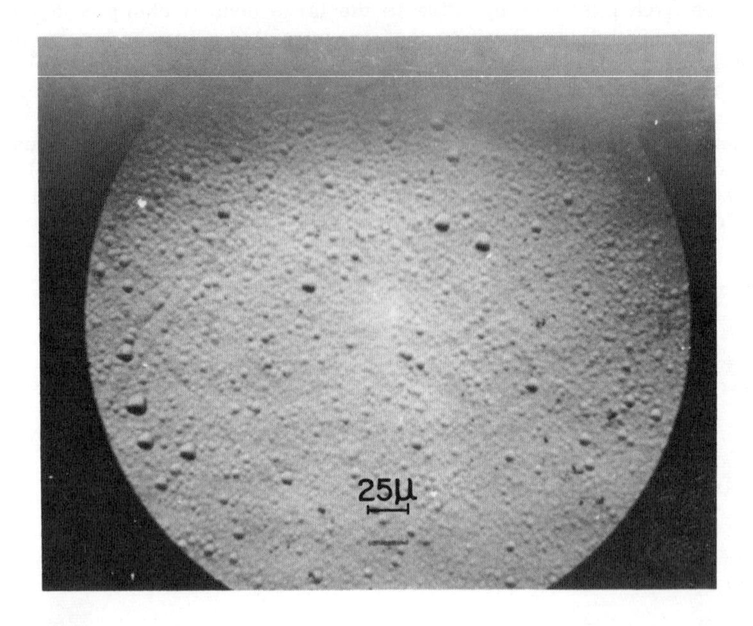

Fig. 4. Germanium, Linde B polished surface heated to 700°C in vacuum and cooled initially at 133°C per minute followed by WAg etch.

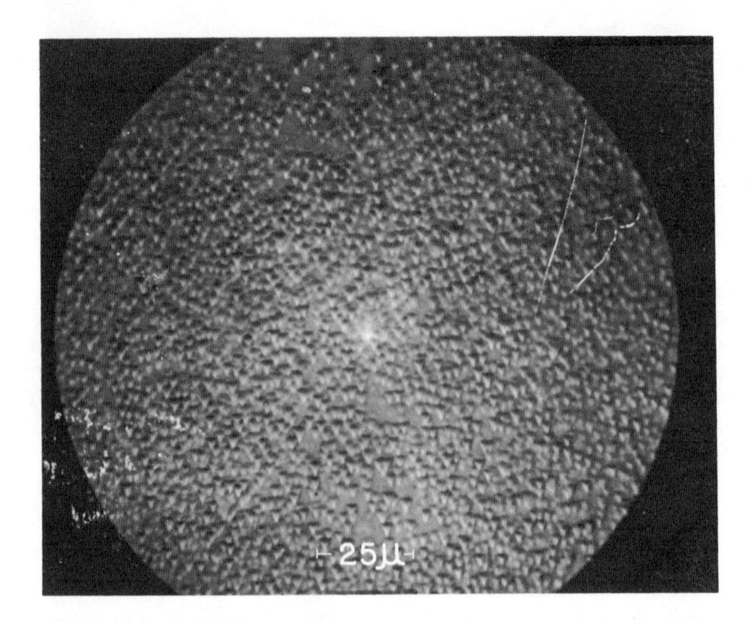

Fig. 5. Germanium, Linde B polished surface heated to 700°C in vacuum with surface stressed and cooled initially at 133°C per minute—followed by WAg etch.

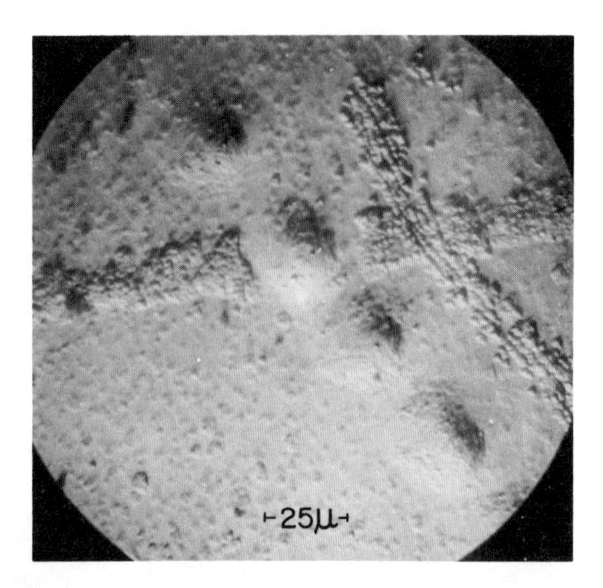

Fig. 6. Germanium substrate with CP-4 etched low angle boundary (large conical pits)—epitaxial film (1.4 μ) deposited, followed by WAg etch (small triangular pits near center of conical pits). Also note surface damage due to handling with tweezers after CP-4 etch (2 wide tracks with high density of triangular pits).

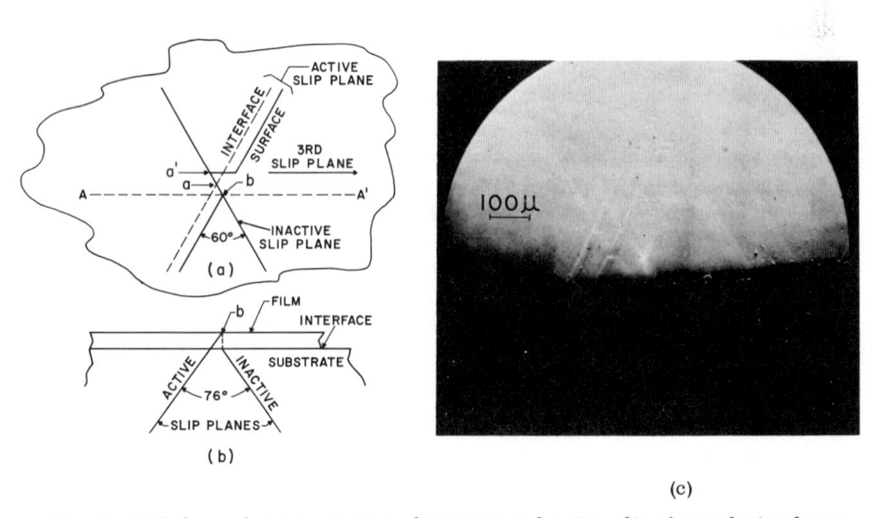

Fig. 7. Kink formed at intersection of inactive and active slip planes during formation of silicon epitaxial film. (a) Plan view of (111) surface showing geometry of interaction. (b) Section at AA' of Fig. 7a. (c) Silicon epitaxial film showing kinks at intersection of slip planes.

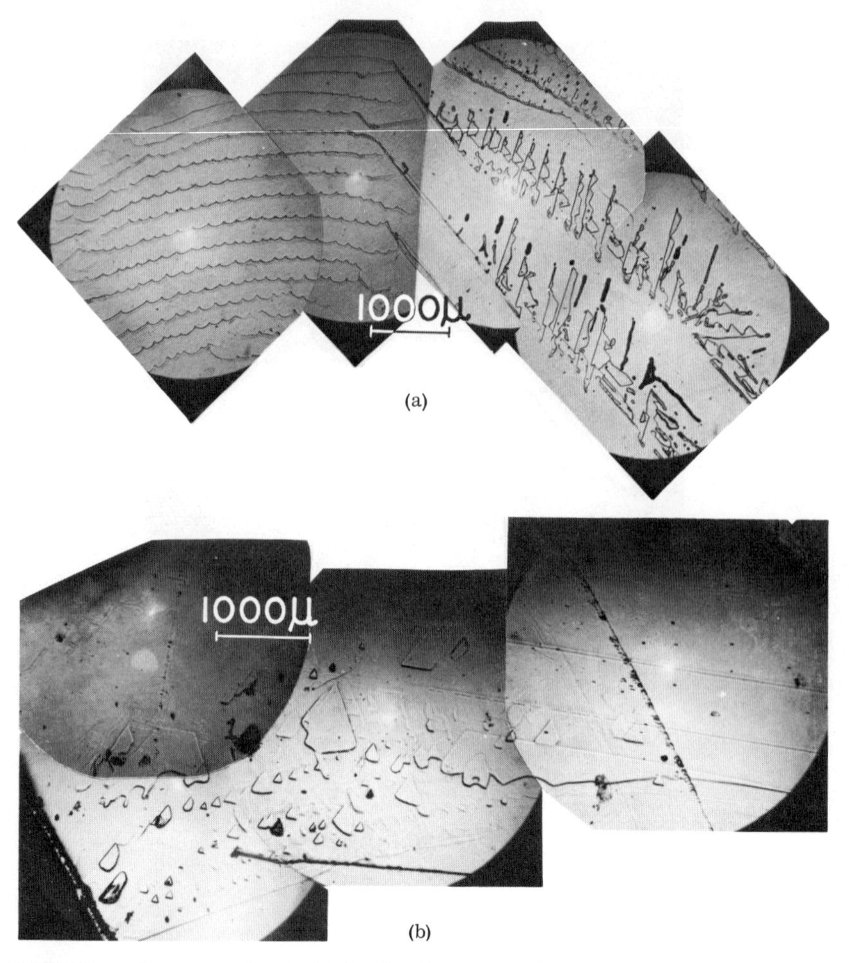

Fig. 8. Impurity segregation in highly doped (~0.0008 ohm cm) silicon substrate material. (a) Substrate material after WAg etch. (b) Epitaxial silicon film with no etching.

of curvature small compared to the film thickness are left behind as the surface advances and those with larger radius of curvature will show up at the final surface. Dislocations that do not intersect the film surface can still have adverse effects on devices.

Figure 6 shows a CP-4 etched germanium substrate with a 3-5 μ epitaxial film. The low angle boundary originally present is extended to the surface of the film as shown by the WAg etch pits. Also to be noted is the large number (10^7 cm^{-2}) of defects in the film and the two bands of etch pits resulting from tweezer damage. These will be treated later in this paper.

Slip boundaries that have been caused by plastic flow during the predeposit heat treatment of silicon substrates extend into the film as well. We can define an active slip plane as a slip plane where dislocations are in motion during epitaxial film deposition. An inactive slip plane is a slip plane where there is no dislocation motion during epitaxial film deposition or where the dislocations have been eliminated from the crystal leaving a step at the surface. During epitaxial film deposition on a $\{111\}$ surface, an inactive slip plane would probably be replicated showing up at the epitaxial film surface in register with its position on the substrate surface. An active slip plane would be offset

Fig. 9. Epitaxial germanium film (3–5 μ) with silicon chip used to mask one part of substrate—etched with WAg etch—left side, surface damage on substrate—right side, epitaxial film.

by an amount proportional to the film thickness. If an active and an inactive slip plane interesect on the substrate surface at position a (Fig. 7a) and an epitaxial layer of thickness t (Fig. 7b) is grown, the active slip plane will intersect the inactive slip plane at position b (Fig. 7a) and a kink occurs in the active slip plane extending from position a′ in the third slip plane. The kink goes from a to a′ because of the 19.4° tilt off normal of the third slip plane. The length of this kink is 2t tan 38° where t is the epitaxial film thickness or the thickness from the point where one of two intersecting slip planes changes from like to unlike behavior. The existence of such kinks as in Figure 7c implies that one slip plane was active during epitaxial film growth.

The slip planes would have to be activated during the predeposit heat treatment or the epitaxial deposition because the substrate wafers have chemically polished smooth surfaces when they are put into the deposition apparatus.

Impurity segregation in heavily doped (<0.001 ohm cm) silicon has been revealed by the WAg etch. This is usually not present near the seed end but appears and gets progressively more noticeable toward the pot end of the crystal. This is probably caused by precipitation of the doping impurity on dislocations. These precipitates cause the formation of complex phases of silicon impurity. An example of this is shown in Figure 8a, and a 7.4 μ epitaxial film grown on such a substrate appears as in Figure 8b. In Figure 8a the wavy lines on the left probably represent a region of lighter precipitation on dislocations, and on the right side the precipitation is sufficient to supply the concentration of impurity (arsenic) needed to form silicon-impurity phases such as SiAs which have been identified in such samples (10). The elongated black areas on the right appear to be inclusions. They are heavily dislocated which could indicate that they are under high compressive stress. The small black triangular areas on Figure 8b such as the one marked "1" are inclusions in the substrate where the epitaxial film on top has a very high degree of lattice disorder. This material is "dislocation free" according to normal dislocation etching procedures, and it is believed that the presence of

Fig. 10a. Germanium substrate with stain resulting from de-ionized water rinse and drying.

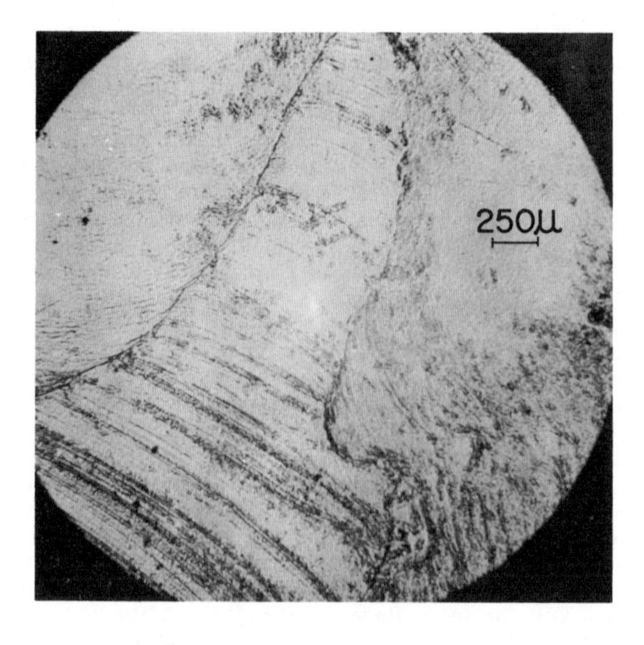

Fig. 10b. Same area as in Fig. 10a after 2 μ film deposition.

Fig. 10c. Same area as in Fig. 10a after WAg etch.

some precipitates on dislocations masks the selective etching action of some dislocation etches.

Surface damage resulting from polishing, handling by tweezers, etc., involves dislocations with very short loop lengths and small radii of curvature. Some of this is annealed out during predeposition heat treatment but much of it still shows up as defects in the epitaxial film. Figure 9 shows a region of germanium substrate that was masked during film deposition and etched with the WAg etch. Surface damage shows up in the masked area to a greater extent than where the film was deposited over the damage. The growing surface of the film would have fewer of the short radii of curvature dislocations intersecting it as the film increases in thickness. The dislocations would still be present as completed loops in the interface region.

Defects Originating At Interface

There are four causes for formation of defects that are present on the substrate surfaces: (1) general surface contamination, (2) oxide film, (3) scratches (polishing), and (4) small cleaved particles of substrate material.

The first three causes will give rise to the same type of defect which for convenience will be called the "open triangle" defect (see Fig. 23). These have been seen only in epitaxial films. The properties of this defect will be discussed further in a later section of this paper.

Heavy surface contamination such as residues left when droplets of deionized water have evaporated from the surface of substrate material have gross effects on the epitaxial layer and in some regions seem to inhibit the formation of the film. Figure 10a shows such a region on germanium where a projection from a droplet of de-ionized rinse water (used after light H_2O_2 etching) has drawn back step by step leaving a line of residue and light etching at each position of the meniscus. Figure 10b shows the same region with a

Fig. 11. (a) Germanium epitaxial film after 2° angle lap and polish (parallel lines and at epitaxial film surface) and WAg etch. Band of etch pits parallel to edge of surface marks interface. (b) Etch pits at interface—higher magnification.

2μ epitaxial germanium film, and Figure 10c shows the same region after the WAg etch. The WAg etch brings out high densities of triangular etch pits and open triangle defects, especially as shown by the broader bands near the base of the projection in Figure 10c. These same regions also initially had areas of varying oxide thickness showing brown and characteristic interference colors.

Light general surface contamination such as from organic vapors may have a negligible affect on the formation of defects compared to oxides because they are removed during predeposit heating, but it can give rise to residual impurity atoms (including carbon) in the initial layers of the epitaxial film which can reduce the stacking fault energy (11,12) and thereby make the formation of defects easier.

The presence of defects at the interface has been observed before (13). Figure 11a shows a 2° angle lap and polish on a germanium wafer with a 2.6μ film that has been etched with the WAg etch. The parallel polishing scratches end at the intersection of the angle lap, and the original suface and the interface show as a band of etch pits (5×10^{-7} cm^{-2}) which are also shown at higher magnification in Figure 11b.

Some germanium epitaxial films when etched with the WAg etch show a region of high stacking fault defect density (triangular pits) and a region of lower vacancy cluster defect density (round pits). A 2° angle lap and polish through these two regions with the WAg etch show that the "bad" region has stacking fault defects through the entire epitaxial film thickness (Figure 12a) while the "better" region shows stacking fault defects only at the interface region (Fig. 12b). The best regions show no defects at the interface (Fig. 12c).

The germanium substrates in the preceding discussion were given a light H_2O_2 etch and de-ionized water rinse before epitaxial film deposition. This is an oxidizing etch and it will leave an oxide film. There is good reason to believe that this oxide film is not uniformly or completely removed during predeposition heat treatment in hydrogen at near atmospheric pressure. Impurities trapped in the oxide film could serve to stabilize the film during this treatment.

The observed variation of epitaxial film perfection with substrate predeposition and deposition temperatures and the observed variation of evaporated germanium film perfection with substrate temperature (14) are probably directly related to the relative stability of the oxide film at these temperatures. Stacking fault defects have been observed in these evaporated germanium films by transmission electron microscopy (15).

Oxide films and the presence of adsorbed oxygen on the surface or oxygen in the film deposition system can give rise to a large number of overlapping open triangle defects such as shown in Figure 13. In this case oxygen was known to be present in the film deposition system.

Scratches also give rise to open triangle defects but only by providing nucleation sites for the formation of these defects. Figure 14 shows how defects form at random intervals along a polishing scratch.

Many silicon epitaxial films on $\{111\}$ faces have raised tri-pyramid defects such as shown in Figure 15. These defects are found in many cases uniformly distributed over wafers, but they have also been found preferentially associated with diamond scribed edges of substrates as shown in Figure 16. Etching studies show no dislocations in the substrates associated with these defects. Most scribing or abrasion processes would result in a small percentage of cleavage tetrahedra among the resulting particles. The air and de-ionized rinse water of most processing areas probably carry these particles. Any of the four sides of these would stick to the $\{111\}$ substrate face by electrostatic attraction and there are three ways a $\langle 110 \rangle$ type edge of the tetrahedra could line up with the $\langle 110 \rangle$ directions of the substrate. The other three $\{111\}$ faces would develop giving the 3-fold symmetry of the tri-pyramid defect with the depression at the center. These defects are oriented uniformly with the substrate which indicates that the nonoriented particles will not withstand the cleaning, etching, and heating treatments.

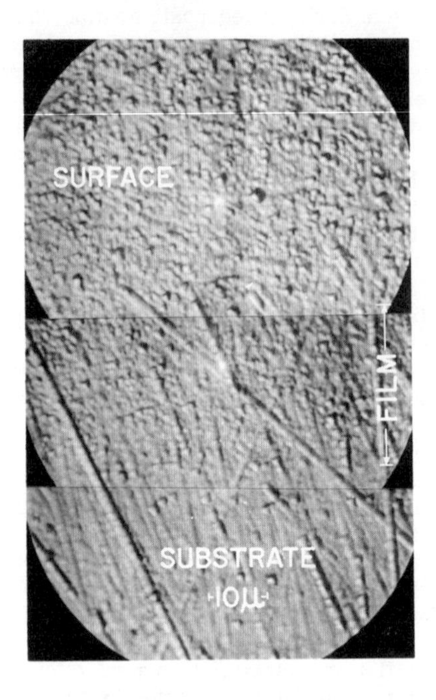

Fig. 12a. Germanium epitaxial film after 2° angle lap and polish (parallel lines, angle lap damage, end at epitaxial film surface). Defects at surface (top of photo) extend through film to interface.

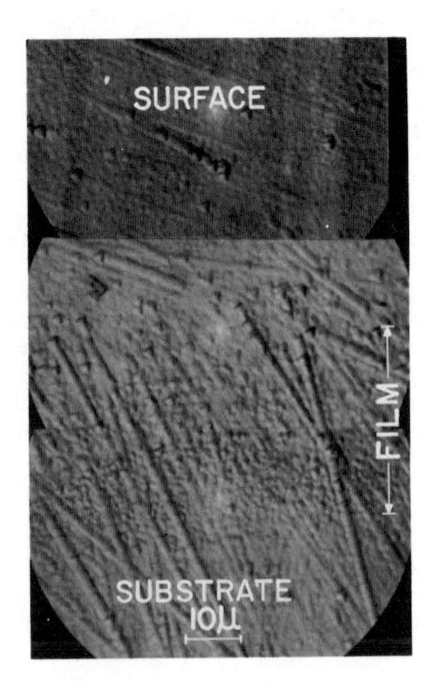

Fig. 12b. Germanium epitaxial film as in Figure 12a. Band of etch pits at interface. Defects at surface show as circular etch pits.

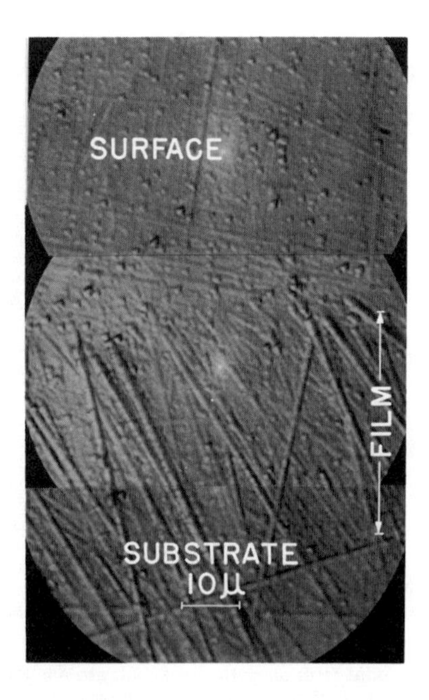

Fig. 12c. Germanium epitaxial film as in Figure 12a. No band of etch pits visible at interface and surface has least density of defects.

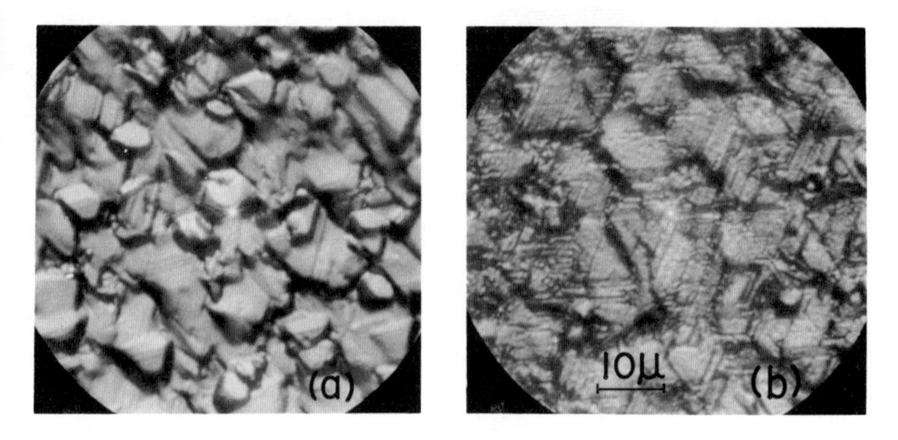

Fig. 13. Silicon epitaxial film inadvertently exposed to O_2 during film growth. (a) Unetched. (b) Etched with WAg etch.

Defects Originating in Film

The defects originating within the film show up in two forms. The difference may be only the presence or absence of stress while the defect is forming. In germanium films the etch pits range in shape from almost round to well-defined triangles, and in analogy with the results of the quenching experiments the defects may range from vacancy clusters to well-defined stacking faults formed from vacancy clusters in stressed regions. An example of this is Figure 17 where the defect etch pits in the three rays extending from the edge

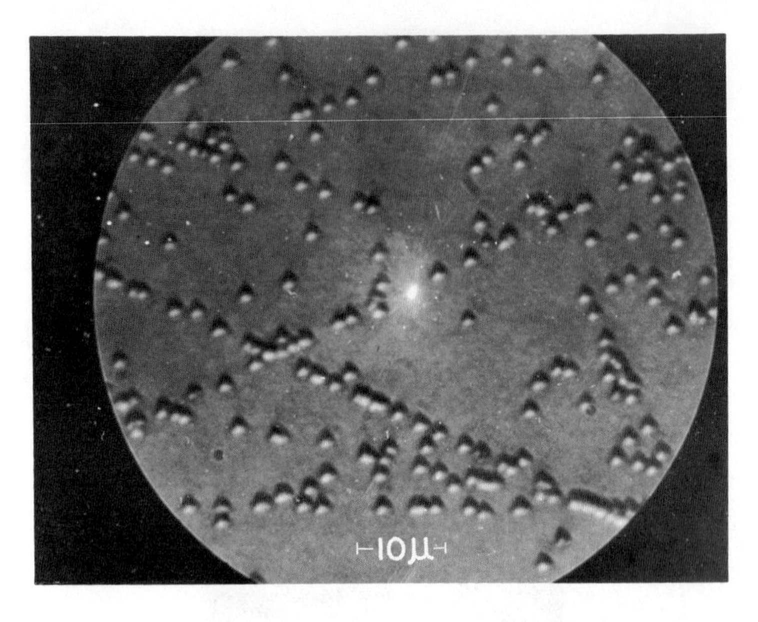

Fig. 14. Germanium epitaxial film on polished substrate showing defects nucleated at polishing scratches—WAg etched.

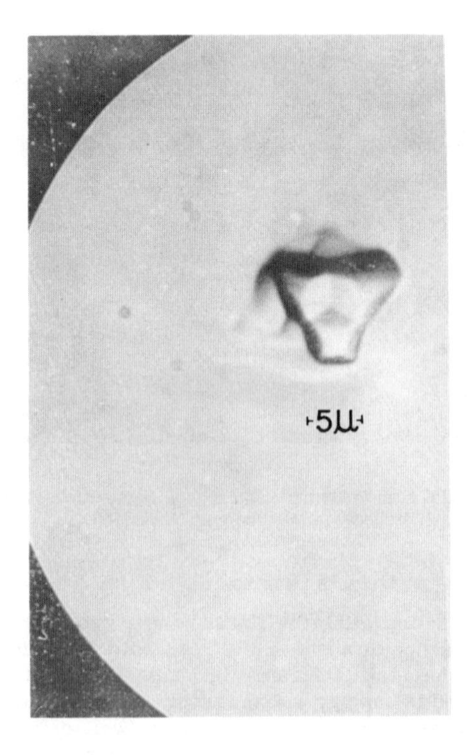

Fig. 15. Silicon epitaxial film with tri-pyramid defect showing depression at center.

Fig. 16. Silicon epitaxial film. Diamond scribed edge on substrate. Tri-pyramid defects form preferentially along diamond scribed edge.

dislocation (dark triangular etch pit) at 120° angles are geometrically oriented with the lattice while the defects randomly spaced in the rest of the area are less regular in shape. These defects exist in surface concentrations as high as 3×10^8 cm^{-2} in germanium epitaxial films. In Figure 18 one can follow the geometry of the etched defects in an epitaxial film as they become larger. The smallest pits are triangular and larger defects are found in areas where the smaller ones are not present. The larger defects range from larger flat bottom triangles to open triangles to rounded triangles to circular loops to larger irregular loops. All of these have formed at the expense of the smallest defects.

The defects that form on scratches such as shown in Figure 14 increase in size with increasing film thickness. The film deposited on the sample of Figure 14 varied in thickness from very thin to 6 μ. Some of the polishing scratches on this sample go from the thin to several micron thickness regions and the defects along the scratch range continuously from small triangular pits of Figure 14 to the larger open triangular defects of Figure 19 thus indicating the same basic defect with the manifestation depending on film thickness. The transmission electron micrograph (Fig. 20) taken by J. Drobek shows an open triangle defect in this same germanium epitaxial film.

Geometrical Properties of Defects

Among the possible Burgers vectors of 60° dislocations (16) are the vectors $1/2[01\bar{1}]$, $1/2[10\bar{1}]$, and $1/2[\bar{1}10]$ which are in the (111) plane. Some of the

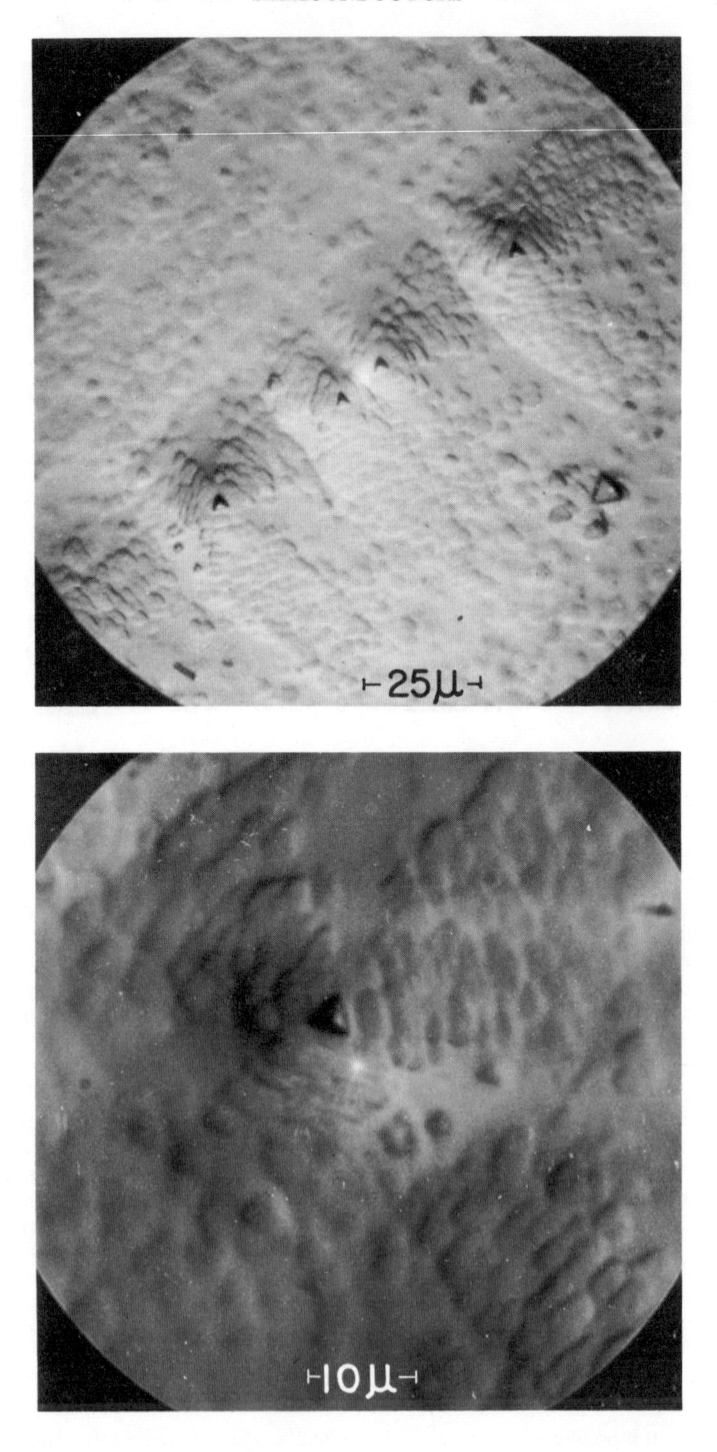

Fig. 17. Edge dislocation in germanium showing defects arrayed by stress field into three rays.

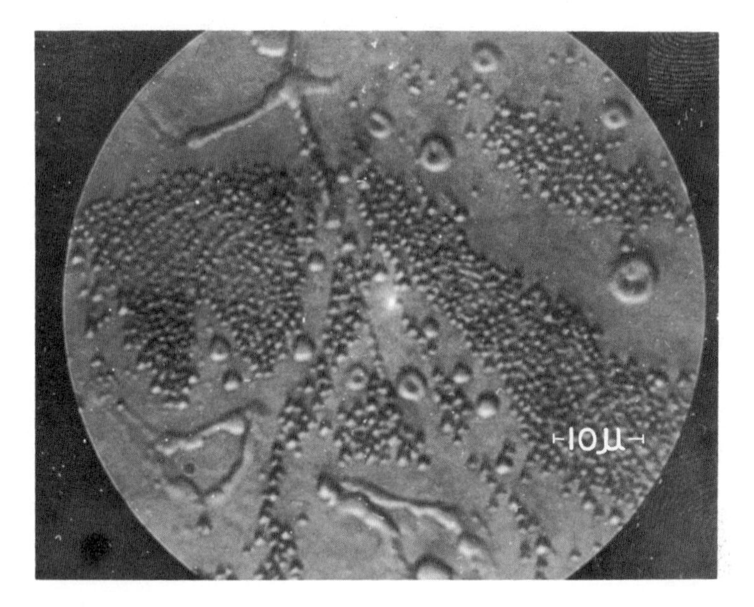

Fig. 18. Germanium epitaxial film after WAg etch showing small triangular pits with larger triangles, open triangles, circular loops, and irregular loops formed at expense of smallest defects.

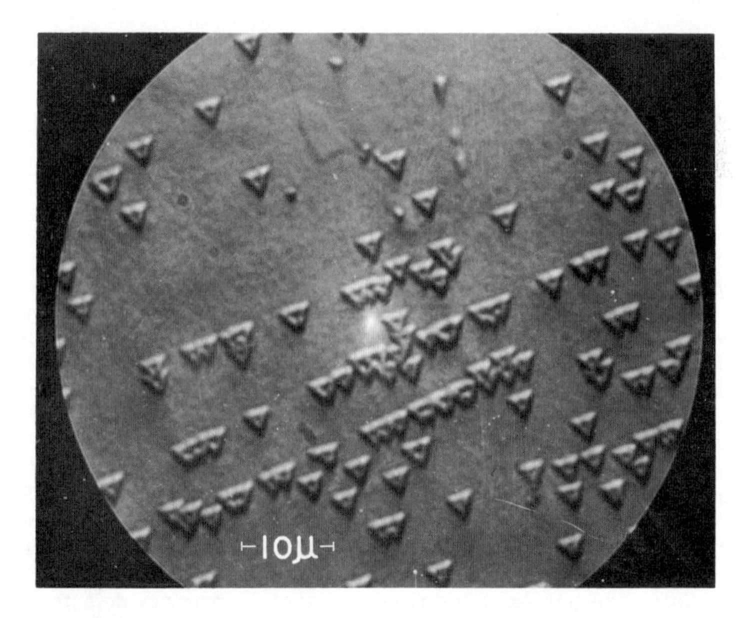

Fig. 19. Germanium epitaxial film after WAg etch showing open triangles formed along polishing scratches when film is thicker than in Fig. 14.

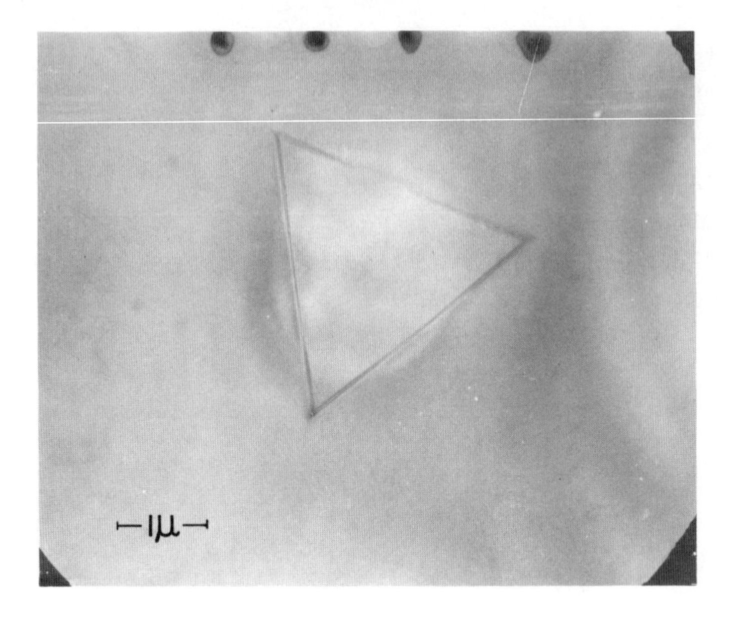

Fig. 20. Electron transmission micrograph of germanium epitaxial film shown in Figs. 18 and 19. Micrograph taken by J. Drobek.

combination rules of these vectors are summarized as follows: If the following definitions are used

$$AB \longrightarrow 1/2\,[10\bar{1}],\ BA \longrightarrow 1/2[\bar{1}01]$$
$$BC \longrightarrow 1/2\,[\bar{1}10],\ CB \longrightarrow 1/2[1\bar{1}0]$$
$$AC \longrightarrow 1/2\,[01\bar{1}],\ CA \longrightarrow 1/2\,[0\bar{1}1]$$

Then

$$AB + BC = 1/2\,[01\bar{1}] = AC$$
$$BC + CA = 1/2\,[\bar{1}01] = BA$$
$$CA + AB = 1/2\,[1\bar{1}0] = CB$$
$$AB + BA = 0 = BC + CB = AC + CA$$

These relationships apply to the dislocations that make up the three sides of the open triangle defects. The relation CA + AB = $1/2\,[1\bar{1}0]$ = CB allows the open triangle defects to combine as in Figure 21a to form the complex defects shown. The relations CA + BC = $1/2\,[\bar{1}01]$ = BA and AB + BA = 0 allow the "open triangle" defects to combine as in Figure 21b to form the complex defects shown. All of these complex defects have been observed in germanium and silicon (Figs. 22 and 23). Some other complex configurations such as near the center of Figure 23 are a combination of those illustrated. Occasionally in a field of open triangle defects there are observed defects that are identical to one side of the open triangle defects in nature and size (Fig. 23). If one postulates a primary triangular defect with Burgers vectors of the same type as previously mentioned, one can use the same set of relations to form the single sides from a row of the primary defects and one can fit the sides together using the same rules to obtain the open triangle defects as illustrated in Figure 24.

These primary defects may be formed originally from vacancy clusters in stressed regions and they may be the same as those formed in the germanium quenching experiments with stress present.

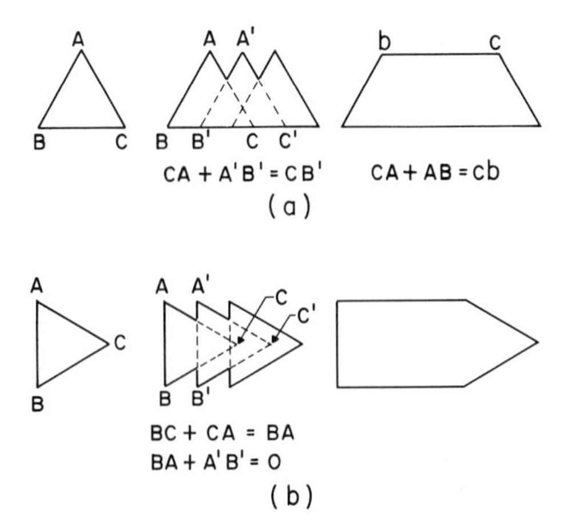

Fig. 21. Illustration of formation of complex defects from simple defects following rules for combination of Burgers vectors of dislocations. (a) Forming along <110> type directions. (b) Forming along <112> type directions.

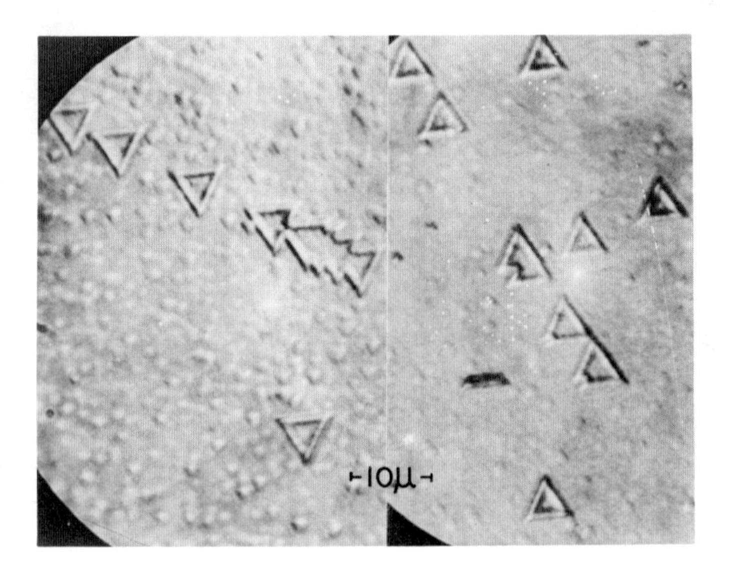

Fig. 22. Silicon epitaxial film—open triangle defects combining in <110> direction and <112> direction. WAg etch.

The evidence (such as uniformity in size) seems to show that open triangle defects originate at the interface. One can then calculate the relation between the length of the side of the open triangle defect and the thickness of the epitaxial film:

$$\text{Film thickness} = \text{Side of triangle } (\sin 54.7^\circ)$$

This has been checked against angle lapping and staining data.

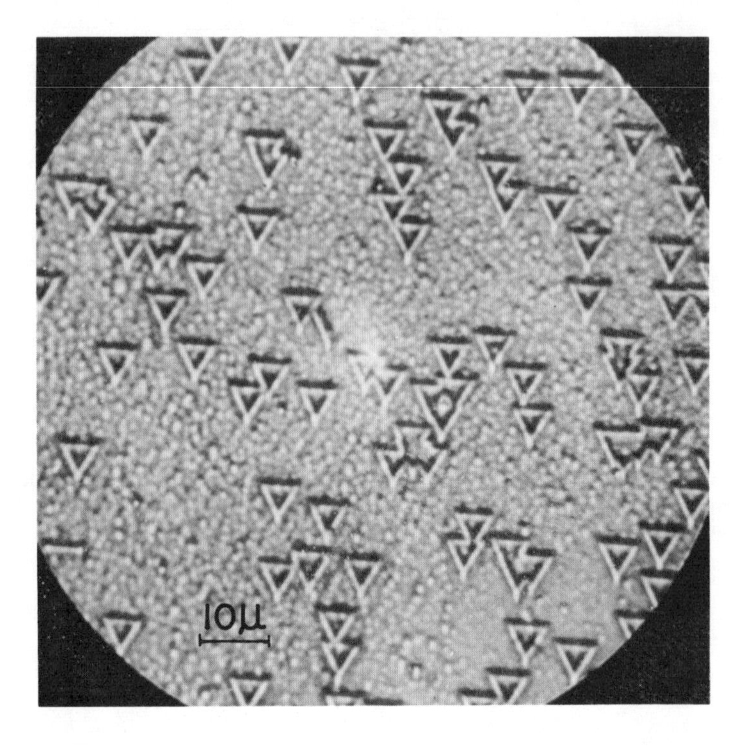

Fig. 23. Germanium epitaxial film after WAg etch showing complex defects and line segments in two of the three <110> type directions.

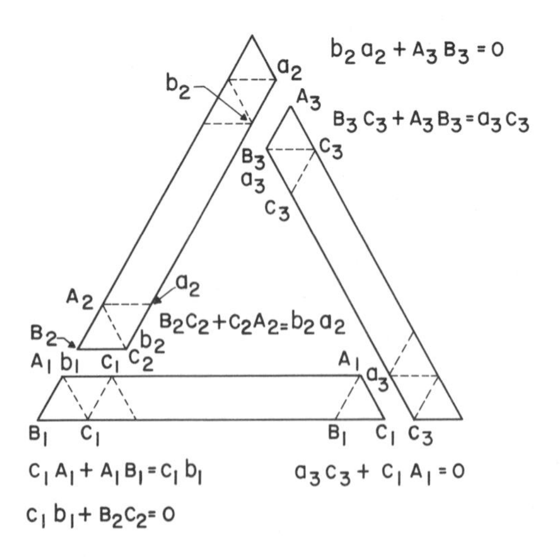

Fig. 24. Illustration of formation of open triangle defects from line segments following rules for combination of Burgers vectors of dislocations.

CONSIDERATIONS FOR MINIMIZING DEFECTS IN FILMS

It is evident that an epitaxial film will not be dislocation free if the substrate is not dislocation free. There will be a one-to-one correspondence in dislocation density between film and substrate for dislocations with radius of curvature large compared to the film thickness. Heavily doped substrates with impurity segregation and second phases will affect the film. Slip due to plastic deformation during epitaxial deposition has been observed.

Light surface damage is masked somewhat by the film, but the defects are nevertheless still present in a critical region. Light surface damage also will cause the nucleation of open triangle defects. Particulate surface contamination can cause gross defects. Surface cleanliness and atmosphere control in process areas should minimize these defects.

Thin film contamination—and especially oxide films—are the most significant causes of defects in epitaxial film work (13). This is probably the major cause of the higher concentration of disorder at the film-substrate interface. It is known that the stacking fault energy of some pure materials is lowered by the addition of impurity elements (11,12). This would enhance the formation of stacking fault type of defects. The presence of impurities in the film growth system could greatly increase the defects in the film, especially near the interface.

References

1. Shockley, W., Czechoslov J. Phys., 11, 81 (1961).
2. Kikuchi, M., J. Phys. Soc. Japan, 15, 1822 (1960).
Moore, R. T., and J. L. Moll, "Studies of Microplasmas and High Field Effects in Silicon," Final Report Signal Corps Contract DA36(039)SC78272, Feb. 15, 1960; Solid State Electronics Lab., Stanford Electronics Lab., Stanford Univ., Stanford, Calif.
4. Bernard, M., and B. Leduc, J. Phys. Chem. Solids, 13, 168 (1960).
5. Batdorf, R. L., A. G. Chynoweth, G. C. Dacey, and P. W. Foy, J. Appl. Phys., 31, 1153 (1960).
6. Schottky, W., Halbleiter Probleme, Vol. IV, 1958, pp. 68-118.
7. Gibson, A. F., F. A. Kröger, and R. E. Burgess, Progress in Semiconductors, Vol. IV, Wiley, New York, 1960, pp. 155-203.
8. Wynne, R. H., and C. Goldberg, Trans. AIME, 197, 436 (1953).
9. Wagner, R. S., and B. Chalmers, J. Appl. Phys., 31, 581 (1960).
10. Jaccodine, R. J., and P. D. Zavitsanos, private communication.
11. Haasen, P., and A. King, Z. Metallk., 51, 722 (1960).
12. Warren, B. E., and E. P. Warekois, Acta Met., 3, 473 (1955).
13. Ingham, H. S., Jr., and P. J. McDade, I.B.M. J. Research Development, 4, 302 (1960).
14. Riesz, R. P., and L. V. Sharp, Solid State Device Research Conference June 1961.
15. Haase, O. H. A., this volume, p. 159.
16. Hornstra, J., J. Phys. Chem. Solids, 5, 129 (1958).

DISCUSSION

E. G. BYLANDER (Texas Instruments): I have observed triangles in silicon along scratches before etching similar to those you showed. Were yours visible before etching?

T. B. LIGHT: Yes.

E. G. BYLANDER: Miller, of our laboratories, estimated dislocation densities of 10^8 cm^{-2} are required to explain a microtwinning interface that he observed. I wonder if your dislocation densities of 5×10^7 cm^{-2} would be a reasonable value to explain stress relief by microtwinning at the interface.

T. B. LIGHT: The number 5×10^7 cm^{-2} is not very different from an estimated number of 10^8 cm^{-2}. I do not believe though that it is a matter of stress relief at the interface through slip processes. Whether the defects are formed during film deposition and the ease of formation depends on the stacking fault energy which is controlled by the impurities present at the interface and in the system.

E. G. BYLANDER: Can this be interpreted as stress relief by slip?

T. B. LIGHT: No.

S. O'HARA (Westinghouse Electric Co.): It is accepted rather widely that both free surfaces and dislocations act as sinks for vacancies. Do you see the small triangular pits, which you believe to result from etching at dislocation loops, very close to actual dislocations?

T. B. LIGHT: Yes.

S. O'HARA: Then if you have line dislocations in your material, you would not expect to see these small triangular pits in the immediate vicinity of a line dislocation. Do you see these small pits?

T. B. LIGHT: You do see them. In Figure 1 there is a very small region that is defect-free around some of the dislocations brought out by the first etch, but this is not observed in all cases. In some cases it is difficult to see if there is a free region.

S. O'HARA: How much material do you remove from the surface of the sample during the etching which shows the small pits?

T. B. LIGHT: Fractions of a micron, except that the pits themselves are a little deeper, but in general it is fractions of a micron; very short etching times.

S. O'HARA: It seems rather surprising that you would see these small loops so close to the surface. The work of Tweet indicates that the surfaces of Czochralski crystals act as sinks for vacancies. In his work there is evidence that dislocation loops exist in the bulk of otherwise dislocation-free crystals; however, there is a region of about 1 mm thickness at the surface which does not appear to contain loops. This is attributed to any excess vacancies, which have been grown in, diffusing to the surface.

T. B. LIGHT: I believe that a so-called perfect, damage-free surface is not as good a sink as a damaged surface, and I feel the surface of a grown crystal would not be as perfect as a chemically polished surface.

S. O'HARA: Is there any theoretical reason for believing that a perfect surface would not act as a sink any more than a rough surface?

T. B. LIGHT: The lattice disorder and damage at and near the surface may be the major sink for vacancies at surfaces, or to look at it in another way, the roughness, disorder, and damage at a surface greatly increase the effective surface area. The effective surface has been extended into the material for a distance corresponding to the depth of damage.

W. D. BAKER (Rheem Semiconductor Corp.): On your angle lap section, where you showed dislocation array, there appeared to be two interfaces on that slide. Could you explain the two interfaces in Figure 11a?

T. B. LIGHT: That is an overlap of two pictures. One region is out of focus when the other is in focus, and I had to match two pictures in focus to cover both regions.

Transmission Electron Microscopy of Evaporated Germanium Films

O. HAASE*

Bell Telephone Laboratories, Inc., Murray Hill, New Jersey

Abstract

Germanium films evaporated on single crystal germanium surfaces were found to be single crystal with imperfections which are thought to be dislocations, on {111} and {100} substrates, and stacking faults, on {111} substrates. Their density depended on substrate surface treatment prior to evaporation, up to 10^{10} cm^{-2} for etched substrates and 10^8 cm^{-2} and less for sputtered and cleaved, {111} only, substrates. The imperfections appear when individual crystals formed on the substrate surface started to intergrow. It is suggested that imperfections found in thicker layers ($1\,\mu$) have their origin near the interface and propagate through the growing layer.

INTRODUCTION

The perfection of semiconductor crystals has been studied by a number of methods (1). The etching technique makes use of the fact that the etching attack near an imperfection differs from that in the perfect regions. The etch pit is subsequently observed with the light microscope or a replica is made and studied with the electron microscope. Another method has been used for silicon. Copper and other atoms diffused into the silicon will precipitate near the crystal imperfections and cause contrast in infrared light. Still another method makes use of the fact that x-rays diffract differently from an imperfect region. These methods are especially useful for crystals with low imperfection densities, lower than 10^7 cm^{-2} for the etch-pit method, 10^5 cm^{-2} for x-rays and infrared.

For higher imperfection densities, transmission electron microscopy has been successful in resolving individual defects in a number of materials (2). In a standard magnification of 20,000×, which is usually necessary to reveal the imperfections, the field of vision covers an area of the order of 10^{-7} cm^2. Therefore, the density of imperfections must be 10^7 cm^{-2} before one may expect to find one in the field of view. Thus, electron transmission microscopy complements the standard methods of detecting imperfections in germanium crystals and is especially useful for imperfection densities of 10^7 cm^{-2} and higher.

---

* Visiting Scholar 1960-61.

Description of Problem

The problem we faced was the determination of the structural perfection of thermally evaporated germanium layers of a thickness of 1 μ and less deposited onto germanium single crystals pulled from the melt.

The aim of the investigation was to determine the effect of the following parameters on the structural perfection of the layer: (a) substrate surface treatment, (b) orientation of substrate crystal, and (c) thickness of the evaporated layer.

EXPERIMENTAL

Deposition of the Films

The samples were taken from a series of experiments by Riesz and Sharp, of our laboratory, who wanted to measure the electrical properties of evaporated germanium films (3).

The evaporation was made from a carbon crucible at a rate of 100 A/sec. The thickness of the deposit was measured by optical interferometry and for thin films extrapolated from evaporation time. The germanium slabs, several mils thick, were held against a jig and the temperature was measured in the jig by a thermocouple. The temperature was 700°C. The actual temperature of the germanium was lower. The cooling rate after deposition was 3 min from 700 to 500°C. The vacuum was 10^{-6} mm Hg generated by an oil diffusion pump and a liquid nitrogen trap.

Substrate crystal faces used were $\{111\}$ and $\{100\}$. The surfaces were (a) CP4 etched and (b) etched and cleaned by sputtering in an argon-gas discharge. Additional $\{111\}$ faces were formed by cleaving crystals *in vacuo.* Both the sputtering and the cleaving was done in the vessel and evaporation was started immediately afterwards.

Method of Making Sections

The method of making the sections from these layers for microscope investigation was developed by Riesz and Bjorling of our laboratory (4).

Its principle is to etch away the substrate and leave the layer untouched. This method enabled us to obtain sections from layers of varying thickness. In a 1-μ layer, the etching was continued through the evaporated layer until a 1000-A section of the outermost part of the film was left. In layers as thin as 100 A, sufficient substrate material was retained to support the layer. Thus, the section thickness could be kept thin enough for transmission microscopy while the film thickness was varied.

Layers on $\{111\}$ Substrates

The micrograph from a layer a few thousand angstroms thick (Fig. 1) shows two types of features:

1. Bands several thousand angstroms long and several hundred angstroms angstroms in width, which are parallel to each other or form angles of 120°. This corresponds to the threefold symmetry of the $\{111\}$ face. Comparison with transmission electron diffraction patterns shows that the lines are $\langle 110\rangle$ directions. Electron diffraction pictures taken in reflection showed intensity lines and additional spots on $\langle 111\rangle$ lines indicating a stacking disorder of $\{111\}$ planes. These cut the top and bottom surface along $\langle 110\rangle$ lines. It is thought that the bands observed are stacking faults or repeated stacking faults (which includes microtwins) on $\{111\}$ planes, which have been observed in face-centered metals by other authors (5).

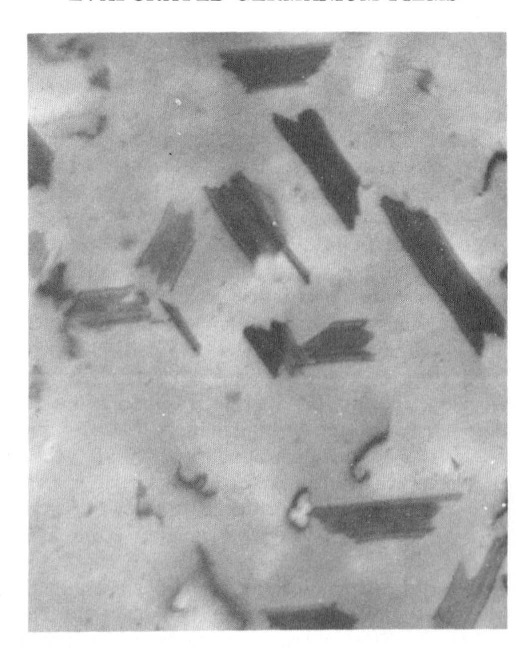

Fig. 1. Germanium layer 4000 A thick on an etched (111) germanium substrate. It shows imperfections which are presumably stacking faults and microtwins on the three {111} planes that intersect the (111) surface, and dislocations.

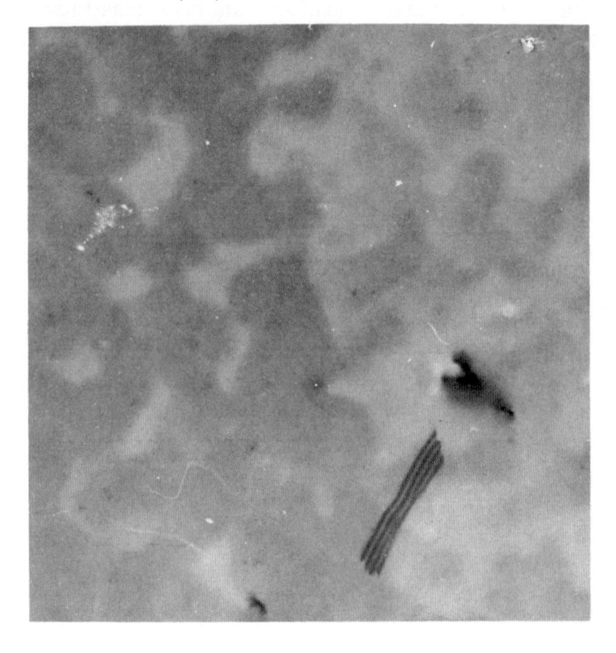

Fig. 2. Micrograph of a section from a 4000 A layer on a germanium substrate cleaned by sputtering showing fringes inside a stacking fault. The imperfection density is of the order of 10^8 cm^{-2}.

2. The other features observable in the pattern are presumably dislocation lines that have been reported in deformed germanium (6). The substrate surface treatment was of considerable influence on the imperfection density in the film. Layers of several thousand angstroms on etched surfaces showed 10^9 to 10^{10} stacking faults and microtwins per cm^2. Surfaces that were sputtered prior to evaporation showed densities of the order 10^8 cm^{-2} (Fig. 2). The density of the dislocations also decreased with the presumed order of surface cleanliness.

The lines visible in an individual stacking fault (Fig. 2) arise from the fact that the lattices above and below the stacking fault are displaced relative to each other. Such fringes which also appear on wedges are thickness extinction contours arising from the dynamical interaction between the primary beam and the diffracted beam. The bright field image shows a bright line when the thickness of the wedge is such that the maximum is in the primary beam (7). Such individual stacking faults can be used to determine the thickness of the section which is otherwise not easily obtained. Since the angle between the $\{111\}$ planes is approximately 70°, the thickness of the section is tan 70° times the width of the pattern, which is the projection of the stacking fault cut out by the section. If this is, for example, 500 A, the section thickness is 1400 A. Thicknesses ranging from 700 to 2000 A were found.

The thickness extention effect also allows us to see the surface topography. The effect is pronounced in a layer of 500 A average thickness formed on a cleaved substrate (Fig. 3). The contrast varies from black to white, depending on the local thickness of the section. The extinction depth varies with operating plane and tilt and was in the case of Figure 2 approximately 500 A. Thus, the thickness extinction effect explains why thickness variations small compared with the total section thickness can give rise to strong contrast. The micrograph shows that the crystals, although of irregular shape, are

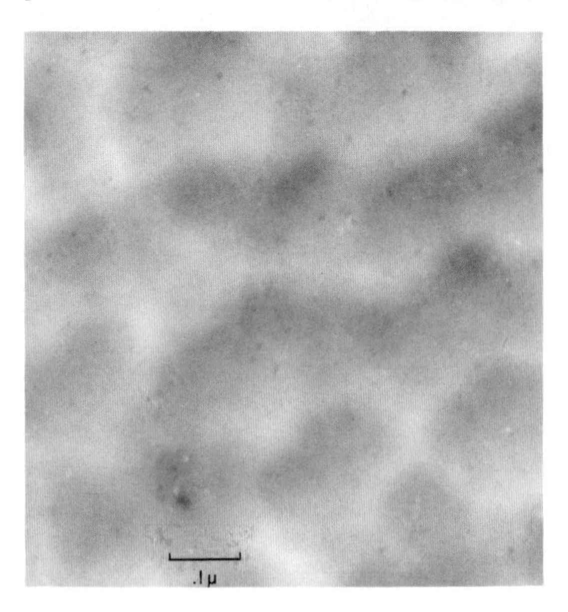

Fig. 3. Germanium layer of 500 A average thickness showing individual crystals of irregular shape about 3000 A in lateral extension. The crystals are in single crystal orientation with the substrate, with a misorientation smaller than 1° around all three axes.

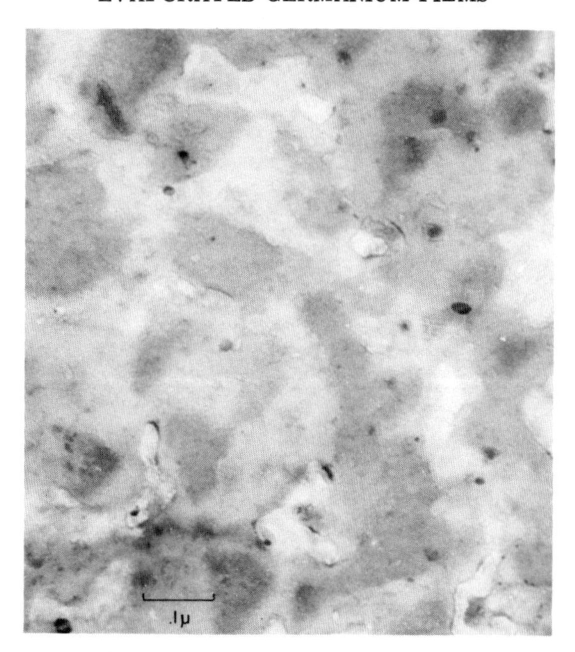

Fig. 4. Germanium layer of 4000 A average thickness. The individual crystals start to intergrow. The stacking faults, if present, are not yet sufficiently developed.

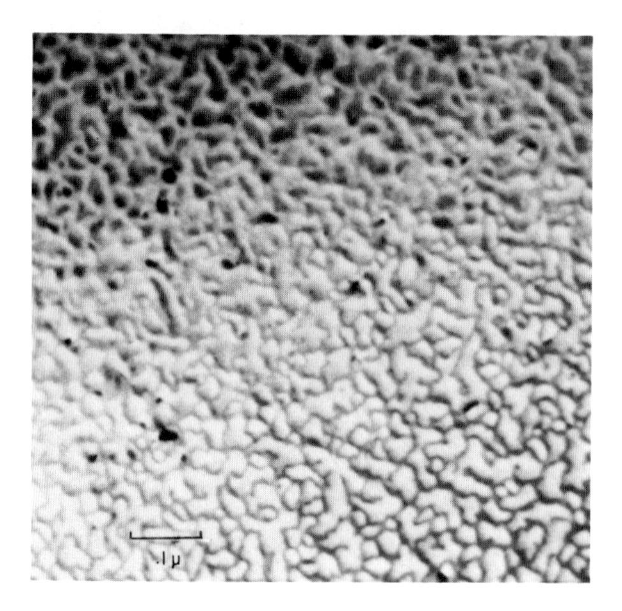

Fig. 5. Germanium layer of 150 A average thickness evaporated on an etched (100) substrate. The average lateral extension of the crystallites is 300 A. The contrast arises from the thickness extinction effect.

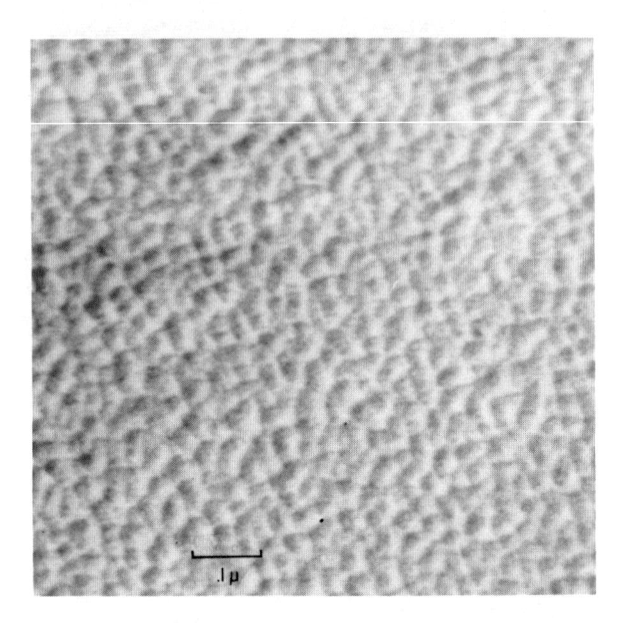

Fig. 6. Replica of the sample from Fig. 5 to show that the surface structure consists of individual crystallites with "ditches" in-between.

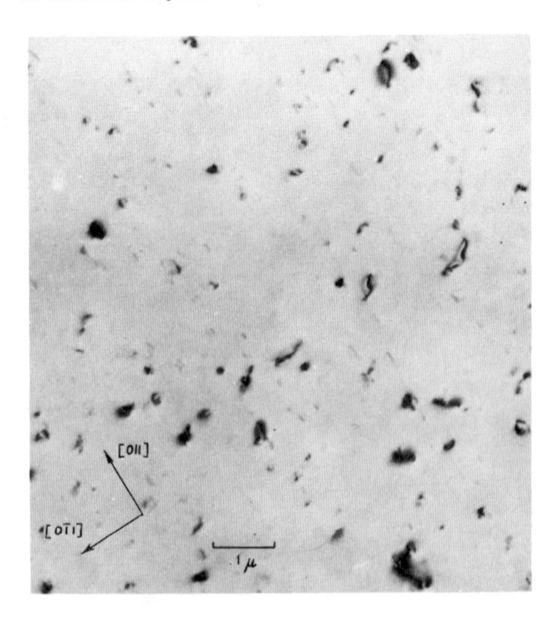

Fig. 7. Transmission micrograph of a film prepared under the same conditions as sample of Fig. 5 but with average film thickness of 300 A. The individual crystallites have intergrown, the "ditches" have disappeared. Many of the features may be inclusions of foreign matter. Note faint lines which are [011] direction. The lines in the [0$\bar{1}$1] direction perpendicular to it are not visible, and the operating diffraction beam is (022).

well oriented. Misorientation around an axis normal to the surface would give rise to Moirée patterns; tilt around an axis in the surface would cause contrast differences between individual crystallites. Both effects are not observed. The orientation of the crystallites must, therefore, be better than one degree.

How the film grows after the substrate surface is covered is shown in Figure 4. This is a section of a 5000 A film evaporated also on a cleaved substrate. At this thickness the individual crystallites intergrow and some unidentified imperfections become visible. Stacking faults, if present, are not sufficiently extended to be recognized. It should be noted that the substrates themselves were highly perfect crystals and did not show any imperfections in the micrographs. The imperfections found in the layers are, therefore, not inherited but generated during the formation of the film. The dependence of substrate surface treatment suggests that foreign matter on the substrate surface may be the cause for the generation of imperfection.

Layers on {100} Substrates

Analogous experiments were made on the {100} face of the substrate. Etched and sputtered surfaces were used; the cleaved surfaces were not smooth enough to make good, thin sections.

Figures 5, 7, 8, and 9 show secctions from germanium layers 150, 300, 2000, and 10,000 A thick, which were deposited on the etched {100} surface. The growth of the films also starts with oriented individual crystals of irregular shape (Fig. 5), as can also be seen from a replica made from the same sample (Fig. 6). The crystals are again well oriented, their misorientation with regard to all three axes being better than one degree.

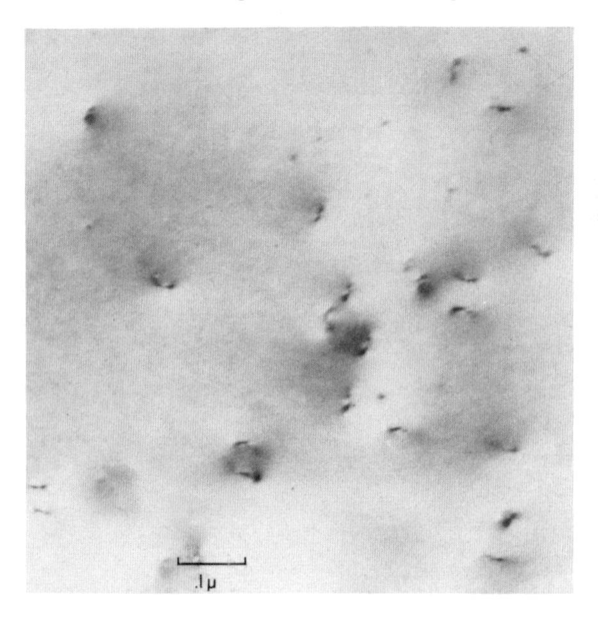

Fig. 8. Film prepared under the same conditions as samples from Figs. 5 and 7 but with film thickness 2000 A. Presumably, some of the dislocations of Fig. 7 persist through the growing layer. Contrast between some of the dislocations may be strained regions caused by the presence of the dislocations.

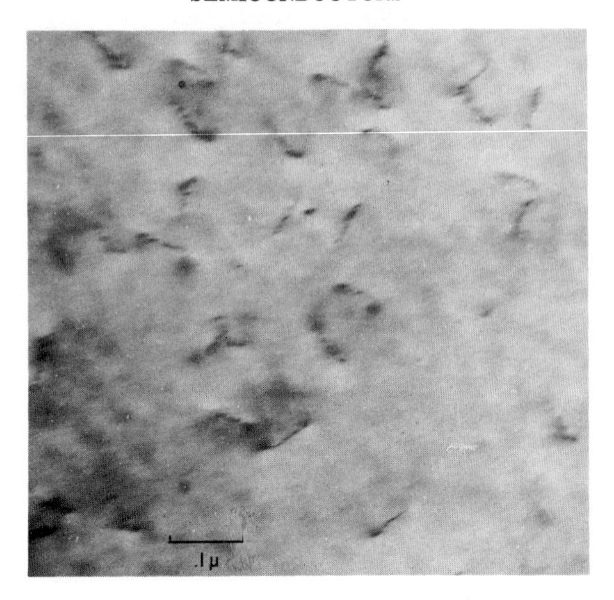

Fig. 9. Film prepared under the same conditions as samples from Figs. 5, 7, and 8 but with film thickness 10,000 A. The form of the dislocations has changed. It is not known how these features interconnect the dislocations found in the 2000 A layer of Fig. 8. There is also no explanation at present for the dotted appearance of the dislocation lines. The change in dislocation density compared with Fig. 8 may be due to the poor reproducibility of the etched substrate surface.

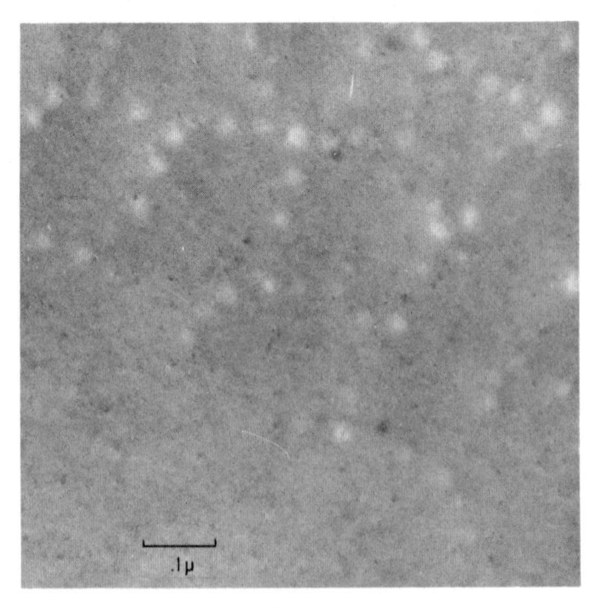

Fig. 10. Germanium layer 4000 A thick deposited onto a germanium substrate cleaned by sputtering and presumably cleaner than the etched substrate. No imperfections in these films were detected. The background contrast visible in the micrograph is presumably surface topography.

At 300 A thickness (Fig. 7) the individual crystals have grown together and form a uniform film. At this stage, a large number of small details can be observed which might partly be due to inclusions of foreign matter from the original surface and partly due to structural imperfections. A large number of faint lines about 200 A long can be observed running along the [011] direction in the (100) surface. These are considered to be [011] components of dislocation lines. Information concerning the Burgers vector of these dislocations can be obtained from the fact that there are no lines visible in the $[0\bar{1}1]$ direction, which is the other <110> direction in the (100) surface. The contrast mechanism is such that a dislocation is not visible when the Burgers vector is in the operating plane, which in Figure 7 is (011). It means that the traces of the dislocation lines and the Burgers vectors are perpendicular. Moreover, the Burgers vector lying in the (011) plane must have a component in the $[0\bar{1}1]$ direction, otherwise no contrast would be observed at all with the beam perpendicular to the surface. The observations would be satisfied with the assumption that the Burgers vector of the dislocations visible in Figure 7 is in the $[0\bar{1}1]$ direction. They would then be partly edge dislocations.

Now we turn to the question where the dislocation line ends. There are the following cases to be considered.

1. The dislocation is connected with one or both ends to the interface between film and substrate.
2. The dislocation line is closed in the film.
3. The dislocation ends at the surface away from the substrate.

The first case is probably not realized. One should be able to see the dislocation connecting to others, since a film of 300 A thickness should be completely embedded in the section.

The second case will probably really happen. In this case, we would only see the trace of the loop, which may be a line. (Occasional double lines may not be real but due to double images.)

The third case means that the dislocation can climb away from the substrate and end in the outer surface, thereby persisting through the layer if more germanium is evaporated on top. The micrograph of a layer 2000 A thick (Fig. 8) supports this view. The dislocations seem to come in pairs and sometimes have a strain field between them. In thicker layers (Fig. 9, 10,000 A) the pattern becomes more complicated and pairs cannot be recognized any more.

The substrate treatment in the case of the {100} substrate was again of strong influence. While layers on etched substrates showed dislocation densities of the order 10^9 cm^{-2}, layers deposited on substrates cleaned by sputtering showed no dislocations but only detail which we attribute to surface topography of the sputtered surface (Fig. 10).

CONCLUSIONS

1. The growth of the layer starts with individual crystals on the surface.
2. The crystals are completely oriented with regard to the substrate. The angular misorientation is certainly smaller than 1°.
3. Imperfections appear when the film reaches a thickness where individual crystal intergrows.
4. The imperfections which are considered to be dislocations and stacking faults cannot be inherited from the substrate. Compared with the imperfection densities in the film, the substrate is highly perfect.
5. Imperfections have been found even in the thickest films investigated. It is suggested that they have their origin near the interface and propagate through the growing layer.

6. The dislocation density in the film depends strongly on the cleanliness of the substrate surface prior to evaporation.

ACKNOWLEDGMENTS

The author is pleased to acknowledge the collaboration of R. P. Riesz and Miss L. V. Sharp and helpful discussions with W. C. Ellis, R. D. Heidenreich, J. R. Patel, C. D. Thurmond, and Mrs. E. A. Wood.

References

1. Pearson, G. L., and F. L. Vogel, "Plastic Deformation of Semiconductors," Progr. in Semiconductor, to be published.
2. As a review see Hirsch, P. B., A. Howie, and M. J. Whelan, Phil. Trans. Roy. Soc., London, 252, 499 (1960).
3. Riesz, R. P., Solid State Device Research Conference, Stanford, California, 1961.
4. Bjorling, C. G., and R. P. Riesz, Rev. Sci. Instr., to be published.
5. Burbank, R. D., and R. D. Heidenreich, Phil. Mag., 5, 373 (1960).
6. Geach, G. A., B. A. Irving, and R. Phillips, Research (London), 10, 411 (1957).
7. Heidenreich, R. D., J. Appl. Phys., 20, 993 (1949).

DISCUSSION

G. R. BOOKER (Westinghouse): Were any tilting experiments tried, while examining specimens? Normally when examining dislocations by the transmission electron microscope method one tilts the specimen and observes the changes in the contrast of the dislocation, and unless one does this one cannot always be quite sure that they are dislocations.

O. HAASE: Yes, the samples were tilted. We could see individual dislocations and stacking faults appear and disappear while the specimen was tilted.

G. R. BOOKER: Did you deduce any values for the stacking fault energy from these configurations?

O. HAASE: No, we did not.

Epitaxial Germanium Layers by Cathodic Sputtering

F. REIZMAN and H. BASSECHES

Bell Telephone Laboratories, Inc., Allentown, Pennsylvania

Abstract

Films less than 1 μ thick can be produced with a control of $\pm 10\%$ on the thickness. The sputtering rate under the experimental conditions was 1.5 μ/hr. Electrically the films were always found to be p-type, regardless of the doping of the cathode. Hall effect measurements on isolated films show the mobility to be similar to bulk germanium of the same resistivity. Interfacial diodes have been made between the film and the substrate and indicate the absence of an intermediate layer.

INTRODUCTION

Several means of depositing epitaxial films are known. For germanium these are chemical deposition, for which several processes are used; evaporation; and cathode sputtering.

Chemical means for depositing germanium films are in use in several laboratories (1,2). Various gaseous compounds may be employed as the source of the germanium. Chemical methods in general are quite successful and were the first to yield usable epitaxial devices. Their principal drawback is the great deposition rate (of the order of thousands of angstroms per minute) which makes it hard to control thickness exactly. Thickness variations may also be due to convection currents in the dense atmosphere from which the material is deposited.

Evaporation of germanium in high vacuum provides slower deposition and easier control. There are fewer variables to manipulate than in any of the chemical processes. The problems relate mostly to finding a suitable crucible which will not dissolve or react with the evaporant. Also, large amounts of power are needed to evaporate germanium, at a usable rate over a large area, and heat radiated from the crucible causes uncontrolled substrate heating. In spite of these problems a few workers (3,4) have reported good epitaxial deposits by evaporation.

The films reported on here were obtained by cathodic sputtering. So far as we know, this is the first reported instance of epitaxial germanium films made in this way. Sputtering is a process in which the source material is vaporized by ion bombardment rather than high temperature. It is thus suited to materials of high melting point or low vapor pressure. The source material is made the cathode in a glow discharge and slowly disintegrates under positive ion bombardment. There is no crucible and no necessity for high temperatures. Although a low pressure gas atmosphere is needed to maintain the discharge, this may be argon or neon and so need not introduce any contamination.

The material ejected from the cathode probably consists mostly of single neutral atoms with energies much greater than thermal. We would expect these energetic atoms to require several collisions on the surface to become

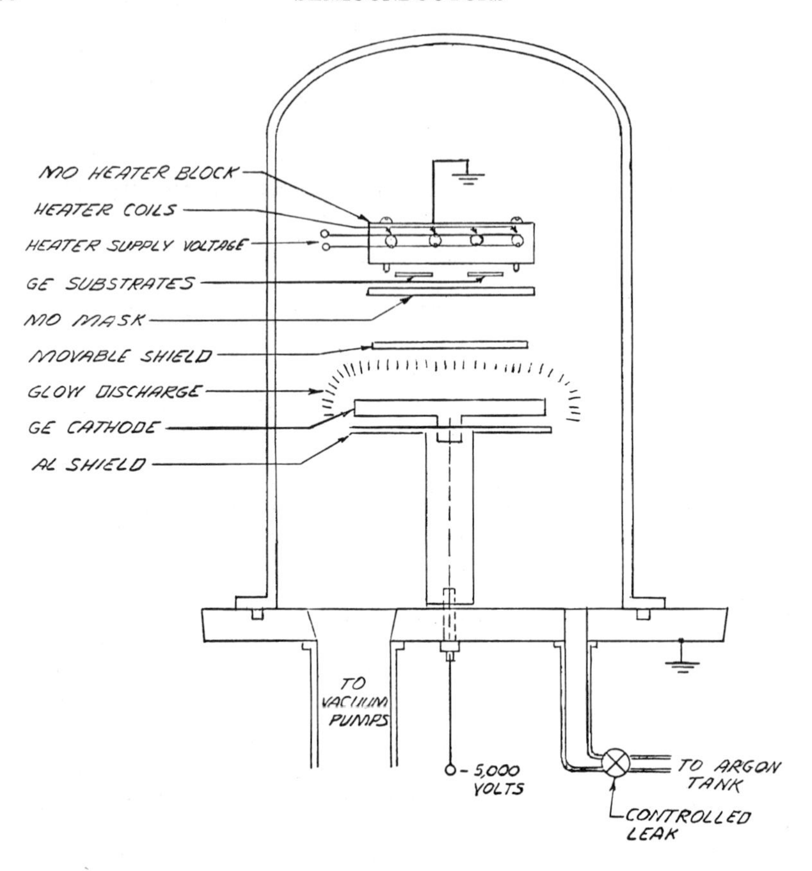

Fig. 1. Schematic of deposition apparatus showing vacuum chamber and exploded view of heater and substrate assembly.

completely thermalized. Thus, there would be a greater mobility of atoms on the surface than would be expected from the substrate temperature. Epitaxial growth depends on a surface migration of atoms, which hit the surface haphazardly and must "find their seats" in the crystal lattice before the next layer buries them. Thus, for a given substrate temperature, anything which enhances surface mobility will probably promote epitaxial growth and crystal perfection.

APPARATUS

The experimental equipment shown in Figure 1 fits inside a bell-jar type of evaporator. The cathode is below the substrate and consists of a slab of germanium about 4.5×6.5 cm. Several interchangeable cathodes of different type and resistivity are available.

The cathode operates at -5000 v and is shielded from below by a grounded metal plate which confines the glow to the top surface. The grounded anode consists of an electrically heated molybdenum block and a mask which accommodates two germanium slices. In use the slices (substrates) are held to the underside of the block by a retainer plate which serves as a mask and germanium is sputtered upward from the cathode. The first material sputtered is caught on a movable shutter to avoid depositing dirt from the cathode surface. The gas is argon at a pressure of about 40 μ, giving a cathode dark space

TABLE I
Deposition Rate in Microns per Milliampere Hour

Run	Mask Position		
	No. 1	No. 2	No. 3
	Cathode-Anode Spacing 6 cm		
FR-8	0.042	0.05	0.045
FR-10	0.036	0.041	0.039
FR-12	0.046	0.056	0.05
FR-14	0.0455	0.05	0.053
Av	0.042 ± 0.0033	0.049 ± 0.004	0.047 ± 0.005
	Cathode-Anode Spacing 5 cm		
FR-16	0.053	0.061	0.064
FR-18	0.045	0.059	0.055
FR-20	0.048	0.068	—
FR-22	0.054	0.058	0.063
FR-24	0.078	0.096	0.089
Av	0.056 ± 0.009	0.068 ± 0.011	0.068 ± 0.011

somewhat more than half the cathode-anode distance of 5 cm. Substrate temperature is monitored by a thermocouple in a groove between mask and block. The gas is continuously pumped out again. Pressure is regulated by varying the leak rate.

DEPOSITION CONDITIONS

Under the conditions used (40 μ argon, 5000 v, 20 ma, 5 cm cathode-anode distance) the deposition rate is about 1.5 μ/hr. By controlling the current and time, the final thickness can be controlled to about 10% of a predetermined value. Table I shows the results of a series of runs aimed at evaluating reproducibility of sputtering rate. The greatest thickness which can be made is limited by the time which the machine can be run uninterrupted. Films 100 μ thick have been made.

The temperature of deposition is usually from 550-850°C, depending on the purpose of the run. Although epitaxial growth is probably possible as low as 400°C, no work was carried out below 500°C.

APPEARANCE OF FILMS

The films look bright and metallic to the naked eye. A slight haze is sometimes seen, especially on films formed at a lower temperature. This is insoluble in HF and is caused by the scattering of light from a surface texture.

Only films on {111} surfaces have been investigated. Under the microscope these films usually show fields of oriented triangular figures, or lines at 60 and 120°. The most striking thing about these triangular figures is their uniform size (of the order of film thickness) and common orientation. The figures are of several types whose relative abundance depends on the temperature of formation.

Low temperatures (~500°C) give faint triangles or 60-120° linear markings. Somewhat higher temperatures give simple pyramids or Y-shaped markings in addition. These features are shown, perhaps indistinctly, in Figure 2a. They are best visible under dark field at high magnification, and are easily overlooked in films less than 3 μ thick. Figure 2b is an electron micrograph by

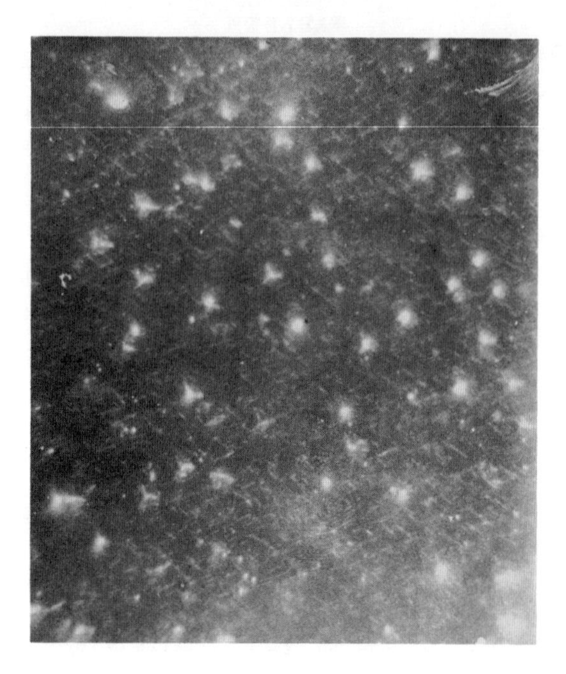

Fig. 2a. Micrograph of as-grown surface showing simple pyramids (bright) and faint triangles. 285×.

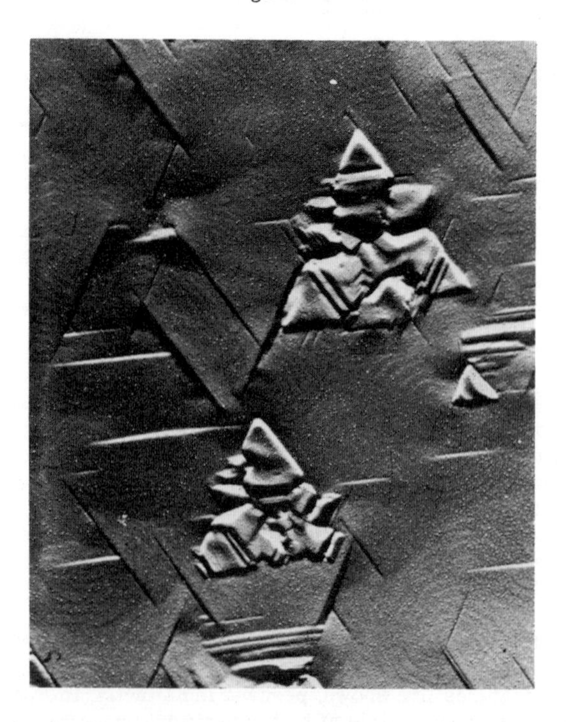

Fig. 2b. Electron micrograph showing faint triangles enlarged (shingle structure) and a feature suggesting a tripyramid, as well as several spiral structures. 2600×.

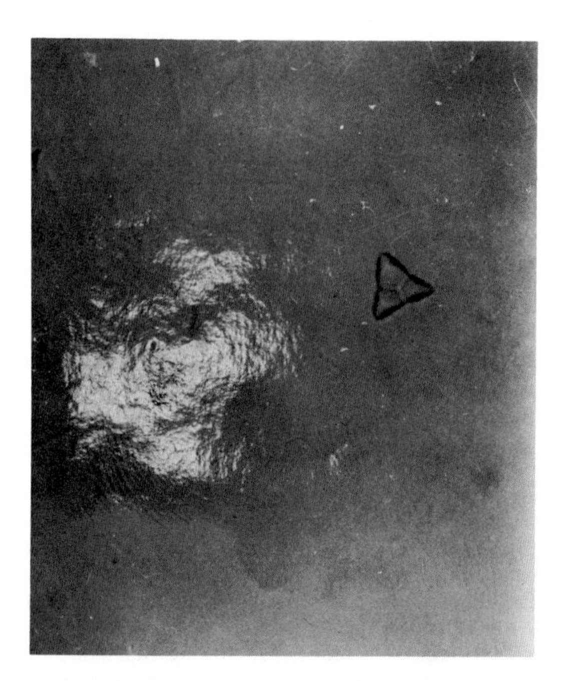

Fig. 3a. Micrograph showing tripyramid on sputtered surface. 285×.

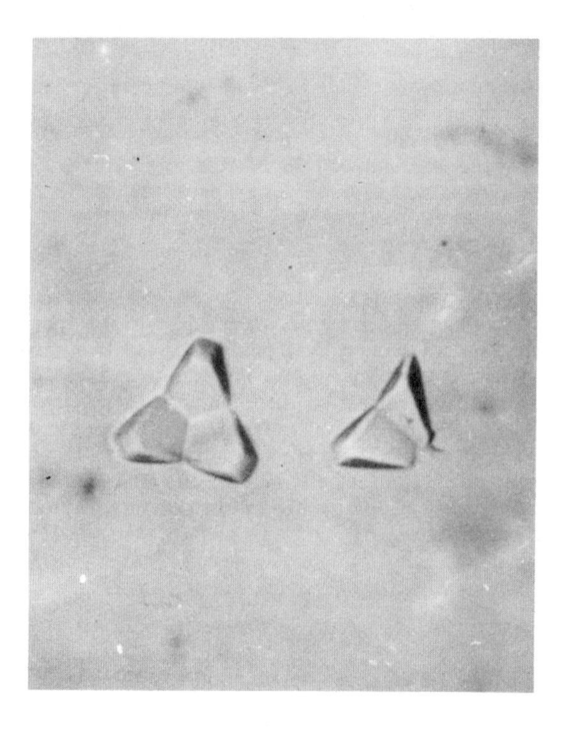

Fig. 3b. Micrograph by Blakeslee showing tripyramid on vapor-deposited epitaxial silicon.

TABLE II
Representative Series of Films on n-Type Substrates

Slice	Deposition temp., °C	Film thickness, μ	Film resistivity, ohm cm	Slice resistivity, ohm cm
FR-60-1	840	4.5	0.071	0.37 n
FR-60-2	840	6.85	0.244	0.37 n
FR-66-1	780	5.65	0.075	0.37 n
FR-66-2	780	5.28	0.073	0.37 n
FR-68	750	6.27	0.078	0.37 n
FR-74	780	0.55	0.031	0.5 n
FR-76	790	5.07	0.089	0.5 n
FR-78	530	5.2	0.018	0.5 n
FR-80-2	820	6.42	0.13	0.5 n

Calbick on a surface showing these features. Here the 60-120° linear markings are plainly brought out, and a new feature is shown in addition, looking like an attempt at a 6-pointed star. This may be related to the feature of Figure 3a, called a tripyramid, which sometimes appears with 3 extra points in the middle of its sides. The tripyramid is seen only on samples made above ~650°C and also occurs on some silicon films made by the $SiCl_4$-H_2 reduction process (6). Figure 3b shows one from a silicon film made by Blakeslee of this laboratory.

In Figure 2b indistinct spiral formations in the lower right corner, and elsewhere may be noted. This is not an artifact of the replica process, but we cannot positively identify it as a true growth spiral.

The exact significance of these markings is not known. However, they do justify two conclusions:

1. Their triangularity partakes of the 3-fold symmetry of the {111} plane, and by their common orientation all over the surface shows that the film in which they appear is epitaxial.

2. The occurrence of identical tripyramids in sputtered germanium layers and chemically deposited silicon layers probably points to some common mechanism in the growth process, in spite of their different composition and different conditions of growth.

Riesz and Sharp report (3) that their evaporated films on {111} substrates are not truly monocrystalline but "may consist of a few large single-crystal areas not having the same orientation with respect to the substrate." Although evidence of twinning was found by Read (5) by electron diffraction in some early sputtered films, untwinned {111} films can be made. This was shown by the following test: a back-reflection Laue photograph was made of a sputtered {111} film 9 μ thick. In this thickness film and substrate make roughly equal contributions to spot density. The photograph was indistinguishable from one taken of an unfilmed {111} slice. Twinning in the illuminated area would have shown up as an extra set of spots while misfits or misorientations would have caused split or broadened spots.

ELECTRICAL PROPERTIES

The films are always p-type, even those sputtered from heavily doped n-type cathodes. The nature of the acceptor involved is at present not known. This invariable p-typeness is a characteristic shared with the evaporated films of Riesz and Sharp and others (3). It can be converted only by a prolonged treatment with phosphorus vapor in a sealed-ampule diffusion.

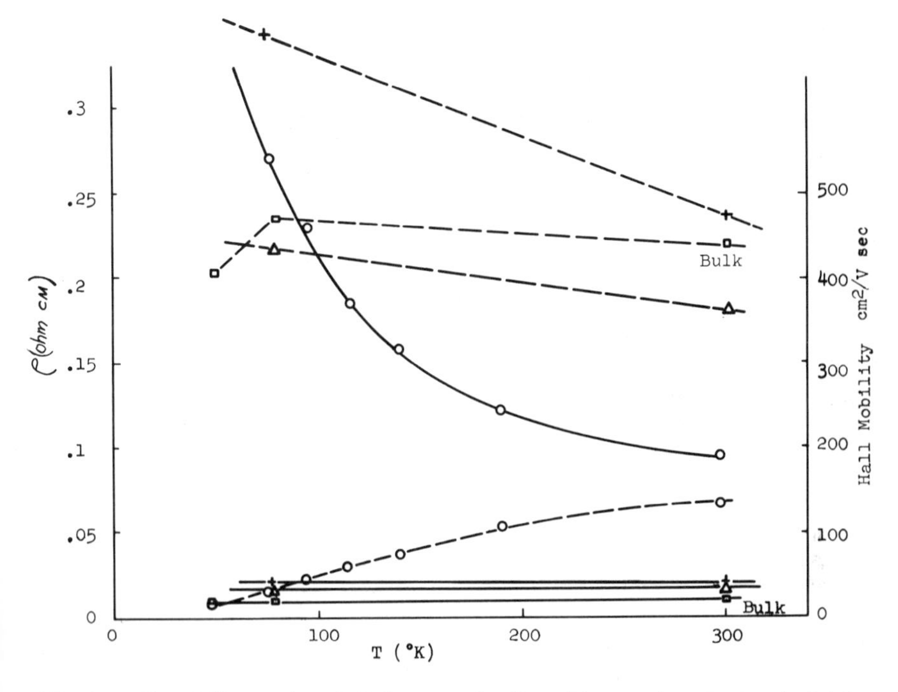

Fig. 4. Hall mobility—and resistivity—as a function of temperature for two isolated sputtered films and a bulk sample.

The resistivity, as measured by 4 point probes in films on n-type substrates is of the order of 0.05 ohm cm (within a factor of 10), indicating that the resistivity is obviously not under experimental control at this time. Table II gives the resistivity, along with other experimental conditions, for a representative series of films on n-type substrates.

Attempts to find occluded argon in the films by x-ray fluorescence have given negative results. This means there cannot be more than one part of argon in several hundred of germanium. Although smaller concentrations of argon (or, more precisely, structural defects associated with interstitial argon) may still be responsible, one cannot well blame argon for Riesz' p-type films, deposited in high vacuum.

Isolated films have been prepared by etching away the substrates. These films were necessarily very thick (several mils) to withstand handling after isolation. With these films, Hall effect studies could be carried out without interference from the substrate. These studies, which were done by L. P. Adda of this laboratory, are summarized in the curves of Figure 4. The carrier density, which is in the range of 5×10^{17} to 1×10^{18} cm^{-3}, puts the samples into the degenerate range, and the resistivity and mobility change little from 45°K to room temperature. The exception is the sample designated 200-1, which behaves anomalously, having higher resistivity and lower mobility at low temperatures. The significance of this behavior, which was quite reproducible in repeated measurements on this sample, is not known. It may result from inhomogeneity in the sample (e.g., a stratum of nondegenerate material) or it may be related to impurity band conduction, only associated with a deep-lying level.

In any case, the measurements do not resolve the problem of identifying the acceptors responsible for the invariable p-typeness of the sputtered layers.

Fig. 5. Angle-lapped and stained germanium layer, showing fringes in sodium light. Beveled surface is below, p-type layer shows lighter.

It may certainly be said, however, that the sputtered material behaves essentially like melt-grown material of the same resistivity (assuming that the one anomalous sample is due to some spurious effect).

The acceptors are probably not any usual chemical impurity. It is hard to see what impurity could be present in such large amounts as to completely swamp intentional dopants from the source material. The usual contaminant, copper, has too low a solubility limit. Single vacancies are too mobile to persist in films formed at 600-800°C; they would immediately diffuse into the substrate or surface. Besides the equilibrium concentration is too low to account for the observed resistivity (7). One is driven to postulating as the acceptor some kind of structural imperfection having too high an activation energy to anneal out and too low a diffusivity to move away during deposition.

JUNCTIONS

If the p-type film is deposited on an n-type substrate, a p-n junction will be formed at the interface. These junctions have been demonstrated by angle lapping and staining (Fig. 5). Two staining techniques have been used successfully. One involves applying a reverse bias to the junction while applying a nitric-hydrofluoric etch. The junction is defined by a discontinuity in the etching action. The other method is the copper plating of the junction by a $CuSO_4$-HF solution in a strong light. Figure 5 is an example of the first type of stain. Minor irregularities in the junction line have no significance in germanium junction stain. The true junction is actually quite planar, and coincides with the original substrate surface. Junction depths, as read from interference fringes, agree with the average thickness of the layers as determined from weight gains.

These junctions have acceptable rectifying properties which may be demonstrated in two ways.

1. Mesas are etched on the film side by masking with wax and etching through the film between the wax dots. Then contact is made to the mesa top

with an indium-doped gold wire and to the n-type surface between mesas with an antimony-doped gold wire. After slight bonding or forming, to remove point-contact effects, the rectification characteristics of the junction can be seen on a curve tracer.

2. By an essentially similar process, diodes were made and encapsulated in standard cans for permanence. The mesa and base contacts are both made by alloying. Five diodes, of very similar characteristics, were made from the same slice. A typical characteristic is shown in Figure 6. Reverse breakdown is about 26 v. The resistivity of this film is of the order of 0.005 ohm cm and its thickness about 5 μ. Substrate resistivity is 0.37 ohm cm. Note, therefore, that the substrate is the high resistivity side of the junction and the breakdown is really characteristic of the substrate.

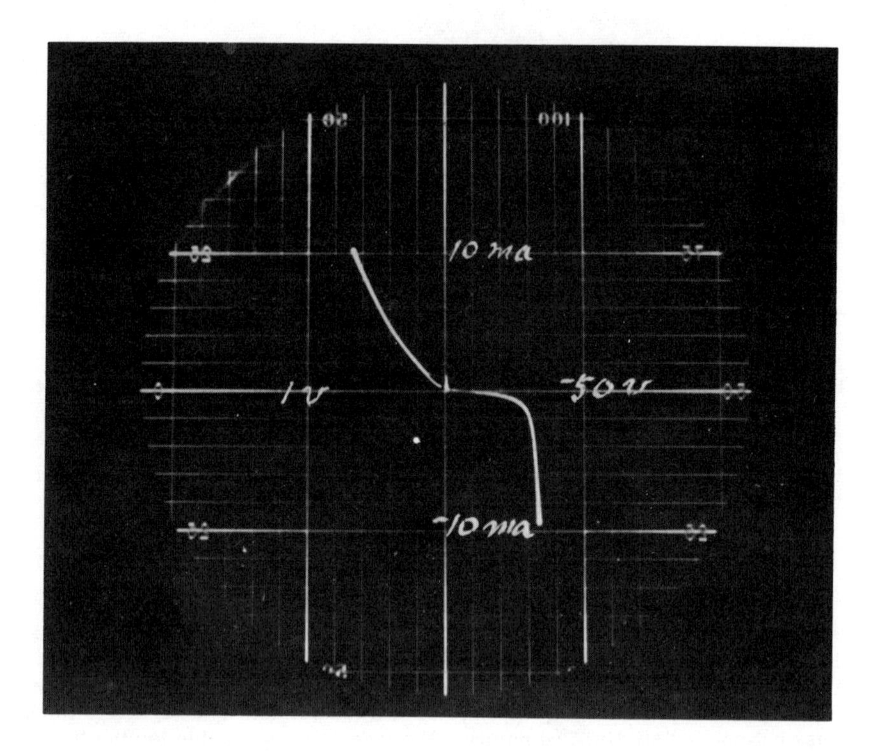

Fig. 6. Curve of encapsulated diode, 27 v breakdown.

These junctions are of the "semideposited" type, which other workers have found to have very inferior properties. These junctions, formed at the film-substrate interface, have generally showed very soft breakdown and high forward resistance. Examples can be seen in ref. 1, p. 288. Christensen and Kuper have had similar experiences (private communication). The poor properties of semideposited junctions have been attributed by the IBM workers to an imperfect interfacial layer of high dislocation count. In an effort to find such a layer we have angle lapped across the interface and etched briefly with silver etch (Fig. 7). Notice that the film is much more imperfect than the substrate but there is no evidence of a third layer, at the interface between them,

Fig. 7. Angle-lapped section etched with silver etch, showing substrate (small pit density) and layer (many pits).

which is less perfect than either. The heavy pitting by silver etch supports the hypothesis of a structural defect as the acceptor.

CONCLUSIONS

It has been shown that epitaxial germanium layers can be made by cathodic sputtering. These layers are analogous in electrical properties to epitaxial layers made by thermal evaporation; i.e., they are p-type and of low resistivity, and probably quite imperfect on the atomic scale although monocrystalline. The deposition process is quite uniform and controllable with regard to thickness but not resistivity. The nature of the acceptor responsible for the invariable p-typeness is uncertain but may be structural. As an application, diodes were made and their characteristics measured. If control of resistivity could be obtained, the method may offer promise as a technique in the preparation of actual devices for microminiature or integrated circuits.

ACKNOWLEDGMENTS

Thanks are due Mr. Conrad Clark for his help in the production and evaluation of films, Mr. A. E. Blakeslee for his permission to use the photograph of Figure 3b, and Mr. L. P. Adda for the Hall effect measurements.

References

1. IBM J. Research, 4, (July 1960).
2. Christensen, H., and A. Kuper, unpublished.

3. Riesz, R. P., and L. V. Sharp, Report presented at IRE-AIEE Solid State Research Conference, Stanford, Calif., June 1961.

4. Kurov, Semiletov, and Pinsker, Kristallografiya, 2, 59 (1957) (in Russian). Translated in Soviet Phys. Crystallography, 2, 53.

5. Read, M. H., unpublished.

6. Light, T. B., this volume, p. 137.

7. Tweet, A. G., J. Appl. Phys., 30, 2002 (1959).

8. Ruth, R. P., J. C. Marinace, and W. C. Dunlap, J. Appl. Phys., 31, 995 (1960).

DISCUSSION

D. M. MATTOX (Sandia Corp.): Were your resistivity and Hall measurements made while the sample was under an inert atmosphere? In other words, were they ever exposed to an oxidizing atmosphere?

F. REIZMAN: They were exposed to air at some time in their history when they were handled and made into samples. I believe that they were under a nitrogen atmosphere while being measured in the Hall apparatus.

D. M. MATTOX: Well, the question boils down to how much oxygen contamination you have. Quite often, especially when you are evaporating it vacuums of 10^{-5} or even 10^{-7} the oxygen in your chamber will prevent the formation of an n-type layer when it is thin, and I was wondering whether you were taking any special precautions for this in your inert atmosphere.

F. REIZMAN: We tried to get a system as leak-tight as we could, but it was not one of the superclean systems. It was exhausted by an oil diffusion pump. When the argon leak was turned off we could get vacuums of 2.5×10^{-6} Torr. While in use the argon was being continuously pumped through and, I hope, diluting leaks. However, there certainly is a possibility that small amounts of oxygen had got in.

V. R. ERDELYI (Knapic Electro-Physics): It has been reported in the past that the sputtering of silicon was especially difficult because of surface oxides. In your apparatus you do not have a means of eliminating the oxides as easily as in the conventional pyrolytic apparatus. In other words, you do not have hydrogen, you have an argon atmosphere. How do you insure that you do not have surface oxide present?

F. REIZMAN: Well, remember, this is germanium, and the oxide is not so stable. It is fairly volatile at the temperatures we are using at the substrate. There are two places where the oxide could be encountered. The oxide which is on the cathode is removed in the first minute or so of sputtering and is caught on the lower side of the shield. This is of no interest. The oxide on the substrate, if it remained there, would probably prevent any epitaxial growth. It probably does not remain there; at least most of it does not, because of the high temperatures.

Also, some of the work was done with an admixture of hydrogen in the argon, just to make sure that the germanium dioxide was reduced. Actually it seemed to make no difference. It was probably being driven off anyway. If there was much of an oxide film sticking to the substrate, it would probably have prevented epitaxial growth, because you need nearest neighbor forces to order the layer as a continuation of the substrate structure.

V. R. ERDELYI: Do you think it is easier in the case of germanium than it is in the case of silicon?

H. BASSECHES: I think it is easier in the case of germanium. We did a little work on silicon sputtering and our rates were certainly a lot less with silicon than with germanium. This is one of the things that induced us to stay with the germanium system.

Microsegregation Phenomena in Semiconductor Crystals

J. W. FAUST, JR., H. F. JOHN, and S. O'HARA

Westinghouse Research Laboratories, Pittsburgh, Pennsylvania

Abstract

Segregation patterns in heavily doped semiconductors were observed by selective etching. The results are discussed in the light of existing theories of crystal growth.

INTRODUCTION

An important by-product of the dendritic crystal program has been the discovery of new microsegregation phenomena during growth of the dendrite. This segregation has been used to deduce crystal growth mechanism. The present work indicates that microsegregation is common to various modes of crystallization and that anodic etching may be employed to study impurity distribution segregation has been used to deduce crystal growth mechanisms (1-4). The present work indicates that microsegregation is common to various modes of crystallization and that anodic etching may be employed to study impurity distribution in semiconductors. It should be emphasized, however, that many of the observations described lack a conclusive explanation.

The microsegregation phenomena described here are important since as device fabrication becomes more sophisticated greater demands are made on the uniformity of the semiconducting materials. Microdevices, tunnel diodes, and certain magnetoresistance devices (5,6) are particularly sensitive to small resistivity variations within the crystal. Techniques for growing crystals have advanced to the point where many of the gross variations in resistivity and quality are no longer troublesome to the device fabricator. More detailed studies on crystal growth and segregation phenomena have made other types of inhomogeneities evident. Resistivity variations within the space of a few microns have been observed in a number of semiconductor crystals grown by various techniques, as described below. Such variations may, in fact, be a feature of all crystal growth. Although the effect of such microresistivity variations on devices have not been assessed completely, it appears likely that this type of inhomogeneity may prove important to the ultimate performance of certain devices.

GENERAL OBSERVATIONS ON SEGREGATION PHENOMENA

Macroscopic Segregation

Several varieties of inhomogeneous impurity distribution have been observed in Czochralski crystals and, to a lesser extent, in float-zone material. Certain types of these inhomogenities can be traced to variable or inadequate stirring at the interface, to radial fluctuations in growth velocity, to extremely non-planar growth interfaces, or to entrapment of impurities between growth projections (7-10). Recently it has been shown that the effective segregation coefficient of certain impurities in indium antimonide (11,12) and germanium (9) is higher for growth on a {111} facet than on other regions. Such crystallographically dependent segregation may lead to marked impurity concentrations in regions of the crystal where faceted growth occurred. Under certain conditions cellular growth may occur (13,14). Rejection of impurities by the growing cells gives rise to a mosaic of higher resistivity material surrounded by regions of lower resistivity.

All of these effects lead to macroscopic impurity variations in the crystal which are easily detectable by conventional measurement techniques.

Microscopic Segregation

The presence of microresistivity variations is a widespread phenomena in crystal growth. They have been revealed recently in the single crystals grown by John (15) from metal solutions using a modified Czochralski technique. Variable segregation in germanium layers regrown from indium and lead alloys have been reported by Tomono (16). Saratovkin (17) has described segregation arising in several types of inorganic salts grown from solutions containing various organic dyes (see Fig. 1). In all the inorganic crystals shown, more dye was incorporated in the re-entrant corner sites than at the edges.

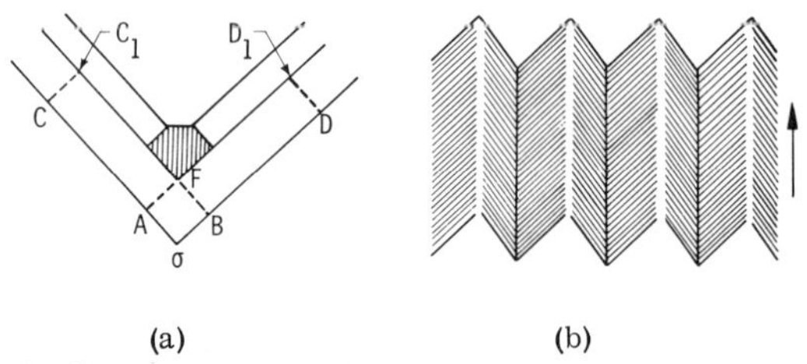

Fig. 1. Impurity distributions revealed by dyes in dendrites of inorganic salts after Saratovkin (27) . (a) How two adjacent growing faces forming a re-entrant angle trap impurities. (b) Distribution of impurities trapped by a growing crystal.

Studies of segregation in dendritic ribbons and in octahedral bodies grown from melts and metal solutions have revealed highly regular microscopic variations in resistivity with a repeat distance between 5 and 50 μ . These microresistivity effects and others described later are probably more closely related to effects described by Saratovkin (17) and Tomono (16), and possibly some of those described by Allred and Bates (18) than the macroeffects outlined earlier.

In its simplest form, segregation may be demonstrated by an octahedron and an idealized tip of a three-twin dendrite. In Figure 2 an octahedron is terminated a {100} surface, while the dendrite tip is a {211} surface. One notices in each, a series of traces, A, parallel to {111} planes and also traces, B, outlining the intersection of {111} planes. It will be shown that B traces are regions of low impurity concentration. These drawings also illustrate that the traces may be continuous around the section.

When cross sections of heavily doped dendrites are etched in selective etchants, segregation traces are revealed. Both A and B traces can be seen in Figure 3 for silicon. That these traces are indeed the result of impurity

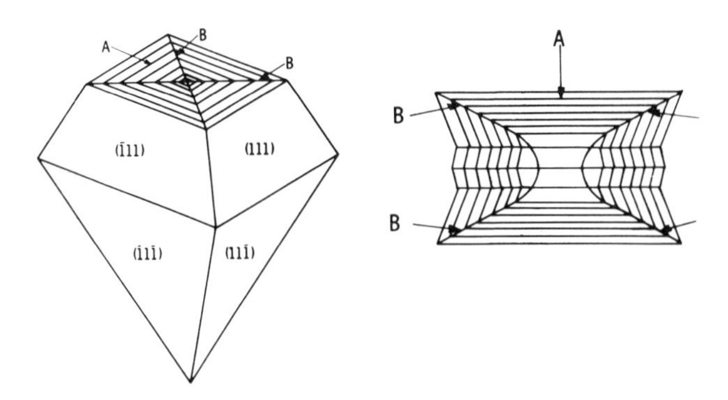

Fig. 2. Segregation traces in an octahedron and in an idealized tip of a three-twin dendrite.

segregation is evident in Figure 4 which shows adjacent samples from an n-type germanium before and after annealing at 800°C for 1 hr. Additional verification that impurities cause the A and B traces stems form the fact that the traces are not observed in undoped material. Although one cannot follow a single A trace completely around the cross section, one can follow numerous traces around several intersections in the unannealed specimen. The silicon dendrite cross section, Figure 3, shows not only the A and B traces but also the continuous curved A traces around the core. The sketch is included to aid interpretation of the cross section.

The basic pattern of A and B traces is found in crystals grown from various metal solutions as shown in Figure 5: silicon grown from a gallium solution;

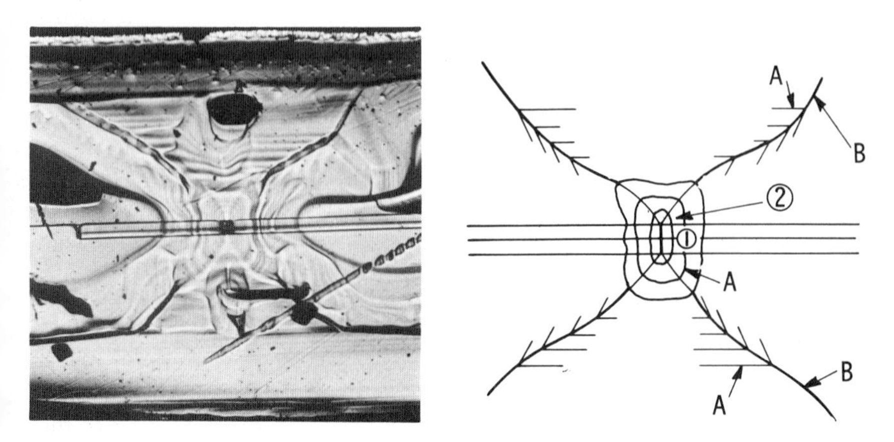

Fig. 3. Cross section of a highly-doped Si dendrite. Etched for 5 min in the Dash etch. 40×.

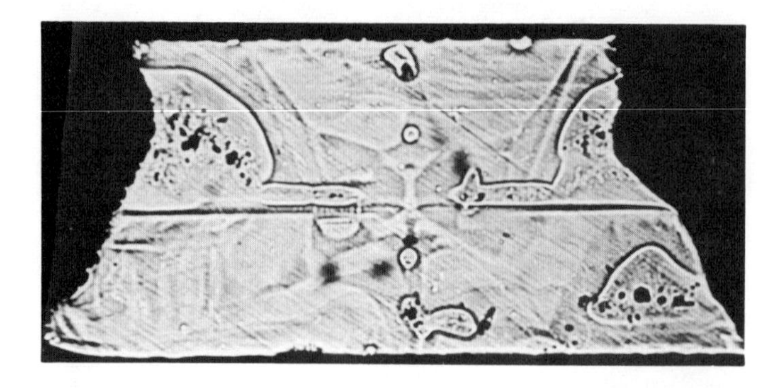

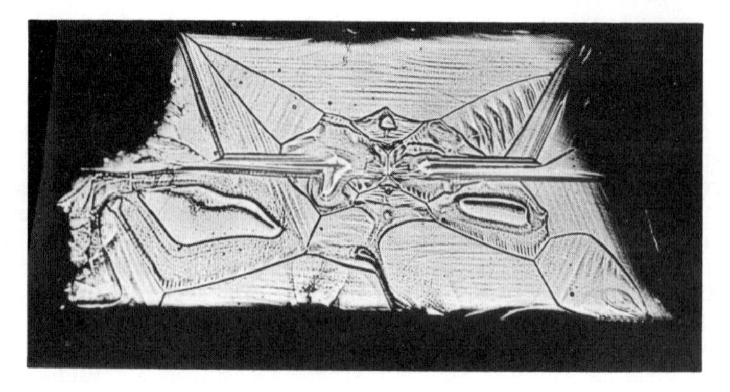

Fig. 4. Annealed and unannealed cross sections of highly As-doped Ge dendrites. Etched for 5 sec in CP-4. 25×.

silicon grown from silver plus 1% antimony solution; and indium antimonide grown from an indium plus 1% tellurium solution. It can be seen that these traces generally conform to the ideal crystallographic traces shown in Figure 2; however, the traces deviate noticeably from perfect crystallographic planes at places, particularly in the indium antimonide specimen. (Note that Figure 5b shows part of a {211} plane on an octahedron.) In the specimens shown in Figure 5, growth from solution is controlled primarily by diffusion of the solute to the growing interface, rather than by the dissipation of the heat of fusion which is the case in dendritic growth from the melt.

THE ANODIC COUPLE ETCHING MECHANISM

It is helpful to know what regions of an etched surface correspond to high and low impurity concentration. By combining selective etching and interference microscopy it has been possible to gain considerable insight into segregation phenomena. When a marked resistivity gradient exists in a semiconductor, the more anodic region will etch faster. Such an effect can be predicted from Turner's data on the variation of electrode potential with resistivity in germanium and silicon (19). The electrode potential becomes more anodic as the equilibrium hole concentration becomes greater, although the increase is not as great for high concentrations of p-type impurities. Thus, the region of highest n-type concentration will etch slowest. In a couple consisting of two p-type regions, the more heavily p-type region will etch faster.

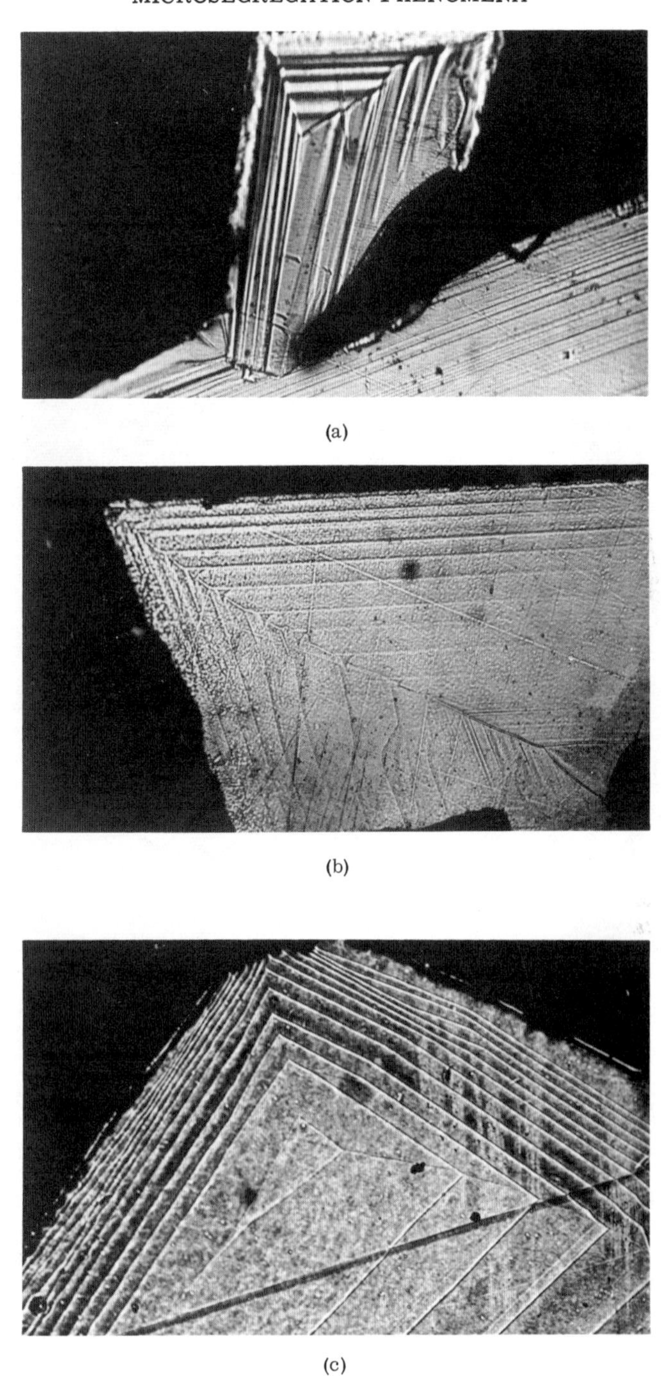

(a)

(b)

(c)

Fig. 5. Sections through crystals grown from metal solutions. a and b etched for 3 sec; c, for 1 sec: all in CP-4. (a) Si-Ga 400×. (b) Si-Ag + Sb 400×. (c) InSb-In + Te 160×.

Because of the smaller change in electrode potential with concentration, the difference in etch rate in the elements of a p-type couple may be less than in an n-type couple.

The effect of impurity concentration upon etch rate can be shown very nicely on etched {111} surfaces of dendrites. Figure 6 shows an interference fringe pattern of an etched, n-type germanium dendrite. For clarity call the intersection of the {1$\bar{1}$1} faces in an acute angle, "valleys" and those intersecting in an obtuse angle, "edges." Examples of B traces which result from valley segregation and edge segregation can be seen, as well as A traces. By moving the interference fringes it can be shown that the B traces outling valley growth are well-developed ridges. The B traces outling the development of the edges, in contrast, lie in grooves.

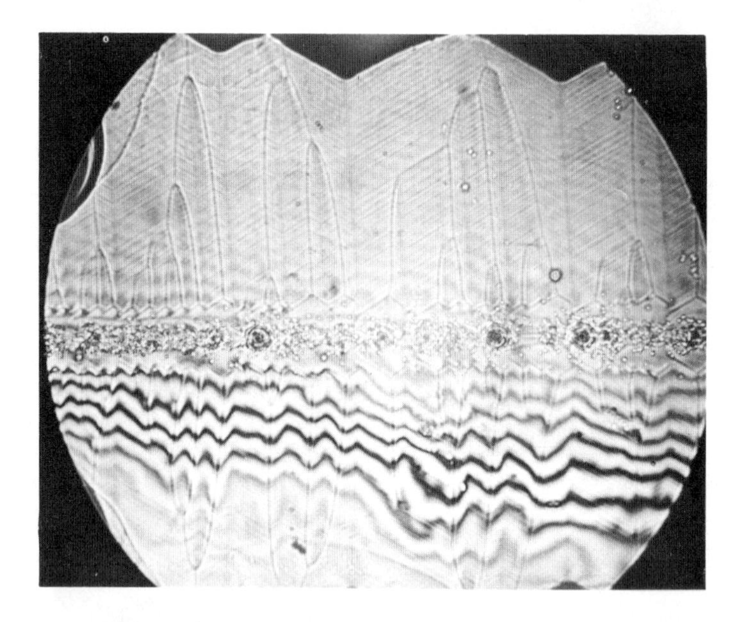

Fig. 6. Segregation traces on {111} face of an As-doped Ge ribbon: interference fringes on one side. Etched for 10 sec in CP-4. 35×.

The traces of the edges must be in a region of lower n-type impurity concentration, since more rapid etching occurs in the anodic regions. The traces of the valleys lie in regions of high donor concentration and thus etch less rapidly. These results are consistent with those reported by Smith (20) using an electroplating technique.

In contrast to this n-type specimen, a heavily doped p-type dendrite showed opposite effects. The B traces resulting from valley segregation gave rise to more rapid etching than the B traces outlining edge growth. It was also noticed that differential etching on heavily doped p-type dendrites is much less marked than in the equivalent n-type case. This is consistent with the proposed anodic couple etching mechanism.

The A traces in heavily doped n-type dendrites etch as grooves between wide plateaus and therefore are regions of low impurity concentration. Information is not yet available on the equivalent A traces in p-type dendrites, because of the less marked etching. It is also interesting to note the similarities in the segregation effects reported by Saratovkin (17) for inorganic salts and dyes (Fig. 1) to those in the dendritic ribbon shown in Figure 6.

DISCUSSION

The dominant feature of the A microsegregation traces, which is common to all the specimens and must be kept in mind in considering possible mechanisms, is the continuity of these traces from one facet system to another with accompanying changes in spacing. The matching of these segregation lines across an edge-type trace is so perfect that it can only be concluded that these traces are an outline of the whole interface at any one instant during growth for the given cross section. Evidence that the A traces outline the solid-liquid interface can be obtained from dendrite tips jerked out of the melt. These show that the A traces are always parallel to the growth surface. The traces are not always straight and may curve following the rounded interface where growth appears to be thermally limited. Figure 3 shows well-defined curved traces which outline the thermally limited core region. As noted previously, segregation traces can be seen in Figure 6 which outline the solid-liquid interface at any instant during growth. These traces also match perfectly across the edge and re-entrant corner traces.

Although we as yet do not have a comprehensive explanation to account for these microsegregation phenomena, some general comments can be made. First, the experimental evidence indicates that edge-type B traces lie in regions of lower impurity concentration. Conversely, valley-type B traces are regions of higher impurity concentration. This is consistent with the idea that the volume of melt into which impurities rejected by the growing crystal can diffuse is greater at an edge than at a surface or in a valley.

No single explanation seems to be entirely satisfactory for the A traces. Theories based on the adsorption of impurities have been considered. One such treatment is that discussed recently by Tainor and Bartlett (21). The principal objection to the adsorption theory is the continuous nature and uniform spacing of the A traces which are observed in dendrite sections. This implies that the event which caused the effect was occurring simultaneously over the whole interface. The existence of curved A traces precludes the explanation that they result from layer growth over facets. Furthermore, the tentative analaysis of A traces by means of anodic etching indicates that they are regions of lower impurity concentration which is the reverse of what would be required by an adsorption mechanism.

Pile-up of impurities in front of layered overgrowth is another possible mechanism. Although layer growth undoubtedly causes pile-up of the rejected impurities and hence diffusion gradients, it is difficult to account for many of the features of A traces on this basis alone. The principal objection is again the continuous nature of the segregation traces.

The nature of the A traces implies an event which occurs simultaneously over the whole interface. An attractive mechanism would be the disturbance of the diffusion front at the liquid-solid interface by vibrations in either the crystal or the melt. Variations in the pull velocity, however, between 0 and 10 in./min, or vibrations deliberately introduced into the crystal melt system, do not appear to produce any detectable changes in the microsegregation phenomena. Furthermore, there are discrepancies when the traces in solution grown bodies are considered. In solution grown materials the traces can be quite variable in spacing, depending on thermal conditions, with no apparent change in vibrational environment.

SUMMARY

In addition to the more familiar macrosegregation and microsegregation traces, new microsegregation traces have been found; in particular, A traces outlining growth fronts and B traces outlining the intersection of growth fronts. These traces were observed in crystals grown by a variety of methods. None of existing theories of segregation seems to account satisfactorily for all of the observed features of the A traces. The B traces can be accounted for qualitatively by simple diffusion considerations.

References

1. Faust, J. W., and H. F. John, J. Electrochem. Soc., 108, 855 (1961).
2. Faust, J. W., and H. F. John, J. Electrochem. Soc., 108, 860 (1961).
3. John, H. F., and J. W. Faust, in Metallurgy of Elemental and Compound Semiconductors, R. O. Grubel, ed. (Metallurgical Society Conferences, Vol. 12) Interscience, New York-London, 1961, p. 127.
4. O'Hara, S., Metallurgy of Elemental and Compound Semiconductors, R. O. Grubel, ed. (Metallurgical Society Conferences, Vol. 12) Interscience, New York-London, 1961, p. 149.
5. Weiss, H., J. Appl. Phys., 32, 2064 (1961)
6. Allred, W. P., and R. K. Willardson, Electrochem. Soc. Abstracts, 9, No. 1, 178 (1960).
7. Burton, J. A., and W. P. Slichter, Transistor Technology I, Van Nostrand, Princeton, New Jersey, 1958, ch. 5.
8. Slichter, W. P., and J. A. Burton, Transistor Technology I, Van Nostrand, Princeton, New Jersey, 1958, ch. 6.
9. J. A. M. Dikhoff, Solid State Electronics, 1, 202 (1960).
10. Camp, P. R., J. Appl. Phys., 25, 459 (1954).
11. Hulme, K. F., and J. B. Mullin, Phil. Mag., 4, 1286 (1959).
12. Mullin, J. B., Electrochem. Soc., Abstracts, 9, No. 1, p. 176 (1960). See also J. Phys. Chem. Solids, 17, 1 (1960).
13. Tiller, W. A., and J. W. Rutter, Can. J. Phys., 34, 96 (1956).
14. Ellis, S. G., J. Appl. Phys., 26, 1140 (1955).
15. John, H. F., J. Electrochem. Soc., 105, 741 (1958).
16. Tomono, M., J. Phys. Soc. Japan, 16, 711 (1961).
17. Saratovkin, D. D., Dendritic Crystallization, Consultants Bureau, New York, 1959, pp. 38-47 (English translation).
18. Allred, W. P., and R. T. Bates, J. Electrochem. Soc., 108, 258 (1961).
19. Turner, D. R., J. Electrochem. Soc., 107, 810 (1960).
20. Smith, R. C., J. Electrochem. Soc., 108, 238 (1961).
21. Trainor and Bartlett, Solid State Electronics, 2, 106 (1961).

DISCUSSION

P. J. SCHLICHTA (Jet Propulsion Lab.): Sodium chloride crystals, when grown in the <111> direction, invariably develop striations whose spacing is independent of rotation or pull rate. We have concluded that they are caused by a slip-strike interaction between the crystal and the meniscus of the molten salt; i.e., the meniscus is carried upward by the edges of the growing crystal until it snaps back to a lower level. Presumably, each time this happens it causes localized stirring of the molten salt which might well result in oscillations of the impurity segregation. Could such a mechanism be operative in your crystals?

S. O'HARA: I did mean to mention that Saratovkin has reported very similar segregation in inorganic salts, and that is what made us say that these features are common to many modes of growth.

Regarding the comment just made, the thickness of these traces is of the order of 5 μ, if you consider the growth velocity of a dendrite, the frequency of these segregation traces comes out to be of the order of, depending upon how you count them, either 80 or 40 per second. We know the meniscus slips; however, this occurs only two or three times a second. Therefore, I do not think these fine traces could be accounted for on the basis of the proposed mechanism.

P. S. FLINT (Fairchild Semiconductor): What are the impurities involved in segregation here? Is this the doping?

S. O'HARA: We doped these crystals deliberately. In the n-type case arsenic was used and in the p-type case gallium. We feel it was the dopant because we did not see the traces in the undoped material. All of the concentrations of the heavily doped material were of about 10^{19} cm^{-3}. But we have seen the traces for much lower doping levels than this. We have seen them at a level of about 10^{16} cm^{-3}, and even in the range of about 10^{15} cm^{-3} the traces were still evident on the flat faces of the dendrite.

P. S. FLINT: At what doping level do you not see these segregations?

S. O'HARA: I cannot give you an accurate answer. It would be somewhere in the region of 10^{15} cm^{-3}.

Fabrication of p-n Junction Structures by Means of Electron Beam Techniques

G. C. DELLA PERGOLA and S. A. ZEITMAN

Westinghouse Research Laboratories, Pittsburgh, Pennsylvania

Abstract

A large number of applications of the electron technique in the field of the semiconductor technology, as well as in many other fields, have been proposed in the past few years [W. Shockley, U.S. Pat. 2, 816,847 (Dec. 1957); M. M. Perogini and N. Lindgren, Electronics, 77, (1960)] In this report the status of an experimental investigation on the use of electron beams to fabricate alloyed p-n junctions will be described.

INTRODUCTION

If a metallic film is deposited, by vacuum evaporation or by plating, onto the surface of a semiconductor wafer, an electron beam can be used to heat both the metal and the substrate locally. The metal can therefore be melted and alloyed to the semiconductor in a small predetermined region. If the metal is properly selected, a p-n junction will result as a consequence of this process.

Electron beam techniques would be expected to have several advantages:

1. These techniques should permit the fabrication of p-n structures having very small sizes and complicated shapes with a high degree of precision and cleanliness.
2. The process is compatible with automatization.
3. The heating can be confined to very small regions; therefore the alloying process can be performed without affecting the other regions of the substrate.
4. The temperature of the bombarded regions can be very high during the electron bombardment, without an excessive increase in the average temperature of the substrate, thus metals that are difficult, and sometimes impossible, to alloy by conventional techniques may be used.

The results obtained showed that the electron beam alloying technique can be successfully used to produce p-n junctions in semiconductors. The technique is, however, fairly complex, because of the large number of parameters involved, each requiring very careful control. Moreover, several difficult problems must be solved to make this technique fully suitable for the production of solid state devices.

DESCRIPTION OF THE EXPERIMENTS

Sample Preparation

Electron beam alloying experiments have been performed on both germanium and silicon samples. The germanium samples were n-type dendrites, having a resistivity of about 2 ohm cm. Indium and aluminum have been used as coating materials, but the best results have been obtained with indium, due to a better adhesion of the film to the semiconductor substrate. The indium film has been generally deposited by vacuum evaporation. A few samples, however, have been electroplated with no appreciable difference in the results. The silicon samples were small bars (approximately $2.5 \times 0.2 \times 0.05$ cm) cut from conventionally grown, n-type, high resistivity (50–100 ohm cm) ingots; evaporated aluminum has been used as a coating material.

All the samples were etched immediately before the introduction in the vacuum chamber for the metal evaporation. Cleaning by ionic bombardment with argon at low pressure was always used before the evaporation to improve the conditions of the surface of the samples. A metallic shield was also used to intercept the contaminants present at the surface of the source at the beginning of the evaporation. The thickness of the film ranged from a fraction of a micron up to several microns. The thickness was controlled by evaporating with a known distance between the source and the sample, a controlled amount of metal. The film thickness was tested by optical interference methods.

Electron Beam Alloying*

The best procedure should be to perform the alloying immediately after the evaporation in the same vacuum chamber to avoid any further contamination of the samples. Such a procedure was not possible during this work. However, to avoid unnecessary handling of the samples between the metal deposition and electron bombardment, which may easily produce some damage to the surface of the samples, the same holder was used for the two operations. The holder was designed in such a way that all the samples were at the same level, as required by the small depth of focus of the electron beam.

The alloying can be made with a continuous or with a pulsed electron beam. In most cases, as it will be discussed later, short heating times are preferable. Therefore, when the beam has to be scanned over the surface of the sample to obtain the desired shape of the alloyed region, one of the two following methods must be used.

1. The beam is deflected electronically. In this case the scanning speed can be very high, and even if a continuous beam is used the heating time at each point of the bombarded surface can be very short.
2. The sample is moved during the electron bombardment by means of simple mechanical translations. In this case the scanning speed cannot be very high. Therefore, to obtain short heating times, a beam of short pulses with a suitable frequency must be used. Under these conditions the alloyed region will be made by a succession of partially overlapping pulses, the degree of overlapping depending on the size of the beam, the frequency of the pulses, and the translation speed.

The pulsed beam method has been selected during the present investigation mainly because of practical considerations, since any desired shape of the junctions is very easily obtained by means of simple mechanical translations of the samples.

The results of the electron beam alloying are strongly dependent on the characteristics of the beam. They are also affected, however, by several other factors such as: the degree of adhesion between the metal film and the semiconductor substrate, which depends on the experimental conditions during

---

*The electron beam alloying was performed with a Zeiss electron beam milling machine (1–3).

the vacuum evaporation (or the electroplating); the thermal conductivity and the thickness of both the metal film and the semiconductor and the presence of a suitable heat sink, which determine the sample capability of dissipating the power received from the electron beam quickly enough to minimize an undesirable rise in the average temperature during a prolonged bombardment.

During the present work, overheating of the entire sample was observed only in a very few cases when high-power high-frequency beams were used. The metallic holder was a satisfactory heat sink in most cases.

DISCUSSION OF THE EFFECTS OF THE ELECTRON BOMBARDMENT

General Considerations

When an electron beam is used to locally alloy a metallic film to a semiconductor substrate, a large variety of results can be obtained depending on the experimental conditions.

Two basically different cases are possible:

1. The power supplied by the electron beam can raise the temperature in the bombarded region up to a value which is higher than the melting point of the metal, or of the metal-semiconductor eutectic, but lower than the melting point of the semiconductor. In this case the junction produced is, in practice, very similar to a conventional alloy junction the main difference being attributable to the presence of large temperature gradients during the alloying process. Therefore in this case the term "electron beam alloyed junction" seems appropriate.*
2. The power supplied by the beam is so large as to locally melt the semiconductor. In this case the junction produced is actually one of the so-called "melt-back," "remelt," or "melt-quench" junctions with the difference that one of the impurities can be added from an external phase ("remelt with additions") instead of being initially present in the parent crystal (4,5). Therefore in this case a better term should be "electron beam remelt junction."

Effects of the Heating Time

In both the above-mentioned cases the junction is generally made with a very short thermal cycle. When a pulsed beam is used, the heating time is determined by the length of the pulses. The cooling rate is, on the other hand, dependent on several factors: the thermal conductivity and the thickness of the metallic film and of the semiconductor, the presence of a heat sink, the size of the heated region, the pulse frequency, and the scanning speed. With a continuous beam, the heating time is given by the ratio between the beam diameter and the scanning speed, the cooling rate being primarily dependent upon the sample geometry.

When a pulsed beam is used, the effects of the electron bombardment are strongly dependent on the pulse length. Figures 1 and 2 show the effects of changing only the pulse length during the electron bombardment for aluminum on silicon. The pulse repetition rate and the scanning speed were kept constant. It is clear that for given values of the accelerating voltage, current density, and beam diameter if the pulse length is shorter than the time required to attain thermal equilibrium within the heated region, the width, the depth, and the maximum temperature of such a region can be increased by increasing the pulse length.

---

*Typical examples of electron beam "alloyed" junctions are shown in the photomicrographs of Figures 3 and 4.

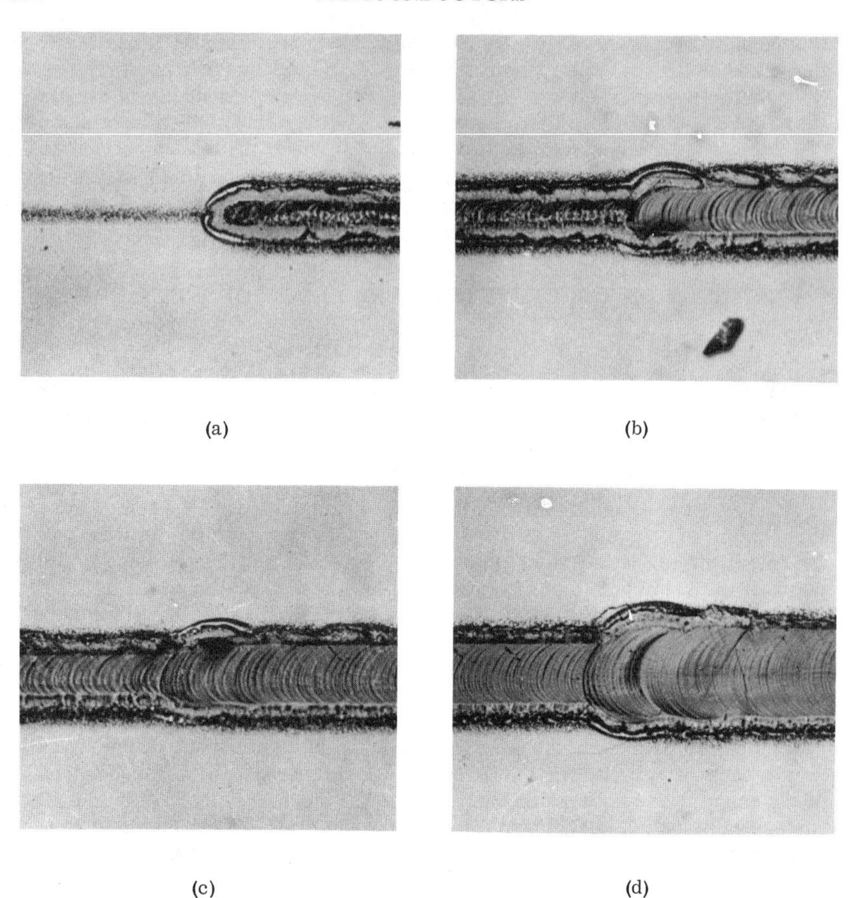

Fig. 1. Effects of changing the pulse width in steps during the electron bombardment on aluminum coated silicon. The sequence of the pulse width is as follows (in μ sec):

	Left	Right
(a)	2	8
(b)	8	13
(c)	13	18
(d)	18	32

From the sample shown in Figure 1, the curves of Figure 2 have been derived by measuring the widths of the regions where the aluminum-silicon eutectic has been melted (upper curve) and where the silicon has been melted (lower curve) as a function of the pulse length. A similar dependence can be expected for the depth of the melted region on the pulse length, but an accurate measurement is much more difficult. By extrapolating the two curves, it can be observed that, with the particular experimental conditions under which the sample has been bombarded, a minimum pulse length of about 5 μsec is required to melt the silicon while approximately 1.5 μsec is the minimum pulse width to make an alloyed junction. Similar considerations, of course, can be applied to the ratio of the beam diameter to the scanning speed in the case of a continuous beam.

The above considerations suggest the possibility that a satisfactory control of the process can be obtained by means of a fine regulation of the pulse length

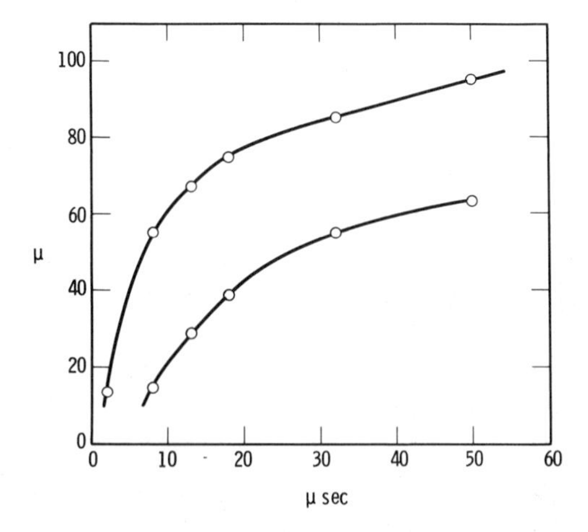

Fig. 2. Experimental curves showing the widths of the regions where the aluminum has been melted (upper curve) and where the silicon has been melted (lower curve) plotted against the width of the electron beam pulses (see Fig. 1).

(for pulsed beams) or of the scanning speed (for continuous beams). In particular it seems possible, by properly adjusting the pulse length (or the scanning speed), to control the junction depth. This is of particular importance in the case of the remelt junctions (case 2, in the preceding section); i.e., when even the semiconductor is melted, since in the case of the alloyed junctions (case 1, preceding section) the alloying depth is mainly determined by a dissolution process just as any conventional alloy junction, and is therefore less directly dependent on the heating time. In general, very short heating times (i.e., shorter than the time required to establish thermal equilibrium in the bombarded region) are needed to take full advantage of the possibility offered by the electron beam technique of producing very shallow junctions. The temperature and the width of the alloyed region, although not completely independent of the heating time, can be varied independently by adjusting the other parameters of the beam; i.e., the accelerating voltage, the current, and the beam diameter.

Properties of the Regrown Regions

Considerable additional work is needed to find the optimum conditions under which the electron beam alloying process must be performed to obtain a defect-free regrowth in the bombarded regions. A very important and undesirable feature of the process is that the crystal near the alloyed regions is often under strain as has been demonstrated by early experiments of Wells (6). Therefore, after the alloying, cracks may appear in the neighborhood of these regions. This problem is mainly important in the case of the remelt junctions, since in the case of the alloyed junctions it does not seem difficult to obtain a reasonably good alloy.

The preliminary results, however, show that it is not impossible to obtain a good quality regrowth even in the case of the remelt junctions, particularly when the alloying depth is kept very small (i.e., of the order of a few microns or less). The example of Figure 1, for instance, shows that with a large pulse length a large alloying depth is obtained where cracks are visible even before etching. With shorter pulses, however, smaller alloying depths are obtained and no cracks occur.

EXPERIMENTAL RESULTS

Germanium

The junctions produced by the electron bombardment on indium coated germanium samples were small strips (Figs. 3 and 4) or relatively large area junctions obtained by several partially overlapping bombardments with a de-

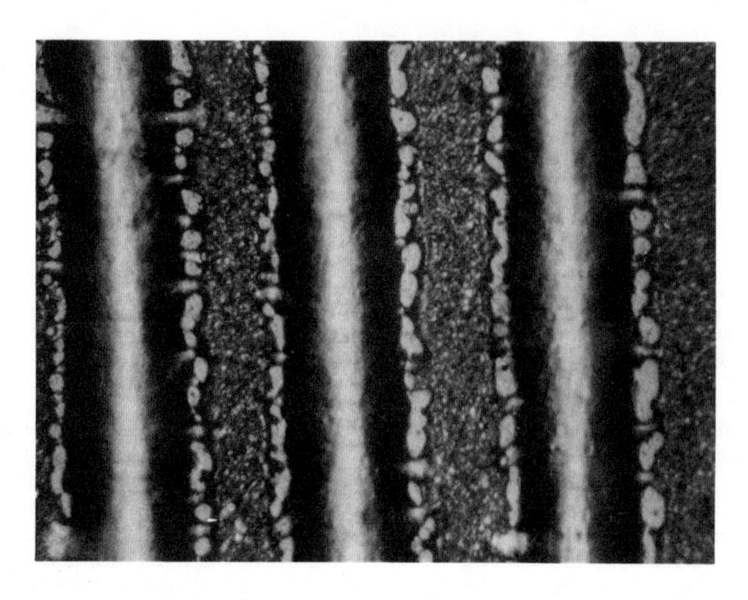

Fig. 3. Electron beam alloyed strips in indium coated germanium thickness of the film about 5 μ. 200×.

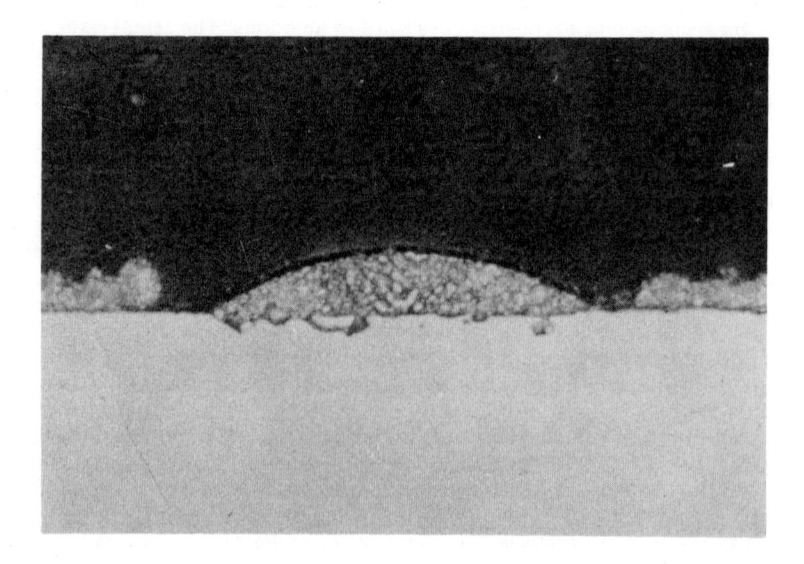

Fig. 4. Cross section of an electron beam alloyed strip, similar to those shown in Fig. 3. 500×.

Fig. 5. Wide area electron beam alloyed region (indium on germanium). 50×.

focused beam (Fig. 5). In both cases they were alloyed junctions, the semiconductor not having been melted. The wide area junctions have been tested electrically in a very straightforward way by etching small mesas in the alloyed regions and measuring the I-V characteristics by means of pressure contacts applied onto the indium-coated surface (Fig. 6). The germanium electron beam alloyed junctions have been found, in general, comparable in reverse breakdown voltage and leakage current to junctions produced by standard alloy techniques. The reverse characteristic of one sample is shown in Figure 7. The results show that the properties of the regrown region, despite the very fast thermal cycle and the high thermal gradients, are normal.

Silicon

A more detailed investigation of the electrical properties of the junctions produced by the electron bombardment has been made on the aluminum-coated silicon samples. In this case the junctions were generally long strips, 0.04 to 0.08 mm wide, parallel to the axis of the samples, and were usually of the remelt type, the silicon substrate having been melted by the beam. To perform the electrical measurements after the electron bombardment the aluminum film was completely removed by etching in HCl.

The problem of making the ohmic contacts to the p-type strips was temporarily solved by alloying onto the surface of the sample a few aluminum dots overlapping the strips. These contacts are rectifying to the bulk n-type silicon and ohmic to the p-type regions. They were necessary in order to eliminate any possible rectification effects which might be produced in attempts to make small probe test contacts to the p-type strips. The ohmic contact to the bulk was made by soldering nickel wires or ribbons with a gold-antimony alloy. The electrical tests were then made by using pressure contacts to the aluminum dots. To avoid the use of such pressure contacts, which imposes several limitations, and to allow a better handling of the specimens, aluminum wires approximately 0.5 mm in diameter were alloyed onto some samples, instead of

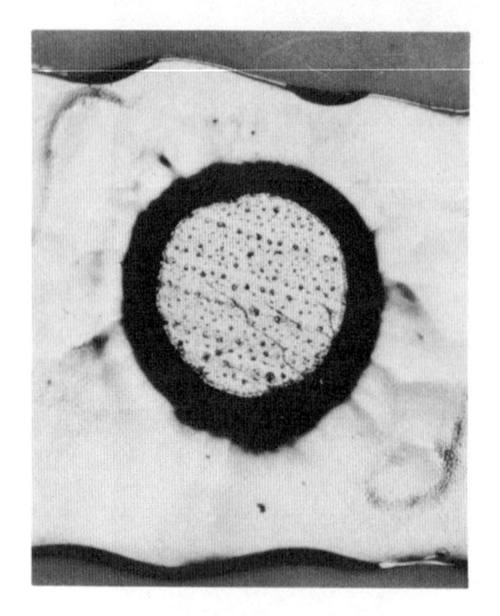

Fig. 6. Mesa etched on a wide area electron beam alloyed region. The original shape of the alloyed region (approximately rectangular) is still visible. 50×.

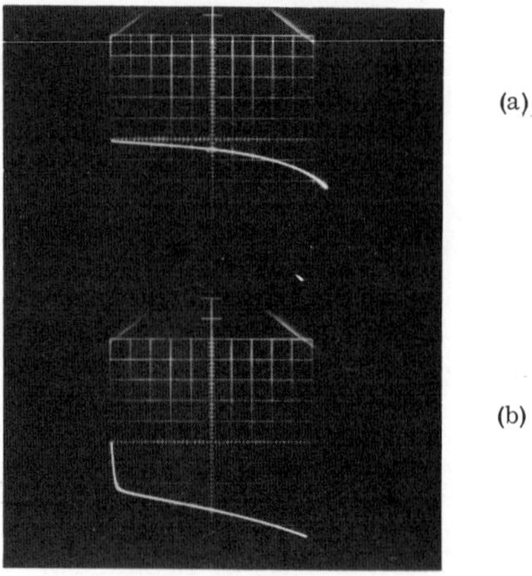

(a)

(b)

Fig. 7. Reverse electrical characteristics of the junction shown in Fig. 6. (a) vertical, 0.5 ma/div; horizontal, 10 v/div. (b) vertical, 10 μa/div; horizontal, 0.5 v/div.

the dots. Nickel wires were then soldered to the aluminum wires by means of a thermosetting conductive cement (Fig. 8).

The electron beam alloyed silicon junctions exhibit useful rectifying properties although definitely inferior to the properties of standard alloyed junctions. It must be noted, however, that the characteristics measured were actually

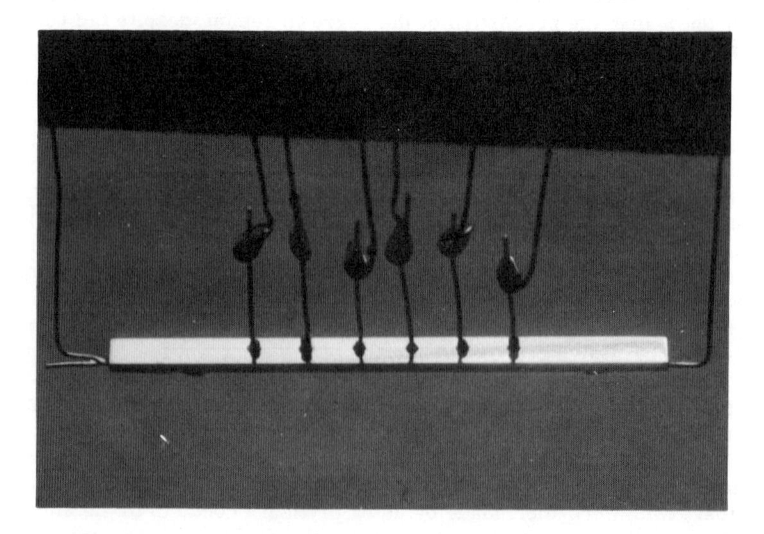

Fig. 8. Assembled silicon sample as used for the electrical evaluation of the junctions made by electron bombardment. The p-type strip is visible at the base of the aluminum wires.

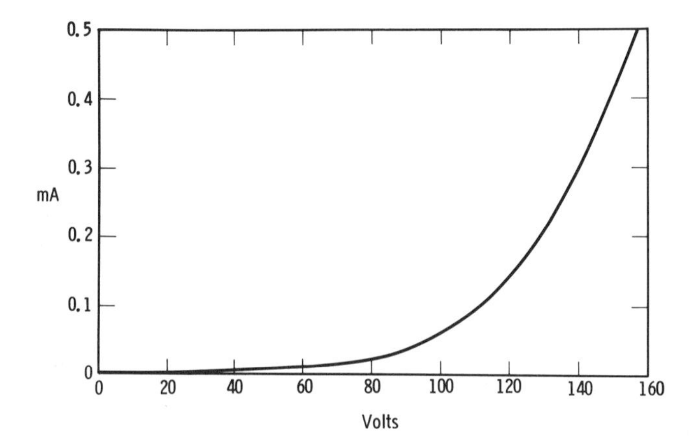

Fig. 9. Reverse electrical characteristics of a p-n junction made by electron bombarding an aluminum-coated n-type silicon sample.

those of a relatively complex system consisting of the electron beam alloyed junction (about 2 cm long and 0.04 or 0.08 mm wide) and of two or more aluminum-alloyed dots. Moreover, the samples were only very lightly etched, or not etched at all, after the application of the contacts. If a satisfactory etching procedure can be found which will allow good cleaning without removing a large amount of the regrown silicon, some substantial improvement may be expected.

The I-V reverse characteristic of one sample prepared on 100 ohm cm silicon is shown in Figure 9. The currents at low voltage (<10) were usually less than 10 μa. The breakdown voltages, however, were always lower than those expected from the resistivity of the n-type bulk material. The junctions tested had usually a "soft" breakdown; nevertheless, in most cases, they were able to withstand a reverse voltage of 120-150 v with currents of the order of 1 ma or less. The forward characteristics of this structure have no significance because the alloyed contact area is larger than the area of the electron beam alloyed strip.

In the tested specimens, the presence of the junctions was also confirmed by applying a bias between two aluminum contacts connected to the same p-type strip, with a suitable bias applied to the base contact in order to insure that the whole junction was reverse biased. The I-V characteristic measured with such arrangement was, as expected, entirely ohmic. It is possible in this way to measure the resistance per unit length of the strips obtaining some information about the resistivity of the p-type regions. The measured resistances were found to be relatively high ranging from a few hundred to several thousand ohm/cm.

From the value of the resistance per unit length of the strips, it is in theory possible to deduce the resistivity of the regrown regions provided that the distribution of the aluminum in these regions is uniform and that the area of the cross section of the p-type strip is known. From a very rough calculation, it may be deduced that the range of resistivity in the regrown regions should be between 0.1 and 0.01 ohm cm (corresponding to an aluminum concentration ranging approximately between 10^{18} and 10^{20} cm^{-3}). These values, however, are only indicative, the aluminum distribution in the regrown regions being almost certainly nonuniform and strongly dependent on the conditions during the electron bombardment.

CONCLUSIONS

1. The results described here show that it is possible to make p-n junctions by locally alloying a metallic film onto a semiconductor substrate.

2. The fact that functional p-n junctions 25 to 50 μ wide over a path of several millimeters have been fabricated gives some encouragement that p-n junction structures having very small sizes and complex shapes can be produced by this method. Therefore, electron beam techniques would seem to be of considerable interest in the field of the semiconductor technology, particularly for the production of molecular circuits.

3. Electron beam techniques, however, are very complex due to the large number of parameters that must be kept under control. To obtain good results with a reasonable reproducibility a large amount of work is still to be done.

4. The production of functional p-n junctions and the amount of control over geometry which has been obtained in these exploratory experiments, however, are such as to encourage further work to find the optimum conditions for the electron bombardment and fabrication of improved p-n junctions.

ACKNOWLEDGEMENTS

The authors wish to acknowledge the cooperation of H. Hosticka, B. Schaaf, and E. Clary, Jr., of the Westinghouse Headquarters Manufacturing Laboratory and of R. Booker and G. A. Kuntz of the Westinghouse Research Laboratory. Thanks are also due H. F. John and R. L. Longini for their helpful discussions and comments. This work was supported in part by the Wright Air Development Division of the Air Force.

References

1. Steigerwald, K. H., *Proc. Third Symposium on Electron Beam Technology*, Alloy Electronic Corp., Boston, Mass., March 1961, p. 269.
2. Opitz, W., *Proc. Second Symposium on Electron Beam Processes*, Alloy Electronic Corp., Boston, Mass., March 1960, p. 32.
3. Rider, M. W., *Proc. Second Symposium on Electron Beam Processes*, Alloy Electronic Corp., Boston, Mass., March 1960, p. 25.
4. Pfann, W. G., *J. Metals*, *6*, 296 (1954).
5. Pankove, J. I., *Transistor I*, RCA, Princeton, N. J., 1956, p. 82.
6. Wells, O. C., *Proc. Third Symposium on Electron Beam Technology*, Alloy Electronic Corp., Boston, Mass., March 1961, p. 291.

Diffusion and Precipitation of Carbon in Silicon

R. C. NEWMAN and J. WAKEFIELD

Associated Electrical Industries, Aldermaston, Berkshire, England

Abstract

Radioactive C^{14} has been diffused into silicon at temperatures in the range 1050 to 1400°C, and the diffusion coefficient found to be

$$D = 1.9 \exp \left\{ \frac{-3.1 \pm 0.2}{kT} \right\} \text{ cm}^2/\text{sec}$$

The surface concentration increases with increasing temperature up to a value of about 10^{18} atoms cm^{-3} at 1400°C. Chemical analysis of some pulled silicon single crystals has shown that the carbon content may be as high as 10^{18} to 10^{19} atoms cm^{-3}. Upon annealing such material at temperatures below about 1300°C, precipitation of the carbon occurs, as shown by infrared microscopy. Precipitates particles can be exposed on the surface in material doubly doped with aluminum and phosphorus by etching, and these particles have been shown to have the structure of β-silicon carbide by reflection electron diffraction.

INTRODUCTION

During the last few years, a considerable amount of work has been done on the behavior of oxygen in silicon (1). In particular, it has been found that dissolved oxygen will precipitate on heating a sample at 1000°C, and the rate of precipitation depends on the dislocation density (2). No corresponding work has been reported for carbon, which might also be expected to be electrically inactive, since up to now there is no known method for detecting its presence other than by destructive chemical analysis. This paper describes measurements of the diffusion coefficient and solubility of carbon in silicon from experiments using C^{14} as a tracer. Chemical analysis of grown crystals not deliberately doped with carbon indicate that carbon is present with a high concentration, and it is shown by reflection electron diffraction that β-silicon carbide precipitates on crystal defects during annealing of each sample.

MEASUREMENT OF THE DIFFUSION COEFFICIENT AND EQUILIBRIUM SURFACE CONCENTRATION

Radioactive C^{14} has been diffused into both oxygen-free floating-zone crystals and crystals pulled from a silica crucible in an argon atmosphere. The samples were sealed into a silica tube together with a source of C^{14}. Three sources have been used: (a) solid barium carbonate which decomposes to give carbon dioxide at high temperatures, (b) carbon dioxide gas, and (c) acetylene

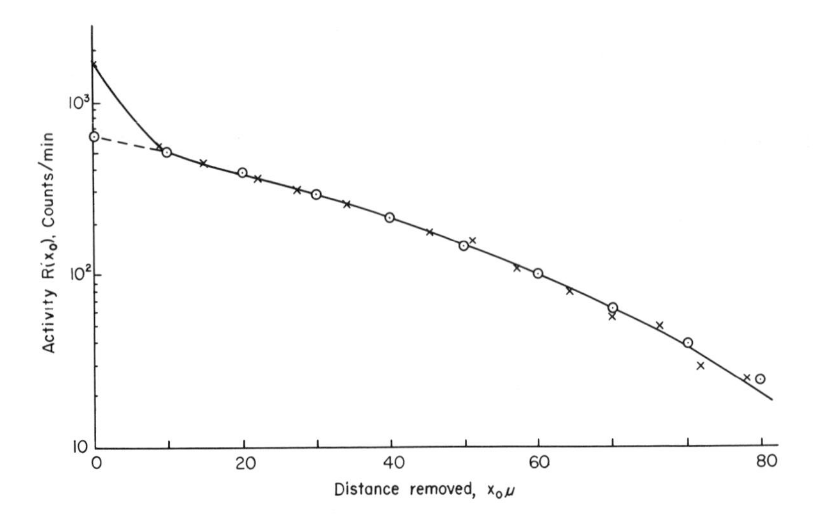

Fig. 1. Variation of counting rate with thickness of material removed by grinding. Specimen heated at 1150°C for 168 hr: (×) Experimental points, (⊙) theory with R(O) = 630 min^{-1}, $\sqrt{Dt}$ = 33 μ.

gas. In all cases a coating of β-silicon carbide, at least 40 A in mean thickness, was formed on the surface of the sample, thereby providing the source of carbon. No significant differences were found for the various possible combinations of samples and sources, and it was concluded that the presence of

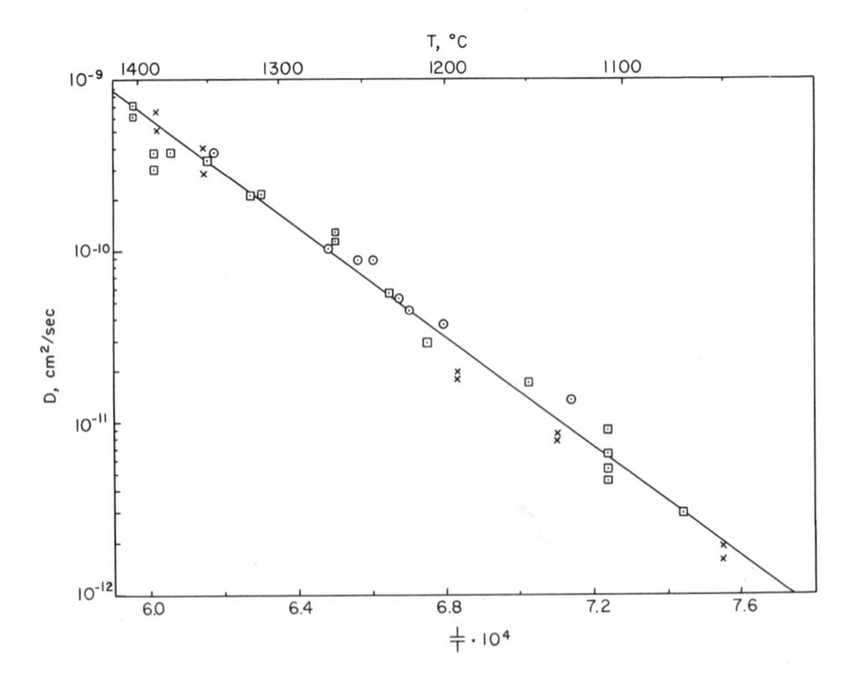

Fig. 2. Variation of D with temperature: (⊙) CO_2 source, (×) C_2H_2 source, (⊡) $BaCO_3$ source.

oxygen in the sample did not affect the rate of diffusion of carbon. Other details of the experimental conditions have been reported elsewhere (3).

If the usual error function compliment distribution of concentration is assumed, then it can be shown (3) that the counting rate, $R(x_o)$, to be expected from the sample of area, A, after grinding off a thickness x_o from the original surface, is:

$$R(x_o) = \frac{3.7 \times 10^{-12}\ Z.A.E.\ c_s}{n\alpha}\left[\operatorname{erfc}\frac{x_o}{2\sqrt{Dt}} - \exp\left\{\alpha^2 Dt + \alpha x_o\right\} \times \operatorname{erfc}\left(\frac{x_o}{2\sqrt{Dt}} + \alpha\sqrt{Dt}\right)\right]$$

where Z is the specific activity of the carbon source in mC/mM, n is the number of carbon atoms per molecule in the source material, α is the absorption coefficient of the β-rays in silicon (equals 600 cm^{-1}), c_s is the surface concentration of the carbon, and E is the efficiency of the counting system.

The measured distributions of activity agree well with the above equation, except close to the surface where some damage occurs, as indicated by photo-

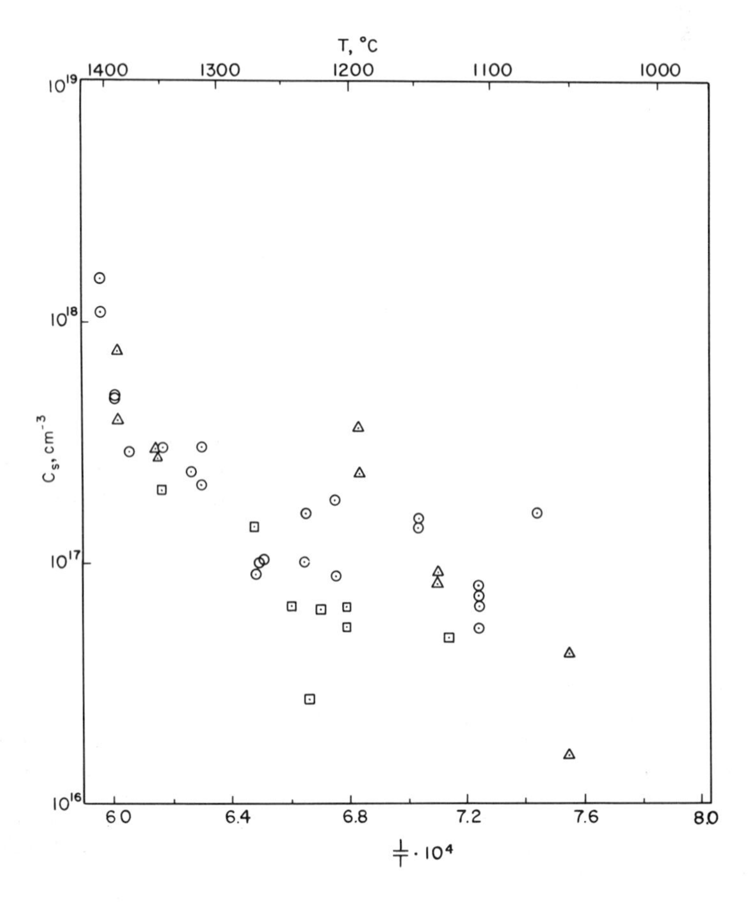

Fig. 3. Observed values of the surface concentration c_s as a function of temperature: (⊙) $BaCO_3$, (△) C_2H_2, (⊡) CO_2.

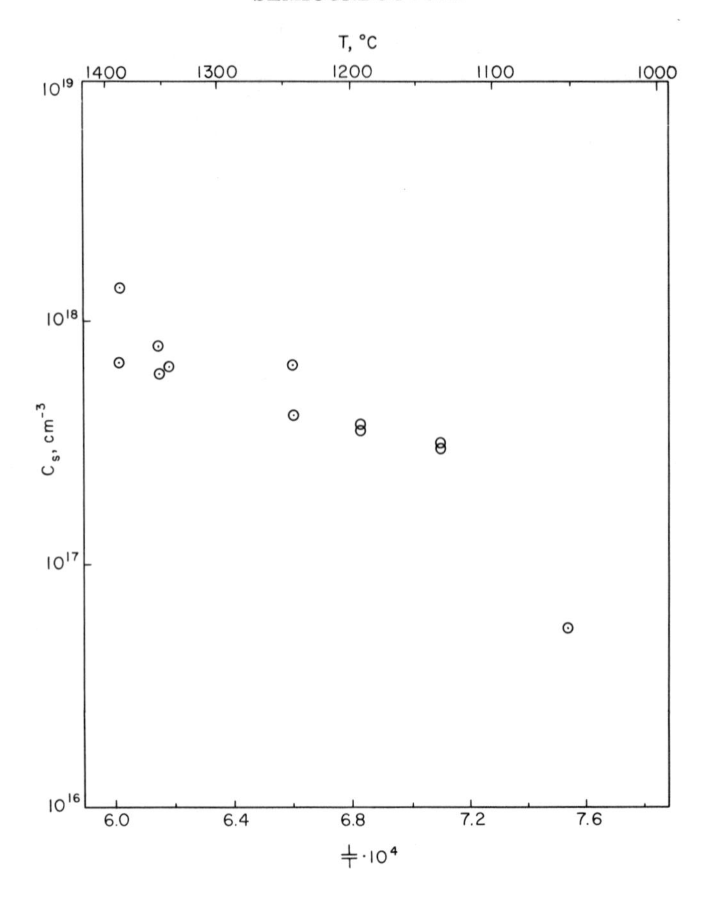

Fig. 4. Variation of c_s normalized to initial surface deposit of $Si^{14}C$.

micrographs; autoradiographs show embedded silicon carbide in these regions. A typical distribution is shown in Figure 1. The values of the diffusion coefficient thus found over the temperature range 1050 to 1400°C are shown in Figure 2, and we obtain:

$$D = 1.9 \exp -\left\{\frac{3.1 \pm 0.2 \text{ ev}}{kT}\right\} \text{cm}^2 \text{ sec}^{-1}$$

The value of the activation energy is somewhat larger than the previously reported (3) value of 2.92 ev but is within the estimated error of 0.25 ev. In any case it is reasonable to infer, from the absolute magnitude of E_D, that carbon occupies substitutional sites in silicon.

Once D is known, it is possible to calculate c_s. The values so obtained are shown in Figure 3, and it is seen that there is a large scatter in the results. It was found that the value obtained for c_s at a given temperature varied with the quantity of silicon carbide present on the surface, even though this was many monolayers in thickness. It is concluded, therefore, that the silicon carbide does not form a uniform coherent film, in agreement with our previous observations by electron microscopy (4), and it is also likely that some non-active carbon is introduced owing to unavoidable contamination. If the measured values of c_s are scaled by a factor proportional to the measured quantity of

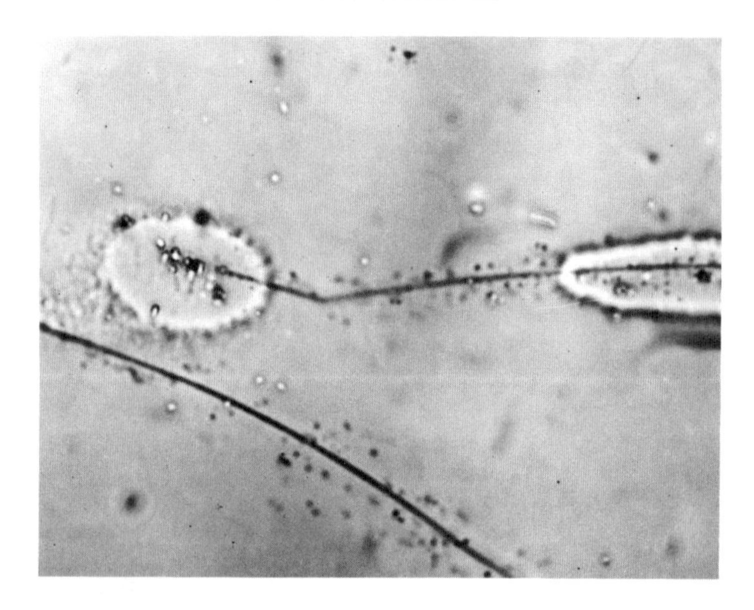

Fig. 5. Transmission infrared micrograph of etch silicon showing p–n junctions where dislocations meet the surface and precipitates around dislocations; sample heated at 1100°C for 90 hr. 1250×.

silicon carbide present, taking the largest quantity of silicon carbide observed as a reference level, we obtain the results shown in Figure 4. It is seen that the absolute values are not changed very much, but the scatter is considerably reduced. The solubility appears to increase with increasing temperature right up to the melting point where the value is about 10^{18} cm^{-3}.

PRECIPITATION OF CARBON

The carbon content of some of our pulled crystals has been determined by vacuum fusion analysis and by mass spectroscopy, and was found to be generally about 10^{18} atoms cm^{-3}. The source of the carbon is not known but may have originated from (a) the raw silicon material used, (b) organic vapors in the crystal puller, and (c) the quartz crucible (synthetic). Hence if such a crystal is annealed at temperatures below the melting point, precipitation would be expected as a result of the observed variation of c_s with temperature. Precipitate particles were, in fact, observed internally by infrared microscopy, after heating crystals in the temperature range 1000 to 1250°C followed by a slow cool. At the higher temperatures the precipitation was fairly light and confined mainly to dislocations, while at the lower temperatures a high density of random particles was also formed.

Similar p-type crystals have also been grown; these were double doped with aluminum and phosphorus. During heating, the aluminum was coprecipitated and regions of n-type conductivity formed around the areas where precipitation occurred (5). By etching these samples with a mixture of hydrofluoric and nitric acids, the pricipitate particles were exposed at the surface in the n-type regions, as shown in Figure 5. The surface was them examined by reflection electron diffraction. The pattern thus obtained from the precipitates was superposed on the silicon pattern as shown in Figure 6. The precipitate pattern was identified as β-silicon carbide. In some cases

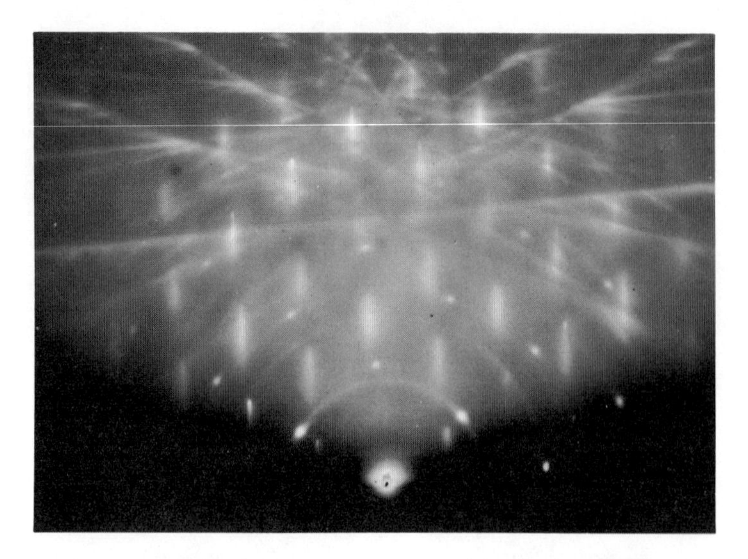

Fig. 6. Reflection electron diffraction pattern from the sample shown in Fig. 5. Streaked spots due to silicon and sharp spots and rings due to SiC precipitate particles. (100) surface; [011] azimuth.

spots attributable to γ-aluminia were found (16). Never was a silica pattern observed, perhaps because the etchant used to prepare the surface would dissolve any such particles.

It is therefore concluded that the precipitates in these crystals consist mainly of silicon carbide. It is interesting to note that if the carbon occupies

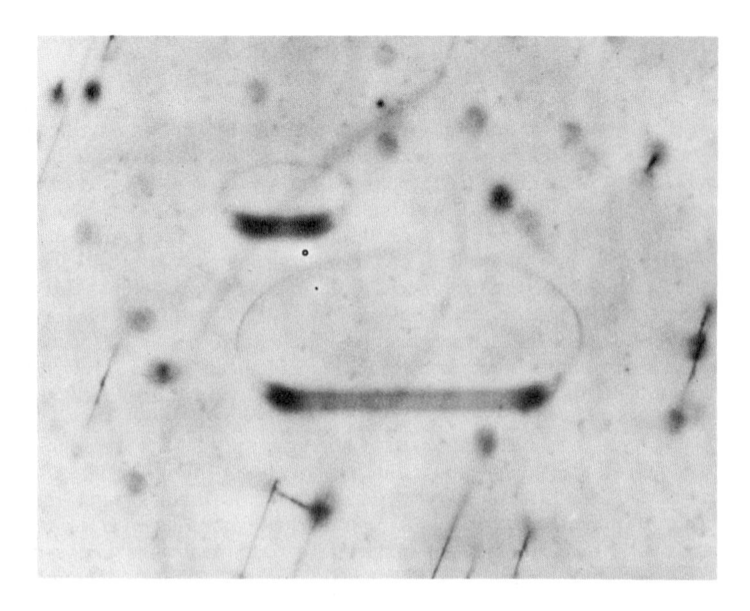

Fig. 7. Transmission infrared micrograph of etched silicon showing dislocation loops and associated etch grooves. Sample heated at 1150°C for 50 hr. 350×.

substitutional sites, as indicated by its activation energy for diffusion, vacancies should be generated during precipitation since a local reduction in volume will occur during this process. The aggregation of these vacancies could provide the explanation for the growth of large dislocation loops on {111} planes about grown-in dislocations in samples heated in the temperature range 1120 to 1200°C. A groove was found where such loops intersected the surface after etching, and it is probable that there is a stacking fault in the plane of the loop (see Fig. 7).

ACKNOWLEDGMENTS

Thanks are due Messrs J. H. Neave and D. J. Maule for experimental assistance, J. B. Willis for taking the infrared micrographs, R. L. Rouse for his encouragement, and G. H. Bush of A.R.D.E., Fort Halstead, for some of the vacuum fusion analyses. The authors also wish to thank Dr. T. E. Allibone, C.B.E., F.R.S., Director of the Laboratory, for permission to publish this paper.

References

1. Reiss, H., and W. Kaiser, in Properties of Elemental and Compound Semiconductors, H. C. Gatos, ed., (Metallurgical Society Conferences, Vol. 5) Interscience, New York-London, 1960, p. 103.
2. Lederhandler, S., and J. R. Patel, Phys. Rev., 108, 239 (1957).
3. Newman, R. C., and J. Wakefield, J. Phys. Chem. Solids, 19, 230 (1961).
4. Newman, R. C., and J. Wakefield, in Solid State Physics in Electronics and Telecommunications, M. Desirant and J. L. Michiels, eds., Academic Press, New York-London, 1960, p. 318.
5. Bullough, R., R. C. Newman, J. Wakefield, and J. B. Willis, J. Appl. Phys., 31, 707 (1960).
6. Newman, R. C., Proc. Phys. Soc. (London), 76, 993 (1960).

DISCUSSION

J. BELOVE (Raytheon Semiconductor): Would you say that any organic compound on the surface of the silicon slice might cause these silicon carbide precipitates if the slice is diffused at some high temperatures?

R. C. NEWMAN: I think it may be possible. Precipitation would not be expected in silicon, initially free of carbon and lattice defects, as a result of heating at a constant temperature in the presence of an organic material. If, however, dislocations are present, then a carbon impurity atmosphere, or Cottrell atmosphere, would be expected to form around them, owing to the large size difference between carbon and silicon, and this atmosphere could condense on cooling the sample.

P. S. FLINT (Fairchild Semiconductor): You mentioned the work that has been done on aluminum precipitation on silicon. Do you believe that carbon is involved in this case, and to what extent?

R. C. NEWMAN: We fine that precipitation of aluminum only occurs in silicon which contains oxygen, and we believe that aluminum oxide complexes form during the high-temperature heat treatment. Carbon is also present in the silicon, and this precipitates simultaneously as β-silicon carbide. It is not clear how the growth of this carbide influences the precipitation of the aluminum.

E. G. BYLANDER (Texas Instruments): For the case of aluminum precipitation, is it not correct that the aluminum precipitates independent of the cooling cycle? How does kinetics of the carbon precipitation depend on the cooling cycle?

R. C. NEWMAN: The degree of precipitation of both aluminum and carbon is independent of the rate of cooling of the samples; samples have been cooled at the rate of one degree per minute or have been quenched into water. The size of the p-n junctions formed and the size of the precipitate particles are the same for both types of sample. No appreciable movement of either aluminum or carbon is to be expected during cooling, because both have small diffusion coefficients and are, hence, relatively immobile, in contrast to interstitial impurities such as copper or iron.

M. TANENBAUM (Bell Telephone Labs.): Have you observed any effect on the infrared transmission of samples containing large concentrations of carbon?

R. C. NEWMAN: Thus far we have not found any infrared absorption band due to carbon in solution in single crystal silicon. If the carbon occupies substitutional sites, then a discrete absoroption band would not be expected to appear, in contrast to interstitial oxygen. Any perturbations of the lattice absorption bands would be expected to be small, with a carbon level of only 10^{18} atoms cm^{-3}.

Impurities in Semiconductors

DAVID A. RESNIK and ROBERT K. WILLARDSON

Bell & Howell Research Center, Pasadena, California

Abstract

Impurity concentrations in semiconductors are being studied using spark source mass spectrographic techniques. The technique used to determine the number of ions which created a given line on the photographic plate considers line density, width, shape and adjacent background density. It was found that cadmium and sulfur ions, which have a mass ratio greater than 3 to 1, cause equal densities on an Ilford QII photographic plate. Anomalous results obtained on zinc in indium antimonide suggest segregation phenomena can occur during sparking.

INTRODUCTION

It has been amply demonstrated (1-5) that a mass spectrograph with the Mattauch-Herzog geometry and utilizing a vacuum spark source and a photographic plate as detector is a suitable instrument for the analysis of impurities in solids in the concentrations of interest in semiconductors. While the qualitative analyses have been excellent because of the high resolution of this type of instrument (with appropriate slit widths), methods of interpreting the data on the photographic plate to give accurate quantitative analyses have not yet been adequately refined. In the work that has been done, two problems have been outstanding. They are (a) an apparent change in sensitivity with mass (6) and (b) several anomalous sensitivities (3), such as that of zinc which has been reported to have a sensitivity of five relative to that of copper and iron in, for example, an aluminum matrix.

A technique which considers both line width and shape, as well as background factors, was employed in the present studies. Abundances of cadmium and sulfur in single crystal cadmium sulfide, chosen because of their large mass ratio, were studied to evaluate the validity of the method. Zinc in a sample of indium antimonide to which 9 ppm (atomic) of zinc had been added was also studied. Measurements of carrier concentration were made to assure that the segregation of zinc was an expected and that the indium antimonide was doped to the proper level.

THE INSTRUMENT

Figure 1 illustrates the basic configuration of the mass resolving system. For the analyses discussed in this paper, the rf sparking voltages were approximately 15 kv (cadmium sulfide) and 21 kv (indium antimonide) repeated at the rate of 100 (cadmium sulfide) and 2000 (indium antimonide) pulses per

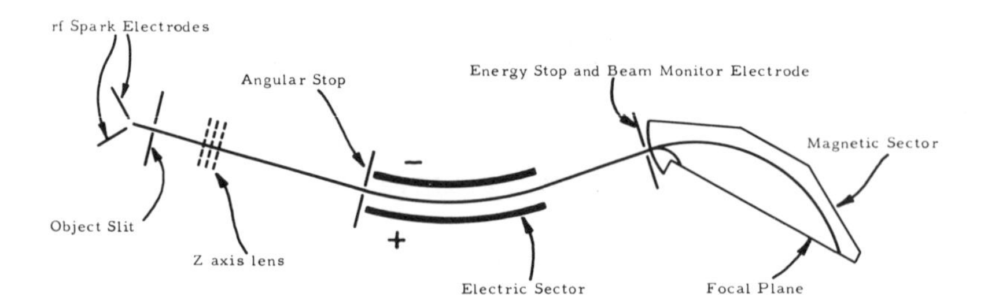

Fig. 1. Basic geometry of mass resolving system.

second. The positive ions are accelerated through a potential of 15 kv and then pass between a pair of oppositely charged plates, the electric focusing section. Only those ions within a certain small range of energies are turned through the critical angle that enables them to enter the magnetic section. Here a magnetic field perpendicular to the ion path sorts the ions according to their ratio of mass to charge. A photographic plate at the boundary of the magnetic field serves as an ion detector and, with it, an entire mass spectrum from mass 6.5 to mass 260 is permanently recorded on 0.19×38 cm strips.

This instrument features a three-element Z-axis lens which focuses the ion beam in the direction normal to the plane of the figure, thus utilizing many ions that would otherwise go astray. This lens increases the number of ions striking the plate by about five times, which is quite important since the concentration of the impurities of interest is often quite low.

Between the electric sector and the magnetic sector is an electrode array which serves simultaneously as energy-limiting stops and as current monitors. Approximately one-half of the ions strike the monitor electrodes. Thus, the current measured by the monitors indicates how many unit charges pass into the magnetic sector. To estimate the number of ions, an allowance must be made for those that are multiply ionized.

The total ion path length varies from 1 to 1-1/2 meters, the higher masses traveling the furthest. Large dimensions are beneficial since they result in high mass dispersion at the focal plane. The particular configuration used here is chosen to nullify some of the second-order aberrations at a selected position on the focal plane, thus giving high mass resolution at that position.

THE QUALITATIVE ANALYSIS

Since the radius of the ion path in the magnetic field varies as the square root of the mass, it is necessary only to identify two lines, measure their positions from the end of the photographic plate, and then the identity of any third line can be judged from its mass, which can be computed from a measurement of its position by the formula

$$\sqrt{m_i} = (k\ x_i + a)$$

$$k = (\sqrt{m_a} - \sqrt{m_b})(x_a - x_b)$$

$$a = (x_a \sqrt{m_b} - x_b \sqrt{m_a}) / (\sqrt{m_a} - \sqrt{m_b})$$

Here x_a, m_a is the position, mass of one known element; x_b, m_b is the position, mass of second known element; and x_i, m_i is the position, mass of any other element. (Due to nonuniformities in the magnetic field and aberrations, precise mass values cannot be calculated if x_a and x_b are separated by more than one or two inches.)

While this is a procedure that is available and is theoretically applicable to every line, it is usually not needed. Generally, the lines of the major constituents (singly, doubly, and triply charged, as well as their dimers and trimers, all easily distinguishable because of their great abundance) serve as sufficiently adequate reference points to enable an experienced person to qualitatively analyze at a glance a photographic plate.

THE QUANTITATIVE ANALYSIS

While qualitative analysis is straightforward, quantitative analysis presents a number of problems. Methods to date have utilized concepts such as "just detectable" lines and the comparison of lines of different elements without regard to their shape or background. The abundance of a minor isotope of a major constituent has been used often as an internal standard. However, in many cases there is not an isotope of the appropriate abundance available to compare with the unknown.

Typical data are indicated in Figure 2. The photographic plate may have as many as 13 exposures. Typically the exposures are between 10^{-12} and 10^{-6} coulombs (10^7 and 10^{13} ions). For each line that is to be analyzed, a micro-

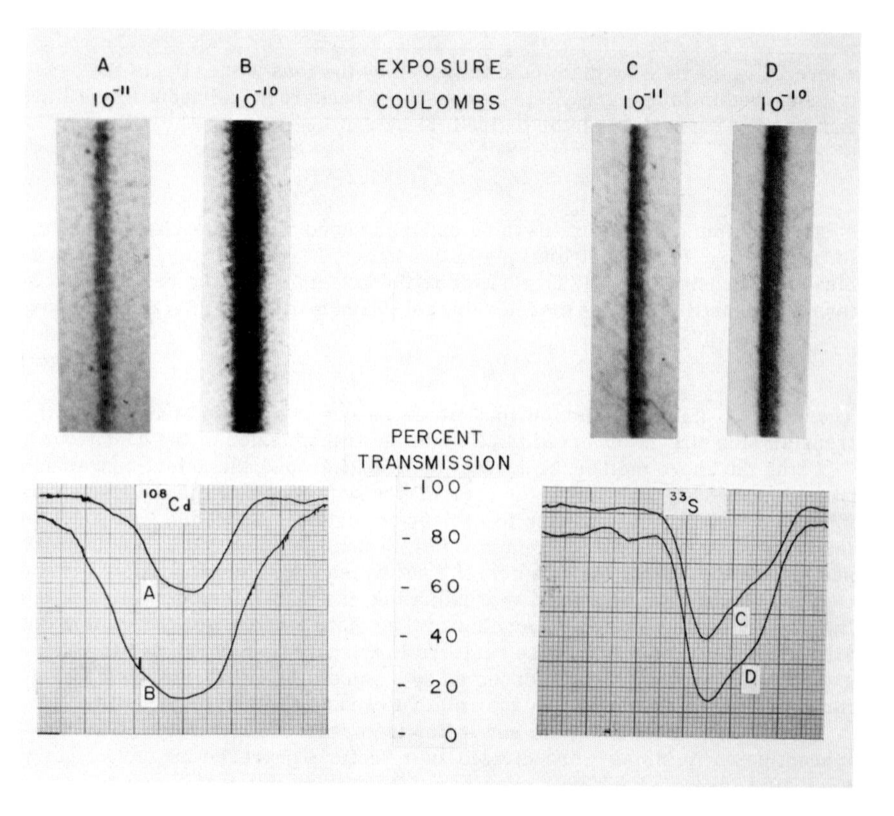

Fig. 2. Spectroscopic lines and corresponding transmission traces.

photometer trace of the transmission versus the distance along the plate is made. The abundances of ^{33}S and ^{108}Cd in CdS are very nearly equal, i.e., 3800 and 4400 ppm, respectively. One would expect, then, that their traces would be quite similar. Such is not the case, however, and this is clearly seen in Figure 2. The cadmium lines tend to be light, wide, and symmetric; whereas the sulfur lines are heavy, narrow, and quite asymmetric. These differences in line shape are due to variations in second-order aberrations along the focal plane and must be taken into consideration. If they are not accounted for, then it will be found that there are apparent variations in sensitivity of as much as 5:1.

In Figure 2 it should be noted that there is on each exposure a different background transmission. Part of this background is due to stray ions and part may be caused by chemical fog. A complete analysis must consider that the background does have a definite effect in darkening even that portion of the plate where the line is found, and must be corrected accordingly. The case of a weak line and a strong background is frequently encountered in semiconductor work.

To correct for background, one must really answer the question: "What would have been the transmission through the plate if there had been no stray ions present and no developmental fog?" If it is assumed that the density of the plate is proportional to the number of developed grains (7), the following relationship is easily developed:

$$D_O = D_I + B\alpha$$

where D_I is the density that would be caused by the ions alone, D_O is the density observed in image area, B is the density of background adjacent to the line, and α is the correction factor defined (8) by

$$\alpha = (D_M - D_O) / (D_M - B)$$

The maximum density achievable under the conditions of development, etc., in use is D_M. In the calculations in this paper, D_M was assumed equal to 3. Since D = density = log (1/T) where T is the transmission, the relationship in terms of densities implies the following relationship in terms of transmissions:

$$T_I = T_O / T_B^{\alpha}$$

where T_I is the transmission that would be due to the ions alone, T_O is the transmission that is observed, and T_B is the transmission of the background.

Using the above method for correcting for background, the actual conversion from line density to number of ions may now proceed. While the line shape may change from one position to another on the focal plane, the line shape is fairly regular in any local region. This, in turn, implies that in a unit cross section of the beam, the number of ions present will be proportional to the current. Using this as a first approximation, the transmission minima of the lines of ^{34}S were plotted as a function of the relative exposure. (From now on, all transmissions not otherwise referred to will have been corrected for background.) These points were connected by a smooth line, and this gave a relationship between transmission and relative exposure in arbitrary units.

Sixteen traces of cadmium and sulfur isotopes were integrated; i.e., the transmission minimas were divided into sections representing unit areas of 3.2×10^{-4} cm^2. Each unit area had a certain transmission which corresponded to a certain number in arbitrary units. Adding these numbers together for each curve constitutes the integration of the trace. The calibration curve

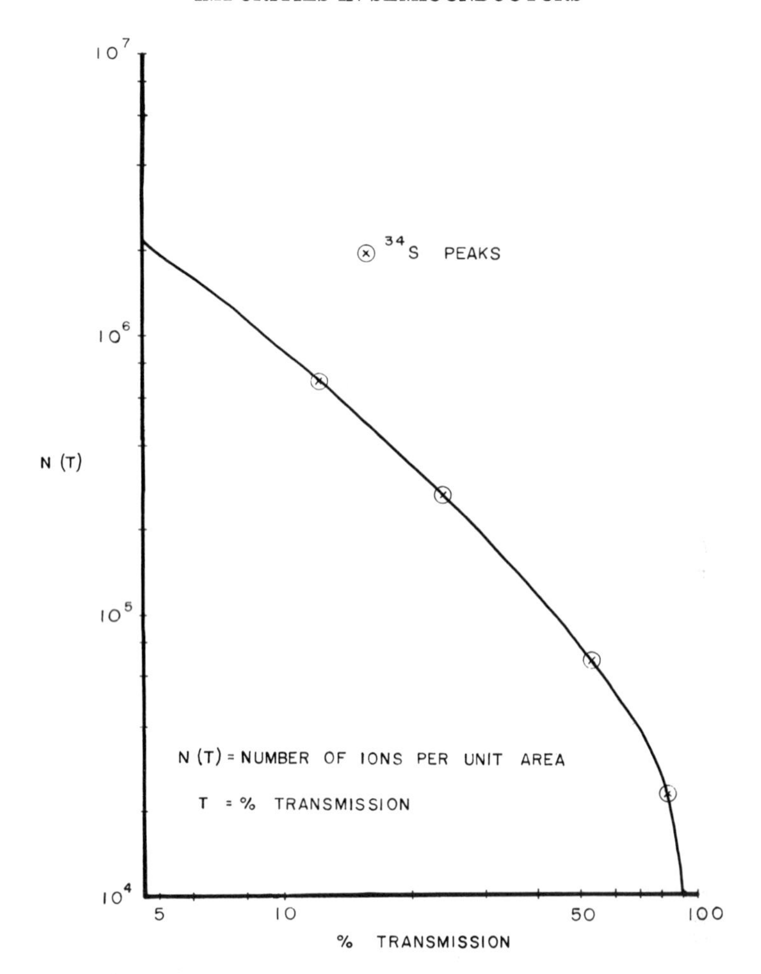

Fig. 3. Number of ions per unit area versus per cent transmission.

was then normalized to the ^{34}S integral on the 3×10^{-12} coulomb exposure. Each arbitrary unit then corresponded to a fixed number of ions per unit area of the line. This calibration curve is presented as Figure 3.

Using the calibration curve in Figure 3, each of the traces could now be integrated and the integral would represent the total number of ions constituting the line. In Figure 4, the number of ions indicated by the current monitors is compared with the integral of the trace. It is seen that there is good agreement between the expected number of ions indicated by the monitors and the values computed from the lines on the photographic plate. This also indicates that the product of the ionization and detection efficiencies for cadmium and sulfur are nearly equal.

The procedure for correcting for background and the calibration curve has been combined into a single chart (Fig. 5) which allows the rapid integration of the transmission trace of any desired line. One notes the background transmission, then the transmission at the point of interest, and from the chart obtains directly the corresponding number of ions per unit area. In the example shown, there was a background transmission of 71%, a transmission of 50% is

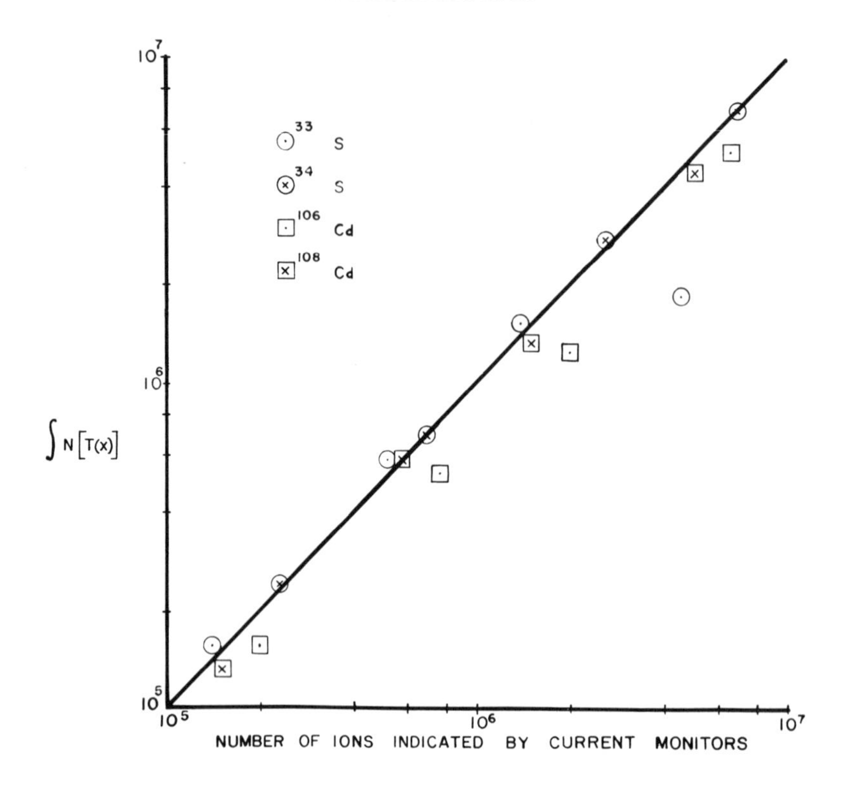

Fig. 4. Integrals of transmission traces versus number of ions indicated by current monitors.

observed, and this corresponds to 3.8×10^4 ions per unit area. The corrected transmission of 72% is not needed in the evaluation of the integrals. To obtain the relative abundance of an impurity, one has only to compute the number of ions causing its line, then divide this number by the number of ions in the total exposure based on the reading of the current monitors.

THE DETERMINATION OF IMPURITY CONCENTRATIONS IN INDIUM ANTIMONIDE

Table I indicates the impurity concentrations of various isotopes of zinc, gallium and arsenic in a sample of indium antimonide. For gallium and arsenic, the standard deviations from the means are 0.3 and 0.14 ppm or 9 and 14%, respectively, of the amounts present, and indicate that the method is at least internally consistent. The overall estimated concentration of zinc is 20% higher than expected from the carrier concentration of 2.72×10^{17} cm^{-3}. The wide variation of concentration with order of exposure and the increase of concentration as time progressed indicate that a heretofore unexpected mechanism may be operative. Melting of indium antimonide at the tip of the electrodes suggests that a segregation phenomenon may be operative. This is made plausible, qualitatively at least, by the fact that the segregation coefficient of zinc in indium antimonide is three.

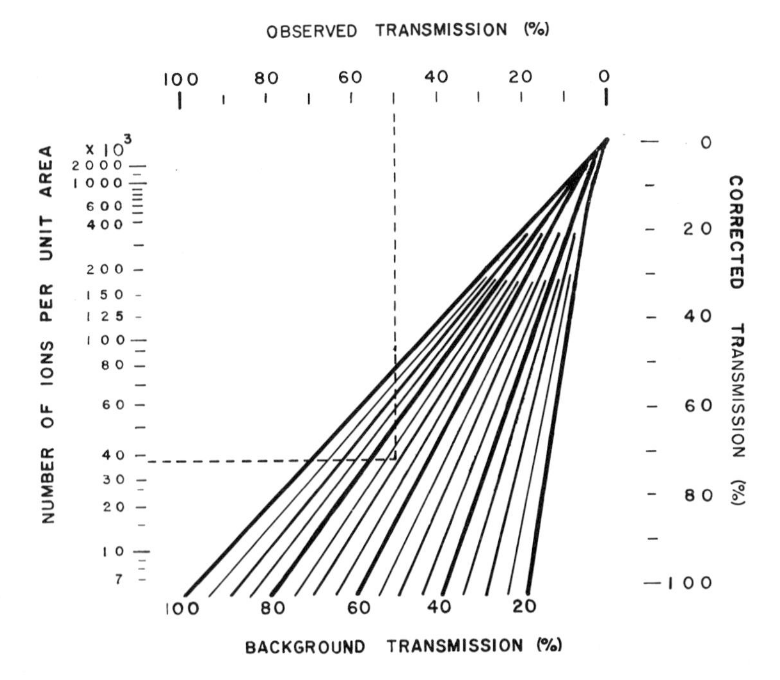

Fig. 5. Transmission observed in image area, corrected for background and related to number of ions per unit area.

TABLE I

Impurity Concentrations in Indium Antimonide[a]

Isotope	Exposure in coulombs							Weighted av.[b]	Estd. concn.
	1×10^{-7}	3×10^{-8}	1×10^{-8}	3×10^{-9}	3×10^{-8}	1×10^{-8}	3×10^{-9}		
^{64}Zn	4.2	19	7.4	9.5	25	23	20	12	
^{64}Zn	4.1	12	9.5	9.5	21	29	27	10	
^{67}Zn	5.4	9.6	12		16	31		10	
^{68}Zn	4.7	12	9.8	5.5	24	29	30	11	
^{70}Zn	11							11	
Weighted av.	4.4	15	8.6	8.7	23	26	24		11
^{69}Ga	3.7	3.5	2.4	2.5	3.6	3.2	3.2	3.5	
^{71}Ga	3.0	3.2	4.2	4.2	3.0	3.5	3.5	3.2	
Weighted av.	3.4	3.4	3.1	3.2	3.4	3.3	3.3		3.4
^{75}As	1.0	.85	.98	.75	1.2	.78	1.6		1.0

[a] Values are given in ppm (atomic).

[b] In the averaging process the values are weighted in proportion to the number of ions of each isotope deposited during the exposure.

Under the sparking conditions that give optimal ion yield for indium antimonide, the initial sparking melts a macroscopic segment of the electrode, and then the spark moves to another segment, allowing refreezing of the first section. On the basis of the expected segregation, the concentration in the last part to freeze would be an order of magnitude less than the mean concentration. The impurity depleted portions of a number of these segments may be expected to supply the major portion of the indium antimonide analyzed in the first exposure (for the first several exposures, depending on their lengths). In subsequent exposures the mean concentration should be approached, but pockets of segregated zinc or depleted regions may cause variations from the mean. If the exposures are short, these variations may be appreciable.

The possibility of the occurrence of such a segregation mechanism suggests further experiments. Such experiments would involve adjusting the sparking conditions in order to avoid macroscopic melting of the sample as well as experiments in which the macroscopic melting and concomitant segregation is carefully controlled.

ACKNOWLEDGMENTS

The authors wish to thank J. R. Shirk for his help in operating the mass spectrometer and express their appreciation to D. M. Pollock, G. D. Perkins, and Dr. C. F. Robinson for valuable discussions and helpful suggestions.

References

1. Hannay, N. B., and A. J. Ahearn, Anal. Chem., 26, 1056 (1954).
2. Ahearn, A. J., Vacuum Technology Transactions, Pergamon Press, New York, 1960, p. 1.
3. Craig, R. D., G. A. Errock, and J. D. Waldron, Advances in Mass Spectrometry, Pergamon Press, 1959, p. 136.
4. Perkins, G. D., C. F. Robinson, and R. K. Willardson J. Electrochem. Soc., 107, 189C (1960).
5. Duke, J. F., "The Application of Solid Source Mass Spectrometry to Determine the Purity of Materials for Semiconductor Purposes"; R. Brown, R. D. Craig, J. A. James, and C. M. Wilson, "Analysis of Trace Impurities by Spark Source Mass Spectrometry"; R. K. Willardson, "Electrical and Mass Spectrographic Analysis of III-V Compounds," all in Ultrapurification of Semiconductor Materials, to be published, Macmillan Co., New York.
6. Owens, E. B., Lincoln Lab., Quarterly Progress Rept., p. 20, March 15, 1961.
7. Mees, C. E. K., The Theory of the Photographic Process, Macmillan Co., New York, p. 169.
8. Meidinger, W., Z. physik. Chem., 114, 89 (1924), (see also reference 7, p. 830).

DISCUSSION

P. S. FLINT (Fairchild Semiconductor): What can you do toward handling the light elements, oxygen, nitrogen, or carbon, in this instrument?

D. A. RESNIK: Carbon, oxygen, nitrogen, and hydrogen can be detected with the solid spark source mass spectrometer. Quantitative studies of their concentrations are now in progress. With the magnetic field set at 9500 gauss the mass range is 7 to 280 atomic mass units (amu). By reducing the field to 3600 gauss, the lower mass range of from 1 to 40 amu is obtained.

P. S. FLINT: Would you expect difficulty in analyzing for such light elements in the presence of fair concentrations of phosphorus or boron?

D. A. RESNIK: No difficulty.

M. TANENBAUM (Bell Telephone Labs.): Do you know what your background level would be for oxygen and nitrogen?

D. A. RESNIK: We are now in the process of attaching a diatron, a second mass spectrometer, to the first mass spectrometer, which will measure the concentrations of background gasses.

The Influence of Surface Damage on the Generation of Dislocations in Lapped Silicon Wafers

S. A. PRUSSIN

MonoSilicon, Inc., Gardena, California

Abstract

Evidence is presented to picture a lapped surface as containing a network of micro cracks. Lapping debris forced into these cracks results in surface tensile stresses and buckling of the silicon wafer. The surface stresses were insufficient for generating dislocations in dislocation-free silicon when the lapped wafers were heated to 1200°C. The significance of microcracks in generating dislocations in vapor-deposited epitaxial layers is noted.

INTRODUCTION

For dislocations to be introduced into a piece of solid silicon, the silicon must be plastically deformed. For this, two conditions are necessary: (1) the silicon must be in the plastic range, i.e., at a temperature of at least 600°C, and (2) the stresses present must exceed the yield point of the silicon. A third condition, although not necessary for plastic deformation, nevertheless plays an important role in determining whether deformation or rupture occurs. This is the presence of surface stress raisers. The role of surface damage in promoting plastic deformation, and thereby the generation of dislocations, would appear to stem from its being a possible source of surface stresses as well as a possible source of surface stress raisers.

Lederhandler (1), in his infrared studies of birefringence in silicon, noted that residual elastic stresses were found along surfaces produced by cutting with a diamond saw. When these as-cut surfaces were lapped, the birefringence patterns disappeared. In this work lapped surfaces were primarily investigated. Silicon wafers with such surfaces are normally heated into the plastic range during device fabrication.

Insight can be gathered into the character of a lapped surface from the appearance of a cleavage plane originating from this surface. Figure 1 shows the cleavage planes originating from surfaces which have been (a) chemically polished, (b) lapped, and (c) produced by a diamond saw. The cleavage plane originating from a chemically etched surface is sharp. In the cleavage plane originating from a lapped surface it can be noted that many subfacets are formed along the lapped surface. This is interpreted as indicating the presence of many fine surface cracks, each nucleating a new fracture and producing its own cleavage plane.* Where the surface is in the as-cut condition this effect is even more pronounced, as seen in Figure 1c. This picture of the lapped surface is confirmed by Buck (2).

*Since the delivery of this paper, the nature of the lapped surface of germanium similarly has been described in a paper by Pugh and Samuels, J. Electrochem. Soc., (1961).

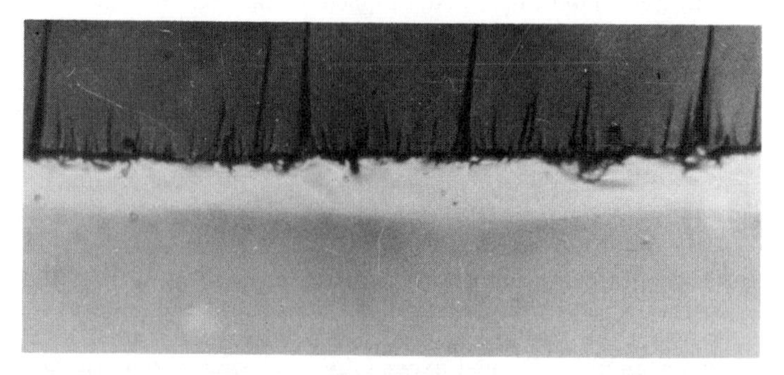

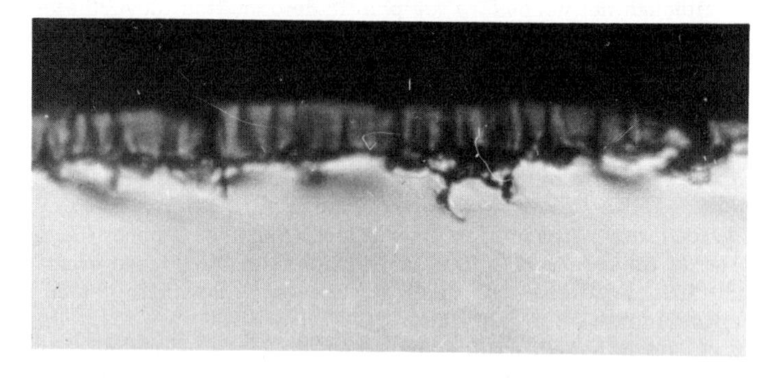

Fig. 1. Cleavage planes originating from differently prepared surfaces: (a) chemically polished, (b) lapped, and (c) as cut by diamond saw.

RESIDUAL SURFACE STRESSES

The following technique was used to detect residual surface stresses in lapped wafers. Silicon slices, lapped parallel on both sides, were chemically polished by submersion in a 1:1:1 hydrofluoric : nitric : acetic acid solution. They were cemented to the bottom of a lapping jig, and one of the surfaces was lapped with a fine garnet abrasive. After lapping, the wafers, still attached to the jig, were stroked several times over a sheet of 4/0 emery paper. This technique, developed by T. J. LaChapelle, brings out a sufficient number of highlights on the lapped surface so that a monochromatic sodium light source and an optical flat can be used to determine surface flatness. In each case, it

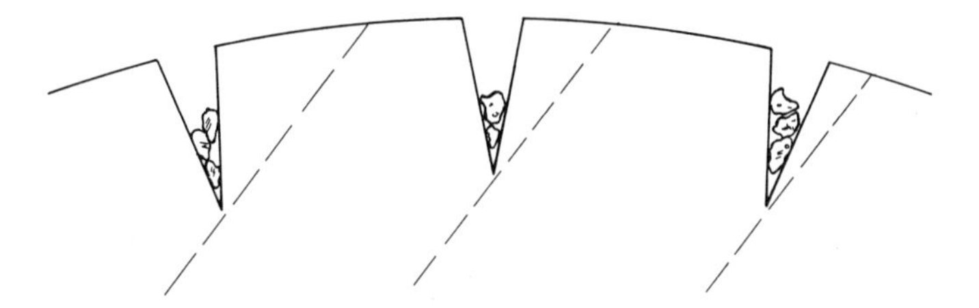

Fig. 2. Schematic representation of a lapped silicon surface.

was found that the surfaces were flat within a wavelength or two. When the lapped wafers were removed from the lapping jig and re-examined for flatness, it was found that in every case, the wafer had buckled, the lapped surface being convex. The radius of curvature was simply and accurately obtained by measuring the spacing between fringes.

Since the lapping operation was performed at room temperature, no plastic deformation could have occurred. The only possible explanation for the tensile stress associated with the lapped surface was that particles of abrasive and abraded particles of silicon had become lodged in the fine crevices associated with the lapped surface. Figure 2 is a schematic representation of this phenomenon. Because wafers of varying thickness were studied, it was desirable to compare the bending moment rather than the radius of curvature alone. The relation between the bending moment per unit edge length, M, and the curvature, R, developed in a circular plate is given by Timoshenko (3):

$$M = \frac{E\ (2a)^3\ \ 1}{12\ (1-\nu)\ R}$$

where 2a is the thickness of the plate, E is Young's modulus, and ν is Poisson's ratio.

A series of wafers were prepared by lapping with W-5 and W-10 garnet abrasives. These were ultrasonically cleaned in aqua regia, a procedure chosen to remove imbedded particles of garnet. The average bending moment for the W-5 lapping 2.0×10^4 dyne-cm/cm. For the finer W-10 abrasive it was 0.66×10^4 dyne-cm/cm. The latter thus produces considerably less buckling.

HEATING OF LAPPED SURFACES

A lapped surface would appear to contain the elements necessary for plastic deformation provided the temperature is raised into the plastic range. The tensile stresses associated with the lapped surface are concentrated at the roots of the surface cracks. It is here that we would expect dislocations to be generated.

The silicon wafers studied were cut from an ingot which was dislocation free. After lapping and cleaning, the radius of curvature of a number of wafers was measured. The wafers were sealed in a helium-filled quartz capsule and heated at 1200°C for 4 hr. After this heat treatment it was found that the curvature of each wafter had decreased significantly. It appeared possible that this stress relief was accomplished by the generation and movement of dislocations in the area of the lapped surface.

A typical wafer, 479/5, was 210 μ thick after being chemically polished and lapped on one side with W-5 garnet abrasive. After being cleaned ultrasonically in an aqua regia solution, it was found to have a radius of curvature of 110 cm. This was determined from the fact that there were 25 circular fringes within a radius of approximately 4.0 mm. The value of 5890 A was used as the wavelength of the light.

After heat treatment, the curvature of the wafer was measured again. Now there were only 14 fringes in the same area, corresponding to the new radius of curvature of 192 cm. As a result of the heat treatment, the bending moment of this wafer was reduced from 1.63 to 0.91 dyne-cm/cm.

Attempts were made to detect such dislocations as might have been introduced during the heat treatment. The use of a Dash etch procedure on the lapped surface was unsuccessful. The general surface attack, nucleated by the presence of work damage, obscured the presence of any existing etch pits. Copper decoration of the heat-treated slices also failed to bring out dislocations. This is understandable since the lapped surface had to be polished away in order to permit satisfactory transmission of the infrared radiation. G. H. Schwuttke has applied his X-ray technique with singular success to the detection of dislocations lying in and just below the surface of a silicon wafer. Slices of dislocation-free silicon were submitted to Schwuttke which were unlapped, lapped, and lapped and heated to 1200°C for 4 hr. No dislocations could be detected.

Dash (4) introduced severe surface damage into dislocation-free silicon by impaction with a diamond point. After heating to 1000°C for a few minutes, the silicon surface was etched and the presence of dislocations in slip arrays was shown to exist. In comparison, it would appear that the stresses associated with lapping surface damage are themselves insufficient to cause the generation of dislocations. Any effect of surface damage in lapping is probably due to the presence of stress raisers. These concentrate externally applied stresses such as thermal stresses and promote the introduction of dislocations in this manner.

The concept of a lapped surface as consisting of microcracks filled with debris has important consequences for epitaxial deposition. Since the deposited layer is simultaneously nucleated at a number of different surface areas, there will be a lattice mismatch when the epitaxial firm crosses a microcrack. Here a low-angle grain boundary will occur and with it the attendant dislocation array.

References

1. Lederhandler, S. R., J. Appl. Phys., 30, 1631 (1959).

2. Buck, The Surface Chemistry of Metals and Semiconductors, H. C. Gatos, ed., Wiley, New York, 1960, p. 107.

3. Timoshenko, S., Theory of Elasticity, McGraw-Hill, New York, 1934, p. 226.

4. Dash, W., in The Properties of Elemental and Compound Semiconductors, H. C. Gatos, ed. (Metallurgical Society Conferences, Vol. 5) Interscience, New York-London, 1960, p. 195.

DISCUSSION

W. V. WRIGHT: (Electro-Optical Systems): I have made microhardness measurements on etched surfaces of both silicon and germanium. Such indentation measurements on silicon usually result in cracking at the corners of the indentations; however, when a very light load (a few grams on a Vicker's diamond pyramid) is used at room temperature, uncracked indentation results. Perhaps the mechanism of dislocation flow at the free surface of silicon is sufficiently different from bulk dislocation flow that surface plasticity is possible at room temperature.

S. A. PRUSSIN: It is a possible hypothesis. I was just going along with what has now become a classical point of view, one which a number of different investigators have looked into, and have set a temperature considerably above room temperature where the first signs of plastic deformation have been noted.

W. V. WRIGHT: I accept the results on single crystal silicon specimens where bulk measurements were made. There may be another mechanism for surface plasticity in silicon, but I am not prepared to explain it.

ANON.: What was your technique of removal of the wafers from the jig?

S. A. PRUSSIN: The wafers were cemented to the jig with a wax and the jig was slightly warmed and the wafer removed.

ANON.: We have been able to polish wafers and then remove them from the jig and they came out dislocation free, but only if the wax is dissolved. If the wax is melted, the wafers buckle.

P. A. ILES (Hoffman Semiconductor Division): Bowden, in England, in his work on friction, showed that you can get very high local temperatures, particularly with good thermal insulators. I wonder if you have looked into this, to see whether you are not getting high local temperature under your lapping particles.

S. A. PRUSSIN: The technique makes use of a liquid vehicle, and the lapping is relatively slow. I happened to be lapping with a Lapmaster, and normally would not imagine that using a liquid vehicle containing this powder, one would expect temperatures high enough to cause plastic deformation to occur. At least there is no sign of that. So I feel that we did not generate such temperatures.

PART II: MATERIALS FOR THERMOELECTRIC ENERGY CONVERSION

The Dependence of the Thermoelectric Figure of Merit on Energy Band Width

R. C. MILLER, R. R. HEIKES, and A. E. FEIN

Westinghouse Research Laboratories, Pittsburgh, Pennsylvania

Abstract

The dependence of the thermoelectric figure of merit, Tz, upon the width of the forbidden band has been calculated assuming the Boltzmann transport equations are valid. It was found that Tz goes to zero as the band width goes to zero and that a maximum Tz occurs for optical mode and ionized impurity scattering when the Fermi level is approximately equal to 4 kT.

The efficiency of a thermoelectric generator depends on the properties of the material through the single parameter

$$Tz = TS^2/\rho\kappa$$

where T is the absolute temperature, S is the Seebeck coefficient, ρ is the electrical resistivity, and κ is the thermal conductivity. The quantity z is usually referred to as the figure of merit. In order to seek out possible new criteria for materials selection, the dependence of Tz on the width of the conduction band was studied. It was assumed that the Boltzmann transport equation was valid at all band widths and that a relaxation time could be defined. This was done in spite of the fact that the applicability of the Boltzmann equation in certain cases has been questioned. From the present point of view, probably the most stringent criterion (1) for its applicability is that the band width multiplied by the relaxation time be greater than $\hbar$. That is

$$E_{max} \cdot \tau > \hbar$$

This is simply a statement that the lifetime of the state allows the error in the definition of the energy to be less than the band width. Thus, it is essentially a statement of the uncertainty principle. It can be easily shown that the above criterion is satisfied in the present model.

In order to obtain specific results, a form for the band shape was chosen and, initially, a one-dimensional model was studied. For ease of calculation, it was assumed that the dependence of electron energy on the wave number is sinusoidal. Thus Tz can be evaluated as a function of the ratio of band width to kT, the position of the Fermi level and a single other parameter, γ, involving the ratio of the mean free path ℓ of the charge carrier to the lattice thermal conductivity, κ_L. Since the position of the Fermi level can, at least in principle, be adjusted by appropriately changing the carrier concentration, we may eliminate this variable by maximizing the figure of merit with respect to the Fermi level. This procedure was carried out numerically by means of a high-speed digital computer. Figure 1 shows Tz, maximized with respect to the position of the Fermi level, plotted against ϵ^*, the ratio of the band width to kT, for one-dimensional conduction. The parameter is defined as

$$\gamma = \ell k^2 T / \hbar \kappa_L$$

where k is the Boltzmann constant. Figure 2 gives the value of ζ, the reduced

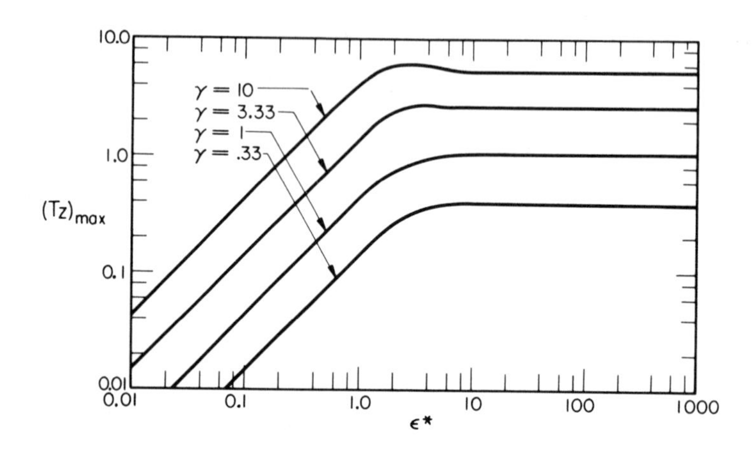

Fig. 1. Tz, maximized with respect to the position of the Fermi level, plotted against ϵ^*, the ratio of the band width to kT, for one-dimensional conduction.

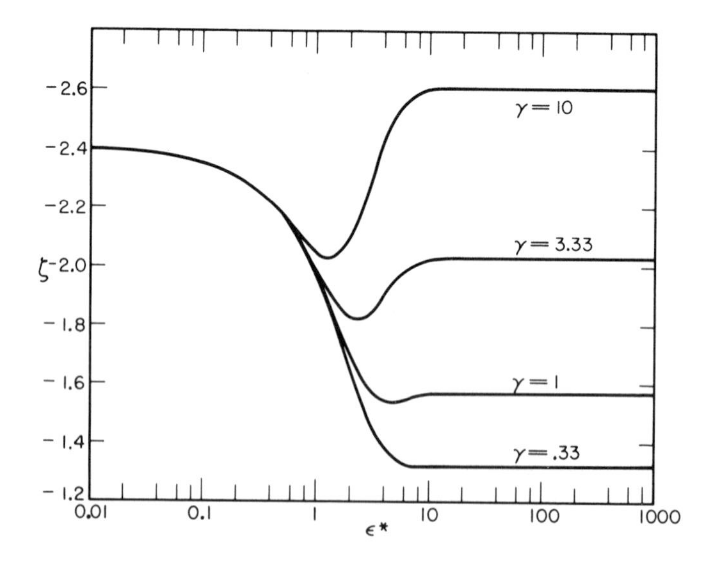

Fig. 2. The value of ζ, the reduced Fermi level, for which Tz is a maximum.

Fermi level, for which Tz is a maximum. It is seen that for $\epsilon^* \gtrsim 2$, Tz is essentially constant while below this value, Tz drops rapidly to zero. Physically, this can be understood by the following argument. If the band width becomes of the order of kT, states of negative mass (states in the upper portion of the band) are populated to a significant degree. As the band width approaches zero, the situation where half the carriers are of negative mass and half of positive mass is approached. Thus, the electrical conductivity goes to zero in the limit of zero band width. Similar arguments show that the electronic thermal conductivity also goes to zero. Furthermore, the calculations show that κ_e/σ, and thereby the Lorentz number, goes to zero at zero band width. Coupling this with the obvious fact that the Seebeck coefficient must remain finite in the limit of zero band width, one then sees the course of Tz versus ϵ^* for small ϵ^*.

Extending the calculation to three dimension, the energy dependence

$$E = \frac{1}{2} E_m \left\{ 1 - \frac{1}{3} \left(\cos ak_x + \cos ak_y + \cos ak_z \right) \right\}$$

was chosen since this combination of cosine functions having cubic symmetry is one of the simplest expressions that can represent a complete band.

The evaluation of the figure of merit for this case can be carried out in the limits of large and small band widths, and it is found that Tz again approaches zero as the band width tends to zero. For large band width, one finds that for optical mode and ionized impurity scattering, Tz decreases monotonically, from a maximum occurring at $\epsilon^* \approx 4$. For acoustical mode scattering, on the other hand, Tz continues to increase as $\epsilon^* \rightarrow \infty$. This arises essentially from the fact that the mean free path and therefore the materials parameter, γ, varies directly with ϵ^*. This does not, however, predict an infinite Tz as the band width increases because a point will eventually be reached where another scattering mechanism will limit the mean free path. From this point, Tz will decrease with increasing band width.

The conclusions developed on this model for a three-dimensional semiconductor are: (a) Tz goes to zero as the band width goes to zero; (b) for optical and ionized impurity scattering, Tz has a maximum at $\epsilon'^* \approx 4$; and (c) acoustic mode scattering in wide band materials does not show a maximum.

Reference

1. Peierls, R. E., Quantum Theory of Solids, Clarendon Press, Oxford, 1956.

DISCUSSION

R. A. CHAPMAN (Texas Instruments): It would seem that a comparison of these results to those obtained with a jump conduction model would be interesting. In the limit when the band width goes to zero, do your results converge to those obtained with the jump conduction model?

R. R. HEIKES: In the limit of zero band width, the results of this calculation do not reduce to the jumping model. The jumping model is, of course, based on a localized picture and such a set of localized states are completely independent of the band states. If the band width becomes smaller, a point will be reached where these localized states are of a lower energy than of the band states. At this point, one will transfer to a hopping-type model. One might estimate that such a transition would take place at band widths less than kT.

R. A. CHAPMAN: So the jumping process results in an inferior-type thermoelectric?

R. R. HEIKES: This is not necessarily so, for one can easily estimate that the maximum zT for a jumping process should be something of the order of 1/2 which will clearly be better than the narrow band materials below a certain band width.

R. A. BERNOFF (Melcor): Can your calculations for zT maximum be modified to cover a material such as silver antimony telluride, where you can have measured thermal conductivities of 0.005?

R. R. HEIKES: If you wish to utilize lattice thermal conductivities of the order of 0.005 in place of 0.01, this would increase the zT by approximately 50%. Thus, you might expect to get zT's of the order of 1.5.

R. A. BERNOFF: Then, on this basis, it would be reasonable to assume that you could exceed a zT of 1.

R. R. HEIKES: Yes. I hope I said at all times "of the order of 0.01," because certainly there is nothing magical about 0.01.

A. J. ROSENBERG (Tyco): The apparent limit on zT, even in the way you have analyzed it, depends finally on the relaxation time of electrons. Do you feel that enough is really understood about scattering mechanisms that one can relate them to the band width in as yet unexplored materials? It seems to me that in your band model the relaxation time does not enter specifically, that its inclusion would require additional assumptions and you have not really treated those.

R. R. HEIKES: As given in my paper, the figure of merit is found in terms of a parameter involving the ratio of the mean free path to the lattice of thermal conductivity. Clearly, the present theory can give no implications as to the value of this ratio.

The principal reason for carrying out this calculation was to determine the dependence of zT on band width; however, as the model chosen allows the effective mass to be disposed of, there is one less independent parameter to be concerned about. In a forthcoming paper, several methods will be presented for estimating an upper limit to zT. Clearly, all of these arguments must be based on a certain amount of empiricism.

H. SAM LEE (Astropower): In the discussion of the improvement of the figure of merit and the effect of carrier mass and mobility, would you be able to make any comment as to whether it was due to the increase in energy gap or the decrease in the lattice thermal conductivity, or perhaps we have to have both to achieve maxium zT.

R. R. HEIKES: The main thermoelectric problem, as we are all aware, is the maximization of the quantity $\mu m^*/\kappa_L$. Although we are all aware that these parameters are not mutually independent, it is frequently forgotten in devising selection criteria. The effective mass and mobility are clearly very intimately related to one another. Even the lattice thermal conductivity has a vague relationship to the electrical properties since it reflects the type of binding in the material. Therefore, in devising selection criteria, one must take simultaneous account of all three parameters.

Some Essentials of Cerium Sulfide

J. APPEL, S. W. KURNICK, and P. H. MILLER, JR.

General Atomic, San Diego, California

Abstract

Electrical, x-ray, and optical data on $Ce_{2+x}S_{3+x}$ $(0 \leq x \leq 1)$ and $Ce_{2+x}Ba_yS_{3+x+y}$ $(0 \leq x+y \leq 1)$ are presented and analyzed in terms of existing polaron theory.

INTRODUCTION

The rare earth sulfides have been under intensive investigation as thermoelectric materials for a number of years. Westinghouse, General Atomic, Radio Corporation of America (RCA), and Research Chemicals, a division of Nuclear Corporation of America, all have programs under way.

The detailed experimental results depend on whether the material is sintered or cast, on the oxygen contamination during processing, and on other details of its preparation. Rather than attempt to give a state-of-the-art review of the field, this paper will concentrate on a more intensive discussion of the physics of the transport processes of cerium sulfide which has been doped with column 2A metals of the periodic table, such as calcium, strontium, and barium. It is believed that the results and interpretations are typical of rare earth sulfides of similar structure.

The general procedure for the synthesis of the semiconductor rare earth sulfides has been along the lines suggested by Eastman and Brewer (1). Additional information on the synthesis of rare earth sulfides is described by Flahaut (2,3), Picon et al. (4), and Banks et al (5). Cerium sesquisulfide was prepared by passing a stream of H_2S over CeO_2 in the presence of carbon at elevated temperatures. Care is taken to avoid the compound of Ce_2O_2S, an intermediate step in the reaction. The final step in coverting the sesquisulfide to a solid solution of $Ce_{2+x}S_{3+x}$, where x is between 0 and 1, was accomplished by melting the compound at approximately 2000°C in a molybdenum boat.

The solid solutions of $Ce_{2+x}S_{3+x}$ and BaS (or any other sulfide of the column 2A metals) were most easily prepared by ball milling the appropriate mixture of CeO_2 and carbonates (or bicarbonates) of the dopant. This mixture was then subjected to H_2S at elevated temperatures (just like the cerium sesquisulfide).

The apparatus used for the melting and controlled freezing of cerium sulfide and cerium-barium sulfides is shown in Figure 1. Sesquisulfides to be melted down in the molybdenum boats are placed horizontally in the graphite susceptor, which in turn is surrounded by a molybdenum radiation shield. Both shield and susceptor have quartz standoffs for thermally isolating the heated objects from the Vycor tube. The Vycor tube is water cooled, not so much to protect it from the intense heating but to condense volatile products. Heating is performed by an induction coil which slides on

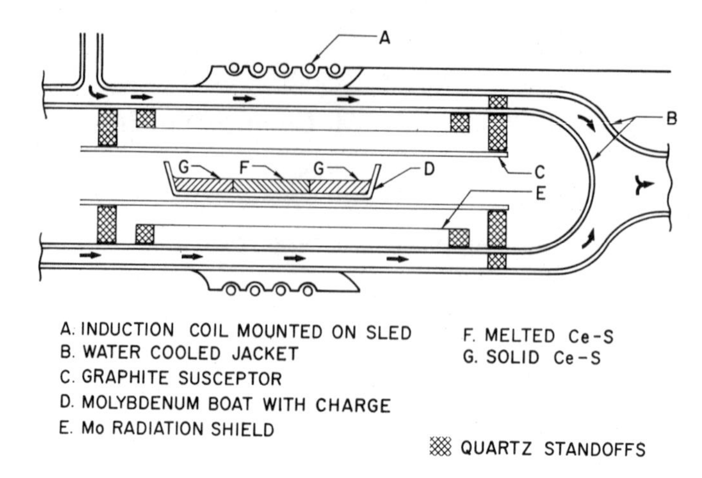

Fig. 1. Meltdown and controlled freezing apparatus.

Teflon runners led by a guide line pulled by a variable-speed motor. The apparatus is quite similar to zone melting, but the impurities and sulfur are boiled off. The chief purpose of the melted zone movements is to control the rapid freezing rate through the large gradient and to avoid the blow holes which are common in sulfide semiconductors. The length of the boats used is approximately 6 in. and the rate of cooling is such that the induction coil is moved along the entire length in approximately 20 min. The resistivity slabs cut from various sections along the length showed uniformity to within some 10% from back to front, in the pure cerium sulfides. However, in the alloy cerium sulfides, particularly with barium sulfide as a doping agent, there were concentration gradients much greater (approximately a factor of 2) down the length of an ingot.

Although no systematic attempts have been made to prepare a single crystal of $Ce_{2+x}S_{3+x}$, it has been noted that in some of the cast material the grains appeared to be about 5×10 mm in size. These apparent grains were cut out and x-ray examination showed them to be single crystals. We hope that larger single crystals can be prepared so that measurement of elastic constants and Debye temperature becomes possible.

GENERAL PROPERTIES OF THE $Ce_{2+x}S_{3+x}$ SEMICONDUCTORS

There are several phases present in the cerium-sulfur system: metallic CeS, the unstable CeS_2, and the γ-phase semiconductors in the range $Ce_{2+x}S_{3+x}$, with x between 0 and 1. The γ-phase is stable at high temperatures and is easily obtained on cooling from the melt. This phase is a semiconductor whose crystal structure is that of Th_3P_4. It is characterized by a cubic unit cell with 16 sulfur ions and a variable cation concentration of from 10-2/3 to 12 cerium ions. In the range where cation vacancies are present the sites are distributed randomly. The $Ce_{2+x}S_{3+x}$ semiconductors range stoichiometrically from Ce_2S_3, a bright red insulator, to Ce_3S_4, which is lustrous, black, and crystalline. It is interesting to note that as cerium ions are added to the vacant sites the lattice actually contracts instead of expanding as might be expected. This is apparently due to the increased bonding resulting from the extra electrons made available from the entering Ce^{3+}. On the other hand, the introduction of Ba^{2+} causes considerable increase in the lattice parameter.' This may be attributed to the weaker polar binding of the Ba^{2+} ion as well as the increased size.

It is a remarkable property of this semiconductor that the range of conductivity varies from 10^{-10} mho cm^{-1} at the Ce_2S_3 composition up to 10^3 mho cm^{-1} at the Ce_3S_4 extreme of composition, a relative change in conductivity of 10^{13} at room temperature. Thus, for the 4/3 vacant cerium sites there are no contributing carriers. The introduction of a divalent cerium ion onto a Ce^{3+} site results in the liberation of one electron. This electron, although tightly bound, is capable of moving to an alternative site. When there are many of these electrons available for conduction the transport mechanism may, of course, be more like that of a semimetal.

BARIUM-DOPED CERIUM SULFIDE

Adding various third ions to the vacant cation sites has effects which alter the size of the unit cell as well as the electrical properties. The addition of SrS to $Ce_{2+x}S_{3+x}$ has been covered previously (5,6). Diluent doping, a term applied by the Westinghouse group, refers to a doping procedure which enchances the conductivity while leaving the Seebeck potential unaltered. The only ion used in their results was Sr^{2+} and the doping concentration was low.

Dissolving BaS in $Ce_{2+x}S_{3+x}$ produces a solid solution $Ce_{2+x}Ba_yS_{3+x+y}$ ($0 \leq y \leq 1; 0 \leq x + y \leq 1$), which is isostructural with Th_3P_4. The results of measurements of electrical conductivity and Seebeck coefficient are shown in Figure 2. Figure 3 is a plot of the relative zT versus T for two samples of cerium-barium sulfide (D-6 and D-23) for comparison with undoped $Ce_{2+x}S_{3+x}$(BaS). This is to be compared with the relative zT's of $CeS_{1.38}$ as measured by the Westinghouse group (7) for various temperatures.

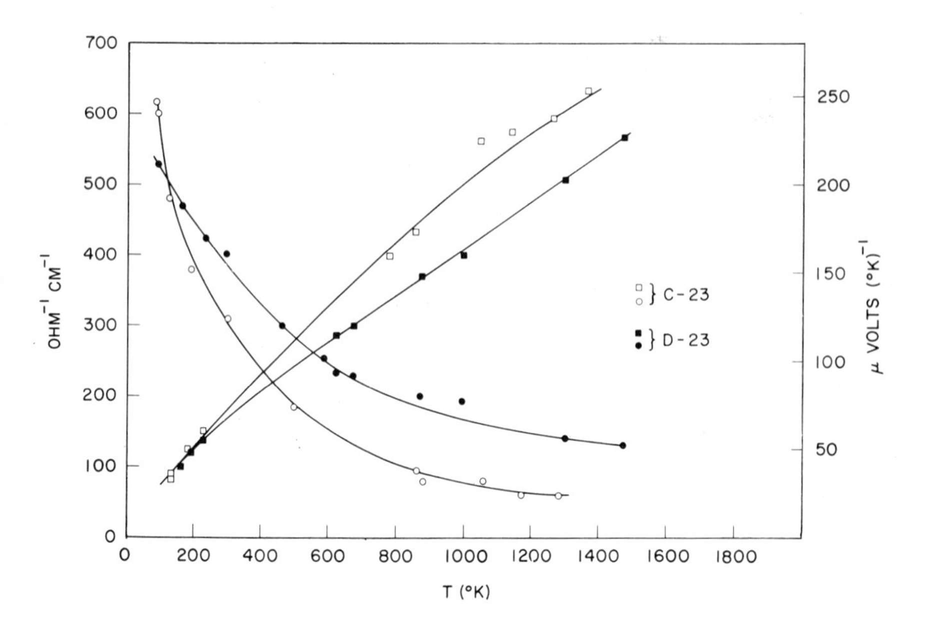

Fig. 2. Behavior of the Seebeck voltage and conductivity at different temperatures for a $Ce_{2+x}S_{3+x}$ semiconductor of higher conductivity, both doped and undoped.

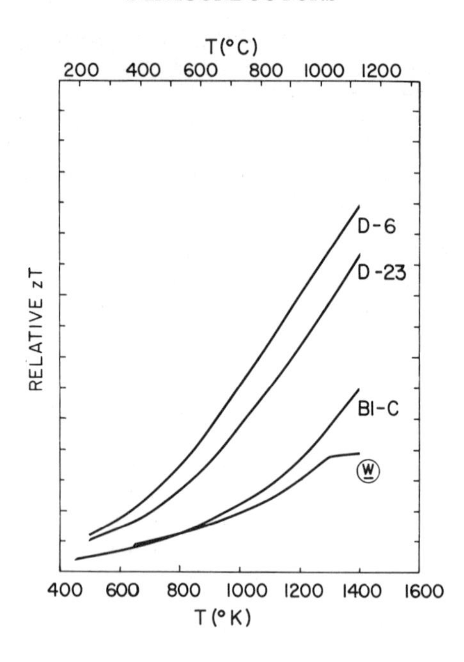

Fig. 3. Relative thermal conversion efficiency ($\alpha^2\sigma$) versus temperature for various samples: D-6 and D-23, cerium-barium sulfide; B1-C, cerium sulfide, prepared by melting; W, cerium sulfide, $CeS_{1.38}$, prepared by pressing and sintering (7).

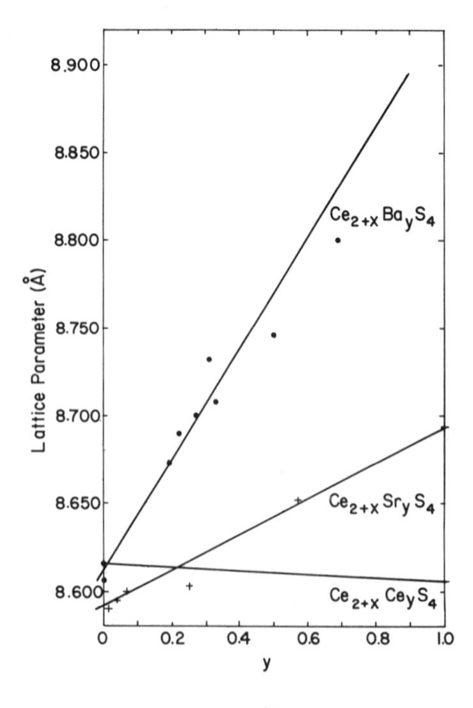

Fig. 4. Lattice parameter as a function of Ba^{2+} content; for comparison, the expansion produced by Sr^{2+} [from Banks et al., (5)] is also shown.

The x-ray powder patterns obtained with barium-doped samples showed a preferential broadening of the high angle reflections. This broadening, which increases with barium concentration, was interpreted as arising from a partial destruction of short-range order in the lattice because of the presence of the barium ions. The barium is present as Ba^{2+} Distributed randomly in a cation sublattice consisting mostly of Ce^{3+}. Since divalent barium is a substantially larger ion than trivalent cerium, localized lattice distortions are to be expected.

In a sample with a few per cent or more of barium, nearly all the ions will see a potential considerably more anharmonic than in an undoped sample, and these additional anharmonicities should increase the thermal expansion coefficient of the crystal. This possibility was checked experimentally. The linear thermal expansion coefficient of the barium-doped sample was 4.5% greater than that of the undoped sample, indicating increased anharmonicities in the doped sample. Whether or not the enhanced anharmonicity might be associated with slightly lower thermal conductivities is to be measured.

In addition to the broadening of the high angle reflections the x-ray data also revealed a change in lattice parameter. The barium ions expand the lattice considerably as shown in Figure 4. This expansion is probably quite closely related to the electrical properties of the material. As noted previously, the lattice of cerium sulfide containing dissolved strontium sulfide expands much less, and its electrical properties are not greatly changed from those of the undoped material. The lattice parameter of $Ce_{2+x}S_{3+x}$ is apparently unchanged by the addition of calcium (5) and the thermoelectric properties are, if anything, less favorable than those of $Ce_{2+x}S_{3+x}$ (see Fig. 5).

From the empirical relation (7) for the thermal conductivity $\kappa \propto a_o / (\alpha\gamma T)$, it can be seen that the change of 2.6% in a_o, the lattice parameter, is more than offset by the increase in α, the thermal expansion coefficient (4.5%). The parameter γ is the Gruneisen constant.

Hall measurements (Fig. 6) show tentatively that a barium-doped sample having the same room-temperature electrical conductivity as an undoped $Ce_{2+x}S_{3+x}$ sample achieves this conductivity by having a higher mobility and a lower carrier concentration than the undoped sample. The improved thermoelectric performance may thus be understood.

It is of interest, therefore, to inquire whether, when $Ce_{2+x}S_{3+x}$ is doped with BaS, the increase in volume of the doped crystal, is greater than the volume of the doping ions themselves. If greater, then the mean distance between ions other than barium has been increased and the overlap thereby decresed; if less, the converse is true. Using the ionic radii of 1.39 A for Ba^{2+} and 1.18 A for Ce^{3+}, and assuming that all the cerium present is characterized by the Ce^{3+} radius, we conclude that the average interionic distance in the crystal at room temperature is probably increased slightly by addition of the barium. In actuality, as cerium ions are added to the vacant sites, the lattice contracts. This is apparently due to the increased bonding resulting from the extra electrons made available from the additional Ce^{3+}. The increase will be greater at higher temperatures because of the larger thermal expansion coefficient of the barium-doped material. A similar estimate for the strontium-doped material, using 1.25 A as the Sr^{2+} radius, indicates essentially little change in the average interionic distance. The limitations of the ionic radius concept preclude further calculations along this line.

POLARON THEORY

Since the electrical properties of this semiconductor cannot be interpreted in terms of the conventional broad-band theory characterized by high

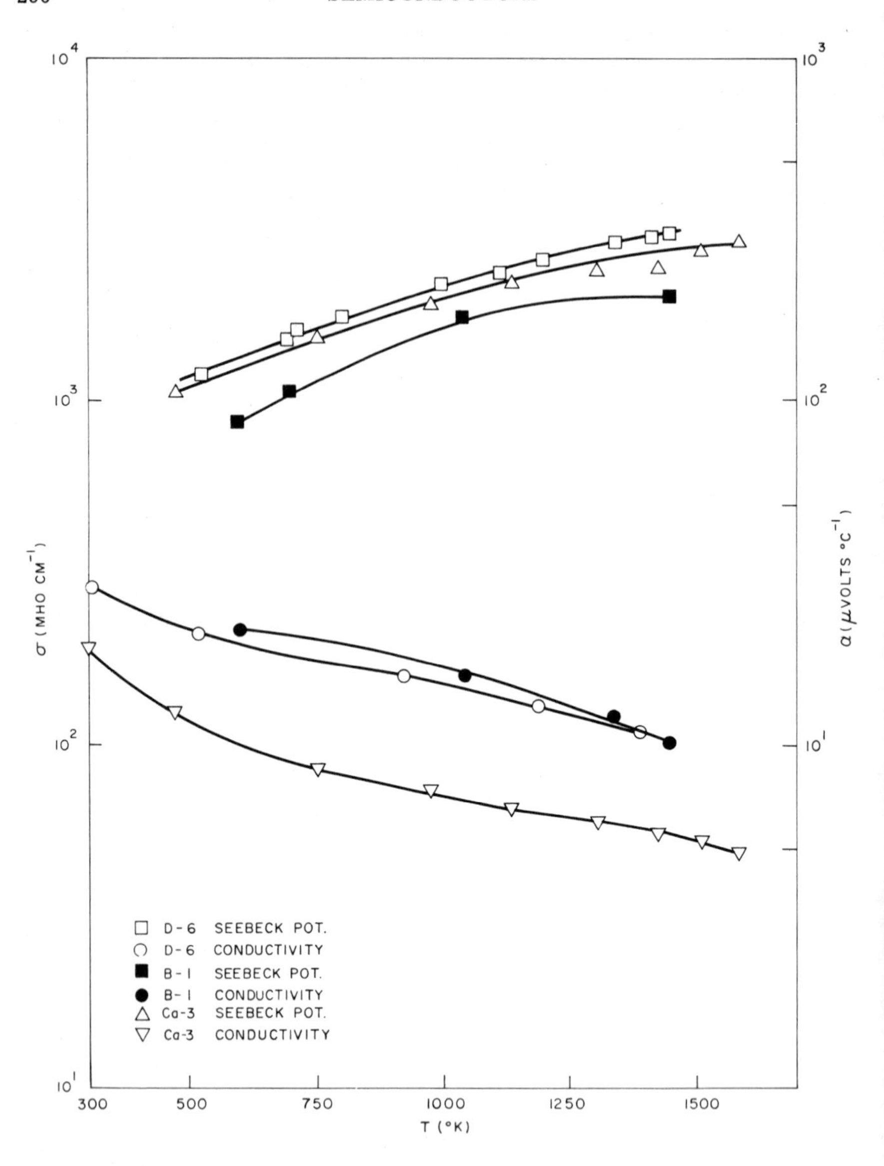

Fig. 5. Comparison of barium-doped sample (D-6) with calcium-doped material (Ca-3) and undoped cerium sulfide (B-1); this shows the enhanced electrical properties of barium-doped material.

mobilities, a new aspect involving conduction in a narrow band of a polar semiconductor applies. With respect to the polaron theory (8,10) and its application to $Ce_{2+x}S_{3+x}$ semiconductors, the question arises: Are the electronic charge carriers significantly different from slow conduction electrons moving in a rigid lattice? This topic has been discussed elsewhere (11), and we can not answer definitively at this time because we need experimental information on the longitudinal polarization modes.

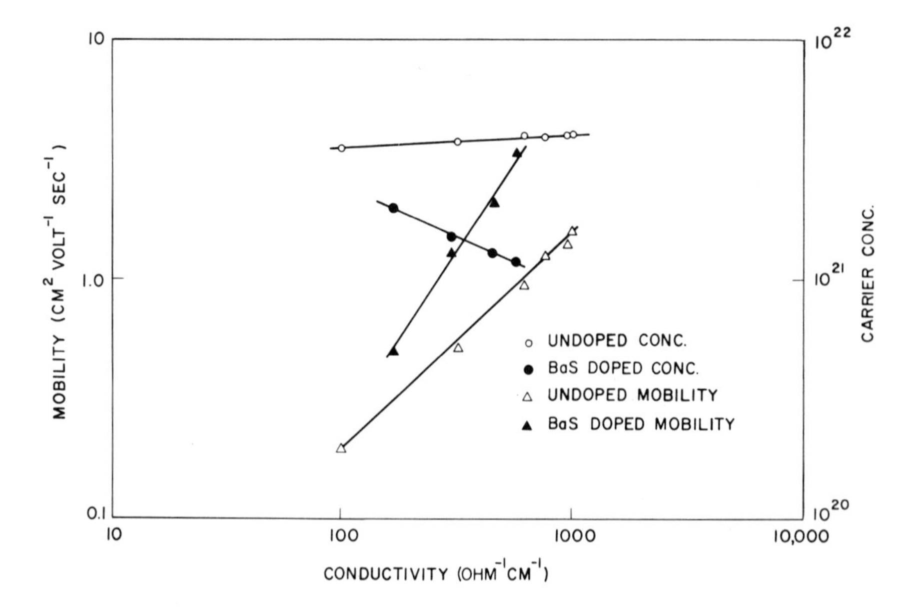

Fig. 6. Hall concentrations and mobilities for both doped and undoped $Ce_{2+x} S_{3+x}$ as a function of conductivity ($T = 300^{0}K$).

Some of the essentials of the polaron theory will be presented and later a review will be given of the experimental evidence which leads to the belief that a narrow polaron band determines the electronic conduction process in the cerium sulfide semiconductors. The Coulomb field of a slow conduction electron moving through a polar crystal displaces the lattice particles. The displacement polarization, in turn, has some influence on the motion of the conduction electron. The electron together with the accompanying self-consistent polarization field can be thought of as a quasi-particle. This quasi-particle is the polaron. The parameter of most interest is the displacement interaction, β, for the polar semiconductor (12). The magnitude of β will determine the strength of the interaction between the free electron and the displacement polarization field. The coupling constant β is defined as

$$\beta = 1/2\ (1/\epsilon_{\infty} - 1/\epsilon_{0})\ e^{2}u/\hbar\omega \qquad (1)$$

with u defined, in turn, as $u = (2m_{0}\omega/h)^{1/2}$, where ϵ_{∞} and ϵ_{0} are the optical and static dielectric constants, respectively, and ω the longitudinal vibration frequency. Then the weak and strong coupling theories for the large polaron, $\beta \ll 1$ and $\beta > 1$, are concerned with the ground state of the system and the low-lying excited states. Thus, the polaron states are determined by virtual emission and reabsorption of longitudinal polarization phonons. An obvious weakness of the strong coupling theories for large polarons lies in the continuum approximation.

The next major subdivision concerns the small polaron picture. According to the weak coupling theory for large polarons, $\beta < 1$, the polaron mass is given by $m^{**} = m_{0}^{*}\ (1 + \beta/6)$. Thus, for small coupling constants, the polaron mass is not much larger than the rigid lattice mass m_{0}^{*}, unless m_{0}^{*} itself is large. Then the continuum approximation breaks down, since it leads to a polaron the dimension of which is comparable to the lattice constant a_{0}. If

$ua_0 \sim 1$, a different kind of approach results in what is called the small polaron picture. The pertinent feature in the new approach is that the atomicity of the lattice is taken into account, and the polaron effective mass increases with rising temperature because the thermal motion of the lattice particle opposes the transfer of the mean positions connected with the motion of a polaron. Thus the polaron band width is a decreasing function of temperature until eventually when the energy uncertainty associated with the finite lifetime of the polaron is of the order of the band width, the band model breaks down.

APPLICATION OF POLARON THEORY TO $Ce_{2+x}S_{3+x}$

For the cerium-sulfide semiconductors, Hall measurements by ac techniques have been made utilizing low impedance transformers, high gain (narrow band pass) amplifiers, and magnetron magnets. To insure low impedance contacts, the Hall sample was first electro-plated in a copper-sulface solution after which the lead wire was soldered with indium. The excess copper coating was then removed by sand blasting. The samples prepared this way gave reproducible Hall voltages which corresponded to concentrations between 10^{21} and 10^{22} cm^{-3} and mobilities between 0.2 and 3 cm^2/volt-sec, depending on the electrical conductivity (see Fig. 6).

The important features of these measurements may be summarized as follows: Since both Hall coefficient and mobilities are low, accuracies of 20% are acceptable for interpretation. It must be stressed that these Hall measurements are open to revision because of the difficulties encountered experimentally. What they purport to show is that the barium-doped samples generally give lower Hall concentrations than those given by the undoped samples. A comparison of undoped $Ce_{2+x}S_{3+x}$ and some samples doped with barium, $Ce_{2+x}Ba_yS_{3+x+y}$, was made to investigate the role of distorting barium ions on the electronic properties. For the barium-sulfide-doped material, the Hall measurements were taken on samples with y = 0.7. Conductivities were altered, i.e., a change in x, by pumping off sulfur first from the melt. It was assumed that negligible barium sulfide was lost. Inasmuch as Hall voltages are quite low and the amounts of conductivity quite insensitive to temperature, as compared with a large conductivity variation of 10^{13} due to the changes in x alone, it was decided to restrict our preliminary measurements to room temperature.

The simplest interpretations of the Hall data have been used; i.e., the Hall coefficient being $R_H = 1/en_o$, where n_o is the carrier concentration and the mobility $\mu = R_H\sigma$. Theoretical interpretation of the Hall coefficient will depend on a more detailed knowledge of the transport theory than is presently available. The use of the equation, $R_H = r/n_oe$ (where $r \neq 1$) is meaningless when compared to the large error in the readings (20%). Also the difference between the usual weak field and strong field approximations may be meaningless for such a narrow band. What the curves of Figure 6 indicate is a general difference between the behavior of $Ce_{2+x}S_{3+x}$ and $Ce_{2+x}Ba_yS_{3+x+y}$. The drop in the Hall concentration is assumed to involve a change in x while y remains constant. This behavior may be questioned since the conductivity increases as x increases in the behavior of $Ce_{2+x}S_{3+x}$. However, the curves may be interpreted generally as a drop in carrier concentration with augmented mobilities as barium is introduced. No further interpretation can be made at this time. More work on the Hall coefficients will be conducted.

Much information on the nature of the carrier transport in $Ce_{2+x}S_{3+x}$ may be gleaned from optical and dielectric measurements on this semiconductor. One such quantity is the coupling constant, which measures the interaction of the polaron with either the acoustical or the optical lattice modes. According to eq. (1), the coupling constant depends on four parameters. Three of these parameters can be measured directly: the frequency of the longitudinal polarization waves, ω, the static dielectric constant, ϵ_0 and the dynamic or optical dielectric constant, ϵ_∞, which in nonabsorbing materials is related to the index of refraction, n, by $\epsilon_\infty = n^2$. This frequency is related to the reststrahlen frequency, ω_r, by Fröhlich's relation:

$$\omega = \left(\frac{\epsilon_0}{\epsilon_\infty}\right)^{1/2} \omega_r \qquad (2)$$

Experimentally, $\omega > \omega_0$ in measuring ϵ_∞ and $\omega < \omega_0$ when determining ϵ_0. The frequency ω is the frequency at which the refractive index and the dielectric constant become infinitely large.

The dielectric constant of several thin plates of Ce_2S_3 was determined by placing them between and in contact with the plates of a condenser. Measurements were made on a bridge at audio frequencies (1 kc). The dielectric constant was then determined from the change in capacity when the sample was introduced. The dielectric constant so measured was $\epsilon_0 = 19$.

The index of refraction may be derived using Fresnel's relations. The sample was prepared by optically polishing a large grain sized block of $Ce_{2+x}S_{3+x}$, whose conductivity was 600 mho cm^{-1}. Figure 7 shows the ratio of reflectivity of the two principal polarizations of incident light as a function of the angle of incidence. At the minimum eq (3) is valid

$$\tan^2 \theta_{min} = n^2 + k^2 \qquad (3)$$

where $n^2 \leq 8$, depending on the absorption of the material. Since in absorbing material $\epsilon_\infty = n^2 - k^2$, we see that the dielectric constant at 0.6 μ is less than half the static value of 19 for the dielectric constant.

To minimize the absorptivity, k, the reflectivity, R, was then measured on Ce_2S_3 cast samples. The procedure followed was to heat treat the optical blanks of $Ce_{2+x}S_{3+x}$ to the red insulating form, Ce_2S_3, in an H_2S atmosphere at 1500°C. The reflectivity measurements were made before and after sulfurization. This processing eliminates the excess carriers and, consequently, gives a reflectivity more like that of an insulator than that due to the many excess electrons whose presence is indicated by high conductivity. It is interesting to note that although the original conductivities may have differed primarily due to the difference in the cerium-sulfur ratio, these have all been cancelled out and all are consistent with one another.

The reflectivity of a sample is given by the general formula,

$$R = \frac{(n-1)^2 + k^2}{(n+1)^2 + k^2} \qquad (4)$$

The absorptivity of one sample of Ce_2S_3 has been estimated from measurements of its absorption coefficient, μ'. Preliminary results give

$$k = \frac{\mu' \lambda}{4\pi} \simeq 10^{-2}$$

indicating that eq. (4) can be simplified to

$$R = \frac{(n-1)^2}{(n+1)^2}$$

This procedure yielded an index of refraction of 2.5, which agrees well with $n \leq 2.8$, derived from the polarized light method described above.

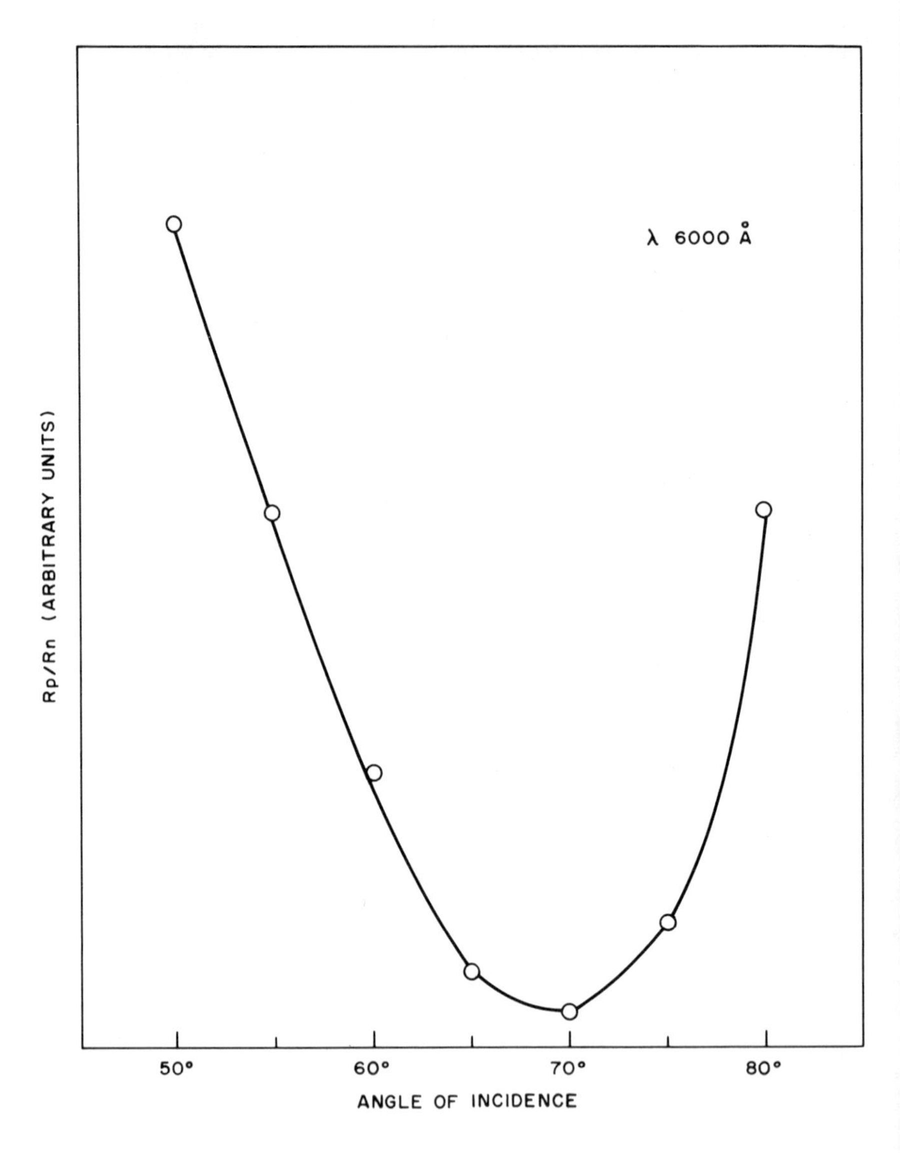

Fig. 7. Ratio of reflected polarized beam consisting of polarization perpendicular and parallel to plane of reflection as a function of angle of incidence.

There was little dispersion in the region of measurement between 1 and 13 μ . At the present moment, ω_r remains undetermined.

On the basis of the experimental results obtained so far, the question arises whether or not the charge carriers in cerium sulfide semiconductors are polarons which are significantly different from slow conduction electrons. Our conclusion is that the electronic charge carriers are slow polarons which can be adequately described by Bloch-type eigenstates which travel through the crystal not by phonon-activated jump processes but by frequent interchange between next nearest cerium ions by tunneling. The probability for such a tunneling process is directly proportional to the value of the appropriate overlap integral, which depends on the lattice displacement and eventually on the temperature. A large difference has been found between the static and optical dielectric constants ϵ_0 and ϵ_∞. This implies that the effective charge of the lattice particle is about the same as in crystalline insulators characterized by predominant polar binding. The difference $1/\epsilon_0 - 1/\epsilon_\infty$ is proportional to the strength of the electron-lattice interaction. However, Frohlich's coupling constant β depends on two additional parameters: the frequency of the longitudinal optical mode ω and the rigid lattice effective mass m*. Substituting the values of ϵ_0 and ϵ_∞ into eq. (1) gives

$$\beta = 1.55 \times 10^7 \left(m_0^*/m_e\right)^{1/2} \omega^{-1/2} \tag{1a}$$

CONCLUSIONS

An appropriate average value for ω and the quantitative value of the effective mass m_0^* is not known, but certainly m_0^* is greater than m_e, the mass of the free electron. This is implied by: (1) the relation between the polaron effective mass and the rigid lattice effective mass for coupling strengths $\beta \sim 1$, and (2) the polaron effective mass obtained from the results of the measured Seebeck voltage versus temperature.

The combined measurement of both the Seebeck effect and the Hall effect has the important advantage that one can measure the charge carrier concentrations as well as their effective mass m*, or m** in the case of polarons. If the Hall effect is interpreted on the basis of the conventional formula $R_H = 1/(en_0)$, one can determine n_0 and μ. With the further assumption that the carrier concentration is independent of temperature, one can obtain the polaron

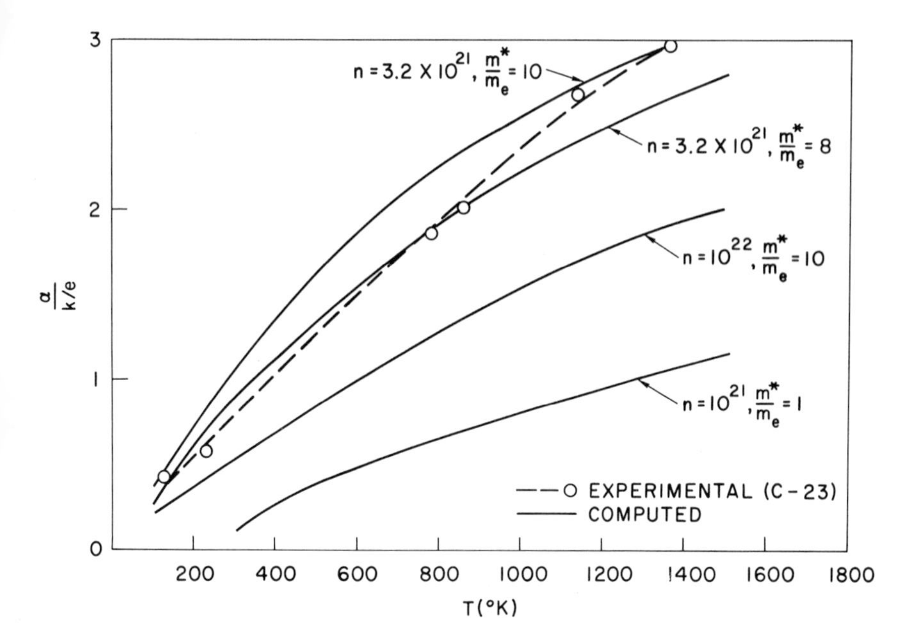

Fig. 8. Fitting of theoretical to experimental thermoelectric power data for sample C-23.

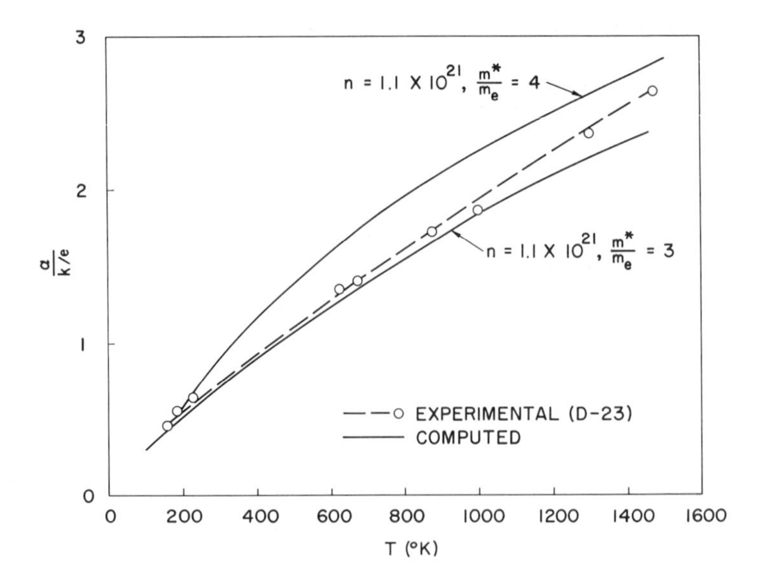

Fig. 9. Fitting of theoretical to experimental thermoelectric power data for sample D-23.

effective mass by analyzing the measured temperature dependence of the Seebeck coefficient with Pisarenko's formula (13)

$$Q = \pm k/e \left\{ (r + 2)/(r + 1) \left[F_{r+1}(\eta) / F_r(\eta) \right] - \eta \right\}$$

where k/e=86 μvolts deg^{-1}, $\eta = \xi/kT$ is the reduced Fermi energy, and $F_r(\eta) = \int_0^\infty x^r dx/\exp(x-\eta)+1$. The scattering index r is taken equal to zero since the electrical conduction depends only slightly on temperature at high temperature. It may well be justified to assume that at high temperatures the free path of the carrier is nearly energy independent. With the value r=0 the curve in Figure 8 has been calculated to best fit the experimental data. In addition, Figure 9 shows the calculated curves for the Hall coefficient concentration of the barium-doped cerium sulfide sample. This shows a much lower effective mass and, in addition, agrees with the higher mobility found in the data of Figure 6.

ACKNOWLEDGMENTS

The work in this paper was the joint accomplishment of many members of the staff of General Atomic. We wish to acknowledge the cooperation of Dr. J. Apfel in the optical measurements, Dr. M. F. Merriam for the dilatometric data, Mr. Lee LaGrange for his considerable help in the chemistry, and Mr. R. L. Fitzpatrick for the electric measurements. Work supported under Contract NObs-77144, Bureau of Ships (Advanced Research Projects Agency, ARPA Order No. 81-59).

References

1. Eastman, E. D., et al., J. Am. Chem. Soc., 72, 2248 (1950); J. Am. Ceram. Soc., 34, 128 (1951); J. Am. Chem. Soc., 72, 4019 (1950).
2. Flahaut, J., and M. Guittard, Compt. rend., 242, 1318 (1956).
3. Flahaut, J., and M. Guittard, Compt. rend., 243, 1419 (1956).
4. Picon, M., et al., Soc. chim. France, Bull., 2, 221 (1960).

5. Banks, E., et al., J. Am. Chem. Soc., 74, 2450 (1952).
6. Westinghouse Thermoelectricity Quarterly Rept. 4, Nov. 18, 1959.
7. Westinghouse, Thermoelectricity Quarterly Rept., 4, B28, Sept. 15, 1960.
8. Sewell, G. L., Phil. Mag., 3, 1361 (1958).
9. Holstein, T., Ann. Physik, 8, 325, 343 (1959).
10. Yamashita, J., and T. Kurosawa, J. Phys. Soc. Japan, 15, 802 (1959).
11. Appel, J., and S. W. Kurnick, J. Appl. Phys., 32, 2206 (1961).
12. Frohlich, H., Advances in Phys., 3, (1954).
13. Ioffe, Semiconductor Thermoelements and Thermoelectricity, Infosearch Ltd., 1957, p. 102.

DISCUSSION

R. K. WILLARDSON (Bell and Howell Research Center): Did you say your Hall effect and conductivity measurements were ordinary dc measurements?

S. W. KURNICK: No, they were ac measurements. We used a low input transformer, preamplifier, and wave analyzer to measure the low level signals. The equipment was checked out on zinc, which has a very low Hall coefficient and mobility. For example, to measure the dc Hall voltage in zinc one generally pushes about two amperes through the sample to obtain a sizable reading. We succeeded in doing it by ac with a few milliamperes. Hall voltages measured were as low as 10^{-10} volt, and the corresponding Hall coefficient agreed well with the dc measurements on zinc. We feel that while the Hall coefficients and mobilities we have are reasonable, we do not yet wish to make an analysis of the absolute Hall coefficients. There appears to be some difference between the doped and undoped samples.

Preparation and Properties of Some Rare Earth Semiconductors

FORREST L. CARTER

Westinghouse Research Laboratories, Pittsburgh, Pennsylvania

Abstract

The preparation of sintered samples of several chalcogenides having the Th_3P_4 structure is described in detail. The lanthanum sulfides and cerium selenides are inferior to the sulfides of cerium and praseodymium for use as thermoelectric elements. The use of the metal hydrides to prepare cerium antimonide and bismuthide at temperatures below 500°C is described.

From $LaH_{0.91}Te$ the maroon, highly conducting LaTe was prepared. The electrical properties of LaTe and $LaTe_3$ are given as a function of temperature. Similarly, the electrical properties of the narrow band-gap ternary CeSSb are shown as a function of temperature, and the preparation of CeSSb and other ternaries is described.

For a few compounds, small single crystals suitable for x-ray analysis have been prepared by various methods. These include: (a) Ce_2S_3 using a vapor transport method at ca. 925°C, (b) $CeSe_2$ lamenae formed by refluxing the selenide for 12 months in excess selenium at ca. 670°C, and (c) gold leaflets of $LaTe_3$ formed simply from reaction of the elements at 750°C in 20 hr. Prepared in the above manner Ce_2S_3 is cubic with the Th_3P_4 structure while the latter two materials are tetragonal. The $LaTe_3$ shows considerable disorder in the c_0 direction.

INTRODUCTION

Methods of preparing and handling rare earth semiconducting compounds are worthy of discussion in view of the increasing interest in these compounds as thermoelectric elements and possible solid state materials. In order to facilitate the work of others in this field, the following is reported on:

1. Useful techniques in the chemical preparation of small lots of materials (ca. 50 g) and, in particular, on the preparation of sintered samples.
2. Various compounds prepared to date, typical methods of preparation, and some known electrical properties.
3. Unit cell data obtained from preliminary single crystal studies of Ce_2S_3, $CeSe_2$, and $LaTe_3$.

TECHNIQUES

The preparative techniques which exist have evolved primarily because either the product and/or its precursors (especially as powders) are subject to degradation from atmospheric moisture and/or oxygen. In general, then, all operations such as weighing, grinding, sieving, pressing, and sintering are performed in a dry atmosphere of argon or argon and hydrogen.

The general sample handling is carried out in an argon-flushed glove box (S. Blickman, Inc. type FLH) with dishes of P_2O_5 present to absorb the moisture which enters through the gloves.* The inclusion of a press and a torsion spring balance in the glove box simplifies much of the handling procedures. Small samples may be ground conveniently in the glove box with mortar and pestal, but larger samples are ground outside in a pica blending mill (Pitchford Scientific Instrument Corp.). In order that the pulverization could be done in an argon atmosphere, the vial clamps were modified so that they could be conveniently attached to the grinder while they held a vial. Thus the ceramic vial would be loaded in the dry box with sample, sealed with the clamp, and after the material was ground outside, then could be reintroduced via the antechamber without air contamination. To further insure air exclusion, the sample vials, after being loaded, were sealed in a polyethylene bag before being placed in the clamps. This method was used to grind the pyrophoric rare earth hydrides.

Rectangular samples ($1 \times 1 \times 3.2$ cm) for electrical measurements were pressed in a short direction in a special double-acting die. The four walls of this die can be released from the sample by means of four large allen head bolts (per side).† Even so, considerable care must be taken to keep the walls clean if several pellets of the same material are to be successively pressed without sample sticking. (A special problem in the case of the tritellurides like $LaTe_3$.) Green pellets destined for electrical measurements were protected against the slightest jar as sample damage might be noted only after the sample had been sintered and trued up for mounting.

Another die of general preparative use was a 2.5 cm diam tungsten carbide die with one end slightly tapered for pellet expulsion without flaking.

Mixtures may be reacted or samples sintered in argon and/or hydrogen in the Vycor apparatus shown in Figure 1. For homogenization of reaction, 2.5 cm diam pellets are loaded into the graphite crucible and placed in the reaction vessel, which may be brought into the glove box via the vacuum antechamber. Even during storage, air is excluded from the interior of the vessel as the boron nitride‡ insulators absorb water and oxygen, both of which during heating may be released and react with the sample. The sample may be rapidly and safely heated to 1500°C using rf techniques. Samples to be sintered for electrical measurements are centered using graphite dishes with rectangular holes in them. Temperatures are determined with an optical pyrometer from a 3-mm thick graphite disc placed over the sample. Rare earth sesquisulfides and sesquiselenides are placed on sheet molybdenum discs for sintering since these do not stick to the samples.

Rectangular pellets have also been sintered in the molybdenum bomb which is shown in Figure 2 with a cap puller used for opening it. The ordinary screw-cap construction of the bomb is obviated by the high sintering temperature (1500°C) of the sesquisulfides and sesquiselenides. Welding occurs at molybdenum-molybdenum contacts when the sintering temperature is above the recrystallization temperature of molybdenum (ca. 1250°C). Two platinum gaskets (0.5 mm thick) are used to separate the cap from the base. The top gas-

---

*An extra pair of rubber gloves is worn over the regular box gloves to protect them from the abrasion and laceration of normal use.

†Designed by Paul Snyder of these laboratories.

‡In a recent talk, Dr. A. Rabenau of Philips Research Laboratory, Aachen, Germany, indicated that boron nitride may be freed from B_2O_3 by heating the material at 2000°C in an ammonia atmosphere.

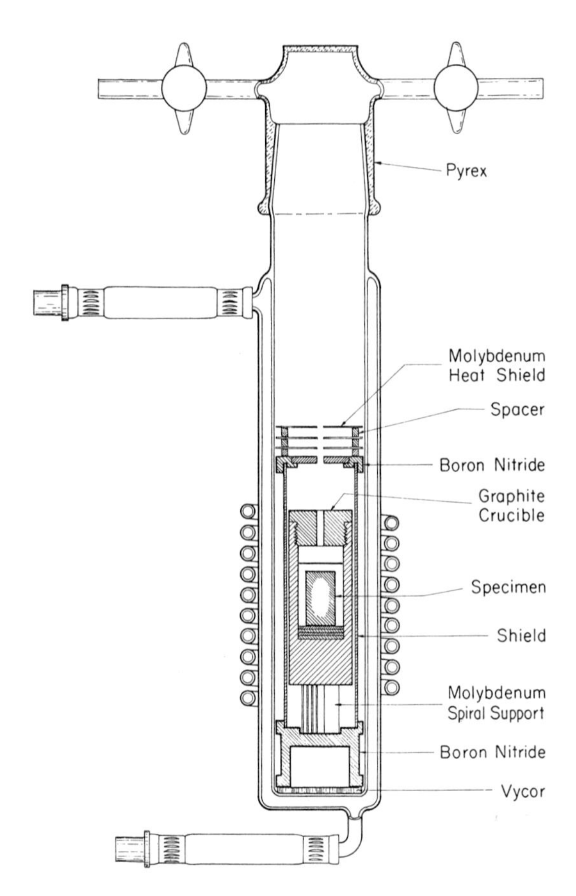

Fig. 1. Vycor apparatus for rf heating.

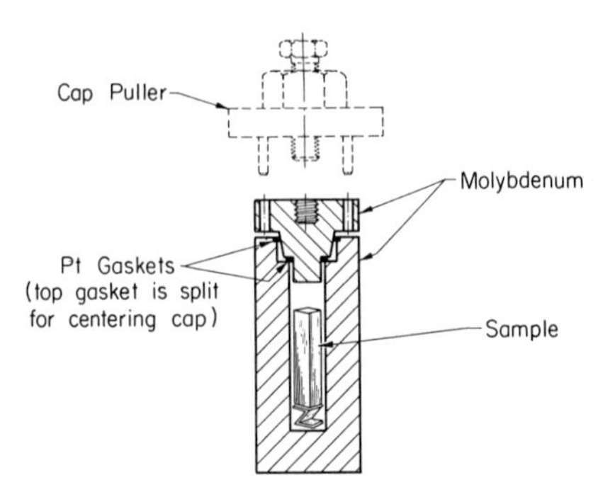

Fig. 2. Molybdenum bomb for sintering cerium sulfide samples.

ket, which is an unclosed ring, centers the tapered cap when it is seated in a press (10,000 lb) against the bottom platinum O-ring. The molybdenum bomb and its contents are then placed in the apparatus of Figure 3. This apparatus includes a special frame which holds the cap in place during firing by means of a large compressed spring (spring rate, 20,000 lb/in.) removed from the rf field. The spring also provides for the differential expansion of the hot bomb and the much cooler supporting frame. While the chalcogen pressure of the sesquisulfides at sintering temperature is quite small (1), such an arrangement as above, when the spring is under compression, would presumably be

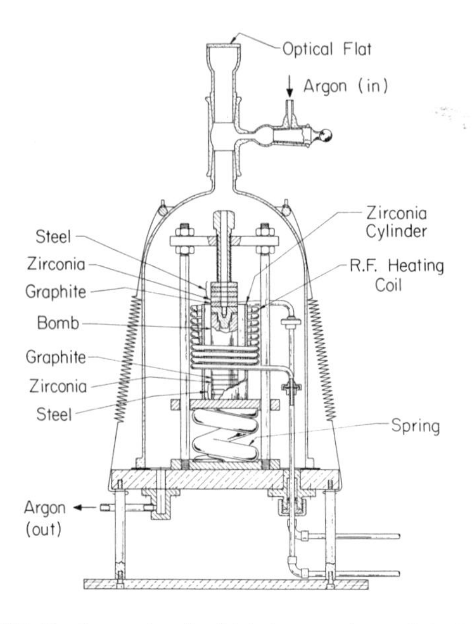

Fig. 3. Apparatus for high temperature sintering.

suitable for an internal pressure of several atmospheres. Due to the increasing embrittlement of the bomb with use at high temperatures it is often necessary to remachine the gasket surfaces.

EXPERIMENTAL DETAILS

Lanthanum and praseodymium metal ingots were obtained from the Lindsay Chemical Division of American Potash and Chemical Co. while the cerium ingots were purchased from Varlacoid Chemical Co. The indicated purities were: lanthanum, 99.9%; cerium, 99.45%; and praseodymium, 99.9%. The lanthanum was found by our analytical laboratory to contain 0.5% yttrium; the cerium contained ca. 0.5% iron. The sulfur, selenium, and tellurium were used as obtained from American Smelting and Refining Co. and had been labeled with the following respective purities: 99.999%, 99.99+%, and 99.99%. Arsenic (99.99+%) and bismuth (99.99+%) were purchased from the same company while antimony (99.997%) was obtained from the Bunker Hill Co. Cerium sesquisulfide (99+%) was sold to us by Union Carbide Corp.

The oxides used were furnished by Lindsay Chemical Division and had the following stated purities: La_2O_3, 99.99%; CeO_2, 99.9+%; and, Pr_6O_{11}, 99.9%. Immediately prior to use they were heated to ca. 850°C and allowed to cool in a desiccator to remove adsorbed water.

Samples prepared for powder diffraction analysis were ground, sieved (150 mesh), loaded into the sample tube, and sealed with Apiezon Q wax in a dry argon atmosphere. A Debye-Scherrer powder camera (114.59 mm diam) was used with filtered copper radiation.

The rare earths were analyzed (by O. H. Kriege) by titration with EDTA solutions and gravimetric procedures while sulfur, selenium, tellurium, bismuth, and antimony were analyzed by known gravimetric procedures. Analysis for hydrogen was performed by burning the sample at ca. 900°C and collecting the water formed.

PREPARATION AND PROPERTIES

In the rare earth elements one finds two features which might be expected to give rise to new and complicated coordination configurations about the rare earth atom or ion. These two features include the larger size of the rare earth and the availability of f-orbitals for bonding as well as the usual s-, p-, and d-orbitals. Hence a larger number of near neighbors might be accommodated and secondly these might be arranged in a greater variety of ways. This is borne out by the number of new structure types reported by Zachariasen (2) in his "5f series." In conjunction with the new structure types, one might also anticipate some interesting chemistry and physics.

Some Compounds with the Th_3P_4 Structure

At the present time, neither the chemistry nor the electrical properties (3, 4) of the rare earth compounds having the Th_3P_4 structure are well understood. The case of cerium sulfide, CeS_x (where $1.5 \geq x \geq 1.33$), will be briefly discussed. This material, in its high temperature or γ- phase (5,6) (Th_3P_4 structure), varies continuously in its electrical resistance from the composition of Ce_2S_3 (or x = 1.5), where it is an insulator, to that of a semimetal (ρ = 5 x 10^{-4} ohm-cm) when the composition is Ce_3S_4 (or x = 1.333). The low temperature phases,* α and β, whose unknown structures appear complicated, have similar electrical resistance properties as a function of composition. The transition temperature between the γ-phase and the others is remarkably sensitive to small percentages of impurities. For example, less than a per cent of oxygen† promotes the decomposition of the γ-phase while SrS is effective in stabilizing the Th_3P_4 structure in an oxygen-containing sample (7).

The pure α-phase may be prepared from the disulfide by removing sulfur at approximately 800°C according to

$$CeS_2 \overset{\Delta}{\rightleftharpoons} CeS_{1.5}\ (\alpha) + \frac{1}{4} S_2 \tag{1}$$

The decomposition of the diselenide, at a lower temperature (760°C), gives the γ-phase instead. This indicates, of course, that the transition is strongly dependent on the size of the anion.

In addition to the reduction of ceria by H_2S at about 1300°C in the presence of graphite, as used by Eastman et al. (1), cerium sesquisulfide may also be

---

*At least two phases in addition to the γ-phase are indicated in powder photographs.

†Just how small an amount of oxygen is effective in causing the transformation is not known since a direct determination of oxygen for these materials has not been available and determination by difference is very uncertain.

prepared from CeO_2 with the use of carbon disulfide and a carrier gas at a much lower temperature.* This reduction reaction may be written as

$$2\,CeO_2 + 4\,CS_2 \overset{\Delta}{\rightleftharpoons} Ce_2S_3 + 4\,CO + \frac{5}{2}\,S_2 \qquad (2)$$

where the driving force is presumably the formation of carbon monoxide. We have prepared Ce_2S_3 in a graphite boat in this manner in an hour at temperatures from about 790 to 860°C. This technique, however, has two purity problems associated with it. The first is the deposition of carbon in the sample from the decomposition of carbon disulfide, and the second is the uncertainty of the oxygen content of the product.

The carbon may be removed by the reversal of the reaction that was its source. Hence, according to eq. (3), the carbon may be removed at approximately 800°C by the formation of carbon disulfide which is swept away

$$C\ (solid) + S_2\ (gas) \overset{\Delta}{\rightleftharpoons} CS_2\ (gas) \qquad (3)$$

by the oxygen-free carrier gas (argon). For this reaction it is better to use a silica boat as graphite boats are attacked.

The first reaction product in the reduction of CeO_2 is a light yellow-brown solid, Ce_2O_2S, which in one case appeared upstream of the hot zone at a temperature of about 650°C. This was also prepared in 48 hr in a Vycor tube at 850°C according to

$$CeS + CeO_2 \longrightarrow Ce_2O_2S \qquad (4)$$

Immediately downstream of the Ce_2O_2S the product was black, but it quickly lightened to a maroon in the hottest zone.

Unfortunately, the maroon color of Ce_2S_3 cannot be used as a reliable index of its oxygen impurity content. Three powdered samples, each prepared in a different manner and each having a different oxygen content, had very similar colors and x-ray patterns.† One was prepared from the decomposition of cerium disulfide [eq. (1)], one from the reduction of CeO_2 with carbon disulfide [eqs. (2) and (3)], and one with the composition of $CeS_{1.2}O_{0.3}$ made from Ce_2S_3 and Ce_2O_3 [eq. (5)] by fusing a mixed pellet at 1500°C:

$$4\,Ce_2S_3 + Ce_2O_3 \overset{\Delta}{\rightleftharpoons} 10\,CeS_{1.2}O_{0.3} \qquad (5)$$

The second sample, prepared from the reaction of CeO_2 and CS_2, had the composition $CeS_{1.18}O_{0.28}$ as determined by chemical analysis. Repeated reduction and/or carefully controlled conditions should give pure Ce_2S_3 by this method.

The Sesquisulfides and Sesquiselenides from the Rare Earth Hydrides

The preparation of the sulfides and selenides of lanthanum, cerium, and praseodymium from the metals is conveniently accomplished through the metal hydrides.‡ The equations involved are the following for the preparation of $CeS_{1.50}$:

$$Ce + \frac{3}{2}\,H_2 \longrightarrow CeH_3 \qquad (6)$$

$$CeH_3 + S\ (xs) \longrightarrow CeS_2 + \frac{3}{2}\,H_2 + H_2S\ (some) \qquad (7)$$

*From a private discussion with Dr. R. Didchenko of Union Carbide Corp.
†These complex photographs contain numerous overlapping and poorly defined lines.
‡For convenience these will be considered as the trihydride (e.g., CeH_3), see ref. 8.

$$CeS_2 \xrightarrow{800^\circ C} CeS_{1.5} + \frac{1}{4} S_2 \qquad (1)$$

The metal hydrides may be prepared by heating metal rods, approximately 1.5 cm^2, in an atmosphere of dry hydrogen over a Bunsen burner in a simple apparatus like that in the right-hand side of Figure 4. Just prior to use the rods should be carefully freed from all visible oxide formation by light filing. When no further hydrogen is absorbed, the expanded and black metal hydride is ground, sieved (150), and stored in a dry argon, or preferably, hydrogen atmosphere.

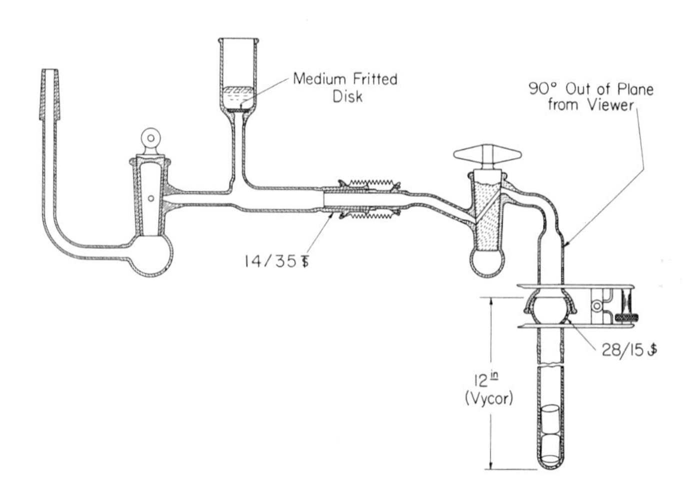

Fig. 4. Apparatus for hydride reaction.

A mixture of the hydride and the chalcogen (say, sulfur) is then spread evenly along a large part of the reaction chamber (Fig. 4) and partially reacted at various places by heating with a Bunsen burner in the hood until exothermic reactions are initiated. Care must be exercised to prevent too rapid an evolution of hydrogen through the mercury gas seal and to prevent the fritted disc from being clogged with sulfur distilled during too vigorous a reaction. The partially reacted mixture is introduced into a furnace at about 350°C and heated slowly to 600°C. During this time the unreacted chalcogen is allowed to reflux by placing the reaction chamber at an angle of about 15°. When the reaction has stopped, as indicated by product appearance and the lack of hydrogen flow, the next step is dependent on what product is desired.

Refluxing is continued until the product has a uniform color if the disulfide is desired, finally distilling off the excess chalcogen at a low temperature (440°C).

The sesquichalcogenides are prepared from the initial reaction product by slowly heating it to 800°C (3 hr) in a horizontal reaction chamber, thereby decomposing the dichalcogenides.

In these reactions excess chalcogen is used (25 to 50% of that calculated for the dichalcogenide). Attempts to use this method to prepare, say, $CeS_{1.40}$ directly result in a product which is inhomogeneous and low in chalcogen. When the chalcogen content is low, the mixture is easily overheated resulting in a violent reaction.

As suggested by Eastman et al. (1) the hydrides may also be used to reduce the sesquichalcogenides according to the example of cerium sulfide, eq.(8).

$$CeS_{1.5} + z\ CeH_3 \rightarrow (1+z)\ CeS_{1.5/(1+z)} + z\ \frac{3}{2} H_2 \quad (8)$$

This is conveniently accomplished by sieving the powders together through a 150 mesh sieve and heating the compacted pellet by rf techniques in the apparatus of Figure 1 to 1200°C for about 15 min. To achieve homogeneity the reacted pellet may be ground and fired a second time.

Inasmuch as the powdered sulfides absorb water and oxygen when exposed to air, the chalcogenides should be freed of these contaminates before use by slowly heating the sample in 1/2 atm of hydrogen to ca. 350°C followed by evacuation.

Sintering Cerium Sulfide for Thermoelectric Measurements

After the sample CeS_x ($1.5 \geq x \geq 1.33$) has been homogenized and pressed with a force of 15,000 lb into a rectangular pellet it is very carefully loaded into either the graphite crucible shown in Figure 1 or the molybdenum bomb shown in Figures 2 and 3. The sample is then heated by rf techniques to the sintering temperature (ca. 1400°C) for a short time and allowed to cool. The heating and cooling rate above 800°C is approximately 800°C per hour. A properly sintered sample should not appear porous and has only a few small gas holes. The shrinkage will amount to approximately 1.6 mm on all dimensions.

Table I lists some typical data for various refractory compounds having the Th_3P_4 structure. The thermoelectric figure of merit, z, for lanthanum sulfide does not vary much with composition. The low z-value shown in Figure 5 for

TABLE I

Sintering Data for Some Compounds Having the Th_3P_4 Structure

Composition	Density, g cm^{-3}	Theoretical density, %	Firing temp., uncor.	Firing time, min
$LaS_{1.38}$	4.29	81	1300 ± 15	15
$CeS_{1.38}$	4.79	87	1380	15
$CeS_{1.42}$ (Mo bomb)	4.12	76	1410	15
$CeSe_{1.43}$	4.97	76	1330	15
$PrS_{1.41}$	4.68	85	1425	15

$LaS_{1.38}$ is fairly typical. The value of z for cerium sulfide (Figs. 5 and 6) is close to that for the optimum composition while no effort has been made to optimize z as a function of composition for praseodymium sulfide (Fig. 5). It appears likely that the praseodymium sulfide is roughly equivalent to cerium sulfide as a thermoelectric material.

For analogous compositions it appears that the selenide ($CeSe_{1.43}$) has a higher resistance than the sulfide ($CeS_{1.42}$) as is indicated in Figure 6. This relationship holds throughout the entire composition range ($1.5 \geq x \geq 1.333$), although the rapid increase in resistivity at 500°C for the selenide is probably due to oxidation or to the development of fine cracks.

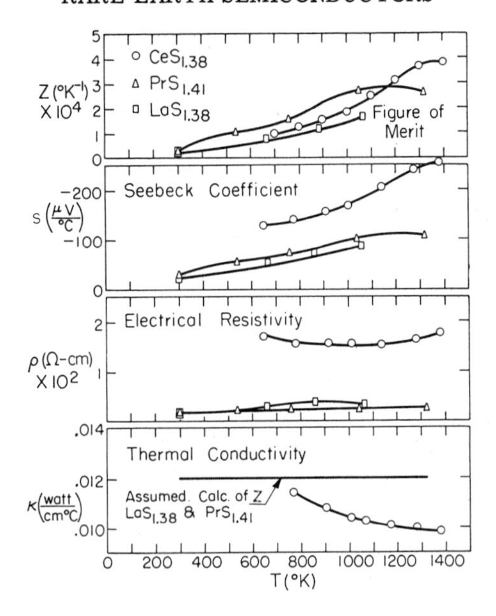

Fig. 5. Thermoelectric properties of $LaS_{1.38}$, $CeS_{1.38}$, and $PrS_{1.41}$.

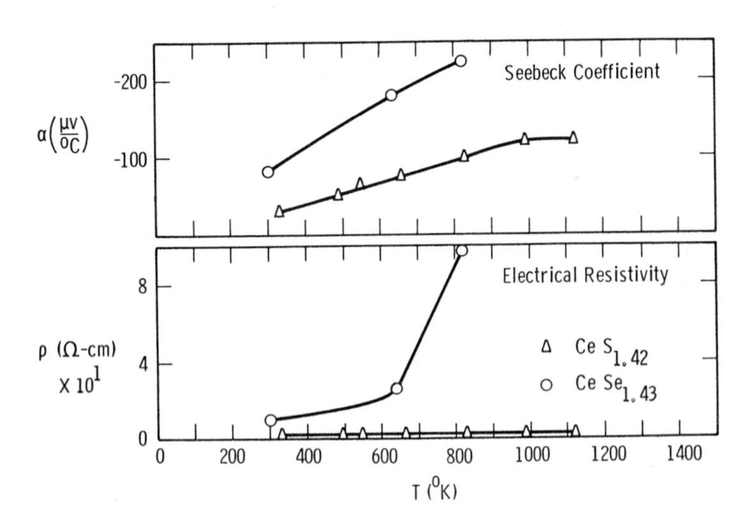

Fig. 6. Electrical properties of $CeS_{1.42}$ (Mo bomb) and $CeSe_{1.43}$.

The molybdenum bomb technique was developed to sinter samples at a high temperature in a closed container, thereby providing a passive atmosphere of their own chalcogen. Presumably such an atmosphere would prevent the formation of a layer of greater conductivity around each grain through loss of sulfur. Such a layer would make a material inferior for thermoelectric devices compared to that of the best pure phase (from a discussion with R. C. Miller). If the molybdenum bomb technique is based on correct premises, then samples of the same overall composition would show different resistivities when fired in a closed molybdenum bomb (high resistance) and when fired in an open crucible (low resistance). Due to experimental difficulties and density and composition variations the question is not settled as yet.

Occasionally samples will be fired at a high temperature (e.g., 1425°) without showing evidence of good sintering. That is, they will be quite porous. It is likely that such samples have an unusually low oxygen impurity content as it is known (1) that the melting point of cerium sesquisulfide is markedly lowered by oxygen. Except for the most impure samples, the high sintering temperatures promote the transition to the Th_3P_4 structure.

The Metal Hydride as an Intermediate

Preparation of several compounds (Ce_2S_3, CeP, CeAs) from the elements is handicapped to some degree by the development of a protective layer of product. This problem is avoided if the metal is present in a fine powder. However, the reduction of the metal by mechanical means is apt to be associated with contamination.

Reduction of lanthanum, cerium, or praseodymium with hydrogen produces a black brittle material which is easily pulverized to a useful size. For making compounds of those rare earth metals which easily form the brittle hydride an excellent method is available for preparing small lots of semiconducting compounds. In some cases the use of the hydride may provide an easy method of preparing the low temperature phases of allotropic compounds.

As an example of a general preparation consider a preparation of cerium antimonide. In this case a pellet of the mixed powders (CeH_3 and antimony) was heated in the reaction chamber of Figure 4 with a Bunsen burner. An exothermic reaction started, and the pellet developed a dull red glow with the evolution of hydrogen. The reaction chamber was then heated in a furnace at 590°C with practically no further hydrogen release. After the sample was heated further at 725°C for 24 hr and allowed to cool, it developed a reddish gray color. Powder photographs showed it to be CeSb, with unit cell edge of 6.422 ± 0.002 A in agreement with Iandelli (9).

The pellet of CeH_3 and bismuth was treated as above to give a dark gray powder which had the x-ray pattern of CeBi (NaCl structure, a_o = 6.504 ± 0.002 A). The initial reaction appeared to be somewhat less exothermic than that of the antimonide.

This simple preparative scheme, however, was not useful in the preparation of cerium arsenide or phosphide in which case the metalloid would distill from the pellet before the reaction temperature was reached. The preparation of cerium carbide by this method has also been unsatisfactory to date. While in one case hydrogen was smoothly evolved up to a temperature of 800°C, the reaction did not go to completion in 6 hr as indicated by the x-ray powder photographs. Some CeC was formed, however. No oil or other hydrocarbon was in evidence.

By cautious heating some unusual compounds can be formed. Table II indicates the conditions of preparation of some rare earth mixed hydride-tellurides. These compounds were prepared in about 6 hr during which time hydrogen gently evolved. During the heating process the initial pellets of the mixed components gradually crumble. If heating is too fast the reaction becomes very vigorous.

Table III gives tentative unit cell dimensions for the first three entries of Table II. The antimonide had the NaCl structure with the unit cell size of 6.42 A, the same as obtained for a preparation from the elements (9). The hydrogen in the tellurides appears to be incorporated into the lattice as part of the struc-

TABLE II

Data for the Preparation of Some Mixed Hydride-Tellurides

Composition	Rare earth, %		Hydrogen, %	x	Preparation temp., °C
	Theory	Determined			
LaH_xTe	51.92	51.22	0.34	0.91	280
CeH_xTe	52.14	51.30	0.53	1.41	250-300
PrH_xTe	52.28	51.20	0.52	1.39	260
CeH_xSb	53.30	51.87	0.08	0.20	500

ture, but for the antimonide the hydrogen might be dissolved interstitally or, more likely, might be in the form of some unreacted hydride.

Lanthanum monotelluride was prepared from the hydride-telluride by simply heating the intermediate in the apparatus of Figure 1 under an argon atmosphere to a temperature of 1200°C. At this temperature, no further hydrogen evolution is noted. The material, as recovered, was a dark maroon mass that turns purple upon powdering, even when it is ground in a dry argon atmosphere. (This color change appears to be limited to a surface phenomenon as the interior of pellets retain their red color even after long exposure to air.) After grinding, the telluride was pressed into a pellet (1×1×3.2 cm) and fired at 1400°C for 15 min under argon. The pellets so treated had a density of 5.6 to 5.7 g cm^{-3} and resumed the dark maroon color.

TABLE III

Preliminary Tetragonal Unit Cell Dimensions

Composition	a_o, A	c_o, A
$LaH_{0.91}Te$	4.50	9.15
$CeH_{1.41}Te$	4.45	9.10
$PrH_{1.39}Te$	4.42	9.05

The lanthanum monotelluride, prepared from the elements below 1000°C, has been reported (10) as a gray semiconductor with a monoclinic structure. As prepared in this laboratory, the LaTe possessed a NaCl structure with a cell edge of 6.43 A in good agreement with that reported by Iandelli (11) who employed the elements in his preparation. His cell edge of 6.42 A gives a good fit to the low-angle lines on powder photographs obtained here. The information given in Figure 7 indicates that the red, NaCl-structure phase of LaTe is semimetallic in character. This might be expected from a comparison with CeS, etc.

The Tritellurides of Lanthanum, Cerium, and Praseodymium

Preparation of semiconducting rare earth compounds directly from the elements is generally unsatisfactory either because a protective coating is formed or the reaction is incomplete, leaving a portion of the unreacted metal ingot intact. This latter problem is much less likely to be important when the compound richest in chalcogen is being formed (e.g., $LaTe_3$, $CeSe_2$, etc.).

The tritellurides were prepared from the reaction of tellurium with freshly cleaned rods of the rare earth metal enclosed in an evacuated Vycor tube. A large tube was used to allow for the excessive expansion of the rod during the formation of the tritelluride. The ampule was then placed in a furnace for about 20 hr at a temperature of 700 to 750°C. The reactions were almost complete under these conditions. The lustrous products were ground under an argon atmosphere, pressed into pellets, and fired as before in a Vycor tube.

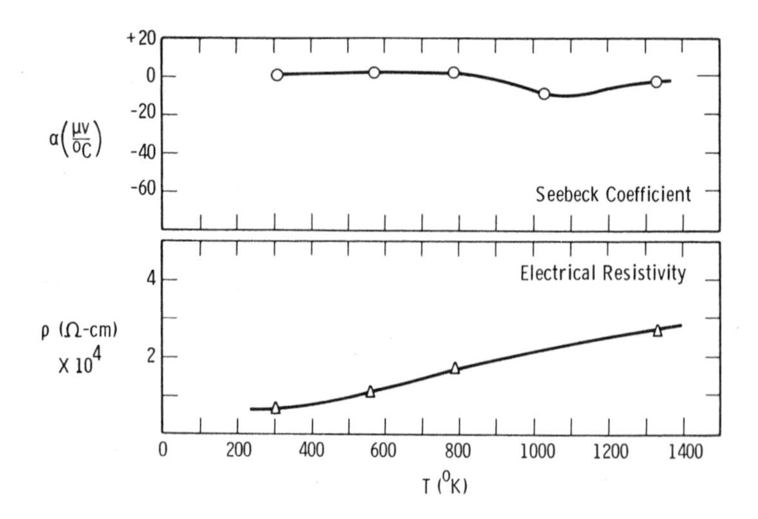

Fig. 7. Properties of LaTe as a function of temperature.

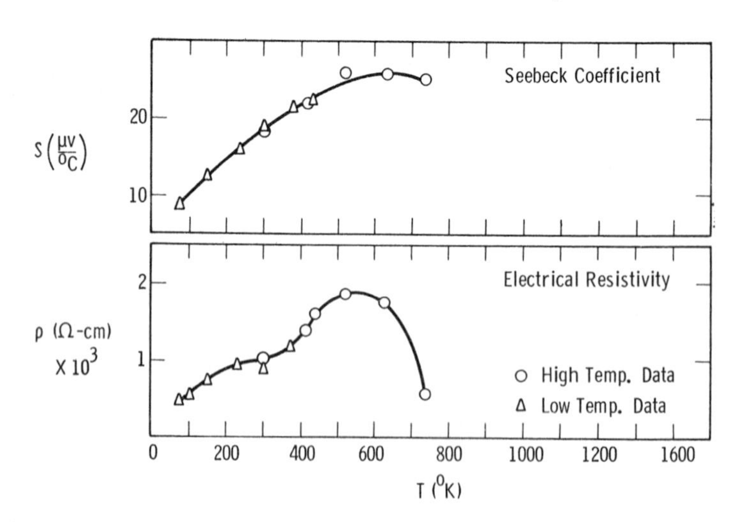

Fig. 8. Properties of $LaTe_3$ as a function of temperature.

As first prepared, the three tritellurides form lustrous light pink-brown platelets. Usually, these are distorted by the expansion during reaction into long serrated arrays. Plates which are formed near the end of the reaction are not subject to such distortion but are found to cleave and bend with extreme ease.

When $LaTe_3$ is exposed to the air, it undergoes a slow decomposition to a gray product. The decomposition of $CeTe_3$ is much more rapid. A platelet of $CeTe_3$ undergoes complete decomposition within a few hours.

These materials are not stable at high temperatures in the absence of excess tellurium vapor. The tritellurides of cerium and lanthanum lose tellurium to a cooler part of the ampule when the sample is near 600°C. At 640°C the loss of tellurium by $CeTe_3$ causes a phase transformation to a gray solid with a composition of $CeTe_{1.85}$ as determined by x-ray diffraction. However, it has been noted that in the presence of considerable excess tellurium $CeTe_3$ does not melt below 1100°C.

A sample of $LaTe_3$ was prepared for thermoelectric measurements by grinding the sintered material and pressing it into a pellet 1 × 1 × 3.2 cm under an argon atmosphere. This sample was then sintered at 610°C for 24 hr. A small amount of tellurium distilled from the sample to the cold end of the Vycor tube. The thermoelectric properties of the rectangular pellet of $LaTe_3$ were measured from liquid nitrogen temperature to 350°C and are shown in Figure 8.

The Reaction of Cerium Sulfide (CeS) with Some Elements

The compound CeS, which has the NaCl structure, cannot be pictured as $Ce^{2+}S^{2-}$ since that does not account for the metallic character of the solid nor does it account for the single unpaired paramegnetic electron per cerium atom (1). However, each cerium can be considered as forming two bonds distributed among the six sulfur neighbors and a single metallic bond distributed among the 12 next nearest neighbors (Ce atoms) to give a total valency of approximately three. The cerium d_{xy}, d_{xz}, and d_{zy} orbitals would give good overlap for the metallic bond. The presence of the free (third) electron in the metallic bond might make CeS susceptible to attack by electron acceptors.

This is shown to be the case by the great ease of reaction indicated in Table IV. The powdered components were mixed, pressed into pellets, and reacted in Vycor or quartz ampules. In no reaction tube was the presence of sulfur noted at any temperature. All these materials except CeSI appear to be semiconductors. The powder patterns of the analogous CeSSb-CeSBi compounds bear no obvious similarity.

This method of making the ternary semiconductors is not unique. Just as CeSSb can be made from the reaction of antimony with CeS, so can it be made by the reaction of sulfur with CeSb [eq. (9)].

$$CeSb + S \xrightarrow{\Delta} CeSSb \tag{9}$$

Both methods of preparation are quite exothermic.

In the latter case, the ignition temperature was probably close to 400°C as sulfur vapor appeared just before the incandescent reaction took place. After the initial reaction, the walls of the tube were coated with an orange-red film (Sb_2S_3?) which was reabsorbed by the bulk material in a few hours at ca. 500°C. The expanded sample, which had a purplish cast, was then ground in argon and heated in pellet form to ca. 1000°C for 15 min. The sintered sample was ground to 150 mesh, pressed, and sintered at 1200°C in preparation for electrical measurements.

The powder pattern indicates the material to be largely single phase and may be tentatively indexed as tetragonal with a_o = 5.86 A and c_o = 8.46 A. The

TABLE IV

Reaction of CeS with Some Elements

Element	Prepared composition	Product color	Reaction temp., °C	Reaction time, hr	Powder pattern	Remarks
I	CeSI	Yellow	430	2	Very complex	Exothermic, lose I_2 reversibly at 650°C to form a green solid: air discolors products and releases I_2.
As	CeAsS	Black	330	1/2	Complex; CeS, CeAs, As, absent	Must melt above 1075°C; loses some As at 1075° but not below 1000°C.
Sb	$(CeS)_3Sb$	Dark grey	450	6	Complex CeS present initially	Initial small amount of silvery third phase present, reacted further between 850 and 1045°C to give a different powder pattern with strong CeSb lines.
	CeSSb	Brown	400	6	Tetragonal	Must melt above 1048°C; loses Sb at 970°C.
Bi	$(CeS)_3Bi$	Dark red	320	6	Complex CeS present initially	Melts above 1073°C; no distillate. Related to $(CeS)_3Sb$.
	CeSBi	Dark red	350	6	Complex	Unstable in air, melts above 1007°C, loses Bi at 1007°C.

low-temperature thermoelectric data indicated in Figure 9 suggests that CeSbS is a small band-gap semiconductor.

SINGLE CRYSTAL STUDIES

Since many of the rare earth semiconductors are either refractory and/or possess high dissociation pressures below their melting points, the prospects of easily obtaining large and varied single crystals are not bright. To date we have not prepared one rare earth semiconducting compound melting below 1000°C. Recently, however, work has begun on several single crystals of chemical interest.

Cerium Sesquisulfide

In an effort to obtain single crystals of one of the low temperature forms of Ce2S3, a transport mechanism of crystal growth was attempted (12). About 10 g of dry Ce_2S_3 was placed in a Vycor ampule with a small crystal of iodine. This was heated for 20 hr at 925 ± 15°C with a drop of about 50°C at the cool end. While practically no transport to the empty cool end was noted, except by crystals of CeI_3, a small cluster of red crystals was observed on the ampule walls in the bulk of the Ce_2S_3.

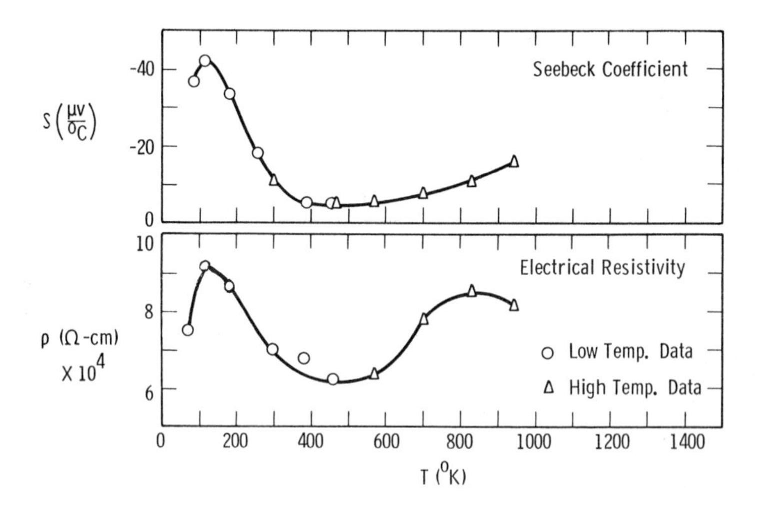

Fig. 9. Properties of CeSSb as a function of temperature.

These crystals, about 0.1 mm on an edge, possessed well-developed faces and were of a ruby red color by transmitted light. The crystal symmetry as determined by Weissenberg camera photographs is that of the Th_3P_4 structure (2, 13). Approximate measurements of the unit cell edges give 8.59 A in good agreement with that obtained by powder photography (8.63 A).

In a second experiment, where the average temperature was 75°C hotter, transport of some red crystals to the cooler walls was noted.

The formation of the high temperature form of Ce_2S_3 at 925°C is quite surprising as the γ-phase has been noted both by the French workers (5,6) and us to occur at a much higher temperature (ca. 1300°C). Of course it is possible that iodine is responsible and is being incorporated into the structure. It is interesting to question its possible site (anionic or cationic) in the structure as well as what influence a smaller halogen might have.

Cerium Disulfide

The structure of CeS_2 (5) was reported to be cubic. However, in the wide angle region there exists two lines which do not fit the cubic index system. Corresponding lines, though still weak, are somewhat more prominent in the case of the diselenide. Since the powder patterns of the disulfides and diselenides (14) of lanthanum, cerium, and praseodymium suggests that these compounds are isostructural, it has long been of interest to do single crystal work on one of them.

Repeated attempts to obtain crystal growth of CeS_2 or $CeSe_2$ in excess chalcogen by various heat treatments followed by extraction with CS_2 were not successful. Finally, however, single crystals of $CeSe_2$ suitable for x-ray analysis were grown during a 12-month period by refluxing excess selenium at 670 ± 15°C in a Vycor ampule tilted about 12° from the horizontal. The lamellar crystals (about 0.5 mm x 0.5 mm x 0.02 mm) were freed from excess selenium by gently distilling it from the crystals at ca. 450°C. When first obtained these crystals had the silvery appearance of germanium, but during a month's time they have tarnished slightly developing a brown tint. The powder pattern of the ground crystals matched that of bulk $CeSe_2$ in the forward angle region; back reflection lines were absent, probably due to the sample being ground too finely.

Weissenberg photographs about the a_o axis show that cubic symmetry is not satisfied while a back reflection Laue photograph with the beam parallel to the unique axis indicates tetragonal symmetry. From these photographs the following approximate tetragonal unit cell dimensions are assigned: a_o = 8.41 A, c_o = 8.39 A. Powder data gives, a_o = 8.44 A for a cubic unit cell.

Lanthanum Tritelluride

Prior efforts to obtain a single crystal of $CeTe_3$ were unsuccessful. X-ray photographs of crystalline $CeTe_3$ always showed multiple twinning. However, several single crystals of $LaTe_3$ have been produced from the elements as described above. These platelets were reduced to a suitable size for x-ray analysis (ca. 0.1 × 0.1 × 0.02 mm) in a solution of HC1 (1 part) and HNO_3 (1 part) in water (4 parts). Back reflection Laue photographs indicate that $LaTe_3$ is tetragonal with the unique axis perpendicular to the cleavage planes. Weissenberg techniques give the following unit cell dimensions: a_o = 4.38 A, c_o = 26.1 A.

It appears likely that there are three $LaTe_3$ molecules per unit cell giving a theoretical density of about 6.6 g cm^{-3}. This gives a sintered sample of $LaTe_3$ with a density of 5.87 g cm^{-3} a value of 89% of theoretical. From the zero layer and first layer Weissenberg photographed, the following systematic absences are observed:

$$
\begin{array}{l}
00L;\ L = 2n + 1 \\
H0L;\ \text{if } H = 2n,\ \text{then } L = 2n + 1 \\
HLL;\ \text{if } H = 2n + 1,\ \text{then } L = 2n
\end{array}
$$

Streaks are very apparent in those lattice rows for which H + K = 2n + 1, however, sharp sports are observed in the lattice rows for which H + K = 2n. The crystal is probably imperfectly periodic in the direction of the c_o axis.

The powder photographs for $LaTe_3$, $CeTe_3$, and $PrTe_3$* appear to be identical except for a very small difference in cell edges.

ACKNOWLEDGMENTS

All electrical measurements as a function of temperature were made by R. C. Miller, Mrs. M. M. Harrison, E. P. A. Metz, and F. M. Ryan. R. A. Jox assisted materially in the preparative work as well as in the powder photography. Finally, the author, would like to acknowledge the financial support of the U. S. Bureau of Ships.

References

1. Eastman, E. D., L. Brewer, L. A. Bromley, P. W. Gilles, and N. L. Lorgren, J. Am. Chem. Soc., 72, 2248 (1950).
2. Zachariasen, W. H., Acta Cryst., 1, 265 (1958).
3. Carter, F. L., R. C. Miller, and F. M. Ryan, J. Advanced Energy Conversion, to be published.
4. Ryan, F. M., I. N. Greenberg, F. L. Carter, and R. C. Miller, J. Appl. Phys., to be published.
5. Picon, M., and M. Patrie, Compt. rend., 243, 1769 (1956).
6. Picon, M., and J. Flahaut, Compt. rend., 243, 2074 (1956).
7. Banks, E., K. F. Stripp, H. W. Newkirk, and R. Ward, J. Am. Chem. Soc., 74, 2450 (1952).
8. Mulford, R. N. R., and C. E. Holley, Jr., J. Phys. Chem., 59, 1222 (1955).
9. Iandelli, A., and E. Botti, Atti accad. naz. Lincei VI, 26, 233 (1937).
10. Brixner, L. H., J. Inorg. & Nuclear Chem., 15, 199 (1960).
11. Iandelli, A., Gazz. chim. ital., 85, 881 (1955).
12. Schafer, H., Angew. Chem., 73, 11 (1961).
13. Zachariasen, W. H., Acta Cryst., 2, 57 (1949).
14. Klemm, W., and A. Koczy, Z. anorg. u. allgem. Chem., 233, 84 (1937).

DISCUSSION

A. C. GLATZ (Carrier Corp.): In the utilization of this hydride technique for the preparation of rare earth selenides and tellurides, there exists the possibility of the formation of extremely toxic decomposition products; namely, hydrogen selenide and hydrogen telluride. Is this a limitation on the use of this technique?

F. L. CARTER: No. It is our technique to use a closed hood for the preparation. Because the reaction may become quite vigorous if large amounts of material are employed, we normally carry out the first preparation of a diselenide or disulfide with less than 50 g of hydride.

---

*$EuTe_3$, with the same structure, probably exists. R. L. Tallman of these laboratories, obtained golden platelets from a reaction of europium with excess tellurium. When the golden product was ground, the powder pattern was that of $LaTe_3$.

Thermoelectric Behavior of the Semiconducting System $Cu_xAg_{1-x}InTe_2$

STOYAN M. ZALAR

Research Division, Raytheon Company, Waltham, Massachusetts

Abstract

By thermal, x-ray diffraction, microscopic and thermoelectric investigations, $Cu_xAg_{1-x}InTe_2$ has been found to be a system of continuous, metastable solid solutions at temperatures above 650°C. Below 650°C the system is unstable and supersaturated. Secondary phases, notably Ag_2Te and InTe, precipitate out of the semiconductor matrix.

Cumulative and continuous annealings at 500 and 300°C, respectively, strongly affect the thermoelectric parameters of the system $Cu_xAg_{1-x}InTe_2$. Due to precipitation phenomena the thermoelectric power increases faster than does the electrical resistivity, resulting in an improved thermoelectric figure of merit.

INTRODUCTION

The investigation of semiconducting solid solutions offers several new research directions in the field of thermoelectric energy conversion. At the present time, at least three mechanisms are known (1) by which semiconductor alloys or solid solutions can and do improve the thermoelectric figure of merit over that of pure compounds and elements.*

The first mechanism is replacement of one type of atoms with atoms of a different size and mass to produce localized short-range lattice distortions. Phonons, which are primary carriers of thermal energy at intermediate temperatures and whose wavelengths are in the order of one interatomic spacing, are strongly scattered by these distortions. Consequently, the phonon thermal conductivity, K_{ph}, is substantially reduced. The charge carriers, however, whose wavelengths are on the order of tens of interatomic distances, are not scattered by short-range lattice deviations and as a result, their mobility, μ, is not greatly reduced.

The second mechanism is based on the changed, i.e., increased, energy gap. Detrimental intrinsic effects are reduced due to the smaller concentration of ambipolar intrinsic carriers. The systems $(Bi, Sb)_2Te_3$, $Bi_2(Se,$

*The thermoelectric figure of merit is defined as:

$$Z = \frac{S^2}{K \cdot \rho} = \frac{S^2 \sigma}{K} \propto \frac{\mu m^{*3/2}}{K_{ph}}$$ (for the nondegenerate state with the optimal carrier concentration 10^{19} cm^{-3})

where S is the Seebeck coefficient, thermoelectric power; K is the total thermal conductivity; ρ is the electrical resistivity; σ is the electrical conductivity; μ is the carrier mobility; m* is the effective mass of carriers; and K_{ph} is the phonon or lattice thermal conductivity.

$Te)_3$, In(P, As), etc. exhibit improvement in the figure of merit due to both of these mechanisms.

A third possible mechanism for an increase of Z by alloying is based on partial solubility of one compound in another, as exemplified by the system $GeTe-Bi_2Te_3$. Here an addition of 5 atomic-% Bi_2Te_3 into GeTe produces the carrier concentration (about $10^{19}cm^{-3}$) required for the optimization of the thermoelectric figure of merit. These alloys are very successfully used in composite thermoelectric structures in the relatively high temperature region of 400-500°C.

It will be seen on the basis of the experimental evidence of this paper that a fourth mechanism probably exists, namely the effect of one or several dispersed second phases or precipitates within the semiconductor matrix.

The system $Cu_xAg_{1-x}InTe_2$ was selected because of preliminary studies at New York University (2) which indicated strong changes of thermoelectric parameters upon annealing. This study also indicated that the band gap of both ternary compounds is approximately 0.95 ev. It was felt that these changes could possibly be due to precipitation phenomena within semiconductor supersaturated solid solutions.

EXPERIMENTAL

Specimens of the system $Cu_xAg_{1-x}InTe_2$ were prepared by direct melting of participating elements in evacuated and sealed quartz tubes. As a rule, they

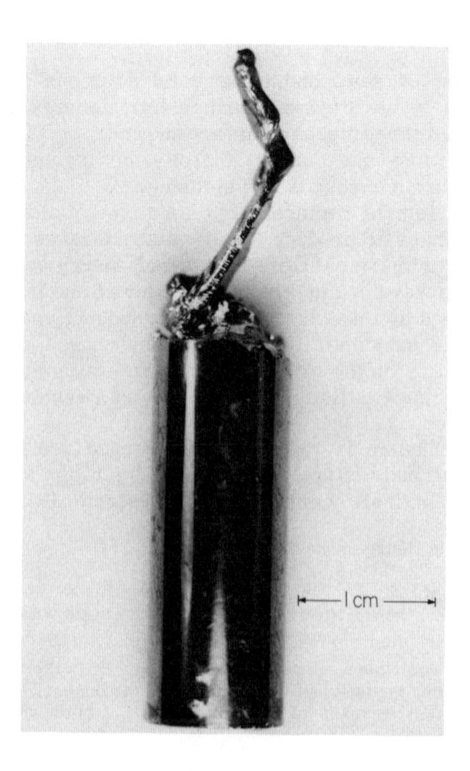

Fig. 1. "Positive solidification pipe" in an ingot of $CuInTe_2$ having been solidified from below by normal freezing.

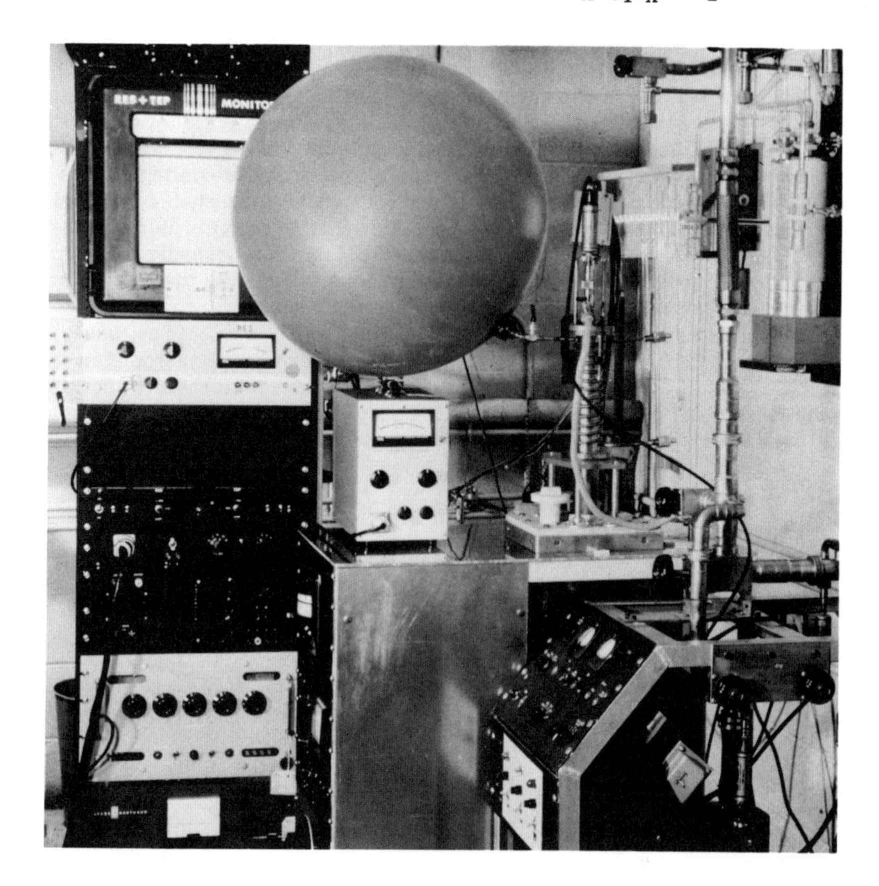

Fig. 2. Experimental setup for the simultaneous measurement of electrical resistivity and thermoelectric power during the continuous annealing of specimens $Cu_xAg_{1-x}InTe_2$.

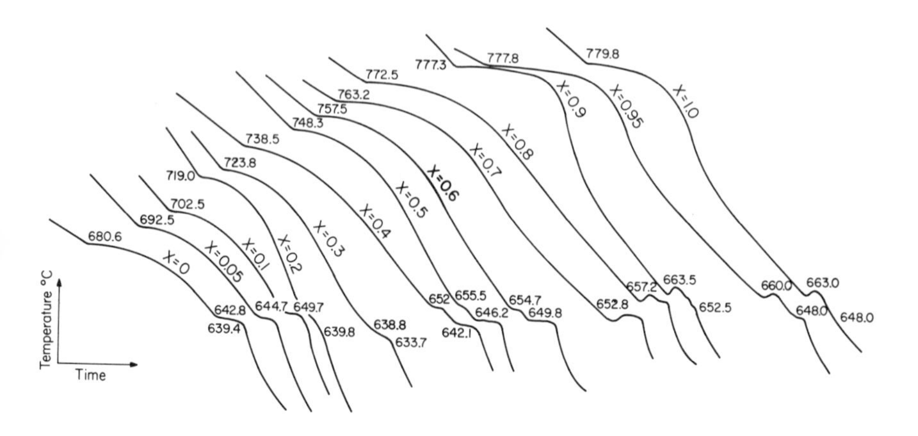

Fig. 3. Cooling curves of the system $Cu_xAg_{1-x}InTe_2$. The accuracy of temperature readings is ±2°C.

were solidified by an accelerated normal freezing technique, so that sound polycrystalline ingots were produced. This Bridgman-type method was necessary because the alloys expanded on solidification, very similar to materials which crystallize with a diamond or zinc blende lattices. A photograph of the "positive solidification pipe" in $CuInTe_2$ is shown in Figure 1.

Thermal analyses were performed by taking continuous cooling curves on samples of about 50 g in weight which were contained in evacuated and sealed quartz capsules. The temperature was monitored with two calibrated, 28-gage, Chromel-Alumel thermocouples (one inserted into the capsule through a closed quartz protection tube, and the second one attached to the outside wall of the capsule) using an automatic, 4-sec interval recorder.

X-ray diffraction data on powdered samples of $Cu_xAg_{1-x}InTe_2$, annealed for one week at 300°C, were obtained using a Norelco diffractometer with filtered copper K_α radiation. The complexity of powder patterns and uncertain resolution of (hkl) peaks restricted lattice parameter measurements to angles less than 60° in 2θ. The (008) and (400) at about 59° were used when the c/a

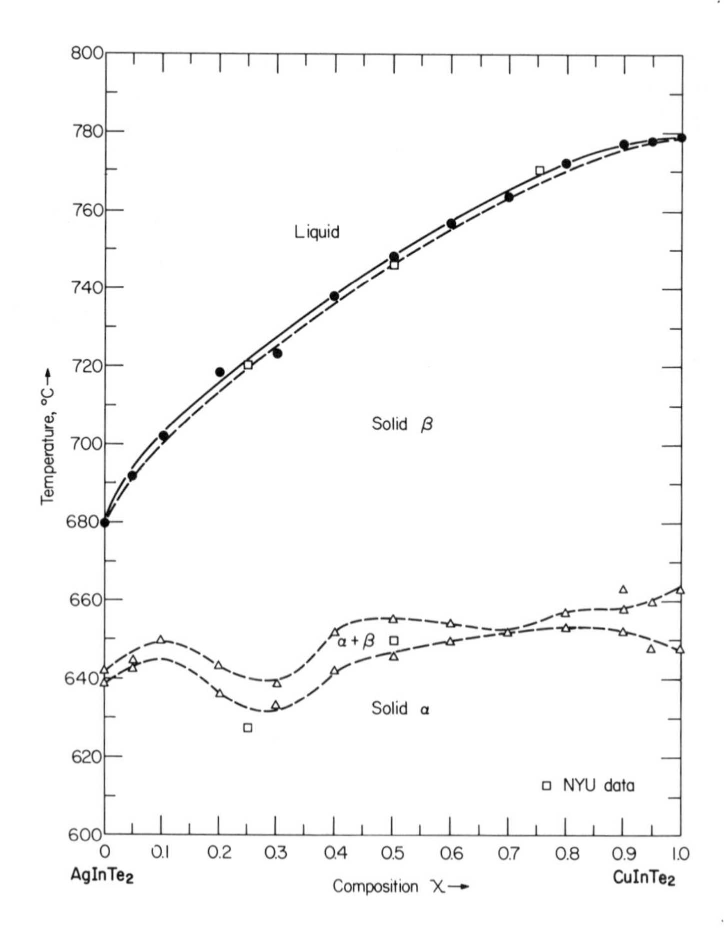

Fig. 4. Quasi-binary phase diagram $AgInTe_2$–$CuInTe_2$ as produced by thermal analyses of the system $Cu_xAg_{1-x}InTe_2$. The NYU data are taken from ref. 2.

ratio was sufficiently different from 2 to permit their resolution. Otherwise the (301) at $2\theta \approx 44°$, the (211) at $2\theta \approx 33°$, and the (112) at $2\theta \approx 24°$ were used. The presence of weak peaks not belonging to the $Cu_xAg_{1-x}InTe_2$ phase was observed.

The thermal conductivity at 0°C of specimens $Cu_xAg_{1-x}InTe_2$ was determined by an improved version of the Putley-Harman technique (3). The specimens (4 mm diam 20 mm long) were hung in vacuum by very thin (0.075 mm) Chromel, Alumel, and copper wires. All results were corrected for radiation from the specimen and for the heat conduction along the lead wires.

In order to follow the annealing effects on the thermoelectric power and electrical resistivity simultaneously during the annealing itself, an experi-

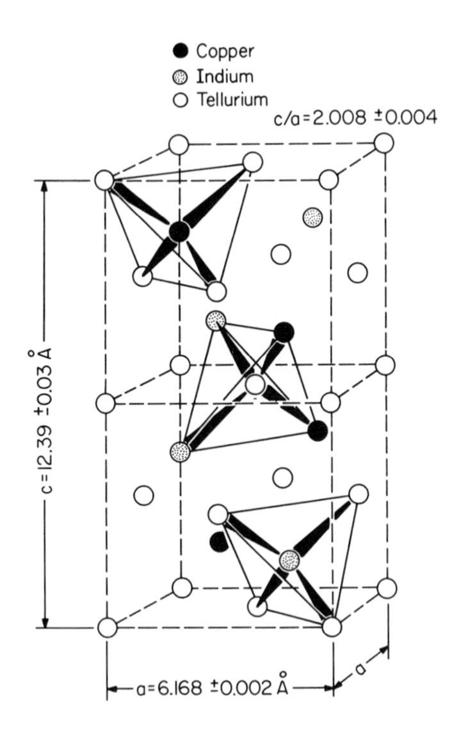

Fig. 5. Chalcopyrite structure of $CuInTe_2$ with its basic covalent tetrahedrons.

mental setup, shown in Figure 2, was built. To the left is the monitor with a 6-point automatic recorder. Two Keithley microvoltmeters, one Leeds & Northrup amplifier, and one attenuator serve to send properly scaled electric signals to the recorder every 4 sec. Thermoelectric power, temperature difference, the lower temperature of the specimen, and resistivities in both directions were recorded on the chart. A North Hill current source in the lower portion of the monitor provided reversible, stabilized dc pulses in the range of 5 to 20 ma.

TABLE I

X-Ray Lattice Parameters of $Cu_xAg_{1-x}InTe_2$

x	a_0, A	c_0, A	c/a	Cell vol., A^3
0	6.401 ± 0.009	12.613 ± 0.004	1.9705	5168 ± 2
0.1	6.371 ± 0.009	12.594 ± 0.004	1.9768	5112 ± 2
0.2	6.350 ± 0.010	12.580 ± 0.004	1.9811	5073 ± 2
0.3	6.318 ± 0.009	12.574 ± 0.003	1.9902	5019 ± 2
0.4	6.294 ± 0.004	12.540 ± 0.003	1.9924	4968 ± 1
0.5	6.279 ± 0.004	12.539 ± 0.004	1.9970	4944 ± 1
0.6	6.275 ± 0.004	12.540 ± 0.030	1.9984	4938 ± 2
0.7	6.256 ± 0.001	12.511 ± 0.006	1.9998	4896 ± 1
0.8	6.236 ± 0.004	12.500 ± 0.030	2.0045	4861 ± 2
0.9	6.204 ± 0.002	12.440 ± 0.030	2.0052	4788 ± 1
1	6.168 ± 0.002	12.390 ± 0.030	2.0081	4744 ± 1

In the center of Figure 2 is the furnace with a stainless steel, vacuum-tight specimen holder. The specimen itself (8 mm diam, 16 mm long) sits in a lavite structure, clamped elastically (by high temperature springs) between two nickel blocks. All values of thermoelectric power measured in this apparatus have nickel as the reference metal. Fourteen measuring wires lead from the specimen pass through a ceramic-to-metal seal and are collected into a flexible cable which goes into the monitor receptacle.

The furnace was heated by direct current to prevent any magnetic induction effects. The balloon, filled with helium, serves to keep an inert atmosphere of fairly constant pressure during the heating, annealing, and cooling periods.

RESULTS

Phase Diagram

For a better understanding of the temperature-dependent phase changes in the system $Cu_xAg_{1-x}InTe_2$, 13 specimens of different compositions were subjected to thermal analysis. The cooling curves are shown in Figure 3. The liquidus point gradually increases with X, from 680.6 ± 2°C for pure $AgInTe_2$ to 779.8 ± 2°C for pure $CuInTe_2$. For all compositions definite thermal effects could be detected in the temperature region of about 650°C. They are believed to be due to the beta to alpha transformation of InTe and are indicative of a general instability of the system below 650°C. It should be noted, however, that these thermal effects are weak, since their time interval in most cases is less than 1 min.

In Figure 4 thermal analysis data are summarized in the form of a binary phase diagram $AgInTe_2$-$CuInTe_2$. Smooth change of the liquidus line suggests a continuous solid solution at temperatures above 650°C. It is somewhat surprising that the course of the solidus line could not be detected accurately.

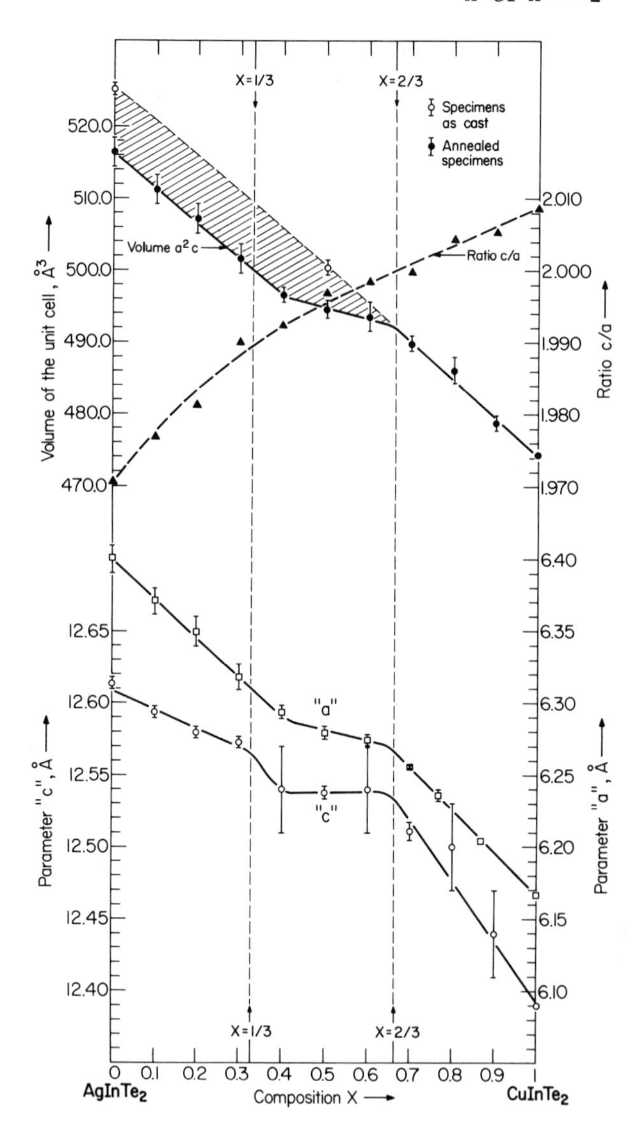

Fig. 6. X-ray lattice parameters of the system $Cu_xAg_{1-x}InTe_2$.

X-Ray Lattice Parameters

X-ray diffraction data from samples annealed at 300°C for one week imply that they all possess a chalcopyrite structure of the form $(A, B)^I C^{III} D_2^{VI}$. The chalcopyrite unit cell (V_d^5) of $CuInTe_2$, representative for all compositions of the system, is shown in Figure 5. This structure is a ternary extension of the elemental diamond (like germanium, silicon, etc.) and binary zinc blende structures (GaAs, InSb, CdTe, etc.). In $CuInTe_2$, each copper or indium atom is tetrahedrally surrounded by four tellurium atoms, while each tellurium atom is bound to two copper and two indium atoms. In the solid solution $Cu_xAg_{1-x}InTe_2$, copper and silver atoms are randomly distributed throughout

TABLE II

Fixed Count Peak Intensities (in counts/sec) of Secondary Phases in Samples $Cu_xAg_{1-x}InTe_2$, Annealed at 300°C for One Week.

Intended Compn., x	d-value Ag_2Te[a]			d-value InTe[a]	
	2.31 A, counts/sec	2.87 A	3.01 A	2.71 A	2.98 A
0	5.2	—[b]	1.5	1.5	4.3
0.1	5.5	5.7	1.6	1.0	4.7
0.5	1.5	2.6	0.5	—[b]	21.0
0.6	0	?	0	—[b]	0

[a] ASTM cards Nos. 4-0795 and 7-112.
[b] Peak was not looked for.

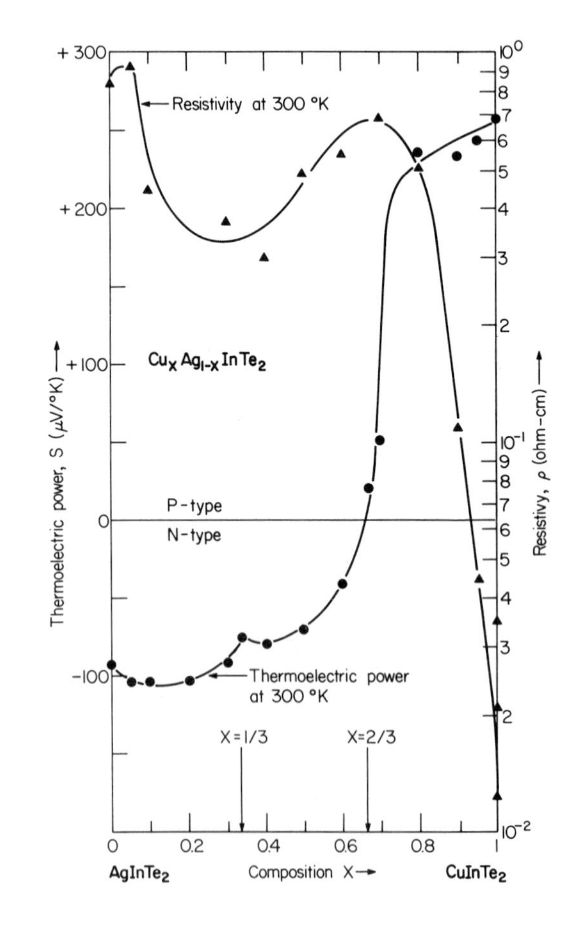

Fig. 7. Thermoelectric power and electrical resistivity of the system $Cu_xAg_{1-x}InTe_2$ measured at 300°C.

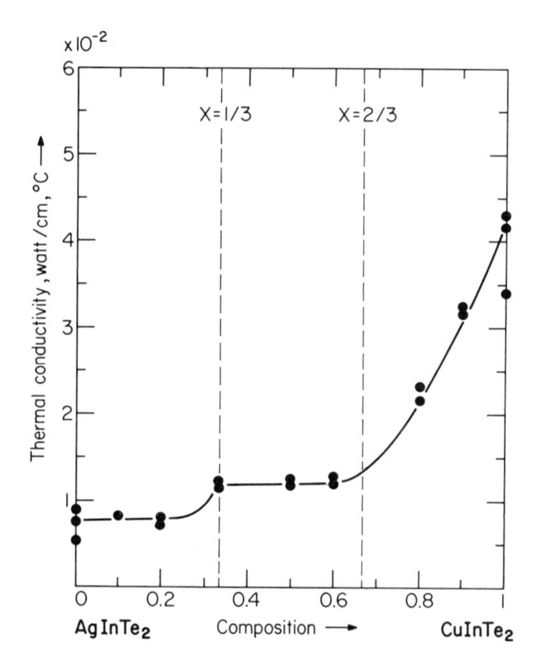

Fig. 8. Total thermal conductivity of the system $Cu_xAg_{1-x}InTe_2$ measured at 0°C by the Putley-Harman technique.

the copper lattice positions. The bonds among participating atoms are strongly covalent, that is, four electrons per atom cooperate in the formation of resonant covalent bonds resulting in the regular tetrahedrons. The chalcopyrite structure deviates from ideal diamond symmetry by showing a slight tetragonality, i.e., the ratio c/a is different from 2.

Hahn and co-workers (4) were the first to perform a systematic x-ray investigation on powdered, synthetic compounds of the composition $A^I B^{III} C^{VI}$. Their lattice parameters for $CuInTe_2$ and $AgInTe_2$ were: $a_0 = 6.16_7$, $c_0 = 12.34$ A, $c/a = 2.00_0$ and $a_0 = 6.40_6$, $c_0 = 12.5_6$ A, $c/a = 1.96_2$, respectively.

Table I and Figure 6 summarize the results of the x-ray diffraction analysis of the system $Cu_xAg_{1-x}InTe_2$. Lattice parameters, a_0 and c_0, do not change smoothly as expected from a true solid solution. The breaking points lie in the vicinity of $x = 1/3$ and $x = 2/3$.

Weak peaks not belonging to the chalcopyrite structure of $Cu_xAg_{1-x}InTe_2$ were seen in both the fast scans and the fixed count runs. Although the powder patterns of secondary phases were too weak to be identified with certainty, there is a good probability that Ag_2Te is present and a possibility that InTe occurs. The peak heights above background, as determined from the fixed count runs of peaks, assumed to belong to Ag_2Te and InTe, are listed in Table II.

The c/a ratio changes smoothly from one end to another, though this fact has no particular physical meaning. It is significant, however, that the results for cast specimens indicate a linear change of the unit cell volume, as expected for ideal solid solutions. This suggests that long annealings precipitate some second phases, such as Ag_2Te and InTe, and shift the intended composition, particularly in the region $0 < x < 2/3$. The region $2/3 < x < 1$ seems to retain the characteristics of a true solid solution.

Thermoelectric Parameters

Figure 7 shows the change of thermoelectric power and electric resistivity, measured at 300°K, with the composition of cast specimens $Cu_xAg_{1-x}InTe_2$. Again the values for x = 1/3 and x = 2/3 look significant. There exists a slight discontinuity of the negative thermoelectric power at x = 1/3, while the resistivity goes through a distinct minimum. At x = 2/3 the thermoelectric power passes through zero and rises rapidly to high positive values of about 230 μv/°K. The resistivity in this composition range goes through a maximum and rapidly decreases afterward.

The change of total thermal conductivity at 0°C, measured by the Putley-Harman technique as a function of composition in the system $Cu_xAg_{1-x}InTe_2$, is shown in Figure 8. The results indicate that the thermal conductivity of the

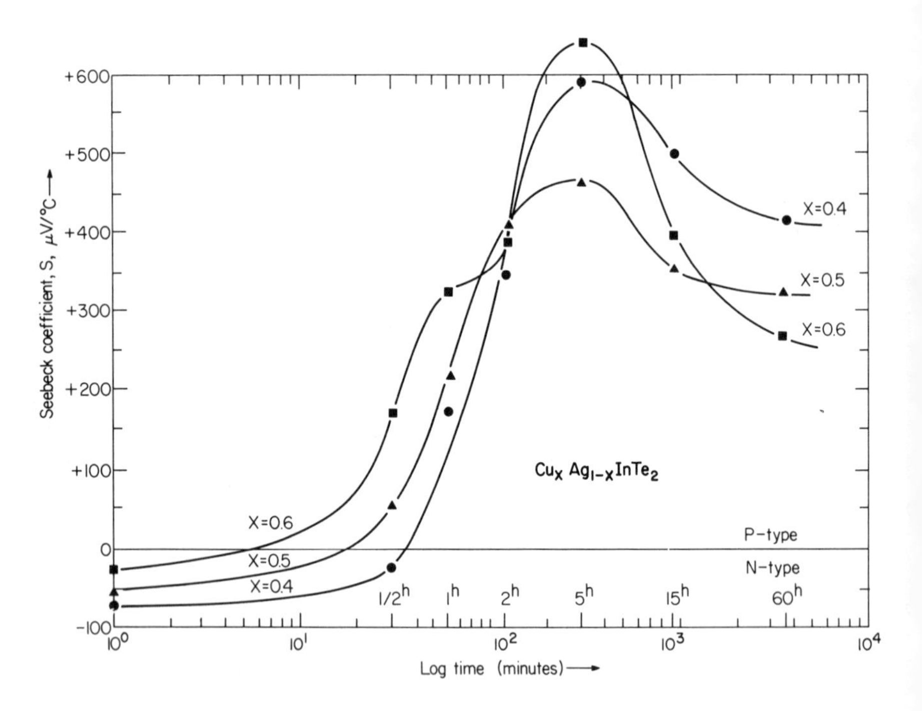

Fig. 9. Effect of cumulative annealing at 500°C on the thermoelectric power of three specimens $Cu_xAg_{1-x}InTe_2$ with the composition x = 0.4, x = 0.5, and x = 0.6.

cast specimens $Cu_xAg_{1-x}InTe_2$ is in the order of 1×10^{-2} to 5×10^{-2}w/cm °C, and thus is much lower than that of elemental semiconductors (silicon: 108.8×10^{-2}; germanium: 53.4×10^{-2} w/cm °C) or that of binary compounds (AlSb: 46.9×10^{-2}, InSb: 15.9×10^{-2} w/cm °C).

Alloys within the composition region of $2/3 < x < 1$ behave normally in the sense that random substitution of silver atoms in the lattice of $CuInTe_2$ produces increased anharmonic phonon-phonon interactions and consequently reduces the thermal conductivity. Thermal conductivity drops from 4×10^{-2} w/cm °C for pure $CuInTe_2$ to 1.5×10^{-2} w/cm °C for $Cu_{0.7}Ag_{0.3}InTe_2$.

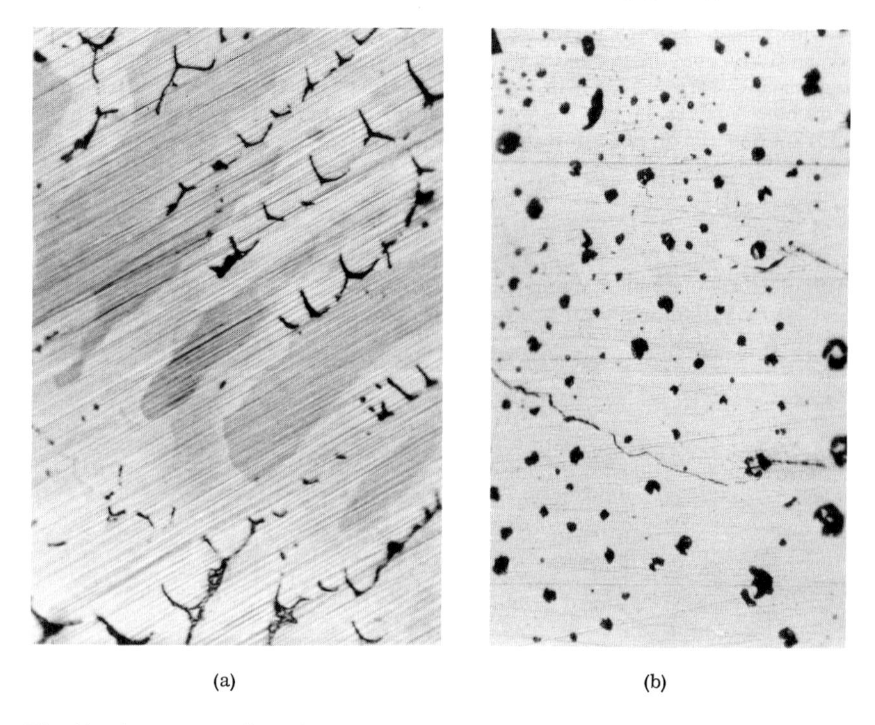

(a) (b)

Fig. 10. Appearance of precipitates in $Cu_{0.6}Ag_{0.4}InTe_2$ before and after the annealing at 500°C. (a) As cast. (b) As annealed, 15 hr at 500°C.180 ×.

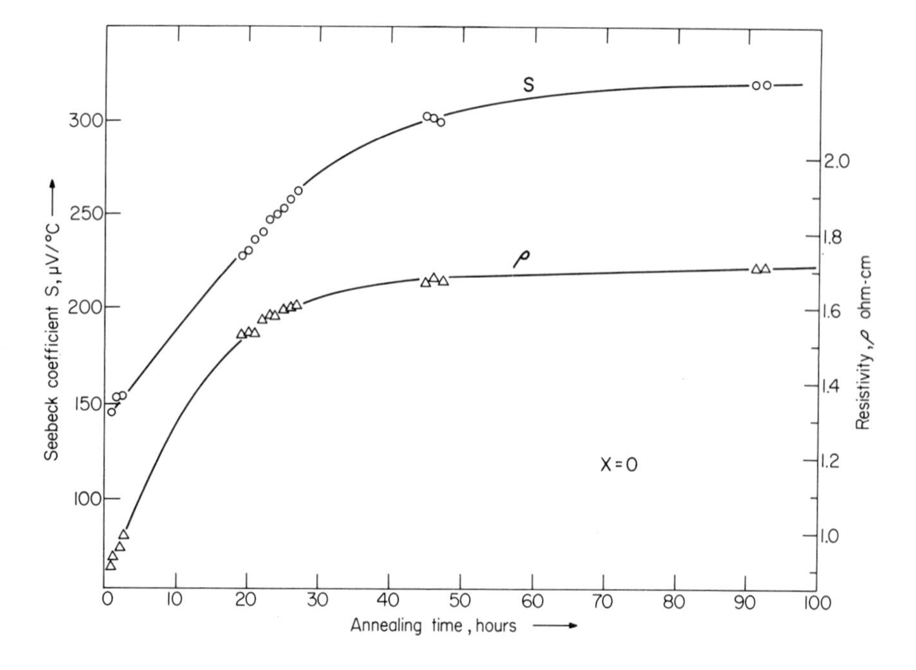

Fig. 11. $AgInTe_2$—change of thermoelectric power and resistivity as a function of time of continuous annealing at 300°C.

The behavior in the region of $0 < x < 2/3$ is much less clear. The thermal conductivity remains practically the same (around 1×10^{-2} w/cm °C) with a discontinuity at x = 1/3. It is very probable that precipitates within the semiconductor matrix in this region influence the mechanism of phonon-phonon interactions and do not allow the curve to raise at the $AgInTe_2$ end of the diagram.

Effects of Cumulative Annealing at 500°C

Figure 9 summarizes the effects of cumulative annealing at 500°C on the thermoelectric power of three $Cu_xAg_{1-x}InTe_2$ specimens with compositions x = 0.4, x = 0.5, and x = 0.6. After each annealing period the specimens, en-

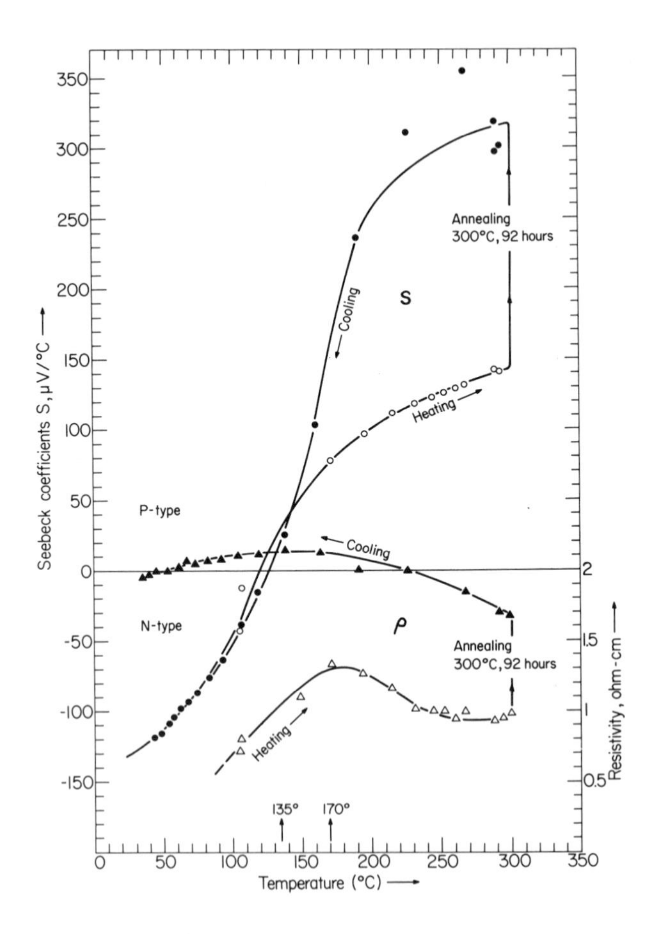

Fig. 12. $AgInTe_2$—effects of the heating-annealing-cooling cycle on thermoelectric power and resistivity as a function of temperature.

capsulated in evacuated and sealed quartz ampules, were taken out of the furnace, cooled, and their thermoelectric power measured at room temperature.

Marked changes in the thermoelectric power are seen as the time at 500°C increases. The initial values of thermoelectric power were on the order of -50 μv/°C. After half an hour of annealing the thermoelectric power changed sign and reached a high peak with values of 450-650 μv/°C in approximately 5 hr. Upon further annealing, the thermoelectric power decreases and levels off at about 300 μv/°C. This behavior very much resembles that of supersaturated metallic solutions, for example, the alloys aluminum—38% silver (5).

Microstructure of $Cu_{0.6}Ag_{0.4}InTe_2$ before and after annealing is shown in Figure 10. The etching agent used in almost all specimens of the system $Cu_xAg_{1-x}InTe_2$ was 1:1 nitric acid-water solution. It is seen that upon annealing new precipitates occur in the microstructure. These precipitates, mostly spot and starlike in form, have been found in practically all compositions of the system.

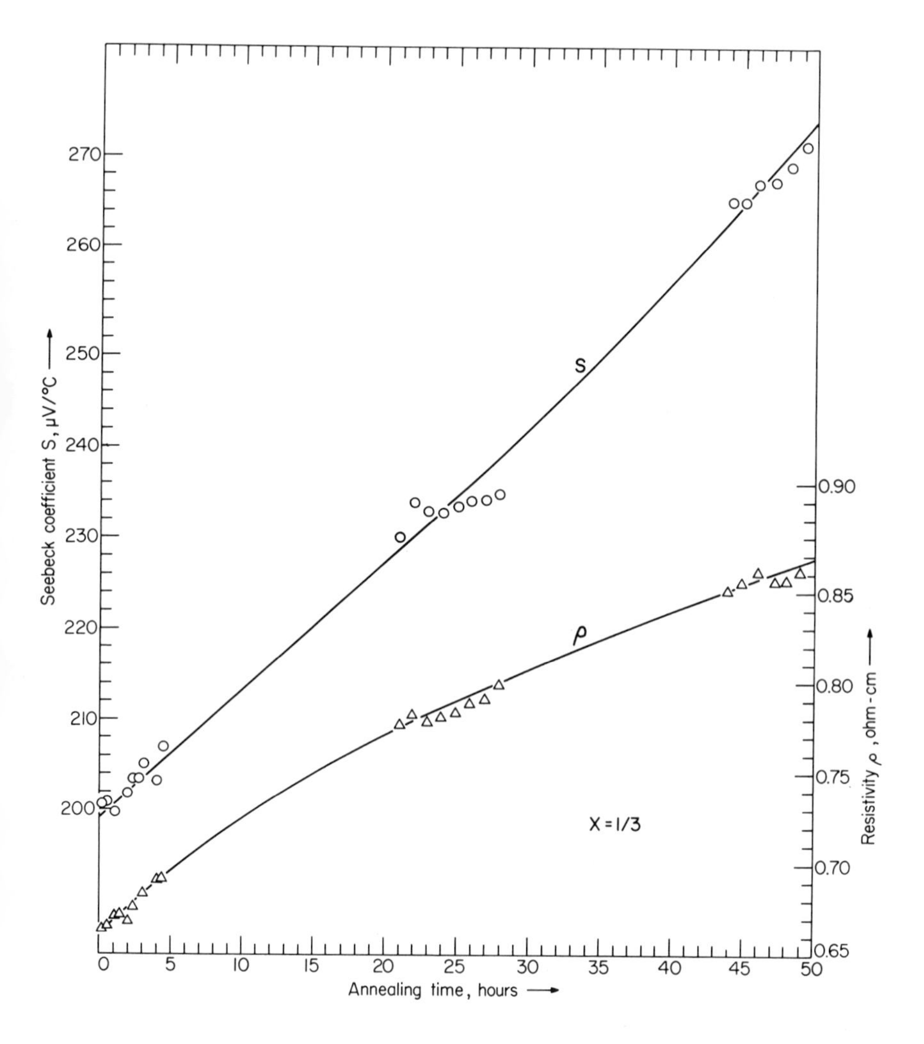

Fig. 13. $Cu_{1/3}Ag_{2/3}InTe_2$—change of thermoelectric power and resistivity as a function of time of continuous annealing at 300°C.

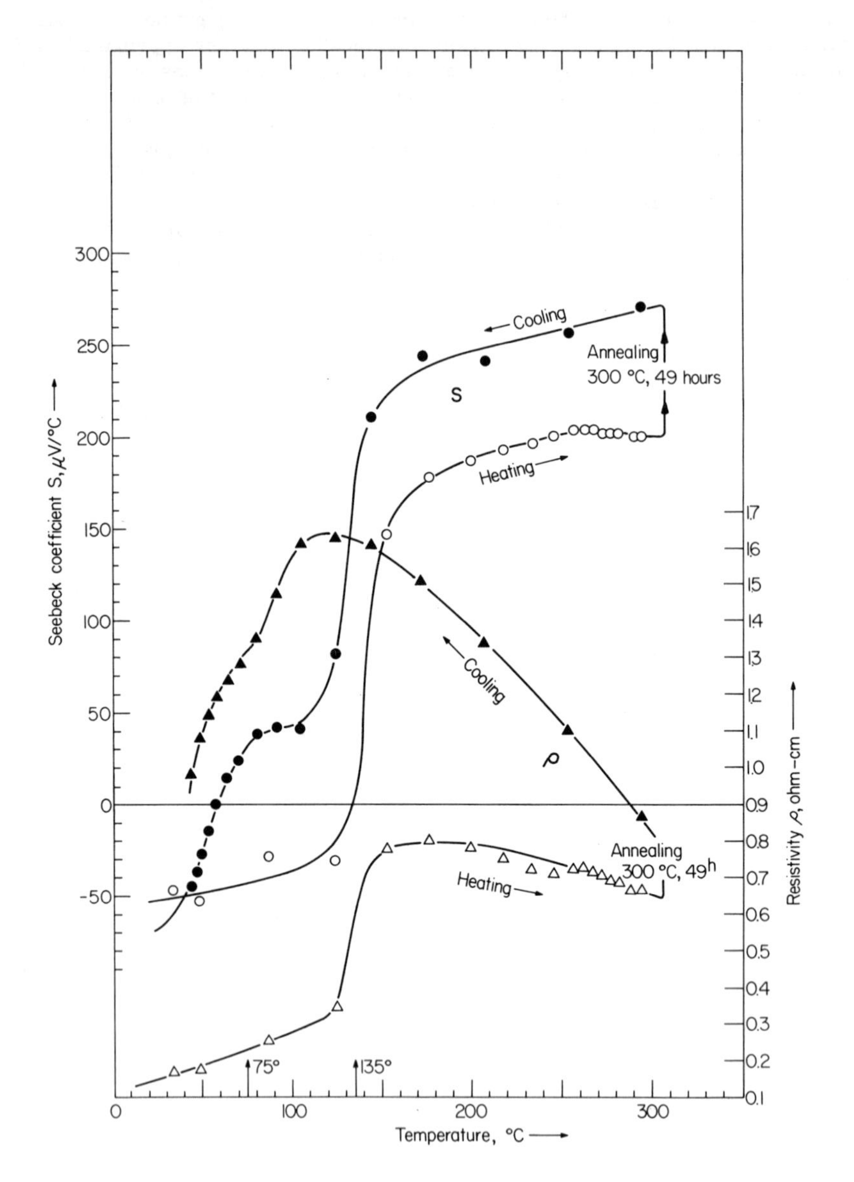

Fig. 14. $Cu_{1/3}Ag_{2/3}InTe_2$—effects of the heating-annealing-cooling cycle on thermoelectric power and resistivity as a function of temperature.

Effects of Continuous Annealing at 300°C

In order to follow the changes of thermoelectric power and resistivity of specimens during the annealing itself, the apparatus shown in Figure 2 was built. The figures which follow show the results of the work with this apparatus.

Figure 11 illustrates the change in thermoelectric power and resistivity of $AgInTe_2$ (x = 0) during continuous annealing at 300°C, as a function of time. The Seebeck coefficient, S, increases 126%, and ρ increases less (about 83%). In Figure 12, the net effect of continuous annealing for 92 hr at 300°C on both thermoelectric parameters is shown as a function of temperature. Note the maxima of resistivity at 170°C on heating and at 135°C on cooling. On cooling, the thermoelectric power of $AgInTe_2$ returns to its initial, negative value.

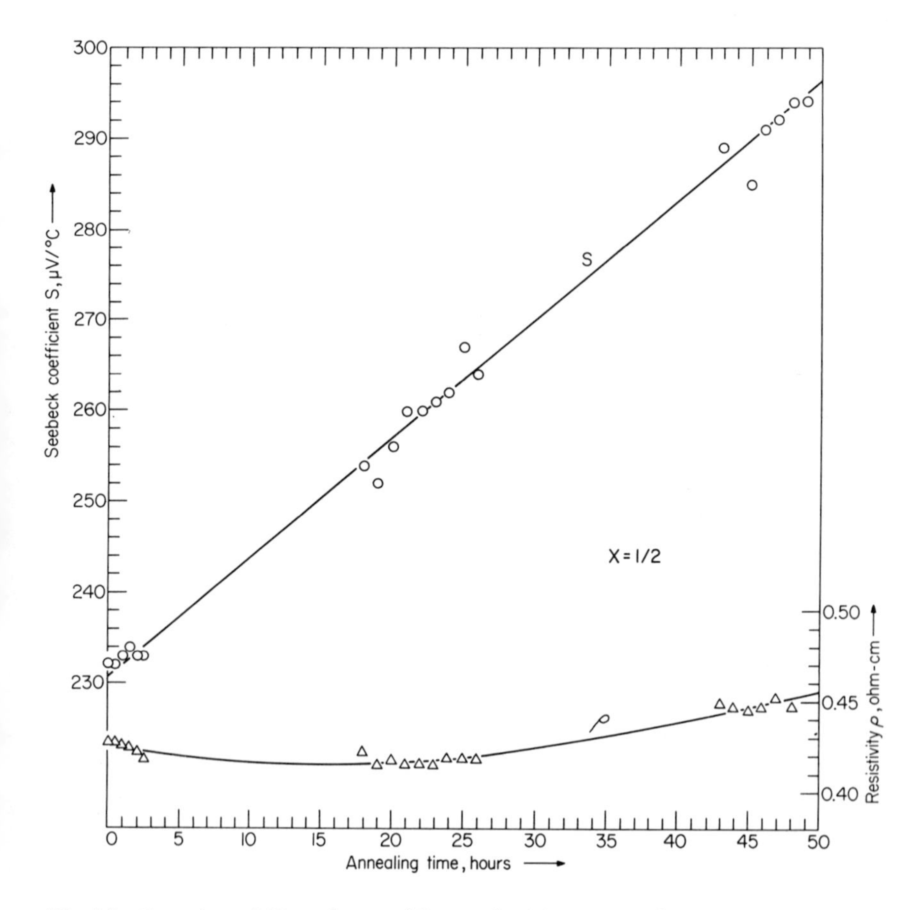

Fig. 15. $Cu_{1/2}Ag_{1/2}InTe_2$—change of thermoelectric power and resistivity as a function of time of continuous annealing at 300°C.

Figure 13 shows the situation for x = 1/3. With an increase in the annealing time both parameters increase (S, 37% and ρ, 30%). The overall effect. shown in Figure 14, indicates a great variety. On heating, the resistivity suddenly increases at 135°C and later tapers off. On cooling, there is a marked increase of resistivity, with the maximum again at 135°C and a possible second but overlapped peak at 75°C. Thermoelectric power, on heating, shifts suddenly into positive values at 135°C and shows, on cooling, an additional knee at 75°C.

It is believed that the effects at 135°C are due to the alpha to beta transition in Ag_2Te.

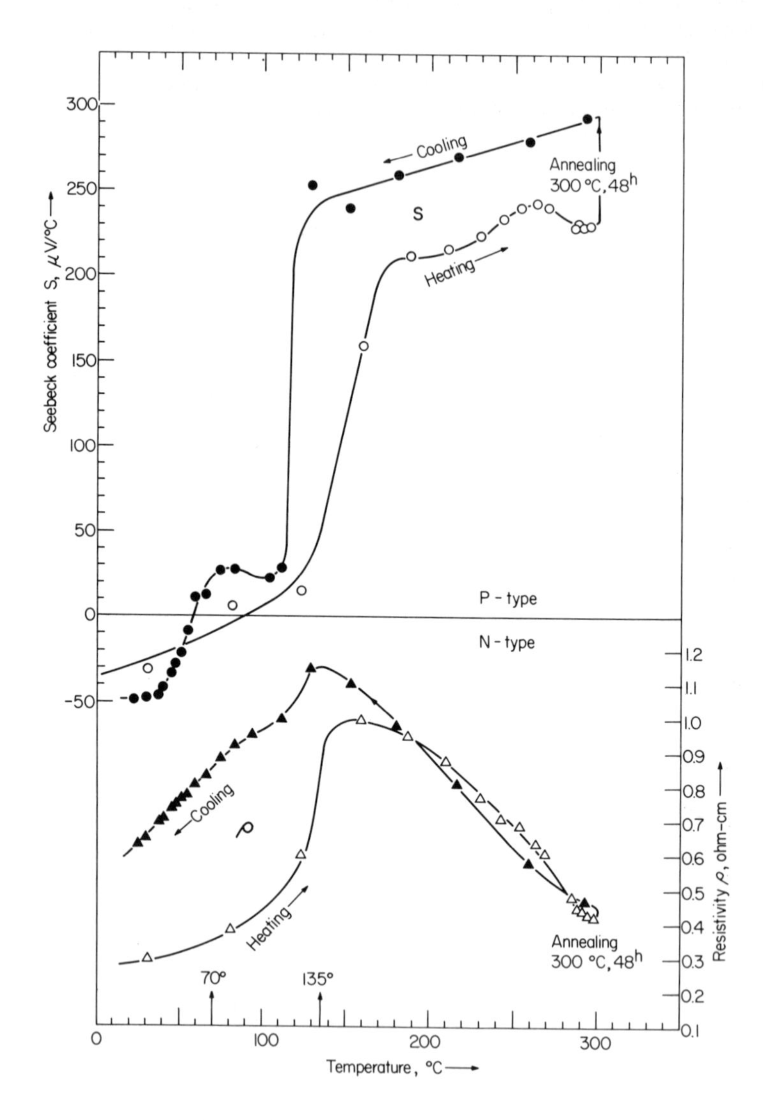

Fig. 16. $Cu_{1/2}Ag_{1/2}InTe_2$—effects of the heating-annealing-cooling cycle on thermoelectric power and resistivity as a function of temperature.

Figure 15, representing the composition x = 0.5 shows a steep increase of thermoelectric power (85%), while the resistivity increases more slowly (only 4% at the end of the annealing). The overall cycle, Figure 16, again indicates a distinct peak at 70°C. Thermoelectric power has the same reversal at about 135°C and on cooling shows a new maximum at 70°C.

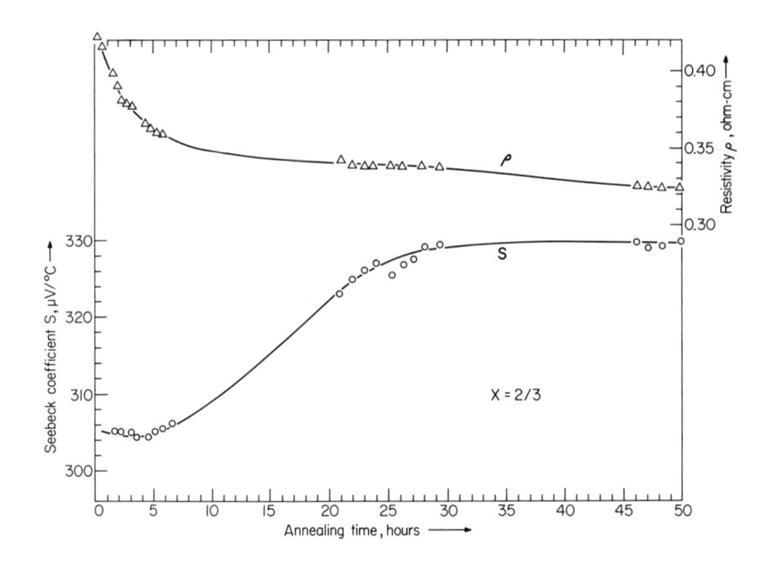

Fig. 17. $Cu_{2/3}Ag_{1/3}InTe_2$—change of thermoelectric power and resistivity as a function of time of continuous annealing at 300°C.

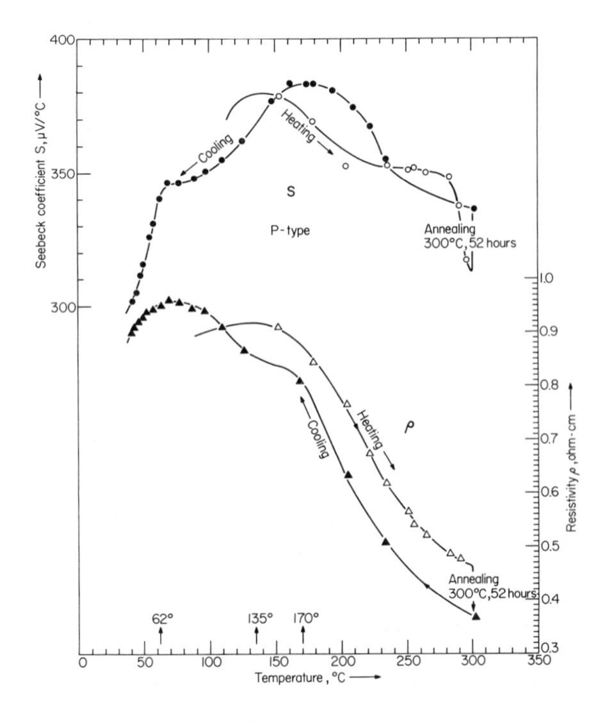

Fig. 18. $Cu_{2/3}Ag_{1/3}InTe_2$—effects of the heating-annealing-cooling cycle on thermoelectric power and resistivity as a function of temperature.

Figure 17 describes the situation for x = 2/3. The results are most significant. Thermoelectric power increases about 6.4%, while resistivity decreases 18.8%. In Figure 18, the resistivity goes through a maximum again at 135°C on heating and has a second maximum on cooling at 62°C. Thermoelectric power goes through a complicated pattern on the level of about 350 μv/°C, and on cooling shows a maximum at 170°C and a second one at 62°C.

In the case of pure $CuInTe_2$, Figure 19, thermoelectric power increases slowly (12%) and resistivity increases more substantially (172%). The effects of the full cycle, Figure 20, are rather unexpected. While heating curves are

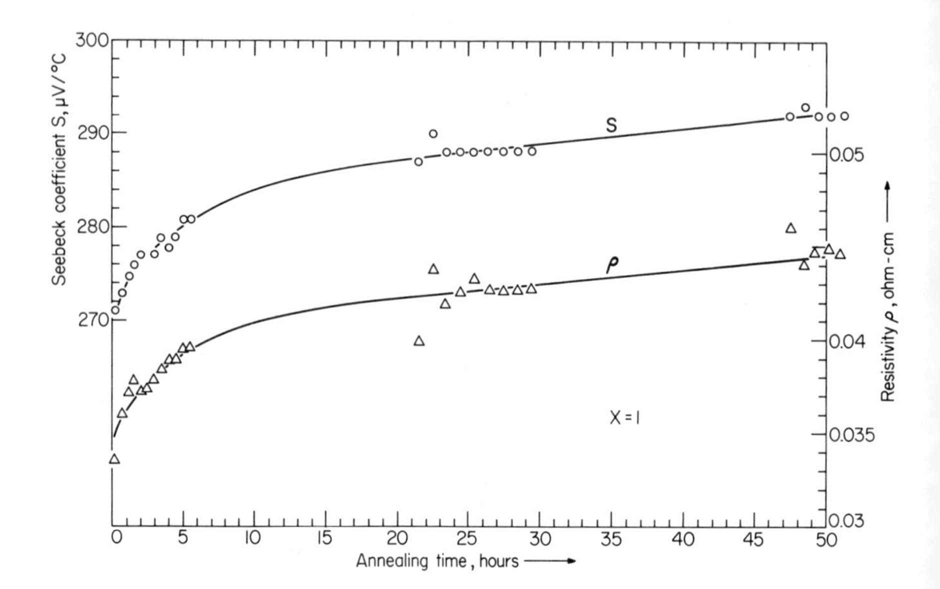

Fig. 19. $CuInTe_2$—change of thermoelectric power and resistivity as a function of time of continuous annealing at 300°C.

linear, the cooling curves show a marked maximum at 170°C for both thermoelectric parameters.

CONCLUSIONS

1. $Cu_xAg_{1-x}InTe_2$ is a system of continuous, metastable solid solutions at temperatures above 650°C.

2. At temperatures below 650°C the system $Cu_xAg_{1-x}InTe_2$ is unstable and supersaturated. Secondary phases, in the amount of less than 4%, precipitate out of the matrix, notably Ag_2Te and InTe. Ag_2Te undergoes an allotropic transformation at 135°C and is probably the basic cause for the formation of cracks in the microstructure of alloys and for the inability to grow their single crystals by zone leveling.

3. Instead of the quasi-binary representation $AgInTe_2$-$CuInTe_2$, the system $Cu_xAg_{1-x}InTe_2$ could probably be better represented as a linear part of the ternary system Ag_2Te-In_2Te_3-Cu_2Te as shown in Figure 21.

4. The chalcopyrite structure is preserved for all compositions of the system $Cu_xAg_{1-x}InTe_2$. In the case of cast specimens, the volume of the unit cell linearly decreases from $AgInTe_2$ to $CuInTe_2$.

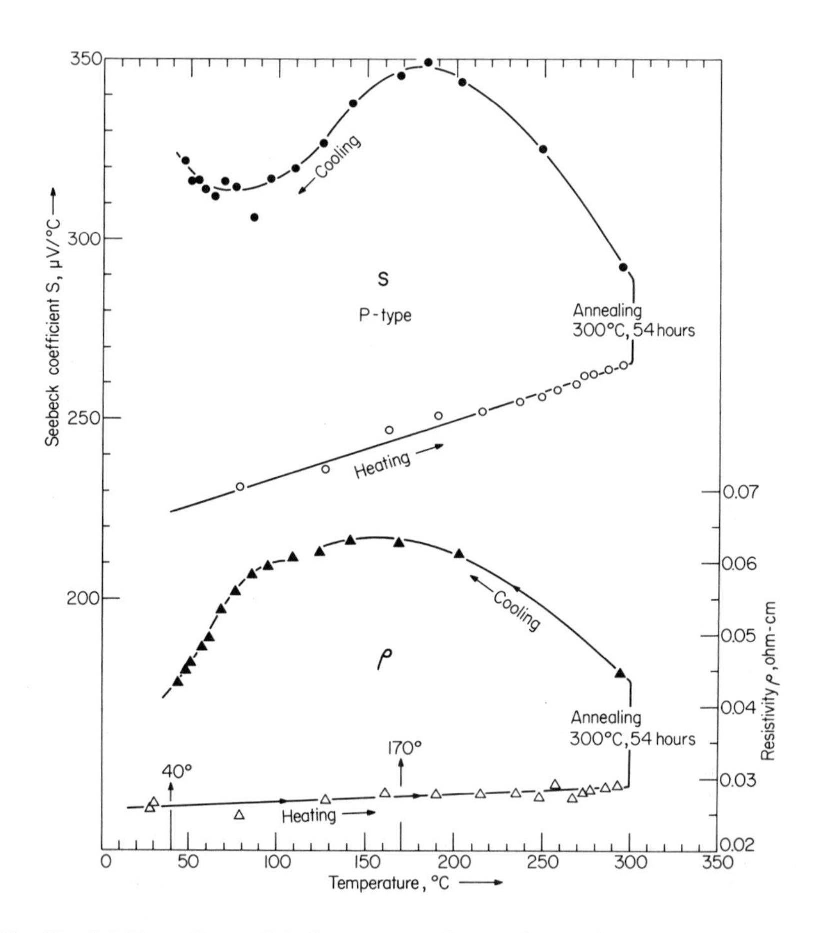

Fig. 20. $CuInTe_2$—effects of the heating-annealing-cooling cycle on the thermoelectric power and resistivity as a function of temperature.

5. The volume of the unit cell in annealed $Cu_xAg_{1-x}InTe_2$ specimens shows a sudden deviation from linearity at x = 2/3. This is believed to be due to the precipitation of secondary phases, mainly Ag_2Te, with the resulting deviation from the intended composition. It is possible that ordering of the copper and silver atoms at the 1/3 and 2/3 compositions could be important even though ordering does not occur in the copper-silver system.

6. Thermoelectric power and resistivity of the system $Cu_xAg_{1-x}InTe_2$ are in the semiconductor range. Up to x = 2/3 the thermoelectric power is n-type and later it abruptly changes into p-type, having the values about 230 $\mu v/^\circ C$.

Resistivity shows a minimum at x = 1/3 ($\rho = 3.5 \times 10^{-1}$ ohm-cm), a maximum at x = 2/3 ($\rho = 7 \times 10^{-1}$ ohm-cm), and later drops sharply, reaching the value of 2×10^{-2} ohm-cm for $CuInTe_2$.

7. Total thermal conductivity of the system $Cu_xAg_{1-x}InTe_2$ ranges from 4×10^{-2} w/cm °C for $CuInTe_2$ to 0.8×10^{-2} w/cm °C for $AgInTe_2$. These values are two orders of magnitude lower than those for elemental semiconductors and one order of magnitude lower than those for III-V compounds.

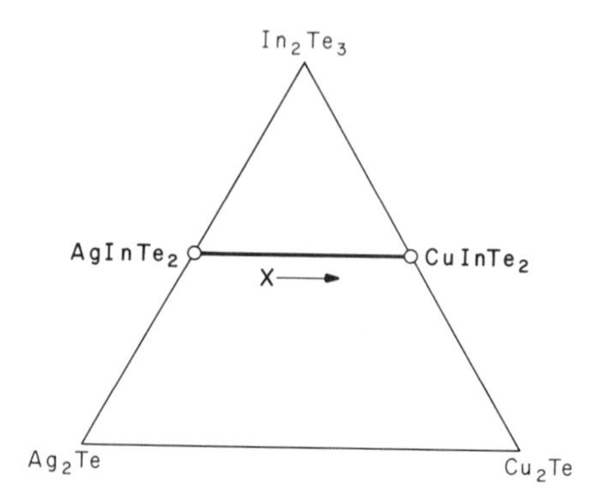

Fig. 21. $Cu_xAg_{1-x}InTe_2$ represented as a linear part of the ternary system Ag_2Te-Cu_2Te-In_2Te_3.

8. In general, annealing of specimens of the system $Cu_xAg_{1-x}InTe_2$ brings about an increase in thermoelectric power and an increase in electrical resistivity. Due to precipitation phenomena, however, the thermoelectric power increases faster than does the resistivity. For certain compositions (for example, x = 2/3) the resistivity even decreases with the time of annealing.

9. Annealing effects in the system $Cu_xAg_{1-x}InTe_2$, based on precipitation of secondary phases within the semiconductor matrix, constitute a new mechanism for the improvement of thermoelectric figure of merit. The product $S^2\sigma$ for the compositions x = 1/3, 1/2, and 2/3 is increased about 47% after annealing of specimens at 300°C for 50 hr.

10. The level of the improved thermoelectric figure of merit in the system $Cu_xAg_{1-x}InTe_2$ is about 10^{-4} °C^{-1}, and is thus one order of magnitude lower than that of economic thermoelectric generators. It is believed, however, that the new principle will prove to be useful in other semiconducting solid solutions and particularly in composite metal-oxide-metal and ceramic systems.

ACKNOWLEDGMENTS

The author wishes to express his appreciation to Mr. R. P. Whelan for highly competent and resourceful experimental assistance, Messrs. R. Hawkes, R. Mozzi and B. Bekebrede for their excellent experimental support in x-ray diffraction studies, Francis Corrigan, a Massachusetts Institute of Technology graduate student, for his careful measurement of the thermal conductivity, and Miss R. A. Brown for her reliable mathematical computations.

References

1. Heikes, R. R., and R. W. Ure, Jr., Thermoelectricity: Science and Engineering, Interscience, New York-London, 1961, ch. 11.3.3.

2. Zalar, S. M., and I. B. Cadoff, Trans. AIME, to be published.
3. Putley, E. H., Proc. Phys. Soc. (London), B68, 35 (1955); T. C. Harman, J. Appl. Phys., 29, 1375 (1958); T. C. Harman, J. H. Cahn, and M. J. Logan, J. Appl. Phys., 30, 1351 (1959); R. Simon, R. T. Bate, and E. H. Longher, J. Appl. Phys., 31, 2160 (1960).
4. Hahn, F. H. W. Klingler, and A. Meyer, and G. Storger, Z. anorg. u. allgem. Chem., 271, 153 (1953).
5. Glocker, R., W. Koster, J. Scherb, and G. Ziegler, Z. Metallk. 43, 208 (1952).

Preparation and Properties of Silver-Antimony-Tellurium Alloys for Thermoelectric Power Generation

R. G. R. JOHNSON and J. T. BROWN

Metallurgical Department, Westinghouse Electric Corporation
Pittsburgh, Pennsylvania

Abstract

The thermoelectric properties of the psuedo-binary system Ag_2Te-Sb_2Te_3 were investigated. The composition range investigated was between 34 and 66 mole-% Ag_2Te. The thermoelectric power, electrical resistivity, and thermal conductivity were measured between room temperature and approximately 800°K. The effect of heat treatment upon the thermoelectric properties was studied.

INTRODUCTION

Other investigators (1) working with the composition $AgSbTe_2$, a material with p-type carrier characteristics, reported promising thermoelectric properties. Of special interest was its unusually low thermal conductivity. On the detrimental side was the reported inability to adjust electrical properties through controlled impurity doping. Recent work (2) at our laboratories revealed the presence of a pseudo-binary section in the ternary diagram, the end members of which are Ag_2Te and Sb_2Te_3. The alloy $AgSbTe_2$ is simply a 1:1 molar mixture of these two binary compounds and is shown as the β-phase in Figure 1. The $AgSbTe_2$ has a disordered NaCl structure with a lattice parameter of 6.078 A (3). Putting this information together seemed to indicate further experimentation was in order. Thus the purpose of this investigation was to examine the electrical properties and microstructure that existed after several methods of preparation were used to make molar mixtures of Sb_2Te_3 and Ag_2Te varying from 2:1 to 0.5:1, respectively.

PREPARATION PROCEDURE AND TESTS PERFORMED

All ingots were made from antimony and tellurium of 99.999% purity, and silver of 99.9% purity. The antimony, tellurium, and silver were placed into 13 mm Vycor tubes. These tubes were charged with 100 g of material, evacuated, and sealed.

The charge was then melted and cast into an ingot within the tube. Melting took place in a small resistance-heated furnace equipped with a proportional controller. The furnace was mechanically rocked for 1 hr at 700°C to insure complete homogenization of the molten charge. Cast alloys were investigated

TABLE I

Compositions Investigated

Mole ratio Sb_2Te_3:Ag_2Te	Formula	Mole per cent		Weight per cent		
		Sb_2Te_3	Ag_2Te	Ag	Sb	Te
2.0:1	$Ag_2Sb_4Te_7$	66.0	34.0	13.84	30.32	55.84
1.7:1	$Ag_2Sb_{3.4}Te_{6.1}$	63.0	37.0	15.30	29.41	55.29
1.6:1	$AgSb_{1.6}Te_{2.9}$	61.5	38.5	16.05	28.95	55.00
1.5:1	$Ag_4Sb_6Te_{11}$	60.0	40.0	16.82	28.47	54.71
1.4:1	$AgSb_{1.4}Te_{2.6}$	58.0	42.0	17.86	27.83	54.31
1.3:1	$Ag_2Sb_{2.6}Te_{4.9}$	56.0	44.0	18.92	27.17	53.91
1.1:1	$Ag_2Sb_{2.2}Te_{4.3}$	53.0	47.0	20.55	26.16	53.29
1.0:1	$AgSbTe_2$	50.0	50.0	22.25	25.11	52.64
0.8:1	$AgSb_{0.8}Te_{1.7}$	44.0	56.0	25.82	22.90	51.28
0.5:1	$Ag_2SbTe_{1.5}$	34.0	66.0	32.39	18.84	48.77

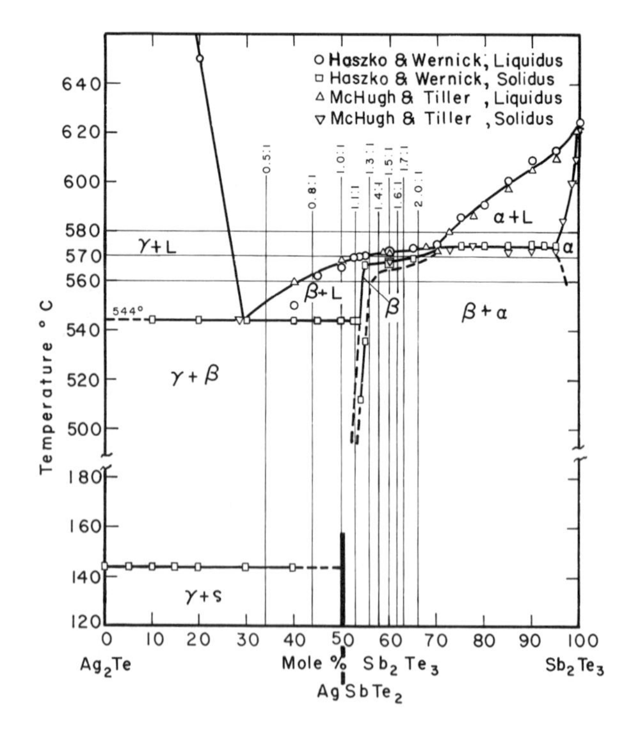

Fig. 1. The equilibrium phase diagram for the pseudo-binary cut, Ag_2Te-Sb_2Te_3, in the Ag-Sb-Te system.

using two different means of solidification—an air quench and a slow cool within the rocking furnace. The air quench consisted of removing the tube from the furnace at 700°C and holding it upright in air during freezing. Approximately 1 min was required for the molten contents to freeze. Furnace cooling was accomplished within the rocking furnace with the furnace stationary and in a vertical position. The charge cooled inside the furnace to 500°C at an average rate of 3°C per minute, whereupon the tube was removed and allowed to cool to room temperature in air. Afterwards the ingots (still within the original tubes) were annealed for 8 hr at 500°C.

Pellets were cut from the ingots for the determination of the thermoelectric properties. For obtaining the room-temperature Seebeck coefficient, electrical resistivity and thermal conductivity, two 1/2-in long pellets were cut from each ingot on a saw equipped with a diamond wheel. These same pellets were also used to determine the average Seebeck coefficient and resistivity over a ΔT of 300°C. Before testing, the pellets were carefully lapped on metallographic paper to remove the traces of the sawcuts and to preserve the parallelism of the cut faces. Data on the Seebeck coefficient and electrical resistivity versus temperature from approximately room temperature up to 800°K was obtained from measurements on a 1.5-in. long pellet in testing equipment especially designed for this purpose.

Metallographic studies were made on all the alloys. A 55% solution of nitric acid was found to be a suitable etchant.

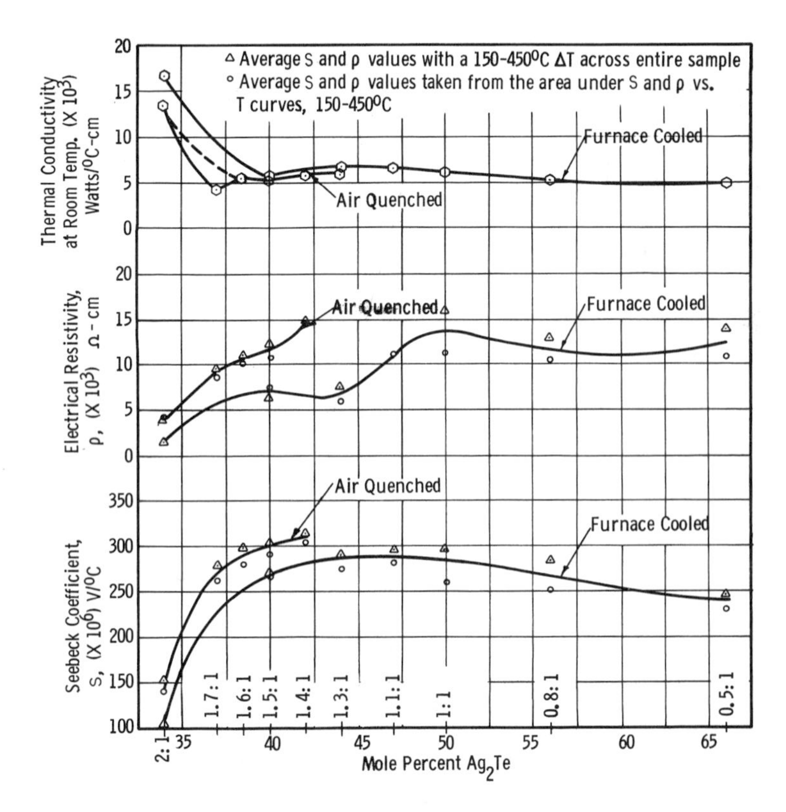

Fig. 2. Thermoelectric properties of silver-antimony-tellurium ternary alloys as a function of composition.

RESULTS AND DISCUSSION

Table I gives the composition of all alloys investigated, expressed in four different ways. Figure 1 (4) illustrates the tentative pseudo-binary phase diagram of the system with the postion of the diagram pertinent to this investigation outlined. Plotted in Figure 2 are the results of the measurements, as a function of composition, of room-temperature thermal conductivity, average Seebeck coefficient between 450 and 150°C, and average electrical resistivity

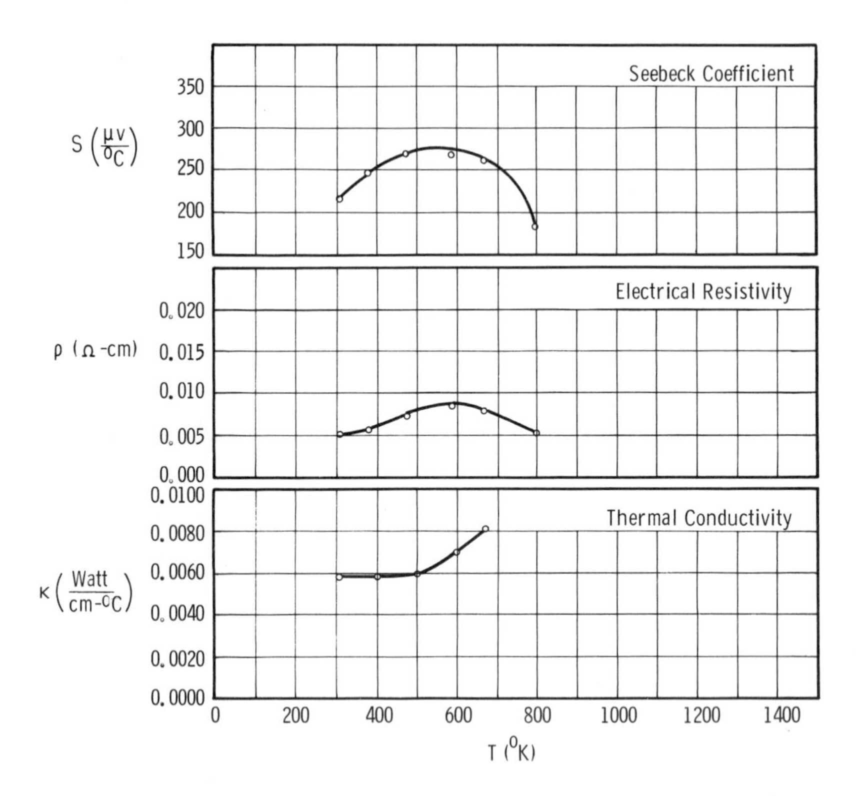

Fig. 3. Thermoelectric properties of furnace-cooled $1.5Sb_2Te_3$:$1Ag_2Te$ as a function of temperature.

between 450 and 150°C. The particular temperature range was chosen because it is of interest for commercial generators. A significant feature of the graphs is that they show the difference between properties obtained from the same composition by the different processing conditions. Air quenching generally produces lower thermal conductivity, higher electrical resistivity, and higher Seebeck coefficient than furnace cooling. Conjecture on the real cause of this behavior is felt out of order at the present time, even though the behavior is consistent with a qualitative simple theory. It can be said, however, that radically different microstructures are exhibited by comparable compositions.

For the furnace-cooled samples in this temperature range, it is seen that the Seebeck coefficient has a rather broad maximum with composition peaking at about 280×10^{-6} v/°C at 47 mole-% Ag_2Te. Electrical resistivity increases with increasing Ag_2Te and levels out at the 50 mole-% composition. Unusually low thermal conductivity is displayed by all furnace-cooled alloys above 38 mole-% Ag_2Te. An anomalously low value of room-temperature thermal conductivity was obtained for the 37 mole-% Ag_2Te air quenched alloy. This result is not readily explainable and perhaps the curve should follow the dotted line indicated in the top graph.

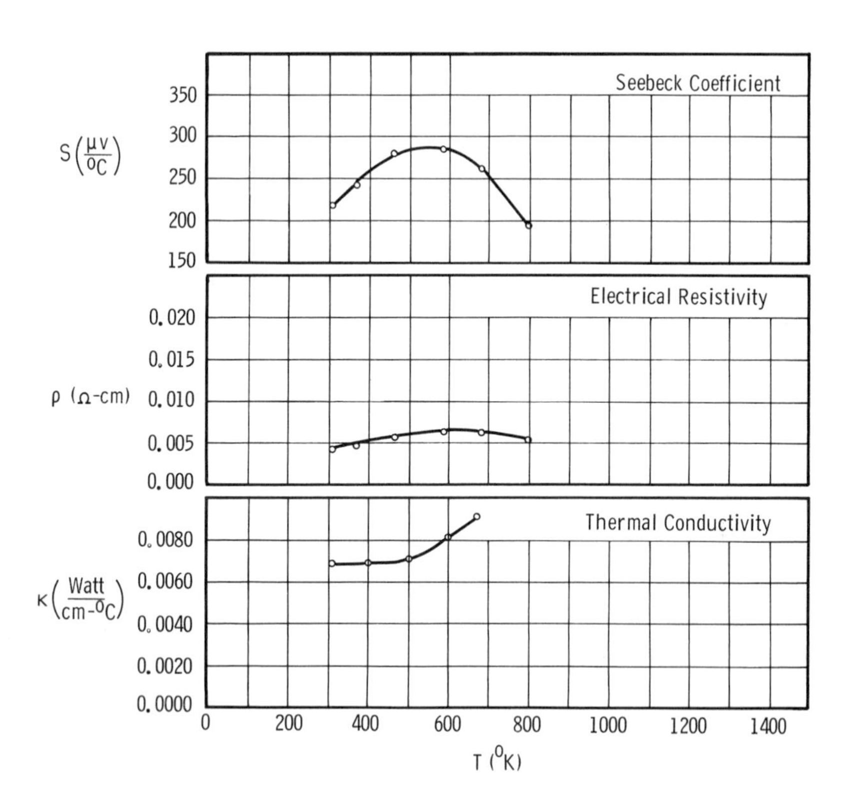

Fig. 4. Thermoelectric properties of furnace-cooled $1.3Sb_2Te_3$:$1Ag_2Te$ as a function of temperature.

Figures 3 to 13 show the data obtained from measurements of Seebeck coefficient thermal conductivity, and electrical resistivity versus temperature obtained on six furnace-cooled alloys, and five air-quenched alloys. Plotted in Figure 14 are curves of thermoelectric figure of merit, $Z = S^2/\rho\kappa$, of the same samples versus temperature. As usual, thermal conductivity versus temperature needs some explanation.

Prior to this work, other investigators (5) had carefully measured on comparative-type apparatus the total thermal conductivity versus temperature of a sample whose composition was $Ag_4Sb_6Te_{11}$. A curve of the lattice thermal conductivity, K_l, versus temperature was established from the relationship $K_l = K_t - K_e$, where K_t is the total thermal conductivity which was measured and K_e is the electronic portion of the thermal conductivity. The relationship between K_e and the electrical resistivity, ρ, is governed by the Wiedemann-Franz law $[K_e = L\,(k/e)^2\,T/\rho]$. Acoustic mode lattice scattering was assumed which means L in the above equation equals 2. It was further assumed that the shape of the K_l versus temperature curve for all the alloys was constant, and

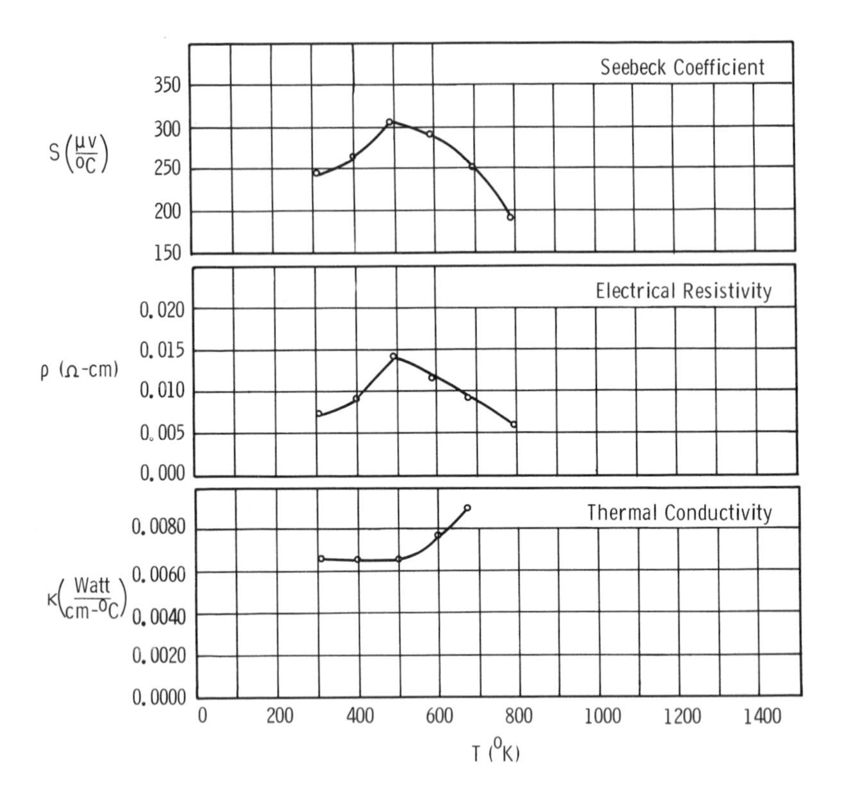

Fig. 5. Thermoelectric properties of furnace-cooled $1.1Sb_2Te_3{:}1Ag_2Te$ as a function of temperature.

thus from our room-temperature thermal conductivity measurements it was possible to calculate the K versus temperature curve for all alloys investigated. To be sure, even though all assumptions are more or less standard, and therefore to be considered reasonable, it does put a "theoretically calculated" label on the figure of merit parameter Z. It should be added that some very recently obtained results on a different sample of the $Ag_4Sb_6Te_{11}$ alloy tested in an absolute type of measuring apparatus, which is capable of independently measuring the Seebeck coefficient, electrical resistivity, and thermal conductivity, have quantitatively agreed with results presented here.

Alloy compositions which lie toward the Sb_2Te_3-rich end of the pseudobinary system exhibit the best properties. Specifically, alloys lying between 40 and 45 mole-% Ag_2Te, which include the 1.5:1, 1.4:1, and 1.3:1 compositions, constitute the most desirable furnace-cooled cast compositions. Figure 3 presents the properties of the 1.5:1 composition indicating a figure of merit of $1.8 \times 10^{-3}/°K$ at 400°K. The 1.3:1 composition shown in Figure 4 has a figure

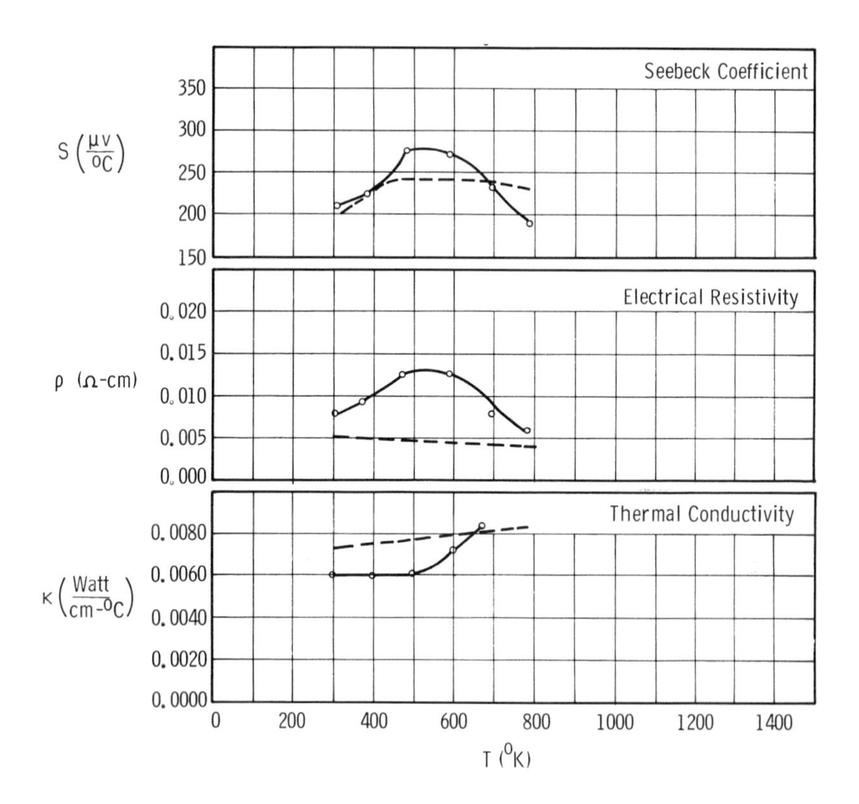

Fig. 6. Thermoelectric properties of furnace-cooled $1.0Sb_2Te_3{:}1Ag_2Te$ as a function of temperature.

of merit at $2.0 \times 10^{-3}/°K$ at 450°K, which is the best obtained using the furnace cool. In spite of higher peak Seebeck coefficients, the figure of merit of alloys richer in Ag_2Te than 45 mole-% decrease, due primarily to their higher resistivities. Although all the alloys feature higher than usual resistivities, when compared with the well-known and used lead-telluride and germanium-bismuth-telluride types (6), their rather low thermal conductivities more than offset this condition and provide a better figure of merit than the above-mentioned alloys.

Further toward the Sb_2Te_3-rich end of the diagram, the Seebeck coefficient and resistivity decrease, while the thermal conductivity increases producing the undesirable condition seen in the 2.0:1 composition of Figure 9. Temperature dependence curves of the thermoelectric properties for the 2.0:1 furnace-cooled heat are not available. This ingot cleaved so readily that it was impossible to cut a large pellet from it.

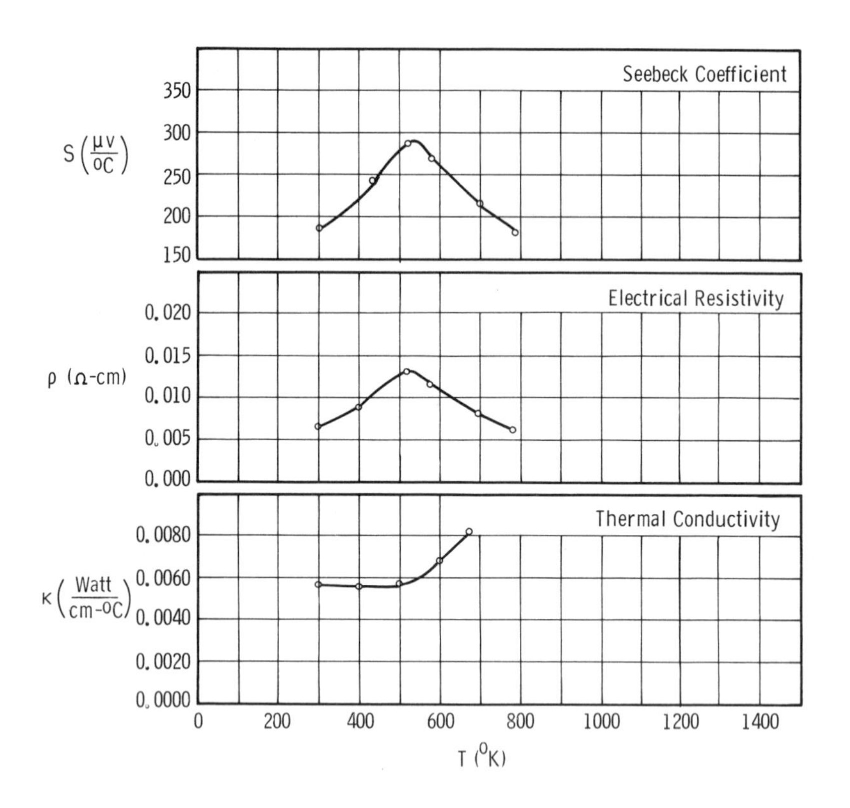

Fig. 7. Thermoelectric properties of furnace-cooled $0.8Sb_2Te_3{:}1Ag_2Te$ as a function of temperature.

Results shown in Figure 6 do not correlate well above room temperature with those of previous investigators (1) of the $AgSbTe_2$ composition, which are shown as the dotted curves.

Figure 14 is included to offer an easy comparison of all alloys' figure of merit versus temperature and processing methods. The rather consistent results indicate increasing figure of merit with increasing Sb_2Te_3 content up to the $1.3Sb_2Te_3$:$1Ag_2Te$ composition and then a decrease. The principal inconsistency is with the $1.7Sb_2Te_3$:$1Ag_2Te$ alloy air quenched, and is probably due, as mentioned earlier, to some questionable thermal conductivity results.

An attempt was made to correlate the alloys' thermoelectric properties with their microstructures. The group of furnace-cooled alloys microstructures appear in Figure 15. A Widmanstatten precipitate of Sb_2Te_3 on the {111}

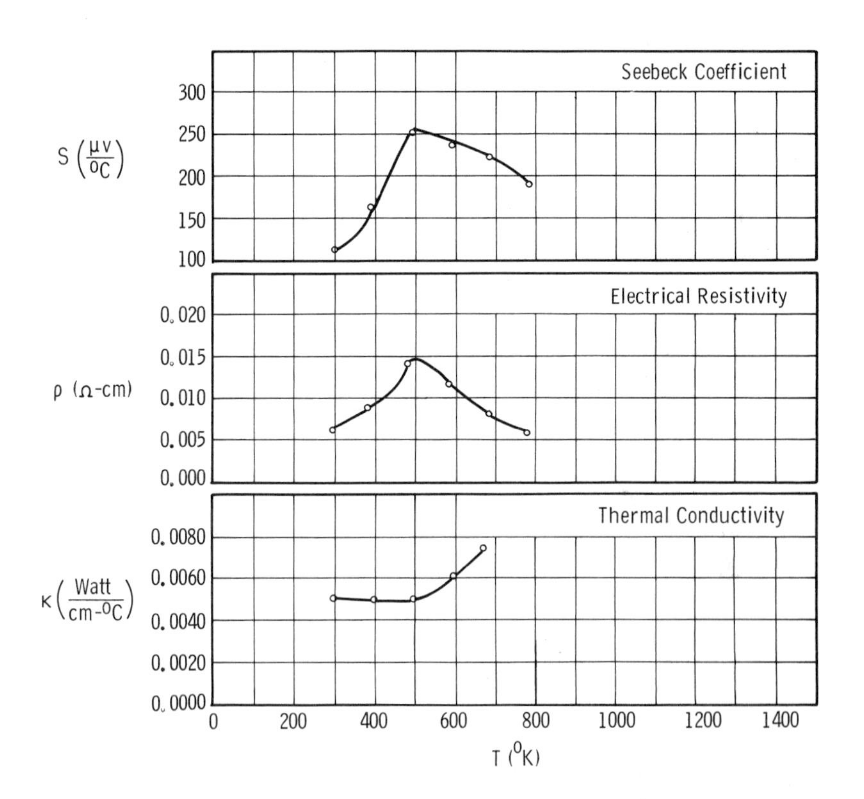

Fig. 8. Thermoelectric properties of furnace-cooled $0.5Sb_2Te_3$:$1Ag_2Te$ as a function of temperature.

planes of the face-centered cubic matrix (the $AgSbTe_2$, β-phase) characterizes the 1.3:1 and 1.5:1 Sb_2Te_3:Ag_2Te composition, Figures 15C and 15B, which also have the best thermoelectric properties. As the alloys become richer in Sb_2Te_3, the Widmanstatten precipitate on the {111} planes becomes heavier until finally the saturation limit is reached and large needles of Sb_2Te_3 appear in the matrix, as can be seen in the 2.0:1 microstructure (Fig. 15A). This leads to poorer properties.

The eutectic structure appears prominently in the microstructure of all compositions 1.1:1 to 0.5:1, Sb_2Te_3:Ag_2Te (Figs. 15D-15G). This is in agreement with the phase diagram shown in Figure 1. The appearance of the eutectic structure is accompanied by an increase in the electrical resistivity and a decrease in the Seebeck voltage resulting in poorer figure of merit alloys.

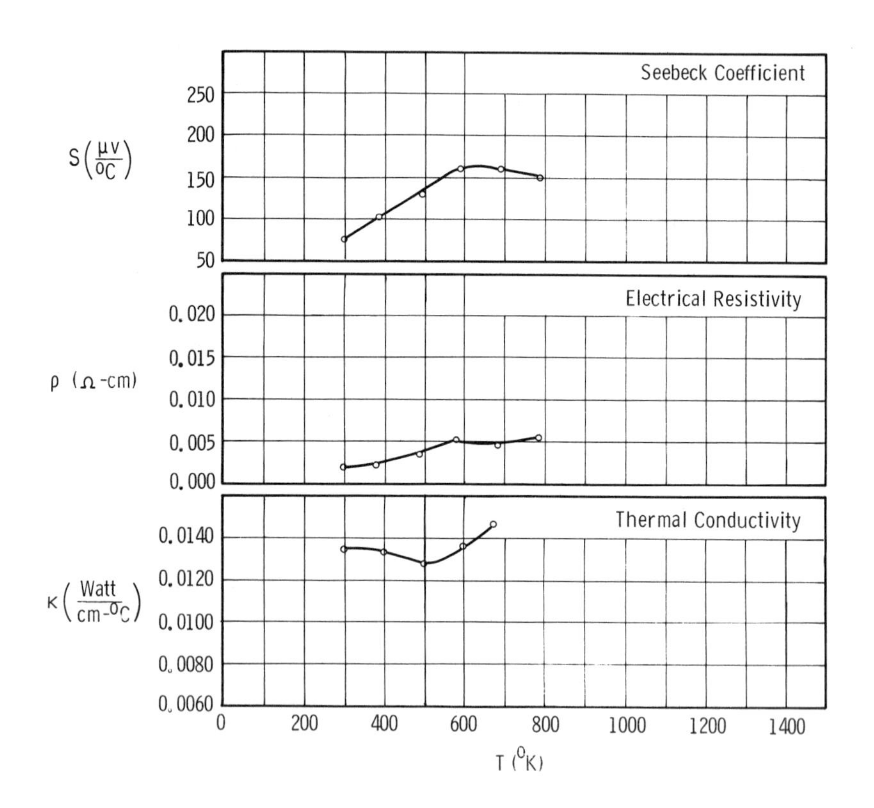

Fig. 9. Thermoelectric properties of air-quenched $2.0Sb_2Te_3$:$1Ag_2Te$ as a function of temperature.

Light-colored β-phase evident in Figure 15D should appear as a single-phased microstructure in one of the alloys somewhere in the region between the eutectic and the peritectic, but it was not seen in this investigation. Some of the eutectic is even visible in the grain boundaries (resolvable at higher magnifications) of the 1.3:1 and traces in the 1.5:1 Sb_2Te_3:Ag_2Te composition. This evidence indicates that even with the slow furnace cool and subsequent anneal, perhaps the equilibrium structures were not obtained.

Apparently, the thermoelectric properties of silver-antimony-tellurium alloys can be adjusted toward optimum by regulating the amount of the Widmanstatten precipitate, since precipitation of Sb_2Te_3 on the {111} planes of the matrix constituted the structure of the alloy with the best properties. At present it appears that the Sb_2Te_3 precipitates at temperatures close to the

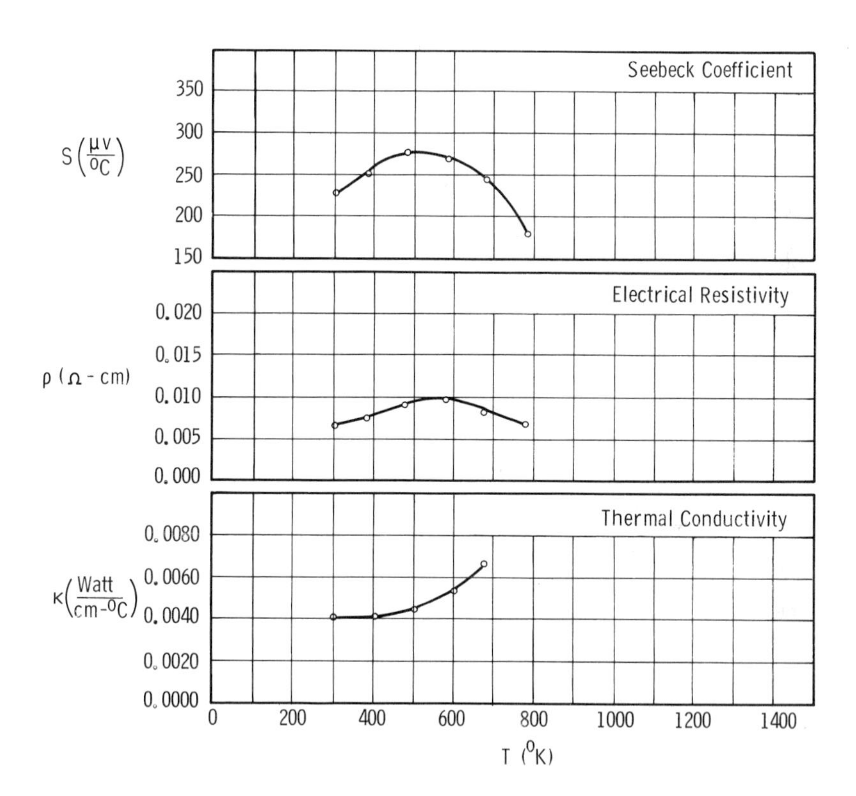

Fig. 10. Thermoelectric properties of air-quenched $1.7Sb_2Te_3{:}1Ag_2Te$ as a function of temperature.

melting point of the alloy. Therefore, the cooling rate during freezing and the annealing temperature are rather critical.

A very apparent macro-difference in appearance exists between alloys made by air quenching and those made by furnace cooling. Air-quenched ingots have a bright silvery metallic luster, while the surface of a furnace-cooled ingot has a soft satin finish which varies in appearance as the ingot is rotated owing to changes in reflectivity from the macroscopic surface grains. All ingots form a solidification shrinkage cavity up to one quarter or more of the total length.

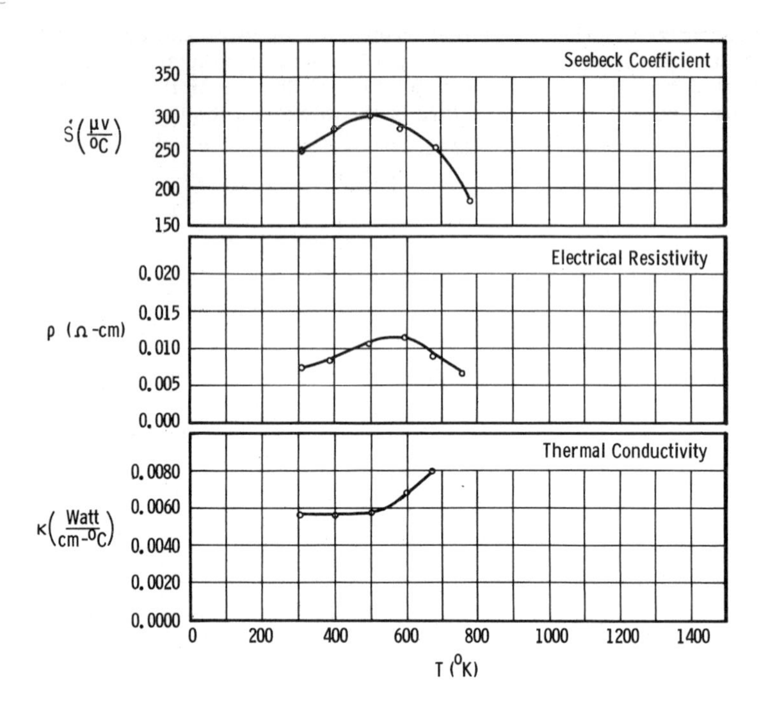

Fig. 11. Thermoelectric properties of air-quenched $1.6Sb_2Te_3:1Ag_2Te$ as a function of temperature.

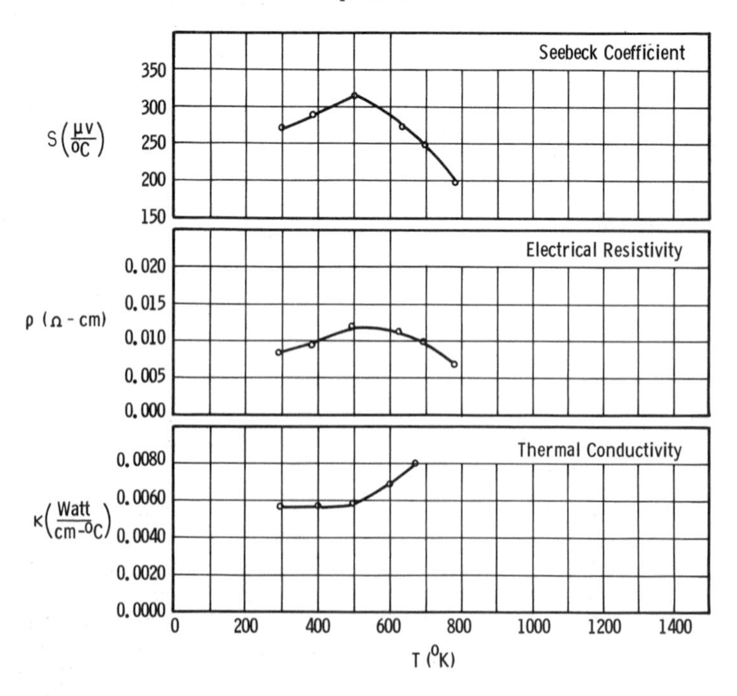

Fig. 12. Thermoelectric properties of air-quenched $1.5Sb_2Te_3:1Ag_2Te$ as a function of temperature.

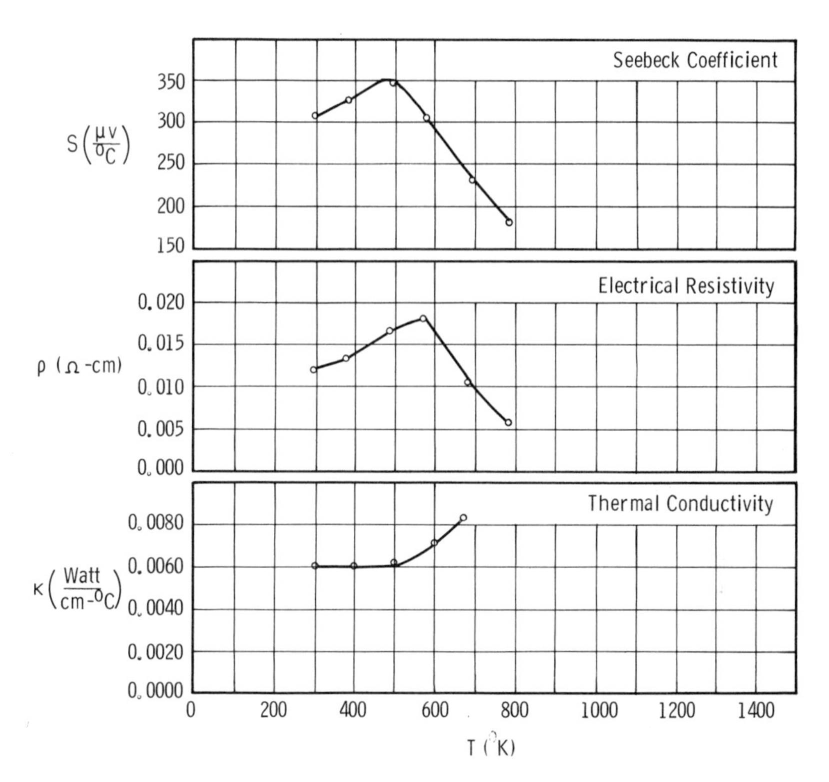

Fig. 13. Thermoelectric properties of air-quenched $1.4Sb_2Te_3$:$1Ag_2Te$ as a function of temperature.

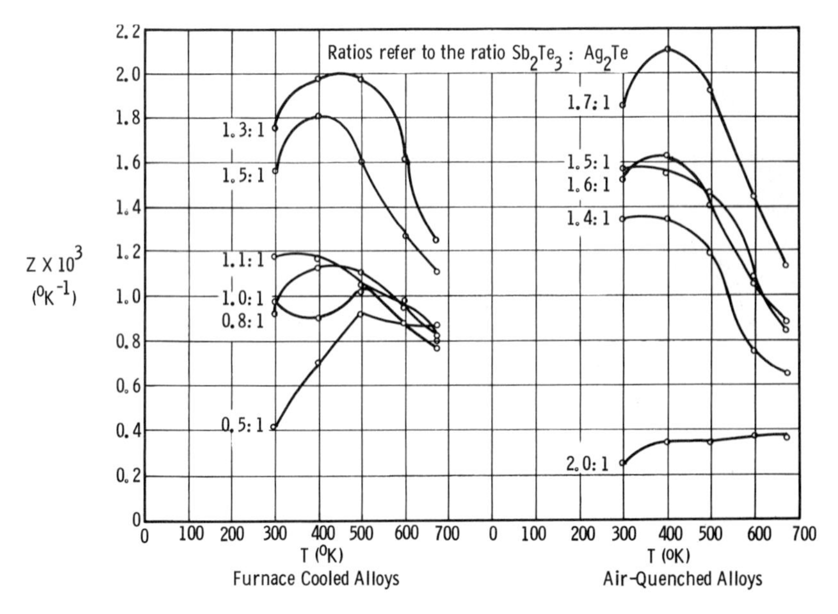

Fig. 14. Temperature dependency of the figure of merit of silver-antimony-tellurium ternary alloys.

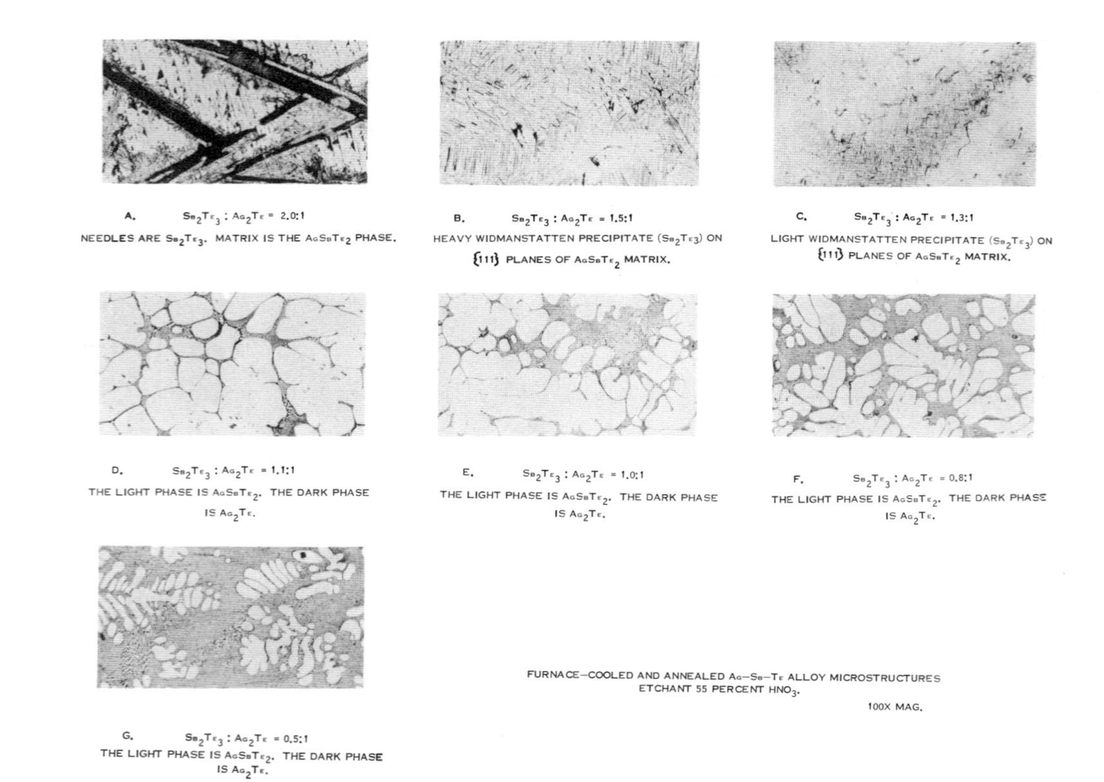

Fig. 15. Furnace-cooled and annealed Ag-Sb-Te alloy microstructures.

Alloys of silver-antimony-telluride perform most efficiently at low temperatures. All the furnace-cooled alloys start to become intrisic conductors between 500 and 600°K with a resulting degradation in the figure of merit. With the exception of the 2.0:1 composition, the air-quenched alloys go intrinsic at 500°K. This behavior naturally somewhat restricts the operating temperature range and maximizes the figure of merit in the vicinity of 400°K; however, useful properties are still anticipated for the alloys up to 750°K.

References

1. Rosi, F. D., J. P. Dismukes, and E. F. Hockings, Elec. Eng., p. 450 (June 1960).
2. Armstrong, R. W., J. W. Faust, Jr., and W. A. Tiller, J. Appl. Phys., 31, (1960).
3. Warnick, J. H., Properties of Elemental and Compound Semiconductors, H. C. Gatos, Ed. (Metallurgical Society Conferences, Vol. 5), Interscience, New York-London 1960, p. 69.
4. McHugh, J. P., W. A. Tiller, S. E., Haszko, and J. H. Wernich, Trans. AIME, to be published.
5. McHugh, J. P., and M. S. Lubell, private communication.
6. Westinghouse Thermoelectricity Quarterly Report. No. 4, U.S. Navy Contract No. NOBS-78365, Feb. 11, 1960.

DISCUSSION

R. A. BERNOFF (Melcor): You mentioned that other people had tried to dope this compound without success. Did you make any attempts at this?

J. T. BROWN: No.

W. T. HICKS (Du Pont): Are these samples harmed by exposure to air at the temperatures where they have good thermoelectric properties?

J. T. BROWN: Yes. We would expect to use silver-antimony-tellurium thermocouples in a closed system.

The Effects of Heavy Deformation and Annealing on the Electrical Properties of Bi_2Te_3

J. M. SCHULTZ, J. P. McHUGH, and W. A. TILLER

Westinghouse Research Laboratories, Pittsburgh, Pennsylvania

Abstract

A mechanism is proposed to account for the observed change in the electrical properties of Bi_2Te_3 as a result of mechanical deformation and post-deformation annealing.

INTRODUCTION

Single crystals of Bi_2Te_3 grown slowly from a stoichiometric melt are found to be p-type at room temperature (1). In fact, p-type behavior is exhibited by single crystals of Bi_2Te_3 grown over a wide range of excess bismuth or tellurium. Polycrystalline samples of Bi_2Te_3 prepared by rapid cooling of a stoichiometric melt are also found to exhibit this p-type behavior (2,3).

Pressed samples of Bi_2Te_3 powder are found to be p-type after pressing but are observed to convert to n-type on annealing (2-4). Other investigators (5) have observed this same inversion and have shown, further, that after sintering at sufficiently high temperature, the pressed material reconverts to p-type.

Quenching experiments performed on single crystal and polycrystalline samples of maximum melting point "Bi_2Te_3" (3,6) have also revealed interesting electrical effects. In both cases, the effect of the quench on these p-type materials was to lower the Seebeck coefficient, S, while increasing the resistivity, ρ, relative to more slowly cooled material. Hall constant and resistivity data on these single crystals (6) showed that the increase in ρ could not have been caused by a carrier mobility decrease alone. The number of carriers in the crystals must have been increased by the quenching operation.

Two main types of defects may be present in the lattice as a result of quenching; i.e., line defects and point defects. The line defects are dislocations and will reduce the mobility of the charge carriers but not their number. By analogy with the charged dislocation considerations carried out for germanium (7), one would expect that only an insignificant number of dislocation sites in Bi_2Te_3 would be charged at 20°C. Further, a simple experiment was performed (8) to determine the effects of dislocations on the electrical behavior of Bi_2Te_3. In this experiment, a single crystal was bent sharply about an axis parallel to the basal plane, plastic flow being primarily by slip along the basal plane. No change in properties was observed.

On the other hand, point defects, which are created in their equilibrium number at high temperatures, may be quenched into the sample at lower temperatures (9). These point defects in an AB-type compound may be of three main types: (a) Frenkel defects (interstitial sites occupied by atoms, together with an equal concentration of vacancies at the proper sites; i.e., $A_i = V_A$ or $B_i = V_B$); (b) Schottky-Wagner defects (vacant sites on both the anion and cation sublattices); and (c) anti-structure defects (anions on cation sites and/or cations on anion sites). These defects may act as donors or acceptors depending upon the energy levels they occupy relative to the Fermi level.

A direct correlation between vacancies and electrical properties of indium antimonide has been previously found by Duga et al. (10) who showed that the effect of vacancies created by plastic deformation in n-type indium antimonide is to decrease the carrier mobility while leaving the number of carriers unaffected. The mobility decrease was attributed to scattering by ionized vacancies, the ionization occurring by a rearrangement of bonding electrons and not from the trapping of conduction electrons and holes as proposed by Crawford and Cleland (11) in their experiments on the neutron irradiation of p-type indium antimonide.

It appears clear from the foregoing, that a wide range of electrical behavior can be imposed upon intermetallic compounds in general and Bi_2Te_3 in particular, depending upon the mechanical and thermal history of the material. The purpose of the present work is to trace the changes in the electrical behavior of Bi_2Te_3 throughout the thermal history of material in which large concentrations of both line and point defects have been introduced by heavy deformation. Finally, a model based on lattice point defect production and annihilation is proposed to explain the experimental observations.

Extrusion was chosen as the mode of deformation; pellets 1 cm in diameter were extruded into rods 0.24 cm in diameter. A description of the apparatus and technique is given elsewhere (12). Annealing was performed in an atmosphere of pure argon at about 2000 psi in order to eliminate vapor loss. Measurements of Seebeck coefficient, S, and resistivity, ρ, as a function of annealing time for a 500°C anneal are shown in Figure 1.

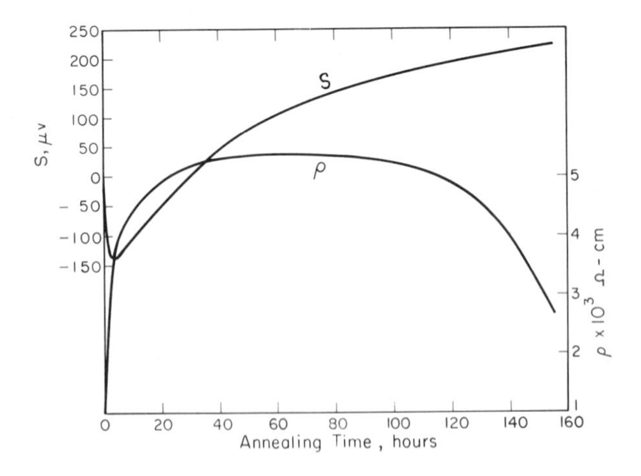

Fig. 1. Measurements of Seebeck coefficient, S, and resistivity, ρ, as a function of annealing time for 500°C anneal.

The following example is proposed to explain the experimental observations. During the deformation process, large numbers of dislocations and point defects are produced. It can be shown that the predominate electrically active point defects are vacancies (12). The ratio of tellurium to bismuth vacancies is greater than 3/2, and the tellurium vacancies must act as donors (12). Thus, upon extrusion, the material is n-type because of the large excess of donors ($\approx 3 \times 10^{20}$, as calculated from Hall measurements).

During the first stages of annealing, dislocations will climb to stable positions, ending at the Van der Waals bonded tellurium-tellurium planes, consuming vacancies in the ratio of 3 tellurium to 2 bismuth in the process. This will cause the negative S to initially become more negative (through a decrease in the number of excess donors) and ρ to increase (through reduction in the total number of carriers), as observed. As the concentration of tellurium vacancies is further reduced, S will become less negative until it is driven positive as the material reverts to the equilibrium state.

The second stage of the annealing process should be the formation of polygonization walls by these stable edge dislocations through glide and climb. This also causes the annihilation of tellurium and bismuth vacancies in the ratio of 3:2. In this region, the carrier mobility should be increasing, thus tending to compensate for the decreasing carrier concentration (flat part of ρ curve in Fig. 1). It is in this region that some grain growth should occur, as we do indeed observe.

The drop in resistivity at the final stage of annealing can only be attributed to an increase in carrier mobility through continued attrition of dislocations at grain boundaries and free surfaces. Such a process requires long range climb of dislocations and hence should occur very slowly. This annealing stage is facilitated by recrystallization and grain growth.

To summarize briefly, heavy plastic deformation produces nonbasal slip in Bi_2Te_3 and introduces large concentrations of point defects into the lattice. The predominant "electrically active" point defects are vacancies. The tellurium vacancies are more plentiful than the bismuth vacancies $\left(V_{Te}/V_{Bi} > 3/2\right)$ and change the extruded rods from p-type toward n-type. Annealing in an inert atmosphere allows dislocations to climb into stable positions, thus annihilating vacancies at a rate determined by the slow migration of tellurium vacancies (12). This causes an initial negative increase in S and an increase in ρ due to donor and acceptor annihilation. At longer times dislocation array formation, recrystallization, and grain growth occur to cause further annihilation of donors and acceptors plus an increase in carrier mobility so that the samples tend toward the properties of carefully grown single crystals.

References

1. Satterthwaite, C. B., and R. W. Ure, Phys. Rev., 108, 1164 (1957).
2. Vasenin, F. I., Zhur. Tekh. Fiz., 25, 397 (1955).
3. Vasenin, F. I., and P. F. Konovalov, Soviet Phys.-Tech. Phys., 1, 1376 (1956).
4. Kokosh, G. V. and S. S. Sinani, Soviet Phys. Solid State, 2, 1012 (1960).
5. George, W. R., R. Sharples, and J. E. Thompson, Proc. Phys. Soc., (London), 74, 768 (1959).
6. Birkholz, V., Z Naturforsch., 13a, 786 (1958).
7. Read, W. T., Jr., Phil. Mag., 45, 775 (1954).
8. Eisner, R. L., Private communication.
9. Kroger, F. A., and H. J. Vink, Solid State Physics, Vol. 3, Academic Press, New York, 1956, p. 352.
10. Duga, J. J., R. K. Willardson, and A. C. Beer, J. Appl. Phys., 30, 1798 (1959).
11. Crawford, J. W., and J. H. Cleland, Phys. Rev., 95, 1177 (1954).
12. Schultz, J. M., J. P. McHugh, and W. A. Tiller, J. Appl. Phys., to be published.

DISCUSSION

J. F. NESTER (Ohio Semiconductors): One of the problems associated with the extrusion of bismuth telluride alloys is, apparently, related to the diameter of the extruded rods. Dr. Tiller said that they extruded samples 0.24 cm. in diameter. Most of our samples have been 0.63 cm. in diameter, and we have found that at this diameter the extruded rod exhibits hot-shortness. This may be due to the presence of a thermal gradient across the larger diameter rod, which causes local variations in the flow rate of material as it passes through the die orifice. This flow differential causes surface cracking and sometimes complete disintegration of the extrusion.

Most of our work has been done with high antimony content alloys, while Dr. Tiller seems to have confined his work to pure bismuth telluride. This could account for some of our difficulties, especially since we have found that temperatures in excess of 500°C are needed to extrude these high antimony content alloys.

One point on which Dr. Tiller did not elaborate is the rate at which he extruded his samples. In light of the previous statements concerning hot-shortness, the extrusion rate could be of great importance.

W. A. TILLER: Our extrusion rates were in the range of 7.5 to 37 cm./min. But we have extruded some alloys of the bismuth telluride type without the attendant difficulties that you seem to have run into. Maybe all your difficulties are size. Certainly when we go to larger size bismuth telluride, it is much, tougher to extrude.

J. F. NESTER: Do you account for this by thermal gradients in the samples?

W. A. TILLER: This is our tentative conclusion in trying to think up why these cracks occur at larger size samples. This seemed to be the most reasonable explanation to us.

R. W. FRITTS (Minnesota Mining & Manufacturing Co.): You mentioned that annealing in the presence of oxygen causes donors to be added to the bismuth telluride. If the oxidized sample is annealed in the presence of hydrogen for a long time, will the original properties of the bismuth telluride be restored?

W. A. TILLER: I certainly do not know the answer to the question, but I would hazard a guess that, yes, you should restore it, but it would take a very, very long time because the point defects that allow the oxygen to go in rapidly may no longer be there and, therefore, the diffusion rate may be cut down by orders of magnitude. I certainly would expect it to go the way you are suggesting.

R. W. FRITTS: Could your raise the annealing temperature beyond 440°C in your hydrogen anneal?

W. A. TILLER: We have annealed these samples up to 550°C. I just happened to show that curve for 480°C.

PART III: MATERIALS FOR THERMIONIC ENERGY CONVERSION

Introduction

J. J. CONNELLY

Office of Naval Research, Washington, D.C.

Since the following group of papers on thermionic energy conversion embrace several technical disciplines, it may be beneficial to review qualitatively the basic concepts of this relatively unknown form of energy conversion.

A thermionic converter can be thought of as a device in which electrons escape from a heated surface (called the emitter) by virtue of their kinetic energy, drift across a gap, and are absorbed on a cooler surface (called the collector) from which heat is removed. The accumulation of electrons on the collector causes a voltage difference to appear between the two surfaces. This voltage is essentially the difference between the work functions of the two surfaces. The work function of a material may be loosely defined as the amount of energy required to remove an electron from the surface of that material.

When the two electrodes are connected through an external matching impedance, current flows, and work can be extracted from the system. Calculations on this simple system, neglecting the space charge and other electrical phenomena, indicate that high operating efficiencies are to be expected.

The way in which the current through the converter effects the efficiency of the device will be described in the following papers. For now it is sufficient to say that it is desirable to have a high current density in the converter, which in turn aggrevates the space-charge problem.

The space-charge effect is simply the electrostatic repulsion between electrons within the gap. Electrons leaving the emitter are influenced by those in transit to such an extent that their motion is retarded, and may, in fact, even be halted. Various techniques have been devised to minimize this effect, and each introduces a major high temperature materials problem.

One method of reducing the space-charge effect is to bring the electrodes very close together, on the order of a tenth of a mil, and thereby, reduce the number of electrons in transit at any instant. Unfortunately, the desired operating lifetime of about two years imposes extremely formidable manufacturing tolerances on such a design, especially in large arrays capable of delivering megawatts of power. The life span of such devices is limited by evaporation, and the subsequent mass transfer across the narrow gap at the high operating temperatures.

An alternative scheme for minimizing the space-charge effect is to introduce positive ions into the interelectrode space. There are several ways of accomplishing this objective.

A gas introduced into the space can be ionized by coming in contact with the surface of a high work function material called the ionizer. If the emitter has a high enough work function to act as the ionizer, the converter must operate at a very high temperature to provide the electrons with sufficient energy to escape from the surface. Failure to do this will curtail emission so that it is inadequate to yield attractive converter power densities.

The problems associated with a high work function emitter may be circumvented by introducing a third electrode somewhere in the system. The emitter may then be fabricated out of a reasonably low work function material while the ionizer has a high work function.

In applications where the converter derives its heat directly from a nuclear reactor, the high energy fission particles can be employed to ionize a noble gas.

Usually, however, it is necessary for the gas to have a low ionization potential and a high vapor pressure. The family of alkali metals, cesium in particular, meet these requirements. Containing alkali metals at high temperatures presents an excellent opportunity to encounter materials problems.

Deposition of some of the alkali metal on the various components of the converter changes the operating conditions. For instance, a monolayer of cesium on the collector surface establishes the work function of the collector as that for cesium.

Solar, nuclear, and chemical combustion all have their advantages and disadvantages as the heat source for a thermionic converter. The solar heating system for space vehicles imposes stringent size and weight requirements, not to mention those associated with reflector orientation and energy storage for dark operation. Neutron capture cross sections limit the variety of materials acceptable when a nuclear heat source is employed. The chemical heating plant introduces a lifetime problem inherent in any fuel-carrying vehicle, and another due to the corrosive nature of the combustion products.

From this brief introduction, and the detailed papers to follow, it is quite apparent that there are numerous, challenging technical problems to be solved before thermionic conversion can be employed in any way approaching routine. However, the potential one can foresee in this area more than justifies the effort.

Considerations in the Selection of Materials For Thermionic Converter Components

PHILIP GOODMAN and HAROLD HOMONOFF

Allied Research Associates, Inc., Boston, Massachusetts

Abstract

Those properties which are of importance to the performance of a material as either the emitter or collector in a thermionic converter are discussed qualitatively. Methods by which the work function for various classes of materials have been calculated and estimated are more extensively discussed in terms of the structural and electronic properties of the atoms in the solid. For selected members of the classes of metals and ionic solids, it has been possible to calculate work functions from a knowledge of the crystalline structure and the nature of the interatomic bonding. For most other materials, including the metalloids, the approximations involved in such calculations are invalid, but correlations of the work function variation within some subclasses have been made with parameters intended to represent the attractive forces of atomic nuclei for the valence electrons. Reported thermionic constants for many nonelemental emitters are tabulated.

INTRODUCTION

The thermionic converter as a means of direct conversion of heat into electricity was first discussed many years ago (1). An emitter to which heat is supplied generates electrons which are received by a collector. Both the emitter and the collector are enclosed in an envelope which may be either evacuated or contain a plasma, such as ionized cesium vapor. Although basically a very simple device, many problems have arisen because of the desirability of obtaining large power outputs from such converters. The electrons leaving an emitter tend to build up a space charge which retards the emission of subsequent electrons. One proposed method to reduce the effects of space charge involves the use of very close spacings between the emitter and collector. Another proposed method requires the presence of plasma; hence, the introduction of cesium vapor which, at the high temperatures present in the thermionic converter, becomes ionized.

The longevity required of useful thermionic converters has also resulted in the introduction of other problems generated by the necessity for use of extremely high temperatures. High temperatures are required because of the desirability of obtaining large currents and also because of the desirability of getting efficient conversion of heat into electricity. The thermionic converter is basically a heat engine in which the thermodynamic efficiency is limited by the efficiency of a Carnot cycle. Thus, the higher the temperature of the heat input stage of the thermodynamic engine, the greater the efficiency which may be theoretically achieved, since this efficiency is, of course, a function of the difference in temperature between the input and output stage. But in a working engine, other loss factors are also introduced. The efficiency of a thermionic converter is defined as the ratio of the output power to the input power. Many authors (2-10) have formulated expressions for this efficiency and for the optimum operating conditions in a converter, taking into account the losses which are associated with both electrical and thermal factors. As a very general statement of the efficiency, η, of such a converter, the expression

$$\eta = (\phi_1 - \phi_2 - V_{ext})/(\phi_1 + P_L/J) \quad (1)$$

can be written where ϕ_1 and ϕ_2 are the work functions of the emitter and collector, respectively, and J is the net current. All the thermal losses inherent in the operation of a thermionic converter are lumped together in P_L. Included in these losses are the radiative energy transfer from the emitter to the collector, the kinetic energies of the electrons which result in a rise in temperature of the collector, the thermal conduction through lead wires, and the gaseous heat conduction. The voltage drop in the external circuit, V_{ext}, determines the electrical loss resulting from the resistance of the electrical circuit. Explicit expressions can be written for these various heat losses and, by assuming arbitrary but reasonable values for certain of the quantities appearing in the resulting expression for the efficiency, this equation can be differentiated and the maximum value of the efficiency can be obtained as one or more of these variables are optimized. Such procedures have been followed by Rasor (2), Houston (3), and Schock (4) among others. The resulting expressions are functions of the dimensions of the components in the thermionic converter, of the emitter and collector temperatures, and of the various material properties of which the components are composed. In the present paper, the various material properties will be discussed qualitatively along with their importance with regard to operation of the thermionic converter. The remainder of the paper will be devoted to a more complete discussion of one of these quantities, the work function.

REQUIREMENTS FOR MATERIALS FOR THERMIONIC CONVERTERS

Those material properties, both physical and chemical, which determine the performance of a given emitter or collector include the following.

The Work Function and Richardson A-Value

The saturation current, J_S, at temperature T°K, generated from an electron emitter is given by the well-known Richardson-Dushman equation:

$$J_S = AT^2 \exp\{-\phi/kT\} \quad (2)$$

where ϕ represents the work function of either the emitter or collector. The constant A is known as the Richardson A and has the theoretical value of 120 amp/cm^{2}deg^2. The output voltage of a thermionic converter is determined by the difference in the work function between the emitter and the collector, as is illustrated in Figure 1. Thus, the output voltage of a thermionic converter becomes greater as the difference between the work function of the emitter and the collector becomes greater. Hence an emitter work function that is too low is undesirable. However, for operation at the same temperature, the output current is determined by the work function according to eq. (2) and is reduced if the work function is too high. The collector work func-

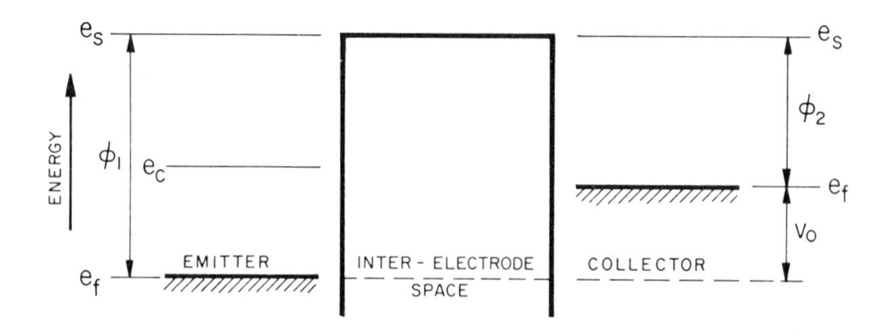

Fig. 1. Energy diagram for thermionic converter, using a semiconducting-type emitter and neglecting space-charge effects.

tion should also not be too low if back emission is to be avoided. The balance between these factors has been discussed by Houston (3) among others.

Vapor Pressure and Rate of Vaporization

Although high emitter temperatures increase the theoretically attainable efficiency, these temperatures also introduce problems with respect to vaporization of the emitter. Initially such vaporization alters the spacial configuration and hence the operating characteristics of the emitter. Eventually the emitter. Eventually the emitter itself disappears as an entity and the converter fails. Therefore, one requisite for a satisfactory emitter is a low vapor pressure at operating temperatures. For cesium vapor converters with electrode spacings on the order of 1 mm, the surface recession rate due to vaporization of the electrode should not exceed 10^{-2} cm per year. A reasonable approximation of the vapor pressure that would permit such a vaporization rate is about 10^{-7} mm of mercury at operating emitter temperatures. Because the vaporization rate is dependent upon the molecular weight of the species being vaporized, and also upon the density and temperature of the solid, it cannot be unambiguously specified for all materials. The value that has been given for the maximum permissible vapor pressure is representative of that for a solid with a density of 4 g cm^{-3} evaporating at 1900°K to yield a vapor with a molecular weight of 100.

One of the simplest optimization criteria which considers both work function and vaporization rate is that given by Wright (11), who proposed that a figure of merit for metallic emitters may be expressed as the ratio of the work function to that temperature at which the vapor pressure is 10^{-5} mm of mercury. Zener et al. (12) have set arbitrary limits with regard to the rates of vaporization of materials and the minimum permissible thermionic emission current. The resulting criteria have been applied to a number of elements and have led to the conclusion that no refractory element will serve the purpose of a long-lived thermionic converter emitter.

Thermodynamic Stability

It is obvious that the materials used for the emitter and collector must be thermodynamically stable at the operating temperatures. If they were not stable then the operating characteristics of the converter would change with time. This stability should exist not only with respect to decomposition of the particular chemical species present but also with respect to the particular crystalline phase that is initially present, since many materials are known to undergo phase changes at high temperatures only very slowly. For converters that are operated with a cesium vapor atmosphere, the thermodynamic stability of the emitter and collector materials must be such that free energy considerations do not favor reaction with the cesium vapor, other impurity gaseous species, or any solid substrates.

Parenthetically, it should be noted that if cesium is adsorbed on surfaces which possess a work function higher than the ionization potential of the cesium atom, the effective work function of the surface is reduced (13-15). Thus the presence of cesium in a thermionic converter makes a twofold contribution to the performance since it also aids in reducing the space-charge problem. However, during long-time exposure to high temperatures, it is conceivable that the cesium atoms may diffuse into the crystalline lattice of the emitter or collector. These diffusing cesium atoms must not exhibit a poisoning effect upon the thermionic emission and work function characteristics of the materials. Experimental studies of such potential poisoning do not appear to have been made except in conjunction with life-tests of cesium vapor thermionic converters.

Emissivity

For efficient conversion of energy, the heat supplied to the emitter should be used exclusively for the energy required to evaporate electrons. However, some of this energy will be radiated by the emitter. The extent to which thermal energy is used for electron evaporation as opposed to being radiated is determined by the emissivity of the emitter. The lower the emissivity, the smaller will be the energy loss resulting from radiation. Radiant energy absorbed by the collector will raise its temperature, thus reducing the conversion efficiency. A low emissivity and a high reflectivity for the collector materials is consequently also desirable.

Other Physical Properties

Other physical properties of emitters and collectors which are important with regard to the efficiency of operation of a thermionic converter include such properties as the thermal conducitivity of the emitter and of the lead going to the external circuit. If the thermal conductivity of this lead is high, then heat will be lost not only through radiation but also by conduction through the lead. The electrical conductivities of the emitter and collector and of the leads are also important since the power losses in the electrical circuit will be high if the electrical conductivity is low.

WORK FUNCTION-DEFINITION

Of those factors which have been discussed as being important to the determination of the requisite properties of materials to be incorporated in thermionic converters, the one which is most critical is concerned with the thermionic emission of the materials. The emission is described by the Richardson-Dushman equation, eq. (2). The presence or absence of emission determines whether the converter works and the emission current determines the magnitude of the output current. Furthermore, differences between the work function of the emitter and collector determine the output voltage. It is thus profitable to examine more closely what can be predicted and what types of correlation are possible among various materials with regard to their thermionic emission. Since, in eq. (2), the exponential term is the one which largely determines the magnitude of the emission, the remainder of this paper will be devoted to a discussion of work function and of various attempts that have been made to predict and correlate work function for both elements and for compounds.

In recent years it has come to be recognized that there are at least three different types of work functions. The first work function, which is the one most clearly defined in therms of theoretical solid state structure, is called the "ture" work function. This work function is defined as the difference in energy between the Fermi energy within a material and the vacuum level outside the surface of the material. Some ambiguity remains in this definition, in that the vacuum level is not precisely defined. However, for most practical purposes, the differences are small between the energy of an electron located just outside the surface of an emitter and that of an electron at a large distance from the emitter. The second work function is the most practical definition of a work function. It has been termed the "apparent" work function and is defined as the value obtained from the slope of a plot of eq. (2); i.e., a plot of log J_s/T_v^2 vs $1/T$. The third work function has been termed the "effective" work function and is obtained by a similar plot of eq. (2), but with the value of the constant A being taken as being the theoretical value, 120 amp/cm^{2}deg^2. Thus values for effective and apparent work functions are experimentally determined. The true work function, however, is basically a theoretical concept. The relations between these three work functions have been admirably discussed by several authors in recent years [see Hensley (16), Cusack (17), Nottingham (18), and Herring and Nichols (19)]. The subsequent discussion shall be essentially concerned only with the true work function. However, it must be emphasized that comparisons made between various materials on the basis of the experimental data must be made on the basis of experimentally determined work functions and not true work functions. In this sense at least, the comparisons will be in error. Other errors arise not only from experimental inaccuracies but also because of the frequent lact of distinction in the literature between apparent and effective work function. The following discussions will be concerned with comparisons that can be made between work functions for a variety of materials, both elements and compounds. They shall not be concerned however, with such factors as patchy surfaces, mixed emitters, surface heterogeneities, grain size and porosity, or gross impurities.

The true work function may be expressed as the difference between two terms, both of which are referred to the energy level of the bottom of the conduction band in the solid (see Fig. 1). The first term, $e_s - e_c$, is called the electron affinity of the solid or the external work function. It is the energy difference between an electron at the surface of the solid and one at the bottom of the conduction band. The second term, $e_c - e_f$, sometimes termed the "internal" work function, is the difference between the energy of the electron at the bottom of the conduction band and its energy at the Fermi level within the solid. The sum of these two terms then is the true work function and is given by the equation

$$\phi_t = (e_s - e_c) + (e_c - e_f) \tag{3}$$

Because both the electron affinity and the Fermi energy are dependent upon temperature, the true work function is therefore dependent upon the temperature. The advantage of expressing the true work function as the sum of two terms lies in the fact that a degree of theoretical correlation and predition is possible with regard to the variation of these two terms from molecule to molecule and as a function of the temperature, etc.

Frequently such comparisons have been made by assuming that one of these two quantities remains constant, and that the other quantity is responsible for the vast majority of the variations observed or predicted. The electron affinity of a solid is primarily a property of the structure of the solid and is largely determined by the magnitude of the dipole layer of charge at the surface of the solid. As such it is influenced by such factors as the interatomic spacings within the lattice of the crystals, the thermal expansion of the crystals, such phase changes as may occur within the crystal as the temperature is changed, and because for a single crystal the interatomic spacing may not be the same for all crystalline faces it is also a function of the crystalline orientation. The Fermi energy is defined as the energy of that level within the crystalline electronic band structure which has an equal probability of being occupied or unoccupied at any given temperature. As such, it is a rather sensitive function of the electronic band structure within the crystal. For extrinsic semiconductors it is also the quantity which is largely responsible for the temperature coefficient of the true work function, since in these materials, it is a function of the effective density of states which, in turn, are sensitive functions of temperature.

WORK FUNCTIONS OF ELEMENTS

Compilations of measured values of work functions for the elements have been given by Michaelson (20) and by Wright (11). A simplified correlation of the work function of a number of elements with the lattice distance within their crystals has been given by Herrmann and Wagener (21) as

$$\phi = \frac{7.2 \times 10^{-10}}{a_o} \tag{4}$$

where ϕ is the work function in electron volts and a_o is the emitter lattice distance in meters. Such a radical simplification, which essentially considers only the electron affinity, would not be expected to provide a satisfactory correlation. Thus it is not surprising to find the correlation partially satisfactory for alkali metals and unsatisfactory for all others.

A number of attempts to theoretically predict the value of the work function for metals have been made. Many of these, however, have been concerned with the value of the Fermi energy within the metal and have ignored the effect of the electron affinity upon the true work function. These approaches have been discussed and summarized by Herring and Nichols (19). That most favored by them, and the one that they have used to make additional calculations, is the approach developed by Wigner and Bardeen (22). It is based on the assumption that the true work function is related to the difference between the actual energy of the ground state of the metal and that of the self-consistent field solution. The self-consistent field solution is then assumed to give essentially the same energy, referred to the ground state of the metal, as that of a gas of free electrons moving in a constant external potential and having the same average electron density as is present in the metal. The assumptions involved in this treatment are valid only for elements on the left side of the periodic table and are perhaps really only valid for alkali metals for which the electronic structure consists of a single electron outside of completely filled shells. The treatment is certainly not valid for transition elements. Although the expression that is given for the work function includes a term representing the electron affinity, the dipole layer of charge on the surface is determined only by taking the difference between the calculated inner work function and the experimentally determined ϕ. A theoretical calculation of the external work function, made by Bardeen for sodium, does agree with the value deduced by difference, but no comparable calculations have been carried out for higher electron densities. Those external work functions, determined by difference for the alkali and alkaline earth metals, appear to increase with the electron density.

Since rigorous theoretical calculations of the true work function for metals are not possible in general; it is therfore desirable to determine what types of correlations can be made with other more qualitative measures of chemical and physical binding of electrons within simple metals. One such correlation (23) is that made by Gordy and Thomas who related the work function of a number of metals to their electronegativity. The electronegativity is a chemical concept first introduced by Pauling (24) and currently used to characterize, in a qualitative fashion, the bonding that exists between unlike atoms in various types of molecules. Pauling originally set up his scale of electronegativity values on the basis of the equation,

$$(x_a - x_b) = 0.208\Delta^2 \tag{5}$$

where x_a and x_b represents the electronegativity of the atoms a and b joined together in a chemical bond and Δ is the energy difference between that bond energy to be anticipated if a pure covalent bond were present between atoms a and b and the observed energy of the bond. The difference, Δ, is attributed to ionic resonance energy. Thus, the scale of electronegativity is related to the attractive forces for a given element for electrons. Gordy (25) originally suggested that electronegativity values could be calculated from the equation

$$x = (Z_{eff})\, e/r \tag{6}$$

where Z_{eff} is the effective nuclear charge of the bonded atom upon the bonding electron when the electron is at a distance from the nucleus equal to the covalent radius r. The calculations for Z_{eff} take into account the screening effect by other electrons present in the unfilled valence shelves. In order to bring the values calculated from eq. (6) into accord with the Pauling scale, a proportionality constant and a shift in origin was required and resulted in

$$x = 0.31\ [(n+1)/r] + 0.50 \tag{7}$$

where n is the number of valence electrons. Still other methods of estimating electronegavity values have been proposed by Gordy (26) who related electronegativity to bond stretching force constants, bond lengths, and bond orders; by Walsh (27) who related electronegativity to the stretching force constant of the elemental bond to hydrogen; and by Milliken (28) who proposed that the electronegativity of an element be determined by the arithmetic mean of its first ionization potential and its electron affinity.

The relationship shown by Gordy and Thomas between the work function and their electronegativities is reproduced in Figure 2. It may be seen that the smaller the electronegativity, the smaller the work function. The equation for the straight line shown in this figure is

$$x = 0.44\,\phi - 0.15 \tag{8}$$

From the concept and formulation of the electronegativity values, it is evident that small electronegativity values are associated with small attractive forces between valence electrons and the effective nuclear charge. Probably the most fundamental atomic parameters related to electronegativity are the interatomic force constant and the interatomic distances. However, these cannot be used directly since the values are modified by such factors as the coordination number of the element in the solid. Gordy's method (25) of calculating electronegativity values is reasonably satisfactory for simple elements.

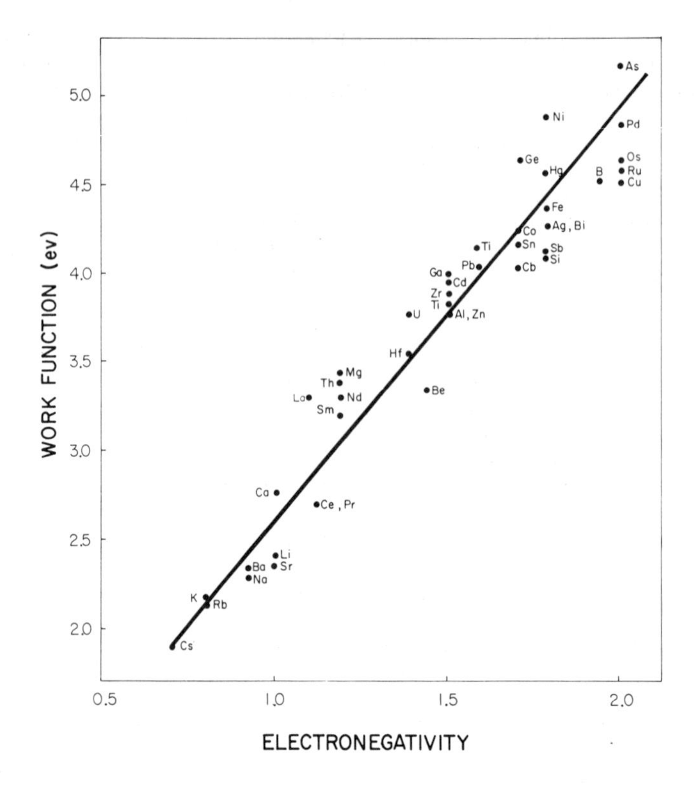

Fig. 2. Variation of work function of metals with electronegativity [after Gordy and Thomas (23)].

However, for transitional elements, there is difficulty in estimating the proper value to be assigned to the screening constants for those electrons located in unfilled inner orbitals.

WORK FUNCTION OF IONIC SOLIDS

Atomic structure and energy levels for isolated elements are reasonably well understood. When interatomic forces between elements must be considered, as in a solid metal, the computation of energy levels is more complicated and can only be done on an approximate basis. However, when elements chemically interact to form chemical compounds in the solid state, calculations become even more complicated, and approximate. One commonly used approximation involves the assignment of valence electrons to individual atomic orbitals, but there is no certainty with regard to this assignment within the solid compound. Once the assignment of electrons to atomic orbitals is made, the calculation of energy levels and band structure in the solid is just as approximate and difficult as it is for elemental solids. Some of the attempts that have been made to predict energy levels and true work functions of different types of compounds in the solid state are discussed below and in the next section. In many cases, the fundamental quantities that are important for the determination of energy levels are also important for the determination of other properties which are of interest with respect to the use of materials in thermionic converters.

An ionic solid is probably the simplest solid compound to visualize. It consists of elements organized together in such a way that the electropositive element has donated one or more of its valence electrons to the electronegative element. It is probable that no real solid is truly ionic and that the bonding that exists within even the most ionic solid has some covalent character to it. Nevertheless, it is possible to treat many compounds as being essentially ionic. The alkali halides are considered to be the most ionic of the known compounds in the solid state, although even within this group there are wide variations in ionicity of the bonds. Various oxides and sulfides are generally also considered to be ionic and it is even conceivable that certain carbides and nitrides possess considerable ionic character. As pure compounds, ionic compounds are insulators. However, they can be converted into semiconductors when interference (impurity or defect) levels are introduced into the crystal, thus disturbing the normal band structure of the insulator.

A method of computing the energy levels, the electron affinity, and the probable maximum value for the work function of alkali halides has been developed by Mott and Gurney (29) and was subsequently modified by Wright (30). For the sodium chloride crystal, it was assumed that the single 3s-electron of the sodium atom has entered the 3p-quantum level of the chlorine atom, thereby completely filling the 3p-band. The 3s-band associated with the sodium atom, considered to have a higher energy than the 3p-band of the chlorine atom, becomes the conduction band in the crystal. It remains empty unless electrons from the filled 3p-band can be given enough energy to enter the higher 3s-band.

A cycle is postulated which involves removal of an ion from the crystal, conversion of the ion to an atom, and replacement of the atom inside the crystalline lattice. The net result is either to add an electron to a positive ion or to remove one electron from a negative ion. Equations can be developed for three quantities: $-\chi$, the work required to introduce an extra electron into the crystal at an energy corresponding to the bottom of the conduction band; ϕ, the work required to extract an electron from the valence band; and θ, the energy required to remove an electron from the valence band but leave it within the field of the positive hole that remains in the valence band. These expresions for $-\chi$, ϕ, and θ involve such fundamental crystalline quantities as the lattice energy per ion pair, the Madelung constant of the crystal, the ionization potential of the electropositive atom, and the electron affinity of the electronegative atom. These equations also contain terms which represent polarization energies, calculable from observed crystalline dielectric constants.

Two additional terms are required to represent the energy required to insert an atom into the lattice in a position which had been occupied by a smaller positive ion and to represent the energy involved with transfer of an electron from a neighboring ion, thus converting it into an atom. These last two terms cannot be calculated and are estimated by intuitive arguments as being either small or large. Wright's (30) contribution to the estimation of maximum work function and electron affinity in alkali halides was to introduce the concept of band width. The allowable energy range within the bands in the alkali halides may be very large, but the distribution of states is such that only a few states are permitted at the extremities of the energy range and the vast majority of the energy states are concentrated around the center of the band. The expressions representing the energies for removal and insertion of electrons remain the same except for the addition of a term representing the energy difference between the bottom of the conduction band and the maximum density of electron states in this band and, similarly, between the top of the valence band and the energy associated with the maximum density of electron states within the valence band. Modern theory would probably relate these energies to the Fermi levels rather than to the position of the maximum of the density of electron states. The values deduced for ϕ are closely related to the thermionic work function if the position of the Fermi level is not too distant from the top of the filled valence band.

This analysis of the energy levels within ionic solids was extended by Wright to divalent oxides and to zinc sulfide. The outer two s-electrons of the electropositive element are assumed to enter the p-shell of the electronegative element and, together with the electrons already present within the p-shell, to form a completely filled valence band. Equations similar to those for the alkali halides can be written and an estimate of the electron affinity, χ, was thus obtained. For BaO this was found to have a maximum value of 0.6 ev, and for the other alkaline earth oxides, the electron affinity did not exceed 1 ev.

Table I gives values for ϕ calculated for the alkaline earth oxides in the manner described. Also given in the last column of Table I are values for experimentally determined work functions for the same compounds. In most

TABLE I

Values of Some Crystalline Parameters for Alkaline Earth Oxides

Compound	Lattice spacing, A	Lattice energy, ev	Vapor pressure, atm., 2000°K	Metal Ionization potential, ev	Calc max work function, ev	Exptl[a] work function, ev
BaO	5.52	32.5	1.0×10^{-6}	10.0	2.6	1.4-1.7
SrO	5.14	34.4	6.9×10^{-7}	11.0	3.3	2.1-2.2
CaO	4.80	36.6	2.1×10^{-7}	11.8	4.8	1.7-2.5
MgO	4.20	40.9	5.0×10^{-6}	15.0	5.9	2.3-4.4
BeO	3.28	49[b]	6.6×10^{-9}	18.1	11.2	3.8-4.7

[a] See Table II for details and references.
[b] Estimated.

cases these are apparent work functions obtained by use of a typical Richardson plot. The trend of the calculated and experimental work functions is the same, but it should be noted that the experimental work function values have generally been measured for solids that contain some impurities in the crystalline lattice which provide donor sites, thus changing the Fermi level and consequently the work function. If the binding energy of the electron in the donor levels is given by W, then the true work function can be approximated by

$$\phi' = \chi + W/2$$

where χ is the electron affinity characteristic of the stoichiometry, and W/2 is an approximation of the energy of the Fermi level relative to the bottom of the conduction band.

Also given in Table I are values for various physical and chemical parameters. Trends in lattice spacing, lattice energy, and the ionization potential of the metal atom can be clearly seen.

The trends of these physical and chemical parameters should be the same for the alkali metal oxides of group I. These should be expected to show smaller work functions than are exhibited by the alkaline earth oxides because of the larger ionic radius and the smaller ionization potential of the alkali metals as compared with the alkaline earth metals. This prediction is verified by a comparison of the experimental work function for cesium oxide, 0.75 ev, and the work function of barium oxide, approximately 1.5 ev. These values will be found listed in Table II which includes most of the experimental values of thermionic emission constants that have been published for solid compounds of interest. Many of the values pertaining to alkaline earth oxide emitters have been omitted, however, because of the large number of conflicting investigations conducted on these emitters during the last 50 years. Also missing from Table II are data on mixed and film-type emitters.

TABLE II

Thermionic Emission Constants for Simple Chemical Compounds

Compound	Work function, ev	Richardson A_2, amp/ cm^2deg^2	Ref. No.	Remarks[a]	
Oxides					
Cs_2O	0.75	0.1	11	ϕ_A	
	0.89-1.17	—	32	ϕ_E, T = 400°K	
BeO	3.8	—	33	ϕ_E, T = 1400°K	
	4.7	—	32	ϕ_E, T = 1400°K	
MgO	3.1	—	33	ϕ_E, T = 1400°K	
	4.4	—	32	ϕ_E, T = 1400°K	
	2.31-3.70	—	34	ϕ_A	Seven types of samples ranging from powder to single crystal
	2.80-3.85	—	34	ϕ_E, T = 1000°K	
CaO	2.50	—	11	ϕ_E, T = 1000°K	
	1.69	3×10^{-4}	36	ϕ_A	
SrO	2.12	—	32	ϕ_E, T = 1050°K	
	2.20	100	11	ϕ_A	
BaO	1.5	0.1	37	ϕ_A	
	1.4	6	11	ϕ_A	
	1.66	—	32	ϕ_E, T = 1050°K	
Al_2O_3	4.4	—	33	ϕ_E, T = 1400°K	
B_2O_3	4.7	—	33	ϕ_E, T = 1400°K	
Sc_2O_3	3.8	—	33	ϕ_E, T = 1400°K	
	4.23	400	38	ϕ_E, T = 1300-1700°K	
	4.04	—	38	ϕ_E, T = 1700°K	
	3.66	9.3	38	ϕ_A, > 1700°K	

(continued)

TABLE II (continued)

Compound	Work function, ev	Richardson A_2, amp/ cm^2deg^2	Ref. No.	Remarks[a]
Y_2O_3	3.16	–	11	ϕ_E, T = 1900°K
	4.14	1000	38	ϕ_A, T = 1300-1600°K
	3.87	–	38	ϕ_E, T = 1600°K
	3.22	1.1	38	ϕ_A, T > 1600°K
	2.0	0.55	39	ϕ_A
La_2O_3	3.25	–	11	ϕ_E, T = 1900°K
	2.50	0.9	11	ϕ_A
	2.80	–	39	ϕ_A
	3.70	6.6	38	ϕ_A, T = 1300-1800°K
	4.20	–	38	ϕ_E, T = 1800°K
	2.95	0.004	38	ϕ_A, T > 1800°K
	3.18	31	40	ϕ_A
CeO_2	2.75	–	11	ϕ_E, T = 1800°K
	2.3	1.0	11	ϕ_A
	2.7	0.005-10	41	ϕ_A ±20%. Richardson A depends upon activation state of cathode.
Pr_2O_3	2.8	0.22	39	ϕ_A
	1.9	0.1	43	ϕ_A
	3.1	–	43	ϕ_E, T = 1850°K
Nd_2O_3	2.0	0.2	43	ϕ_A
	3.0	–	43	ϕ_E, T = 1850°K
	2.3	0.99	39	ϕ_A
	2.96	53	40	ϕ_A
Sm_2O_3	3.32	285	42	ϕ_A Pulsed emission measurements.
	2.38	2.0	43	ϕ_A
	2.8	0.33	39	ϕ_A
	3.1	–	43	ϕ_E, T = 1850°K
	2.98	50	40	ϕ_A
Eu_2O_3	2.48	0.3	43	ϕ_A
	3.50	–	43	ϕ_E, T = 1850°K
	2.6	0.11	39	ϕ_A

(continued)

TABLE II (continued)

Compound	Work function, ev	Richardson A_2, amp/ $cm^2 deg^2$	Ref. No.	Remarks[a]
Gd_2O_3	2.18	0.5	43	ϕ_A
	3.1	—	43	ϕ_E, T = 1850°K
	2.1	0.66	39	ϕ_A
	2.58	4.2	42	ϕ_A
	3.00	51	40	ϕ_A
Tb_2O_3	2.1	0.22	39	ϕ_A
Dy_2O_3	2.28	1.6	43	ϕ_A
	3.0	—	43	ϕ_E, T = 1850°K
	2.1	0.96	39	ϕ_A
Ho_2O_3	2.3	0.33	39	ϕ_A
Er_2O_3	2.4	0.76	39	ϕ_A
Yb_2O_3	2.75	4	43	ϕ_A
	3.3	—	43	ϕ_E, T = 1850°K
	2.7	1.42	39	ϕ_A
Lu_2O_3	2.3	0.11	39	ϕ_A
ThO_2	2.57	7.9	44	ϕ_A
	2.55	3.3	45	ϕ_A
	2.67	5.6	46	ϕ_A
	2.71	21	38	ϕ_A, T = 1100-1300°K
	3.06	—	38	ϕ_E, T = 1300°K
	1.66	0.001	38	ϕ_A, T $>$ 1300°K
	3.1	—	11	ϕ_E, T = 1900°K
TiO_2	4.7	—	33	ϕ_E, T = 2000°K
	3.7	—	11	ϕ_E, T = 2000°K
	3.87	0.458	38	ϕ_A, T = 1500-2000°K
ZrO_2	4.2	—	33	ϕ_E, T = 2000°K
	3.9	—	11	ϕ_E, T = 2000°K
	4.11	281	38	ϕ_A, T = 1400-1700°K
	3.96	—	38	ϕ_E, T = 1700°K
	3.12	0.363	38	ϕ_A, T $>$ 1700°K

(continued)

TABLE II (continued)

Compound	Work function, ev	Richardson A_2, amp/ $cm^2 deg^2$	Ref. No.	Remarks[a]
HfO_2	2.0	0.02	43	ϕ_A
	3.6	--	43	ϕ_E, T = 1900°K
	3.76	381	38	ϕ_A, T = 1300-1500°K
	3.60	—	38	ϕ_E, T = 1500°K
	2.82	0.49	38	ϕ_A, T > 1500°K
SiO_2	5.0	—	33	ϕ_E, T = 2000°K
Borides				
TiB_2	3.95	—	47	ϕ_E (?)
VB_2	3.88	—	47	ϕ_E (?)
CrB_2	3.36	—	47	ϕ_E (?)
MnB_2	4.14	—	47	ϕ_E (?)
ZrB_2	3.70	—	47	ϕ_E (?)
ZrB	4.48	3×10^4	45	ϕ_A, Stoichiometry not certain.
NbB_2	3.65	—	47	ϕ_E (?)
MoB_2	3.38	—	47	ϕ_E (?) Stoichiometry may be Mo_2B_5.
TaB	2.89	10	45, 47	ϕ_A, Several per cent TaB_2 probably present.
W_2B_5	2.62	—	47	ϕ_E (?) Ref. 47 lists compounds as WB_2, which probably does not exist at room temperature.
CaB_6	2.86	2.6	48	ϕ_A
SrB_6	2.67	0.14	48	ϕ_A
BaB_6	3.45	16	48	ϕ_A
ScB_6	2.96	4.6	49	ϕ_A
YB_6	2.20	15	50, 51	ϕ_A
LaB_6	2.66	29	48	ϕ_A
	2.68	73	51	ϕ_A
CeB_6	2.59	3.6	48	ϕ_A
	2.93	~580	51	ϕ_A
PrB_6	3.46	~300	51	ϕ_A
NdB_6	3.97	~420	51	ϕ_A

(continued)

TABLE II (continued)

Compound	Work function, ev	Richardson A_2, amp/ cm^2deg^2	Ref. No.		Remarks[a]
SmB_6	4.4	–	52	ϕ_A (?)	
EuB_6	4.9	–	52	ϕ_A (?)	
GdB_6	2.05	0.81	51	ϕ_A	
TbB_6	6.0	–	53	ϕ_A	
DyB_6	3.53	25.1	51	ϕ_A	
HoB_6	3.42	13.9	51	ϕ_A	
ErB_6	3.37	9.9	51	ϕ_A	
YbB_6	3.13	~2.5	51	ϕ_A	
LuB_6	3.0	0.36	51	ϕ_A	
ThB_6	2.92	0.5	48	ϕ_A	
Carbides					
ThC_2	3.5	550	45	ϕ_A	
	3.2	100	11	ϕ_A	Exact sample stoichiometry not known.
UC	3.62	90	54, 55	ϕ_A	Fused UC, $T < 1590°K$.
	2.67	3.5	54, 55	ϕ_A	Fused UC, $T > 1610°K$.
	4.57	7.2×10^5	54	ϕ_A	From $\log_e J_{sat}$ vs T plot.
	2.94	33	56	ϕ_A	Powdered UC.
$ZrC_{0.8}UC_{0.2}$	4.3	6.6×10^4	54	ϕ_A	From $\log_e J_{sat}$ vs T plot
	3.1	12	57	ϕ_A	Hot-pressed sample.
	3.53	55	55	ϕ_A	Fused sample, $T < 1610°K$.
	2.53	0.3	55	ϕ_A	Fused sample, $T > 1650°K$.
TiC	3.35	25	45	ϕ_A	
ZrC	2.18	0.31	45	ϕ_A	Powdered ZrC
	2.3	0.2	11	ϕ_A	Powdered ZrC
	3.8	134	54	ϕ_A	Fused ZrC
	4.0	140	57	ϕ_A	Fused ZrC
SiC	3.5	64	11	ϕ_A	
TaC	3.14	0.3	45	ϕ_A	

(continued)

TABLE II (continued)

Compound	Work function, ev	Richardson A_2, amp/ cm^2deg^2	Ref. No.	Remarks[a]
Sulfides				
CaS	2.61	–	59	ϕ_E, T = 1113-1313°K
SrS	2.54	–	59	ϕ_E, T = 1113-1313°K
BaS	2.1	–	59	ϕ_E, T = 1113-1313°K
	2.38	0.06-4400	58	ϕ_A, Richardson A depends upon activation of cathode and method of preparation.
Selenides				
CaSe	2.70	–	59	ϕ_E, T = 1043-1373°K
SrSe	2.57	–	59	ϕ_E, T = 1043-1373°K
BaSe	2.07	–	59	ϕ_E, T = 1043-1373°K

[a] Here ϕ_A is the apparent work function and ϕ_E is the effective work function.

WORK FUNCTION OF METALLOIDS

Another major group of materials which is sufficiently refractory for incorporation into thermionic converters can be classed under the heading of metalloids. The term refers to chemical compounds in the crystalline state in which the metalloid atoms are bonded together to form a continuous network extending throughout the crystal. A number of different metallic elements can be incorporated into the crystal which exhibits many properties similar to those of metals. The type of chemical bonding that exists between the metallic atoms and the network is not at all clear. In fact, it may be that a variety of bonding types exist within the various metalloids. Included among the groups of metalloids are such compounds as the borides, carbides, nitrides, beryilides, and silicides. Only for the borides have the thermionic emission characteristics for a relatively large number of compounds been investigated. Despite the fact that uranium carbide and uranium carbide-zirconium carbide mixtures are presently considered to be among the most promising thermionic emitters, very few other carbides have been investigated. No values for the work function for nitrides, beryllides, or silicides appear to have been published. The discussion that follows will therefore be largely concerned with the borides. However, attention will be drawn to resemblances to other types of metalloids where appropriate and possible. As opposed to the semiquantitative calculations for work function and other quantities closely related to work function made by Wigner and Bardeen (22) and Herring and Nichols (19) for metals and by Mott and Guerney (29) and Wright (30) for ionic solids, the treatment of the work function of metalloids has been attempted only on a purely qualitative basis, largely by Samsonov and Neshpov (47) and co-workers.

Diborides

In attempting to relate a number of different physical and electrical properties of solids containing transition elements to their electronic structure, Samsonov et al. (47, 60-67) have suggested the use of the quantity 1/Nn, where N is the principle quantum number of the incomplete d-shell of the transition metal, and n is the number of electrons contained within this shell in the free metal atom. This quantity is considered to be indicative of the extent of the screening of the outer s-electrons from the nuclear charge. Thus, the smaller the value of 1/Nn, and the larger the value of Nn itself, the smaller is the attraction between the nucleus and the valence electrons. This quantity does not, however, take into account the particular stability of the half-filled d-shell, nor does it consider the small difference in energy that exists between the d-shell and the next higher s-shell. Thus, Samsonov is required to make distinctions between the trends observed and predicted for groups IV, V, and VIB compounds and the trends observable for groups VIIB and VIII compounds. Comparisons cannot be extended because the considerations used by Samsonov primarily apply to the position of the Fermi level within the solid. The variation of the electron affinity of the solid with structure and interatomic spacings present in the solid is not considered.

The bonding that exists within the transition metal diboride is not clearly established. The earlier papers of Samsonov (60-64) considered that the bonding between the metal and nonmetal atoms in the structure was largely covalent, especially for groups IVB and VB. The electrons of both the metalloid and metal atoms were considered to contribute to the electronic collective, an s-d band in the crystalline lattice which is some type of unspecified hybridized orbital. The unused d-orbitals remain empty and play the role of the electron acceptor while the two boron atoms per unit cell act as the donors. Despite the presence of strong covalent bonds in this picture, the metallic character of the compounds is maintained. The assumption is made that the work function is a measure of the energy required to excite electrons from lower occupied levels in the conduction band to higher unoccupied levels within the same conduction band. For large values of 1/Nn, the valence electrons are more strongly bound to the nucleus. Thus the bands associated with the valence electrons are narrower and the density of states on the Fermi surface is greater. Since narrower bands result in smaller energy differences between the lower and upper states within the bands, the conclusion reached is that for large values in pure metals of 1/Nn the work function should be smaller. The correlation obtained for transition metals is shown in Figure 3 and that for transition metal diborides in Figure 4. For the diborides, it is reasoned that an increase in the value of 1/Nn for the transition metal atom is accompanied by an increased probability of transfer of the valence electrons from the metalloid atoms to the free, unoccupied levels in the s-d band. This results in a widening of the s-d band and, in accordance with the above reasoning, in an increase in the work function.

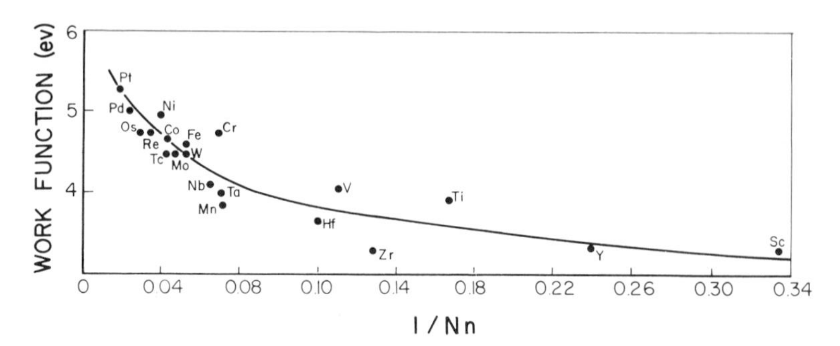

Fig. 3. Dependence of work function of transition metals on 1/Nn [after Samsonov and Neshpor (47)].

In Figure 4, there is some question with regard to inclusion of the work functions for WB_2 and TaB_2. The W_2B_5 is the only established stoichiometry, but there is some evidence that WB_2 is the predominant phase at high temperatures (68). The stoichiometry of tantalum boride is apparently TaB (45), and the crystal structure is NaCl-type, whereas the other diborides in Figure 4 posses simple hexagonal crystal lattices. The method by which the work functions for most of the borides shown in Figure 4 was determined was not given, but it is thought that these are effective work functions. (Using the same experimental emission data, the effective work function will probably be higher than the apparent work function.)

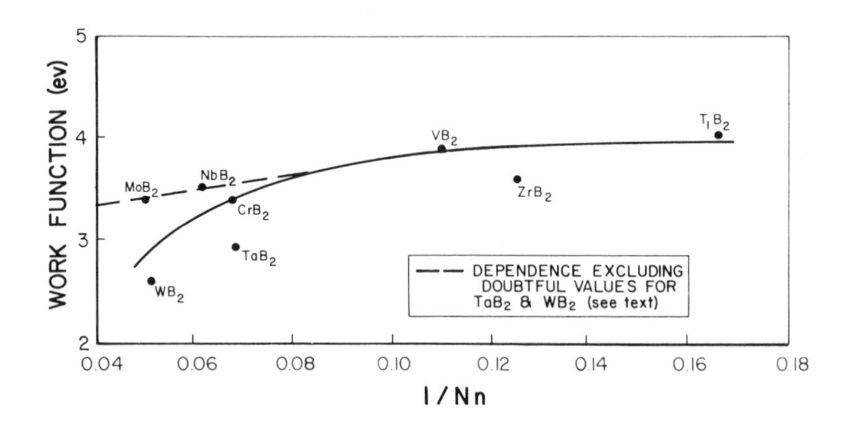

Fig. 4. Dependence of work function of transition metal diborides on 1/Nn [after Samsonov and Neshpor (47)].

The reasoning applied by Samsonov and co-workers (47,66) to the prediction of values of work functions for diborides has been extended (47,66) to provide a comparison between the work functions for corresponding transition element borides, carbides and nitrides. Since the probability of transfer of valence electrons from the metalloid atom into the s-d band should vary with the inverse of the ionization potential of the metalloid atom, the higher the ionization potential, the narrower should be the s-d band and, as a result, the lower should be the work function. Consequently, the work function is predicted to decrease as one goes from borides to carbides to nitrides. The ionization potential of these metalloid atoms are 8.30, 11.26, and 14.54 ev, respectively. If this same reasoning were applied to the beryllides and silicides, whose ionization potentials are 9.32 and 8.15 ev, respectively, then the silicides would be expected to have the highest work function whereas the beryllides should have a work function intermediate between that of the borides and the carbides.

The above view of the nature of the bonding orbitals in diborides is similar to that proposed by Rundle (69) and Hume-Rothery (70). More recently, Kiessling (71) and Juretschke and Steinitz (72) have proposed that the d-band in all transition diborides is filled. Additional electrons enter the s-band. Samsonov et al. (66) have adopted this latter view in explanation of various electrical and optical properties of the diborides. This picture of the disposition of electrons within the bonding orbitals was suggested by, and is in conformity with, various electrical measurements which show that the group IV diborides have higher Hall coefficients and lower electrical resistivities than do the groups V and VI diborides. The greater mobility is attributed to an increase in the screening of the transition metal atoms by the filled d-band which causes a decrease in electron scattering and hence, a higher electrical conductivity. The widening of the conduction band as a result of the presence of electrons associated with the metalloid atoms in the d-band could be considered to predict a higher work function for the group IV diborides as compared with the work function of the pure transition metals. This prediction is borne out by the comparison of the work functions for ZrB_2 and zirconium although no data exists for HfB_2, and the measured work functions for titanium and for TiB_2 are the same. On the other hand, the groups V and VI diborides possess electrons in the s-band and, hence, their work functions should be lower than that of the corresponding metal. Available data are in agreement with this conclusion. Applying similar reasoning to a comparison of the work functions for the diborides for groups IVB, VB, and VIB, the resistivity is found to increase as the group number increases. If this increasing resistance is attributed to a narrowing conduction band, then following Samsonov's reasoning, one would predict that the work function should decrease as one goes from group IV to group VI. The measured values of the work functions, as given by Samsonov and Neshpov (47), actually do follow this prediction.

Following this newer view, the resistivity of the transition metal carbides is greater than that of the borides because the d-band in the carbides is not filled, and therefore, electron scattering does exist to a large extent. Since the conduction band in the carbides should be narrower than it is for the diborides, the work function of a carbide would be predicted to be lower than the work function for the corresponding diboride. Experimental data confirm this prediction, but it should be noted that very few work functions for carbides have been determined and that, of those which have been measured, the accuracy is not very great.

The preceding discussion with regard to diborides is extremely qualitative and concerns itself largely with what, in more conventional terms, would be called the position of the Fermi level with respect to the bottom of the conduction band. It does not concern itself in any way with the electron affinity of the solid. In this connection, it is interesting to note that, although the diborides crystallize in hexagonal structures, most of the transition carbides crystallize in a NaCl-type cubic lattice. Nevertheless, the distinction between the lattice types makes it seem probable that the electron affinities of the carbides and borides cannot be equated.

Hexaborides

The alkaline earth and rare earth hexaborides have received considerable attention in recent years because of their relatively low work functions accompanied by their high temperature stability. The unit cell of a crystalline hexaboride is a body-centered cubic structure with each corner of the cube occupied by six boron atoms, positioned in such a way as to form a regular octahedron. These octahedra are located at the corners of a simple cube, in the center of which is a metal atom. Thus, the crystal structure may be considered as two interpenetrating simple cubic lattices.

The hexaborides possess metallic properties and it is for this reason that Lafferty (48), during the course of his study of thermionic emission from hexaborides, postulated that each boron atom was bonded to five other boron atoms, and that the three valence electrons of the boron atom were distributed over these five covalent boron-boron bonds. But no valence bonds were assumed to exist between the metal atoms and any of the surrounding boron atoms. This structure permitted the outer valence electrons of the metal atoms to remain associated with the metal and thus become contributors to a metallic type of conduction.

An almost completely opposing point of view was adopted by Longuet-Higgins and Roberts (73), who made a quantum mechanical calculation of the bonding orbitals used for the hexaborides. They assumed that all the valence electrons of both the metal and boron atoms were confined to the boron network, which is equivalent to neglecting all structures involving covalent metal-boron bonds. This approximation becomes more valid as a difference in electronegativity between the metal and the boron atoms becomes greater and the bond approximates that of an ionic bond. The band structure, resulting from the LCAO molecular orbital approximation applied by Longuet-Higgins and Roberts, predicted a valence band capable of containing 20 electrons, 18 of which would be contributed by the 6 boron atoms plus another 2 from the divalent metal atom. As a result of this calculation, they concluded that divalent metal hexaborides should be insulators and that any conduction observed for them is the result of impurities or lattice defects. However, the extra valence electron present in trivalent metal hexaborides is assumed by Longuet-Higgins and Roberts to be associated with the metal atom itself. In this case, the crystal should act as a metallic conductor.

This latter view of Longuet-Higgins and Roberts is essentially the same as that adopted by Samsonov and Neshpov (74). However, they permit the insertion, within the boron octahedron, of between 1.6 and 2 electrons from the metal atoms. The evidence for the insertion of less than two electrons from a divalent metal comes from the work of Bertaut and Blum (75), who replaced some of the metal atoms in the hexaboride with monovalent sodium atoms, forming compounds of the type $(Na, M)B_6$. The limit of their substitution corresponded to 1.6 electrons per metal atom. The remaining 0.4 electrons per divalent metal atom, and 1.4 electrons per trivalent metal atom, are associated with the metal atom and responsible for the metallic-type conduction of the hexaborides. For the rare earth hexaborides, Samsonov and Neshpor (74) consider that the energy band system is composed of "a narrow, weakly excited 4f-band," presumably the valence band, and "a broad, hybrid 5d-6s band." The latter band is considered to determine both the electrical conductivity and the work function of a rare earth hexaboride. The work functions have been plotted as in Figure 5, which also shows a plot by Neshpor and Samsonov of the number of possible states that exist for the particular rare earth element. The assumption is made that Hund's rule, relating the degree of stability of an element to the multiplicity, is applicable to this particular electronic structure. Although the method by which the number of possible states for the 4f-electrons has been calculated is unclear, it is true that greater stability in rare earth elements is associated with the presence of a half-filled shell, as in europium and gadolinium both of which possess in their normal states seven electrons in the 4f-shell. The asymmetry of the work function versus atomic number curve is explained by the high degree of screening present in the higher atomic number rare earth hexaborides, which results in an increase in the work function according to the discussion given above for the diborides. Therefore, the decrease in work function on the right-hand side of Figure 5 is less rapid than is the rise in work function on the left-hand side. Gadolinium hexaboride has a low work function which is attributed to the presence of a single 5d-electron in the electronic configuration of the element. This 5d-electron is considered to be substantially free and thus results in a small work function for the compound. There are two possible electronic configurations for terbium: $4f^8\,5d^1\,6s^2$ or $4f^9\,6s^2$. Corresponding to these two possible electronic configurations, a work function between 2.06 and 3.53 or 4.9 and 3.53 ev is predicted. An approximate value for this work function is 6.0 ev (53). Although the accuracy of this work function measurement is not very great, it is probably sufficient to indicate that the

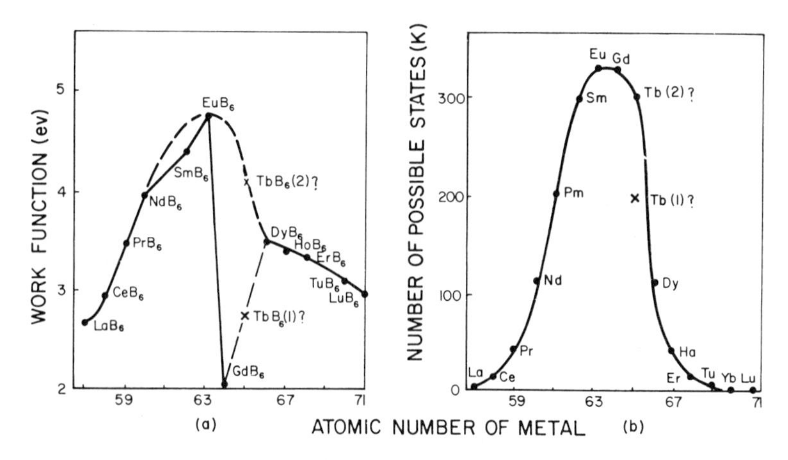

Fig. 5. Variation with atomic number of work function and number of possible states for rare earth hexaborides [after Samsonov and Neshpor (74)].

latter configuration of terbium is the correct one if the assumptions of Neshpor and Samsonov are considered to be completely valid.

SUMMARY

Those properties of materials which are important with regard to the performance of the material in a thermionic converter have been discussed. To the extent that these properties may be discussed as operating independently of each other, the work functions of the emitter and collector are considered to have the greatest bearing upon the potential performance of a thermionic converter. Several attempts to theoretically calculate and to correlate and predict true work functions for metallic elements, ionic solids, and metalloids have been discussed. The work function is determined by the energy of the Fermi level referred to the bottom of the conduction band and is also a function of the electron affinity of the solid. The Fermi energy is dependent upon the presence of absence of impurities and is determined by the type of electronic bonding that exists between the elements in compounds. The electron affinity is primarily a function of the dipole layer present at the surface of the solid and is therefore a function of the interatomic spacings and of the electronic configuration of elements composing the solid. The calculation of work functions from a knowledge of the electronic configuration and crystalline habit of the atoms has not been attempted except for the simplest solids, and the present ability of theory to predict values of work functions from datum on chemically similar solids is severely limited.

ACKNOWLEDGMENT

This work was supported by the Office of Naval Research.

References

1. Schlichter, W., dissertation, Univ. of Goettingen, 1915.
2. Rasor, N. S., J. Appl. Phys., 31, 163 (1960).
3. Houston, J. M., J. Appl. Phys., 30, 481 (1959).

4. Schock, A., Republic Aviation Corp., Rept. PPL-TR-61-3A(259) Contract Nonr-3285(00), Feb 1961.
5. Hatsopoulos, G. N., and J. Kaye, J. Appl. Phys., 29, 1124 (1958).
6. Wilson, V. C., J. Appl. Phys., 30, 475 (1959).
7. Webster, H. F., J. Appl. Phys., 30, 488 (1959).
8. Moss, H., J. Electronics, 2, 305 (1957).
9. Hernquist, K. G., Nucleonics, 17, 49 (1959).
10. Nottingham, W. B., J. Appl. Phys., 30, 413 (1959).
11. Wright, D. A., Proc. Inst. Elec. Engrs. Proc. (London), 100, 125 (1953).
12. Zener, C., R. D. Baker, and H. Thurnauer, Report to NASA Material Committee (ad hoc Committee on Direct Power Conversion) Aug. 4, 1960.
13. Taylor, J. B., and I. Langmuir, Phys. Rev., 44, 423 (1933).
14. Steele, H. L., Report on the 20th Annual Conference on Physical Electronics at Massachusetts Institute of Technology, Nov., 1960, p. 76.
15. Kaye, J. and J. A. Welch, eds., Direct Conversion of Heat to Electricity, Wiley, New York, 1960.
16. Hensley, E. B., J. Appl. Phys., 32, 301 (1961).
17. Cusack, N., The Electrical and Magnetic Properties of Solids, Longman's Green, New York, 1958, ch. 2.
18. Nottingham, W. B., Handbuch der Physik, Vol. XXI, Springer-Verlag, Berlin, 1956.
19. Herring, C., and M. H. Nichols, Revs. Modern Phys., 21, 185 (1949).
20. Michaelson, H. B., J. Appl. Phys., 21, 536 (1950).
21. Hermann, G., and S. Wagener, The Oxide Coated Cathode, Vol. 2, Chapman and Hall, London, 1951, p. 17.
22. Wigner, E., and J. Bardeen, Phys. Rev., 46, 509 (1934).
23. Gordy, W., and W. J. Thomas, J. Chem. Phys., 24, 439 (1956).
24. Pauling, L., The Nature of the Chemical Bond, 3rd edition, Cornell Univ. Press, Ithaca, N. Y., 1960, ch. 3.
25. Gordy, W., Phys. Rev., 69, 604 (1946).
26. Gordy, W., J. Chem. Phys., 14, 305 (1946).
27. Walsh, A. D., Proc. Roy. Soc. (London), A207, 13 (1951).
28. Mulliken, R. S., J. Chem. Phys., 2, 782, (1934); 3, 573 (1935).
29. Mott, N. F., and R. W. Gurney, Electronic Processes in Ionic Crystals, 2nd edition, Oxford Univ. Press, New York, 1948.
30. Wright, D. A., Proc. Phys. Soc., (London), 60, 13 (1948).
31. Tyler, W. W., Phys. Rev., 76, 1887 (1949).
32. See ref. 21, pp. 216-217.
33. Spanner, H. J., Ann. Physik, 75, 609 (1924).
34. Stevenson, J., and E. Hensley, J. Appl. Phys., 32, 166 (1961).
35. See ref. 21, p. 181.
36. Hopkins, B. J., and F. A. Vick, Brit. J. Appl. Phys., 9, 257 (1958).
37. Fan, H. Y., J. Appl. Phys., 14, 552 (1943).
38. Bondarenho, B. V. and B. M. Tsarev, Radio Engi. and Electronics, 4, 214 (1959).
39. Kulvarshala, B. S., V. B. Marchenko, and G. V. Stepanov, Radio Eng. and Electronics, 3, 40 (1958).
40. Schaefer, D. L., quoted by V. L. Stout in Thermoelectricity, Wiley, New York, 1960, p. 219.
41. Uzan, R., Vide, 2, 1139 (1952).
42. Gaines, G. A., "Investigations of Rare Earth Oxide Cathodes," Battelle Memorial Instute, OTS-PB145734.
43. Wyler, E. H., F. C. Todd, and R. C. McMaster, (1950), quoted by D. A. Wright, ref. 11.
44. Weinreich, M. O., Rev. gen. elec., 14, 243 (1945).

45. Goldwater, D. L., and R. E. Haddad, J. Appl. Phys., 22, 70 (1951).
46. Hanley, T. E., J. Appl. Phys. 19, 583 (1948).
47. Samsonov, G. V., and V. S. Neshpor, Radio Eng. Electronics, 3, 155 (1957).
48. Lafferty, J. M., J. Appl. Phys., 22, 299 (1951).
49. Samsonov, G. V., Proc. Acad. Sci., U.S.S.R., 133, 969 (1960).
50. Kudentseva, G. A., et al., Akad. Nauk. S.S.S.R., Uval Filial, 6, 271 (1958).
51. Kudentseva, G. A., and B. M. Tsarev, Radio Engi. Electronics, 3, 182 (1958).
52. Kudentseva, G. A., and B. M. Tsarev, quoted by G. V. Samsonov and V. S. Neshpor, Soviet Phys. Tech. Phys., 3, 1029 (1958).
53. Internal Communication, Boeing Co., Seattle, Washington.
54. Pidd, R. W., G. Grover et al., J. Appl. Phys., 30, 1575 (1959).
55. Kuczynski, G. C., J. Appl. Phys., 31, 1500 (1960).
56. Haas, G. A., and J. T. Jensen, J. Appl. Phys., 31, 1231 (1960).
57. Danforth, W. E., and A. J. Williams, J. Appl. Phys., 32, 1181 (1961).
58. Grattidge, W., Phys. Rev., 81, 320 (1950).
59. Nikonov, S. P., data reported in Radio Eng. and Electronics, 4, 255 (1959).
60. Samsonov, G. V., Doklady Akad. Nauk, S.S.S.R., 93, 689 (1953).
61. Samsonov, G. V., and V. P. Latysheva, Doklady Akad. Nauk, S.S.S.R., 109, 582 (1956).
62. Samsonov, G. V., Zhur. Fiz. Khim., 30, 2057 (1956).
63. Samsonov, G. V., Uspehki Khim., 25, 190 (1956).
64. Samsonov, G. V., Soviet Phys.-Tech. Phys., 1, 695 (1956).
65. Samsonov, G. V., Proc. Acad. Sci., U.S.S.R., 133, 970 (1960).
66. L'vov, S. N., V. F. Nemchenko, and G. V. Samsonov, Soviet Phys.-Tech. Phys., 5, 1334 (1961).
67. Neshpor, V. S., and G. V. Samsonov, Soviet Phys.-Solid State, 2, 1966 (1961).
68. Aronsson, B., Modern Materials, Vol. 2, Academic Press, New York, 1960, p. 171.
69. Rundle, R. E., Acta. Cryst., 1, 180 (1948).
70. Hume-Rothery, W., Phil. Mag., 44, 1154 (1953).
71. Kiessling, R. J., J. Electrochem. Soc., 98, 166 (1951).
72. Juretschke, H. J., and R. Steinitz, J. Phys. Chem. Solids, 4, 118 (1958).
73. Longuet-Higgins, H. and M. Roberts, Proc. Roy. Soc. (London), A244, 336 (1954).
74. Samsonov, G. V., and G. V. Neshpor, Soviet Phys.-Tech. Phys., 3, 1029 (1958).
75. Bertaut, P., and A. Blum, Acta Cryst, 7, 81 (1954).

The Lifetime and Efficiency of a Thermionic Energy Converter

L. S. RICHARDSON, A. E. FEIN, M. GOTTLIEB, G. A. KEMENY
and R. J. ZOLLWEG

Westinghouse Research Laboratories, Pittsburgh, Pennsylvania

Abstract

The effects of the presence of cesium on the evaporative lifetime and the operating efficiency of a thermionic energy converter have been considered. Experimental measurements of the transport rate of molybdenum through a cesium atmosphere are presented and shown to agree with the theoretical calculations. Calculations of the efficiency show that the maximum obtainable efficiency with a tungsten cathode should be nearly independent of temperature between 1600 and 2200°K.

We discuss: the effects of interelectrode spacing upon lifetime and efficiency; the inter-relationship of lifetime and efficiency; and the use of a power-to-weight criterion for optimizing the operation of a thermionic energy converter.

INTRODUCTION

The design, contruction, and operation of all devices are dictated to some extent by their required lifetimes. This requirement varies from minutes for a rocket motor to years for a commercial power plant. The mechanisms of failure must be considered in order to predict this lifetime. In a thermionic energy converter, the basic cause of failure is the evaporative transfer of cathode materials. Other causes such as corrosion, mechanical failure, or radiation damage can in most cases be prevented by ingenious design. Evaporative loss, however, is an inherent limitation because of the high operating temperatures required. To predict the lifetime of a thermionic energy converter, only the allowable loss and the rate of evaporative transfer must be known. The allowable loss is determined by the exact mechanism of failure. Whisker formation, changes in the thermal emissivities, and changes in the work functions are examples of such mechanisms, any of which may be caused by evaporation, or the subsequent deposition. The rate of evaporative transfer is dependent upon the geometrical configuration, the operating temperatures, the materials of construction, and the internal pressure of the device.

An energy converter must possess a useful energy output as well as a useful lifetime. In order to compare the operation of different energy converters, some criterion of comparison must be chosen. Some possible criteria are the efficiency, the output power, or the power-to-weight ratio. Each of these is important in certain applications; for example, a stationary, land-based power station is usually an efficient low-cost device while a space power plant is usually a light device. In this paper, the efficiency has been taken as the criterion. This choice was made for several reasons: (1) The efficiency is always of interest to the designer regardless of the application of the device. (2) Calculation of the power-to-weight ratio requires a detailed knowledge of the heat source and sink, and is thus dependent upon considerations beyond the scope of this paper. This point will be discussed further in a later section. (3) The output power and other possible criteria are more specifically related to particular applications than is the efficiency.

A number of authors have calculated efficiencies of thermionic energy converters using a number of different approximations (1). While the efficiency is a function of a large number of variables, most of these variables are determined, in principle, by a relatively small number of parameters. If the materials of construction and their properties are known, then the efficiency is a function only of the cathode-to-anode spacing, the cathode and anode temperatures, the means of space-charge neutralization, and the lead and load resistances. In most converters presently under investigation, cesium vapor is used for space charge neutralization, and frequently also for modification of the cathode and anode work functions, as recently patented by Feaster (2). In this case, the efficiency is also a function of the cesium pressure. In all but the most recent calculations of efficiency referred to above, the interrelationships of work functions, emissivities, temperatures, etc., have been largely ignored. Furthermore, few of the calculations have considered variations in load resistance or the presence of back current. None of the calculations have attempted to optimize the values of anode and cesium temperature while simultaneously optimizing lead and load resistance. A recent report (3) and the Appendix summarize a method for this simultaneous optimization.

Both lifetime and efficiency can be shown to be functions of the same design parameters: operating temperatures, internal pressure, materials of construction, and geometrical configurations. For a converter built of a given material and operating in the most efficient manner possible, lifetime can be shown to decrease with increasing efficiency. The balancing of the desires for high efficiency and long lifetime requires the knowledge of how these quantities change as the design parameters are changed. In this paper, a proposed method for calculating these quantities is discussed. The relationship of efficiency and lifetime derivable from these calculations is also presented.

The methods by which lifetime and efficiency can be calculated are described in Sections II and III, respectively. The relationship between these two properties is discussed in Section IV.

II. LIFETIME

A simple approach for estimating the evaporative lifetime is to divide the allowable surface recession by the evaporation rate in vacuum. Since no measurements of the allowable surface recession have appeared in the literature, a value of 2.5 μ was used as a conservative estimate. This value, equivalent to about 10^4 layers of atoms, is small compared to the spacing of most converters presently under consideration. In a later section the exact value used will be shown to be unimportant. Figure 1 shows the results of this calculation for molybdenum, niobium, tantalum, and tungsten (4).

Several important effects occur, particularly in the cesium diode, that are neglected in this calculation. The presence of a partial cesium coverage may increase or decrease the rate of loss. The presence of contaminant gases (such as water vapor) in the device may also change the rate of loss. Experimental measurements under very carefully controlled conditions are necessary to determine the magnitudes of these effects.

Another effect of considerable importance is the reduction of evaporative transfer caused by back-scattering from the cesium atmosphere present in

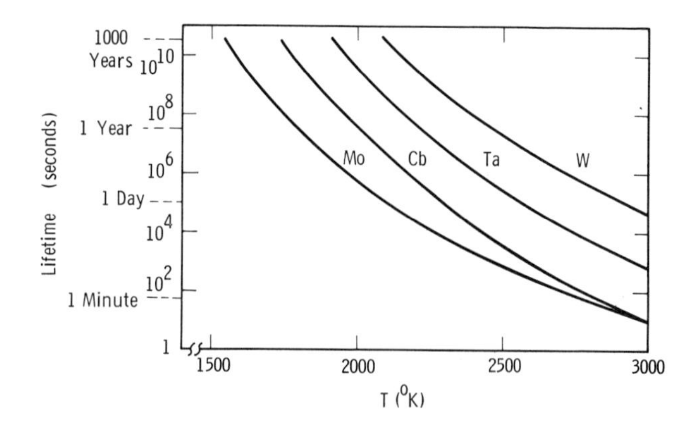

Fig. 1. Time to lose 2.5 μ by evaporation from molybdenum, niobium, tantalum, and tungsten as a function of temperature.

the anode to cathode gap. The magnitude of this effect can be estimated by a relatively simple method.

As a first approximation it was assumed that all of the metallic atoms that strike the cold anode stick, and that the partial coverage of cesium has no effect on the evaporation of atoms from the cathode.

For relatively large spacings, the rate of loss is limited by the rate of diffusion through the cathode-to-anode gap. If the diffusion constant, D, does not change within the gap, then the flux of metal atoms, R, is

$$R = D\frac{\Delta\rho}{d} \tag{1}$$

where $\Delta\rho$ is the difference in density of evaporated atoms at the cathode and anode, and d is the width of the cathode-to-anode gap. The evaporative flux R_V at any temperature T is given by

$$R_V = \rho V/4 \tag{2}$$

where ρ is the density and V is the arithmetic average velocity of the evaporating species (5). Since the pressure and temperature of evaporated atoms adjacent to the electrodes are nearly equal to those characteristic of the electrodes, $\Delta\rho$ is equal to the difference of densities at the cathode and anode. As the cathode temperature is considerably higher than the anode temperature, the density at the cathode ρ_c is much greater than that at the anode ρ_a. Thus $\Delta\rho$ is approximately equal to ρ_c and the transport flux is

$$R = \frac{4DR_{vc}}{Vd} \tag{3}$$

where R_{vc} is the normal evaporative flux at the cathode temperature.

The quantity D/V has the dimensions of length and is proportional to the mean free path λ. Hard sphere calculations (6) show that D/V equals $\lambda/3$.

Thus,

$$R = \frac{4\lambda R_{vc}}{3d} \tag{4}$$

The evaporative transport problem is analogous to the neutron scattering problem in a non-absorbing medium, where R_{vc} is analogous to the incoming neutron intensity and R to the intensity transmitted. Maynard has solved this neutron scattering problem numerically for small spacings (7). Examination of his results between 1/4 and 4 mean free paths shows that the equation

$$R = R_{vc}/(1 + 3d/4\lambda) \quad (5)$$

fits the results very accurately. For large values of d/λ this expression approaches that of eq. (4).

A more general approach is to consider the interelectrode space as a succession of reflecting and transmitting slabs. The transmission through each of these slabs must have the same functional relationship to thickness as the transmission through the assembly. If the multiple reflections are summed up, assuming isotropic behavior, the functional relationship is

$$f\left(\sum_i x_i\right) = \left(1 + \sum_i\left(\frac{1 - f(x_i)}{f(x_i)}\right)\right)^{-1} \quad (6)$$

where x_i is the thickness of the i-th slab and $f(x_i)$ is the fraction transmitted through that slab.

It can easily be shown that the only function which obeys this relationship is

$$f(x) = (1 + ax)^{-1} \quad (7)$$

This implies that $R = R_{vc}/(1 + ad)$ is the general form of the solution to the transport equation. For a sufficiently large d, this equation becomes identical with (4) when a is $3/4\lambda$. The scattering approach of Maynard also led to the same value for the constant. Both approaches are based on all collisions being of the hard sphere variety. Thus eq. (5) is the general solution for transport where the factor 3/4 will be different if the collisions do not obey the hard sphere model.

An experimental verification of eq. (5) was attempted. A diode was constructed using a molybdenum cathode and a nickel anode. A schematic diagram of this cell is shown in Figure 2. Four different anode to cathode spacings were

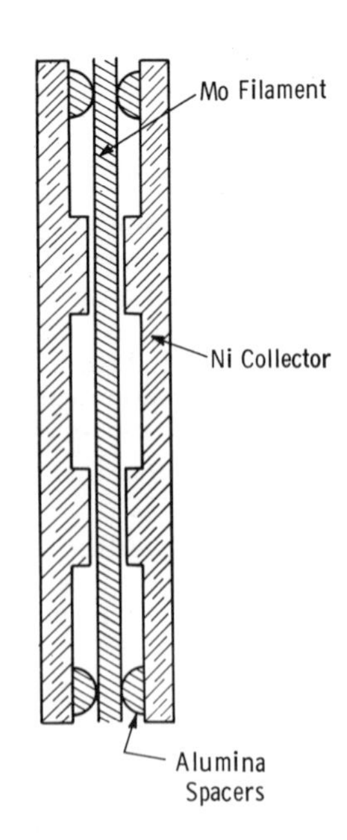

Fig. 2. Schematic diagram of cell for determination of evaporative transport.

TABLE I
Experimental Determination of Molybdenum Transport[a]

Spacing d (μ)	Amount transported G (μg)
155	19
226	12
450	6.2
450	2.6

[a]Temperature = 2100°K; time = 49,200 sec; area of cell = 0.2 cm^2; evaporation in vacuum, G_V = 250 μg (ref. 4); cesium temperature, 552°K; cesium pressure, 1.0 mm of Hg.

used. The cathode was held at a fixed temperature for a measured length of time, after which the nickel anodes were analyzed for total molybdenum content. The spectrographic analysis had a sensitivity of about 2 micrograms of molybdenum. The analytical reproducibility was estimated to be about 15% of the analyzed value. Table I summarizes the results of the first test of this cell. The ratio of the fluxes R_{vc}/R is clearly the same as the ratio of the amounts G_V/G.

If these data are plotted as a function of spacing (Fig. 3), eq. (5) can be used to calculate values of the mean free path. The mean free path deduced is about 8.6 μ at a cesium pressure of 1 mm of Hg.

No other measurement of mean free path under these non-isothermal conditions was found in the literature. Theoretical estimates assuming isothermal conditions were 0.9 to 20 times the experimental value, depending on the assumptions made.

Since the mean free path is inversely proportional to the pressure and is not a sensitive function of the temperatures, the following relationship was

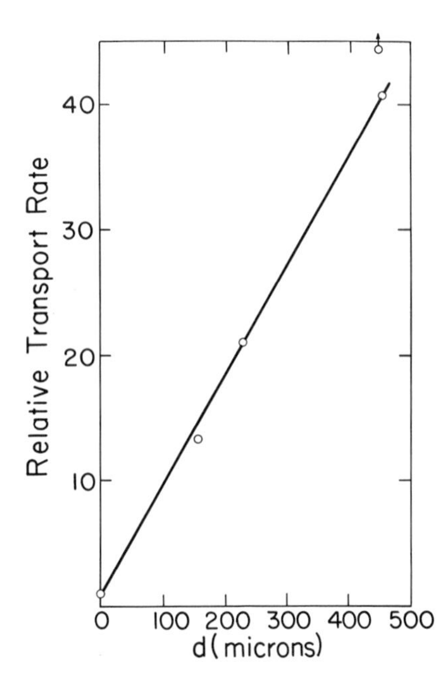

Fig. 3. The relative transport rate as a function of spacing.

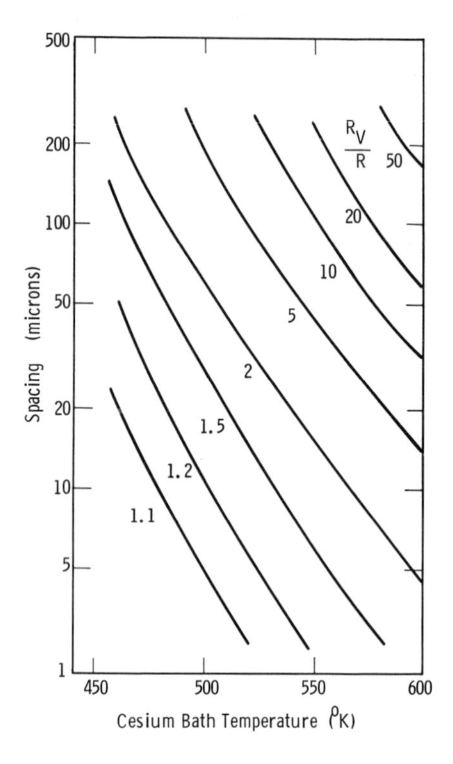

Fig. 4. Relative lifetime of a cathode as a function of cesium bath temperature and spacing.

assumed to hold true for the transport flux

$$R = (1 + 0.087\ pd)^{-1} R_{vc} \tag{8}$$

for the pressure, p, in mm of Hg and the spacing, d, in μ. The value of 0.087 is based upon only one series of experiments and should be accepted only with considerable reservation.

From eq. (8), it is possible to determine lifetime as a function of cathode temperature, spacing, and cesium bath temperature. The relative lifetime, defined as the ratio of the lifetime in a gas-filled diode to the lifetime in vacuum, is equal to R_{vc}/R. In Figure 4 the relative lifetime is plotted as a function of spacing and cesium bath temperature which determines the pressure. Better determinations of the mean free path as a function of pressure may change the values shown in this figure, but the general validity of the figure should not be affected. The lifetime will obviously increase with increasing cesium pressure and spacing.

III. EFFICIENCY

The basic equation used for the calculation of the efficiency, η, of a thermionic diode is

$$\eta = \frac{J\ V_o}{W + H_c - H_a - Q_J + X + H_K} \tag{9}$$

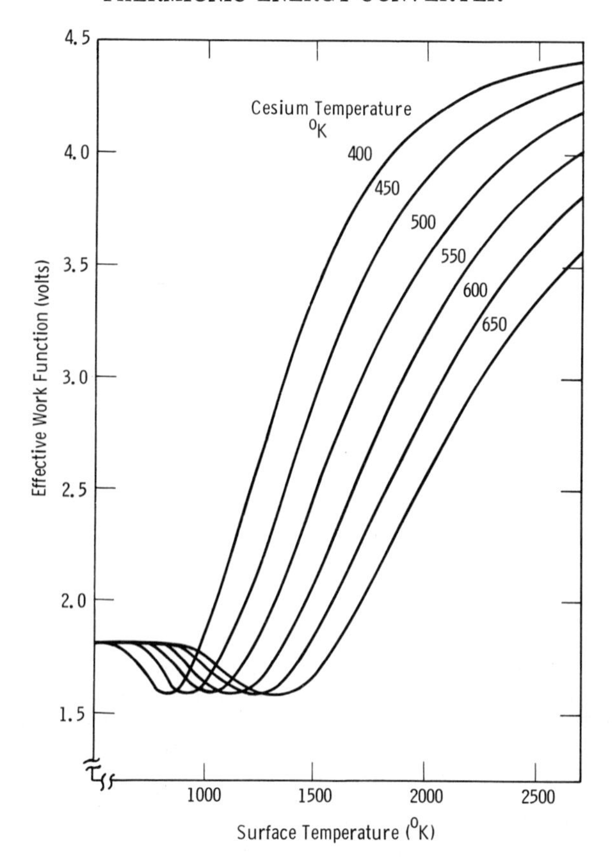

Fig. 5. Work function as a function of cesium bath temperature and surface temperature.

where J is the net current density, V_0 is the output voltage (internal voltage minus the potential drops in leads and plasma), W is the net thermal radiation to the anode, $H_c - H_a$ is the net energy carried by the electrons from the cathode to the anode, H_K is the heat lost by conduction through leads and plasma, Q_J is that part of the Joule heating in leads and plasma that returns to the cathode, and X is the support losses. Minor terms, such as thermoelectric effects, have been neglected. Space charge is assumed to be neutralized, either intrinsically by cesium ionized by the cathode or by some other means.

This expression can be optimized with respect to both lead resistance and load resistance. If this is done, the values of efficiency are functions of the materials of construction and only four parameters—the cathode temperature, the anode temperature, the anode-to-cathode spacing, and the cesium bath temperature. A complete description of the optimization procedure is presented in the appendix. In this calculation, several effects ignored by most other authors are considered and shown to be significant. In particular, the effect of back current, thermal conduction through the plasma, and electrical resistivity of the plasma have been calculated. The effects of varying the load resistance upon the behavior of both the forward and back currents have also been included.

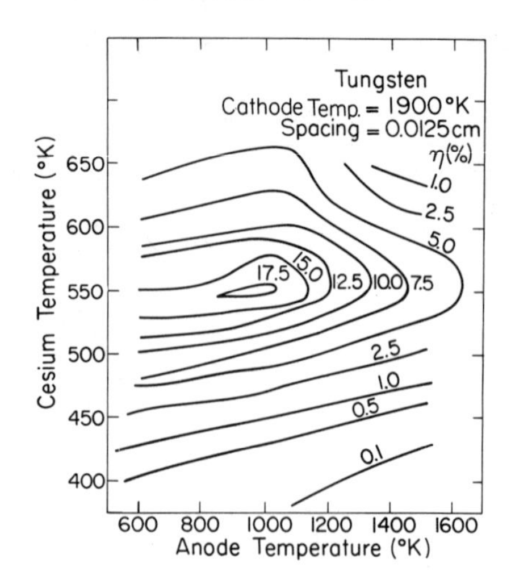

Fig. 6. Efficiency as a function of anode and cesium bath temperatures.

Figure 5 shows the work function as a function of cesium bath temperature and surface temperature.

Calculations were carried out on a high speed digital computer for more than 5000 different combinations of spacing, and cathode, anode, and cesium bath temperatures. At the values of lead and load resistance which yielded the maximum efficiency for each of the given combination of parameters, some forty different properties, such as power delivered to load, external voltage, net current, etc., were tabulated in addition to the efficiency.

Some sample curves of efficiency are given in Figures 6-9. For purposes of discussion a more complete set is presented in the appendix. These were drawn by holding two of the four parameters constant and plotting curves of constant efficiency as a function of the remaining two. These curves represent

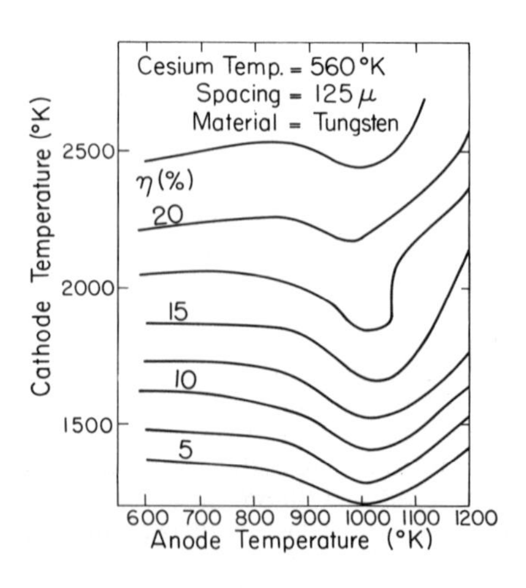

Fig. 7. Efficiency as a function of cathode and anode temperatures.

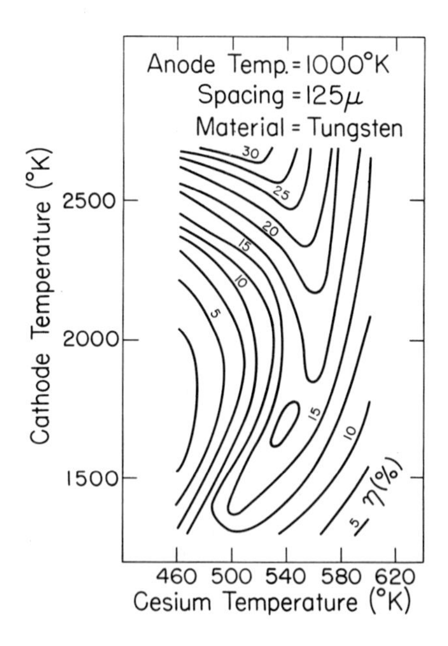

Fig. 8. Efficiency as a function of cathode and cesium temperatures.

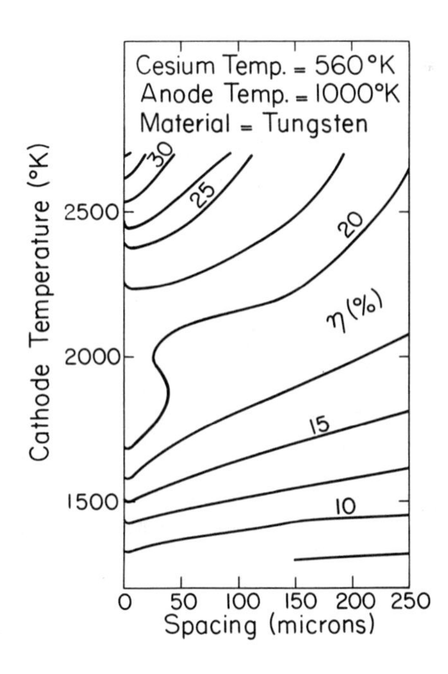

Fig. 9. Efficiency as a function of spacing and cathode temperature.

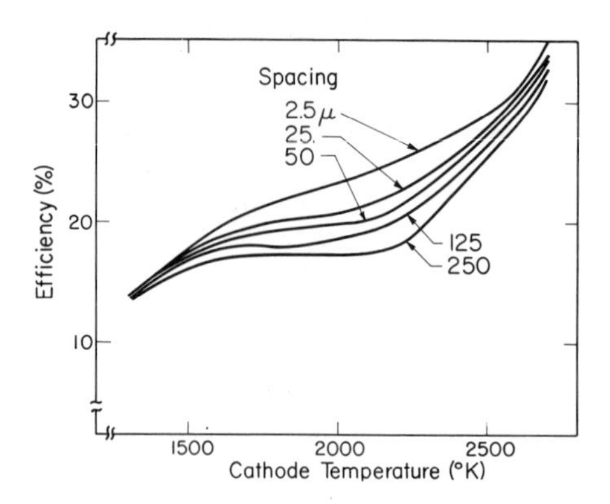

Fig. 10. Optimum efficiency as a function of cathode temperature.

the efficiency of a diode which has been optimized with respect to both lead and load resistances. In Figure 6, the cathode temperature and spacing have been fixed and the efficiency has a maximum occurring at a unique combination of cesium and anode temperatures. This maximum is a general occurrence for all of the possible combinations of spacing and cathode temperature, although its precise location will depend upon the particular values of spacing and cathode temperature considered. It should be mentioned that these maxima may only be local and that higher efficiencies may be obtainable at lower anode temperatures. However, anode temperatures less than 600°K were not considered to be reasonable from a design standpoint. Figures 7 and 8 are drawn similarly to Figure 6 except the temperature fixed was that of the cesium bath or anode, respectively. Curves of the latter two types do not display a maximum of efficiency since higher efficiencies can always be obtained by going to a sufficiently high cathode temperature.

In Figure 9, the anode and cesium temperatures have been held constant and curves of constant efficiency have been drawn as a function of the spacing and cathode temperature. For a fixed value of cathode temperature, it can be shown that an optimum value of spacing exists. However, these spacings are extremely small (less than 2.5 μ) and not realizable. Therefore, for spacings larger than this, it can be seen that efficiency decreases with increasing spacing.

A summary of the maxima obtainable from plots of the type of Figure 6 is shown in Figures 10 and 11. In these figures three sets of curves are shown as functions of cathode temperature—maximum attainable efficiency and the anode and cesium bath temperatures necessary to attain that efficiency. Each set consists of five curves corresponding to five different spacings. Several important features of these curves should be noted. There is, for a fixed spacing, a wide range of cathode temperatures in which the ultimate efficiency is relatively independent of cathode temperature. Thus, a small improvement in the temperature capability of most existing materials does not produce a large improvement in efficiency of operation. Further, the anode temperature required for maximum efficiency is seen to be within the range 900 to 1100°K regardless of the cathode temperature or spacing.

It should be emphasized at this point that the numerical results presented in the figures and in the above discussion refer to a device with tungsten anode and cathode. For other materials, the numerical values might change considerably, but it is believed that the qualitative nature of the results should be the same.

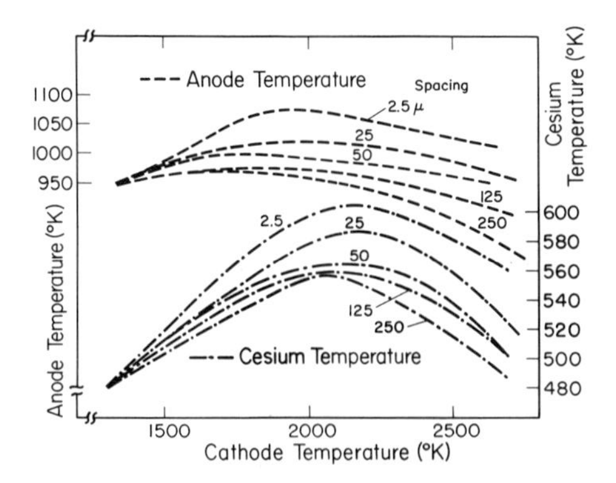

Fig. 11. Optimum anode and cesium temperatures as a function of cathode temperature.

It must also be recalled that these calculations are based on the assumption that space charge has been eliminated. Since there is a minimum cesium temperature at which this is true, certain areas of the curves presented (particularly at the lower cesium temperatures) are inaccurate. Fortunately, from the limited data available in the literature, the optimum operating points do not appear to be in this excluded area.

IV. RELATIONSHIP OF EFFICIENCY AND LIFETIME

With the aid of the calculations and experiments in sections II and III, it is possible to determine efficiency and lifetime as functions of only four parameters—cathode, anode, and cesium bath temperatures and the cathode-anode spacing. If this determination be made, then for a fixed spacing, a maximum efficiency can be found for a given minimum evaporative lifetime. The measurements of mean free path summarized in a previous section were made using a molybdenum cathode. The relative lifetimes deduced from this value should not be seriously different for tungsten, since both atomic diameter and chemical behavior are very similar. A curve of maximum efficiency as a function of the evaporative lifetime for tungsten is shown in Figure 12 for a 125μ spacing diode. Similar curves can be drawn for other spacings. This efficiency has been simultaneously maximized with respect to anode, cathode, and cesium temperatures for a fixed lifetime. In principle, this procedure could be continued to determine an optimum value of spacing. However, impracticably small spacings result.

The extremely large values of lifetime for tungsten shown in Figure 12 are clearly of no practical significance. However, for materials of higher vapor pressure than tungsten, an analogous curve would be useful. In particular, if the electron emission properties of molybdenum in the presence of cesium were the same as those of tungsten, then the lifetime scale would be shifted as indicated in Figure 12. It is clear that under these conditions, a required lifetime of one year or greater will result in an efficiency less than 20% for molybdenum. It must be recalled that this statement is based on the allowable surface recession being 2.5 μ. However, the efficiency in the region of interest for molybdenum is not a sensitive function of lifetime. Allowing 10 times the recession used here does not markedly increase the possible efficiency.

If accurate emission data were available, calculations of the type discussed here could be carried out for other materials. Similarly, lifetime calculations

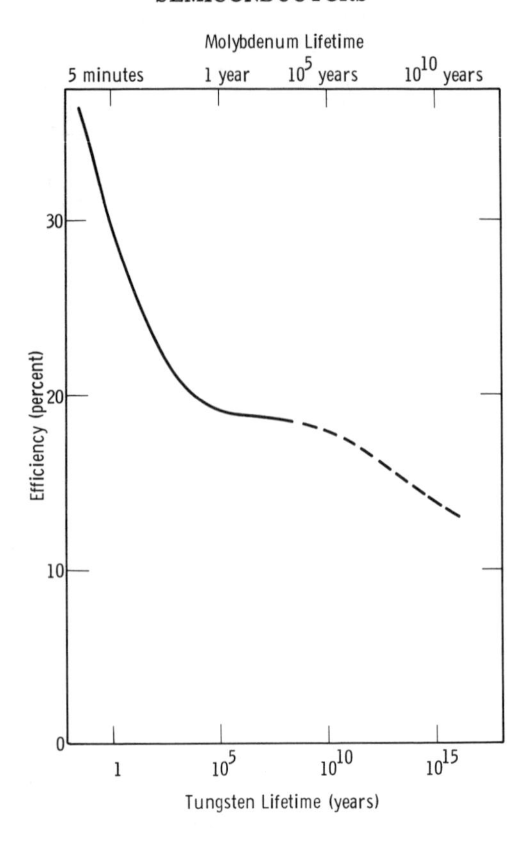

Fig. 12. The relationship between evaporative lifetime and efficiency for tungsten and molybdenum.

could be carried out if measurements of mean free path for other materials were available. While experimental investigations of these data are presently being made by various laboratories, little information is available in the literature.

V. POWER-TO-WEIGHT RATIO

As mentioned in Section I, it may often be more desirable to design a system with a maximum value of the ratio of output power to system weight in lieu of maximum efficiency. This is an extremely difficult problem to solve in general since the weight of the diode itself is negligible, and the weight of the system is the dominant factor. Thus, the construction of the system must be known in great detail. For example, in the case of a reactor-powered thermionic converter, a reactor core, shield, assembly, reflector, and radiator would have to be designed for each choice of spacing and cathode, anode, and cesium temperatures in order to determine the weight of the system.

It is possible, however, to say a few things in a semi-quantitative way about the problem. Consider the radiator, which is responsible for a large part of the weight of a space power system. Let P_o be the power output, P_i the input power, η the efficiency of conversion, and T_a the anode temperature.

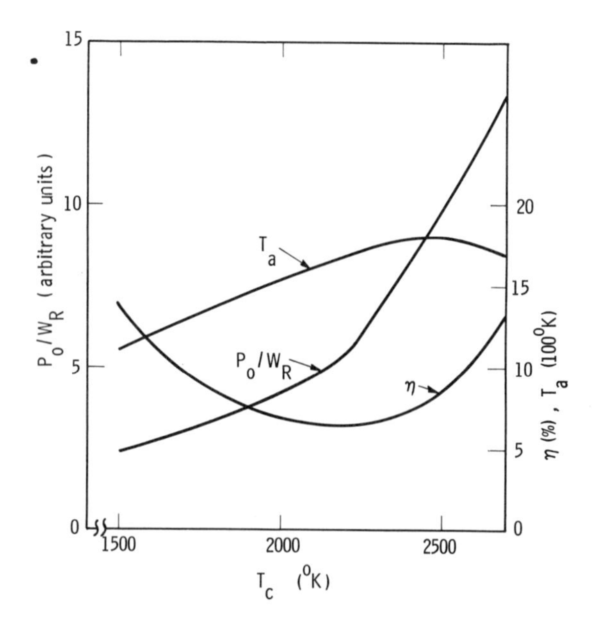

Fig. 13. Power-to-weight ratio, optimum anode temperature, and efficiency at minimum weight as a function of cathode temperature.

Then by definition of η

$$P_o = \eta P_i \tag{10}$$

The amount of power, P_w, which the radiator must dump as waste heat is given by:

$$P_w = P_i - P_o = (1 - \eta) P_i \tag{11}$$

If the waste heat is radiated, the following equation applies,

$$P_w = C_1 A T_a^4 \tag{12}$$

where A is the surface area of the radiator and C_1 is some numerical constant. Further, if the weight W_r of the radiator is proportional to its area,

$$W_r = C_2 A \tag{13}$$

When combined, the above equations yield for the ratio of output power to radiator weight

$$P_o/W_r = C_1/C_2 \left(\frac{\eta}{1 - \eta}\right) T_a^4 \tag{14}$$

The quantity $\eta\ (1 - \eta)^{-1}$ is a monotonically increasing function of η so that for fixed anode temperature, maximizing the efficiency will maximize P_o/W_r as well. However, in many cases, the anode temperatures cannot be considered fixed. A procedure similar to that used in the preceding section can be followed using the maximum power-to-weight ratio, instead of efficiency as the criterion. The results are very similar to those of Section III in that higher cathode temperatures result in higher power-to-weight ratios. Figure 13 shows the results of optimizing P_o/W_r for a tungsten diode with 125 μ spacing.

A comparison of the results using the two different criteria, efficiency (Fig. 10) and power-to-weight ratio (Fig. 13), shows that a significant decrease in efficiency is the price that must be paid for the maximum power-to-weight ratio. This is particularly true in the region between 1600 and 2000°K cathode temperatures where the optimum efficiency (Fig. 10) does not vary rapidly with temperature.

In Figure 13 the relative values of the power-to-weight ratio and the optimum anode temperatures are also plotted. The peak in anode temperatures near a cathode temperature of 2500°K is almost certainly a local maximum, since the power-to-weight ratio can obviously be increased by going to infinite anode temperatures, using a "slightly larger" value of cathode temperature.

VI. CONCLUSIONS

When for a fixed spacing and cathode temperature all other parameters are optimized using efficiency as the criterion, the resulting optimum efficiency is a relatively insensitive function of the cathode temperature in the temperature range 1600 to 2200°K. Clearly, a small improvement in temperature capability will not result in a significant improvement in efficiency. Only a large increase in operating temperature provides a marked improvement. In order to operate a thermionic converter of the type discussed here at 25% efficiency, a cathode temperature above 2200°K is necessary. Only tungsten is capable of a reasonable lifetime at this temperature.

A thermionic converter using tungsten electrodes operating with a cathode temperature of 1600 to 2200°K would have an efficiency between 15 and 20% and an indefinite lifetime. If the emission characteristics of molybdenum are similar to those of tungsten, a molybdenum diode could be constructed with a lifetime greater than one year at an efficiency less than 20%.

The introduction of cesium into a thermionic energy converter has two opposing effects upon the efficiency. Thermal and electrical losses occur in the cesium, while the reduction of cathode and anode work functions allows more efficient operation. As the pressure of cesium is increased from zero, the efficiency is increased reaching a maximum. Above this critical value of cesium pressure, the efficiency decreases because of the increasing electrical and thermal losses in the plasma. The attainment of maximum efficiency depends critically upon establishing the optimum cesium pressure. The use of a cathode material possessing a larger affinity for cesium would make possible lower cesium pressure to obtain a given work function. The net result would be a higher efficiency of operation. In short, the use of a cathode material of higher affinity for cesium will result in increased efficiency. The converse is also true. Thus, consideration of the cesium affinity is of considerable importance in the selection of a cathode material.

From examination of the operating characteristics of thermionic converters (such as Figure 6) it is seen that efficiency is a relatively insensitive function of anode and cathode temperature and of spacing. The cesium pressure, on the other hand, is a quite critical parameter and should be controlled very closely. The lead and load resistances are also critical parameters. Operation away from these critical values can decrease the efficiency by a sizeable fraction. These relationships should be carefully considered in the design of thermionic converters.

The evaporative lifetime of a thermionic converter can be increased by increasing the spacing and the cesium pressure. While the data presented here are only preliminary, the lifetime of a cesium-filled diode will probably be at least 10 times that estimated by the use of evaporation rates into vacuum.

The data necessary for calculations of the type described are extremely scarce and relatively unreliable. The obtaining of more extensive and more accurate information would allow the design of thermionic converters to be made in a relatively routine fashion. For tungsten such a design has been made and the device is presently being built.

The calculations of efficiency have been checked against the results of the few experimental devices available in the literature. Where sufficient information is given to allow comparison, the agreement has been quite reasonable. While the calculations are not to be considered as the exact theoretical results, the values are probably quite close to reality. While the model of a plasma is not entirely realistic and the effect of space charge has been largely ignored, the errors introduced are small in the regions of interest. Devices of 35% efficiency are clearly impractical, while a 20% efficient device can probably be built with present techniques and materials.

As has been already mentioned, the model of an ohmic plasma is crude; it is, however, the only workable model available. It is therefore advisable to derive theoretically and to verify experimentally a more accurate description of the effects of the partially ionized plasma on the transport of electrons. Further, it is important to obtain a better picture of how the pressure of this partially ionized plasma effects the electronic space charge and the final potential distribution in the diode.

Causes, control, and uses of the arcing mode are little understood. In order to intelligently design a thermionic diode, the designer must know how to suppress the arc if it is not desired or how to initiate it if it appears that operation in that mode is desirable.

ACKNOWLEDGEMENTS

Discussions with L. D. Jaffee (Jet Propulsion Lab.) on the type of optimization were most helpful. Some of the work described herein has been performed under Air Force Contract AF33(616)-8292.

APPENDIX

Optimization Procedure for a Thermionic Diode

by A. E. Fein and L. S. Richardson

A. DESCRIPTION OF THE OPERATION OF A THERMIONIC DIODE

In this section, a mathematical model describing the behavior of a thermionic diode will be set up and used to evaluate the efficiency of such a device. For the sake of mathematical convenience, the operation will be expressed in terms of variables which are not all independent. (For example, thermal emissivity is a function of surface temperature.) This will not affect the derivations described below. Before numerical calculations can be performed, however, the actual interrelationships among the physical parameters must be taken into account. A description of the manner in which this will be done will be deferred until the analysis of the operation of the thermionic diode has been presented.

It is known (11) that the reduction of current flow by space charge is very sensitive to the distance between the cathode and anode. For separations of 5 μ or less, the detrimental effects of space charge can be made almost negligible (12). The problems encountered in maintaining such spacing make it advisable to employ some other technique for circumventing the space-charge problem.

In the model, it will be assumed that the space charge in the region between the anode and cathode has been eliminated by using a cesium plasma. It will further be assumed that the sole effect of the plasma on the behavior of the diode is to introduce a "plasma electrical resistance," R_p, and a "plasma

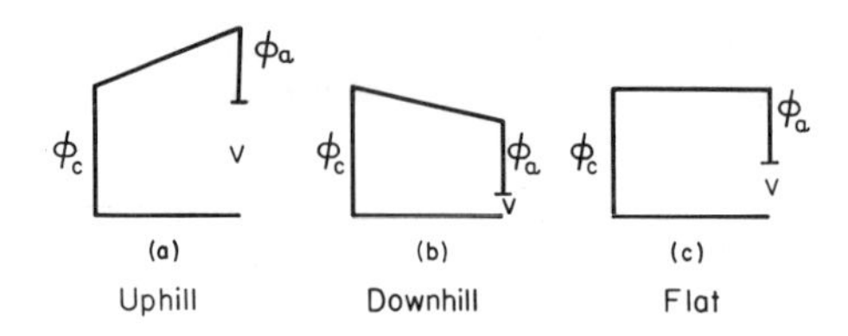

Fig. 14. Modes of operation.

thermal conductance," K_p. The former will contribute to an internal voltage drop inside the diode, and the latter to a thermal conduction of energy across the cathode-anode gap. Also, only the case of plane parallel plates 1 sq cm in area* will be considered.

Let Φ_c and Φ_a be the work functions of the cathode and anode, respectively. (When similar quantities are defined for both the cathode and anode, a subscript "c" will refer to the former and "a" to the latter.) The potential an electron experiences in going from the cathode to the anode will be represented by one of three possible cases depicted by Figure 14. The mode of operation will be classed as uphill if $V > \Phi_c - \Phi_a$ (Fig. 14a), downhill if $V < \Phi_c - \Phi_a$ (Fig. 14b), and flat if $V = \Phi_c - \Phi_a$ (Fig. 14c).

The saturation emission current densities will be denoted by J_{cs} and J_{as}. By Richardson's equation

$$J_{cs} = AT_c^2 \exp\left(-\Phi_c/KT_c\right) \tag{A1}$$

$$J_{as} = AT_a^2 \exp\left(-\Phi_a/KT_a\right) \tag{A2}$$

where A is Richardson's constant which is equal to 120 amp cm^{-2} deg K^{-2}; if k is Boltzmann's constant, and e the electronic charge, $K = k/e = 8.616 \times 10^{-5}$ volts deg K^{-1}. In the flat mode the net current density, J, from the cathode to the anode is simply the net saturation current density, J_s

$$J = J_{cs} - J_{as} = J_s \tag{A3}$$

In general, this is not true, and the net current must be written

$$J = J_c - J_a \tag{A4}$$

where in the uphill mode

$$\begin{aligned} J_c &= J_{cs} \exp\left[-(V - \Phi_c + \Phi_a)/KT_c\right] \\ J_a &= J_{as} \end{aligned} \tag{A5}$$

and in the downhill mode

$$\begin{aligned} J_c &= J_{cs} \\ J_a &= J_{as}\left[\exp -(\Phi_c - V - \Phi_a)/KT_a\right] \end{aligned} \tag{A6}$$

The heat input to the cathode and power output to the load must be known in order to calculate the efficiency of operation of a thermionic generator. The power output is simply the load voltage times the current, where the load voltage is equal to the emf, V, generated by the diode minus the voltage drops

---

* In this regard, it must be remembered that the value of resistance must correspond to electrodes of unit area, and that net current densities calculated will be numerically equal to currents.

due to plasma resistance R_p and lead resistance R_L .

$$P_{out} = J\left[V - J\,(R_L + R_p)\right] \qquad (A7)$$

By the first law of thermodynamics the total heat input to the cathode is equal to the total energy leaving the cathode. The latter can be expressed as the sum of several terms. If such secondary effects as thermoelectric potentials and the energy required to ionize the plasma are ignored, those remaining are:

(1) Heat radiated from cathode to anode, W.
(2) Support and end losses, X.
(3) Heat absorbed by electrons leaving the cathode and reaching the anode, H_c.
(4) Heat conducted from cathode to anode by lead wires and plasma, H_κ.
(5) Heat returned to the cathode by electrons arriving from the anode, H_a.
(6) Heat returned to cathode from the Joule heating of lead wires and plasma, Q_J.

The first four effects carry heat away from the cathode while the last two bring heat to the cathode. The total heat input, H_{in}, to the cathode can be evaluated.

$$H_{in} = W + H_c + H_\kappa - H_a - Q_J + X \qquad (A8)$$

The individual terms are readily calculable. The thermal radiation term W is given by the Stefan-Boltzmann law

$$W = \sigma F \left(T_c{}^4 - T_a{}^4\right) \qquad (A9)$$

where, if ϵ_c and ϵ_a are the total emissivities of a parallel cathode and anode, the view function F is (13)

$$F = \left(\epsilon_c{}^{-1} + \epsilon_a{}^{-1} - 1\right)^{-1} \qquad (A10)$$

and σ is the Stefan-Boltzmann constant (5.67×10^{-12} watts cm^{-2} deg K^{-4}). The heat carried by thermal conduction away from the cathode is

$$H_\kappa = (K_p + K_L)\ (T_c - T_a) \qquad (A11)$$

where K_p and K_L are the thermal conductances of the plasma and the leads, respectively. It will be assumed that the Joule heat returned to the cathode is simply one-half the total Joule heat produced in the leads and in the plasma.* Thus

$$Q_J = (1/2)\ (R_L + R_p)\ J^2 \qquad (A12)$$

In order to compute H_c and H_a, the mode of operation must be considered. From Figure 14, it is seen that in the downhill mode an electron leaving the cathode must have absorbed an amount of heat at least equal to Φ_c. Similarly, an electron arriving at the cathode from the anode must also have absorbed at least that amount of heat. For the uphill mode, this minimum amount of absorbed heat is $V + \Phi_a$. Define

$$\Phi = \max\ (\Phi_c,\ \Phi_a + V) \qquad (A13)$$

The value of Φ is thus the minimum heat absorbed by an electron traversing the diode in any mode. In addition to this minimum heat Φ, there is some

---

* This result is obtained rigorously by Schock (1).

kinetic energy which the electron may possess. If the electrons immediately outside the cathode or anode have a Maxwellian distribution characteristic of the temperature of the surface which they have left, it can be shown that, on the average, this additional kinetic energy will be 2KT where T is the temperature of the emitting surface. Thus

$$H_c = J_c\ (\Phi + 2KT_c) \tag{A14}$$

and

$$H_a = J_a\ (\Phi + 2KT_a) \tag{A15}$$

The efficiency, defined as the ratio of output power to heat in, is then given by

$$\eta = \frac{J\left[V - J\left(R_L + R_p\right)\right]}{W + J_c\left(\Phi + 2KT_c\right) + \left(K_L + K_p\right)\left(T_c - T_a\right) - J_a\left(\Phi + 2KT_2\right) - J^2\left(R_L + R_p\right)/2 + X} \tag{A16}$$

In practice, the load resistance, R_{load}, is a quantity which can be varied at will. However, mathematically the internal voltage

$$V = J\left(R_L + R_p + R_{load}\right) \tag{A17}$$

is much more convenient to use. Hence, in lieu of R_p, R_L, and R_{load} being considered as three independent variables, R_p, R_L, and V will be used, the transformation being non-singular. It must be kept in mind, therefore, in the following analysis that if, for example, a variation of R_L at constant V is considered, a variation in R_p and/or R_{load} is necessary. The value of the variable V is not completely optional as there are physical limits.

Only the load resistance can be physically varied from zero to infinity, short and open circuit respectively. Each case yields an equation which may be solved for the internal voltage. The minimum value for V, V_{sc}, is obtained from the short circuit equation

$$0 = V - J\left(R_L + R_p\right) \text{ at } V = V_{sc} \tag{A18}$$

Equation (18) is transcendental and cannot be solved analytically. The open circuit voltage V_{oc} is determined by the condition

$$J = 0 \text{ at } V = V_{oc} \tag{A19}$$

An analytic solution to (19) for V_{oc} can be obtained from (A4), (A5), and (A6)

$$V_{oc} = \Phi_c - \Phi_a + KT \ln\left(J_{cs}/J_{as}\right) \tag{A20}$$

where

$$T = T_c \text{ if } J_{cs} > J_{as}, \text{ and } T = T_a \text{ if } J_{as} < J_{cs}$$

B. OPTIMIZATION OF LEAD RESISTANCE

The efficiency (16) can be optimized with respect to the lead resistance at fixed V as follows:

If A_c, ℓ, ρ, and κ are, respectively, the cross-sectional area, length, electrical resistivity, and thermal conductivity of the leads

$$R_L = \rho\ell/A_c \tag{A21}$$

$$K_L = \kappa A_c/\ell \tag{A22}$$

The Wiedemann-Franz law relates the electronic portions of ρ and κ for a metallic conductor at a temperature T_o.

$$\rho = LT_o/\kappa,\ L = 2.44 \times 10^{-8} \text{ watt-ohm-deg K}^{-2} \tag{A23}$$

If T_o is assumed to be the average of T_c and T_a

$$K_L = (L/2\,R_L)\,(T_1 + T_2) \tag{A24}$$

Thus K_L can be replaced in (16) by its value in terms of R_L from (24) and obtain an expression which can be reduced to

$$\eta = \beta \eta_o \tag{A25}$$

where η_o is the efficiency of an identical diode with ideal leads (i.e., $R_L = 0$, $K_L = 0$)

$$\eta_o = \frac{J\left[V - JR_p\right]}{W + J_c\,(\Phi + 2KT_c) - J_a\,(\Phi + 2KT_a) + K_p\,(T_c - T_a) - J^2R_p/2 + X} \tag{A26}$$

and

$$\beta = (1 - \gamma)\left\{1 - \frac{\eta_o}{2}\left(\gamma - \frac{1}{G\gamma}\right)\right\}^{-1} \tag{A27}$$

$$G = \left(V - JR_p\right)^2\left[L\left(T_c^{\,2} - T_a^{\,2}\right)\right]^{-1} \tag{A28}$$

$$\gamma = \frac{JR_L}{V - JR_p} \tag{A29}$$

From (A17), therefore,

$$\gamma = \frac{R_L}{R_L + R_{load}} \tag{A30}$$

Thus physically the parameter γ is the fraction of the external emf of the diode which appears across the leads; or equivalently, it is the ratio of the lead resistance to the total external resistance. Its range therefore is from 0 to 1.

The efficiency can then, in principle, be maximized with respect to the lead resistance R_L. Using the representation of (A25) and noting that R_L enters only through the dimensionless parameter γ, it is found that the efficiency optimized with respect to R_L is (quantities evaluated at that value of R_L giving maximum efficiency will be denoted with an overhead bar)

$$\bar{\eta} = \bar{\beta}\eta_o \tag{A31}$$

where

$$\bar{\beta} = \frac{\sqrt{1 + B} - 1}{\sqrt{1 + B} + 1 - \eta_o} \tag{A32}$$

$$\bar{R}_L = \left[(V - JR_p)/J\right]\left[\left(\sqrt{1 + B} - 1\right)/B\right] \tag{A33}$$

$$\bar{\gamma} = J\bar{R}_L/\left(V - JR_p\right) \tag{A34}$$

where

$$B = G(2 - \eta_o)\eta_o \tag{A35}$$

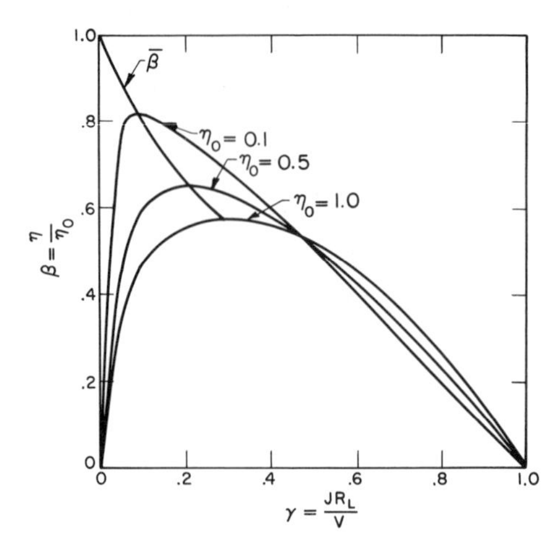

Fig. 15. β as a function of γ and the ideal efficiency, η_0.

As the value of B ranges from zero to infinity, the value of $\bar{\gamma}$ will range from 1/2 to zero. Thus under no conditions will more than half the external emf appear across the leads when they are optimized.

The quantity β is the fraction to which the efficiency of the ideal diode with ideal leads (zero electrical resistance and zero thermal conductivity) has been reduced when the physical leads have been optimized. Considered as a function of η_0 alone (V, T_c, T_a fixed), $\bar{\beta}$ can be shown to be monotonically decreasing.

If non-optimum leads are used, a decrease in efficiency will result. For the values $T_a = 0°K$, $T_c = 3000°K$, and $V - JR_p = 1$ volt, G is equal to 4.55. Figure 15 shows the results of calculations of β as a function of η_0 and γ. Also shown in Figure 15 is the locus of the maximum of the curves, β. Several features of these curves are of interest. At a critical value of γ, equal to $G^{-1/2}$ (0.47 in this case), the curves for different values of η_0 intersect. For γ greater than this critical value, β is a monotonically decreasing function of η_0, but very weakly dependent upon η_0. For γ less than this critical value, β is a strong monotonically increasing function of η_0.

Curves of $\bar{\gamma}$, $\bar{\eta}$, and $\bar{\beta}$ as functions of η_0 are presented in Figure 16 for the values of T_c, T_a, and V used above.

It is clear from Figures 15 and 16 that there is a serious loss in efficiency due to the leads. Even when this loss is minimized by optimizing the value of the lead resistance, it is seen that for the case considered, the ideal efficiency, η, is decreased by amounts ranging from 18% (at $\eta_0 = 0.1$) to 42% (at $\eta_0 = 1.0$). For other cases, the loss may be less severe, but is still appreciable. It should be emphasized that the curves of Figure 16 are for optimized leads. For non-optimized leads, the loss of efficiency is much greater as can be seen from Figure 15.

It is seen from Figure 16 that, for almost the entire range $\bar{\eta}$ is, to a good approximation, a linear function of η_0. This fact may be borne out by expanding eq. (A31) in a series about $\eta_0 = 1$ (see ref. 3).

C. OPTIMIZATION OF LOAD RESISTANCE (INTERNAL VOLTAGE)

The assumption has often been made in the literature that the flat mode of operation yields the maximum attainable efficiency for a given diode. This assumption is not necessarily valid, as a simple numerical example will show.

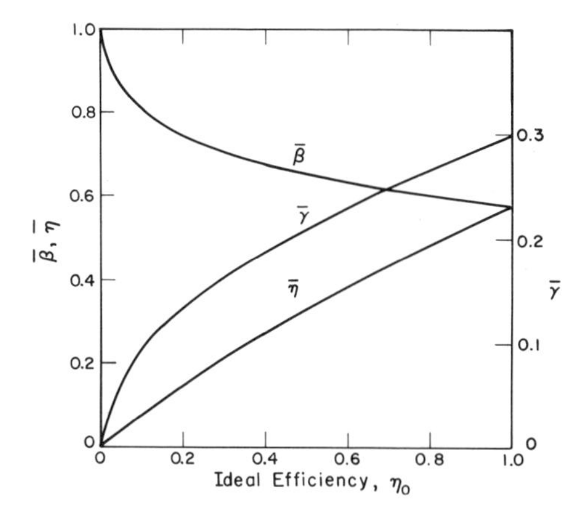

Fig. 16. Optimum values of η, γ, and β as functions of the ideal efficiency, η_0.

If $R_p = 0$, $\Phi_c = 2.5$ volts, $\Phi_a = 1.5$ volts, $T_c = 2200°K$, $T_a = 0°K$, and $\epsilon_c = 0.25$, then $\bar{\eta}_0 = 34.5\%$ when operating in the flat mode. Using the same parameters, but operating in the uphill mode, so that V = 1.5 volts, η_0 is equal to 41.5%. At the same time, a reduction of power density from 1090 watts cm^{-2} to 117 watts cm^{-2} and a reduction of current density from 1090 amps cm^{-2} to 77.8 amps cm^{-2} takes place.

The reduction of the power density is of significant value because a comparable power must be dissipated at the anode, and the dissipation of large quantities of power can pose a serious engineering problem. Similarly, the handling of large current densities is difficult; thus, the reduction of the output current density is also of engineering importance.

The reason for this increase in efficiency in the uphill mode can be seen qualitatively by considering the simple case of ideal leads, no plasma or support losses, and no back current. Then (A16) can be written

$$\eta = \frac{V}{\Phi + 2KT_c + W/J_c} \tag{A36}$$

As long as W is small compared to J_c, the efficiency will increase with increasing V. ($\Phi = V + \Phi_a$ in the uphill mode). Eventually, J_c becomes so small that W/J_c is no longer negligible. At this point, the efficiency begins to decrease with increasing V. When a maximum efficiency occurs in the uphill mode, it can be shown that this efficiency is independent of the cathode work function, Φ_c.

Numerical calculations of (A16) show that this effect exists as well for that more accurate solution. Further, it is possible for the optimum efficiency to occur in the downhill mode. This can easily be explained by considering the case of a diode whose parameters are such that $J_{cs} < J_{as}$. Thus the only possible mode of operation is the downhill mode and the maximum efficiency must occur in that mode.

The efficiency of (A16) can be simultaneously maximized with respect to V and R_L by setting the partial derivatives of η with respect to these variables equal to zero. The differentiation of (A16) with respect to V yields equation (A37).

$$J + \frac{dJ}{dV}\left[V - 2J\left(\bar{R}_L + R_p\right)\right] - \bar{\eta}\,\frac{dJ_c}{dV}\,(\Phi + 2KT_c)$$

$$- \frac{dJ_a}{dV}\,(\Phi + 2KT_a) + \left[J\,\frac{d\Phi}{dV}\left(R_L + R_p\right)\right] = 0 \qquad \text{(A37)}$$

where $\bar{\eta}$ and $\bar{R}_L$ are given as functions of V by (A31) ff. Equation (A37) is actually two equations—one for the downhill and one for the uphill mode of operation. It can be shown that each of these two equations always has one and only one solution for V between V_{sc} and V_{oc}. If the solution to the equation for the downhill mode lies between V_{sc} and $\Phi_c - \Phi_a$, there is a maximum for the efficiency in that mode. Similarly, if the equation for the uphill mode has a solution between $\Phi_c - \Phi_a$ and V_{oc}, there is a maximum in the uphill mode. If neither of these two conditions result, the maximum efficiency results in the flat mode $(V = \Phi_c - \Phi_a)$.*

Once the value of V resulting in the maximum efficiency is found, the value R_L necessary for it to be obtained is given by (A33). From these two quantities all other properties at this optimum point can be found (e.g., power out, net current, etc.).

The model of the diode used for the foregoing calculation possesses several limitations with respect to the manner in which the plasma is included. These are important enough to warrant further discussion. It has been assumed that "sufficient" plasma has been introduced so that the space-charge is eliminated. Although experiments with plasma filled diodes seem to indicate this can be done, only the crudest estimates are available of the conditions under which this is possible. Further, it is known that the introduction of plasma can result in potential sheaths at the electrodes. In lieu of considering this difficult and as yet unsolved problem, the procedure described assumes that the plasma can be considered ohmic and that an effective electrical resistance can be

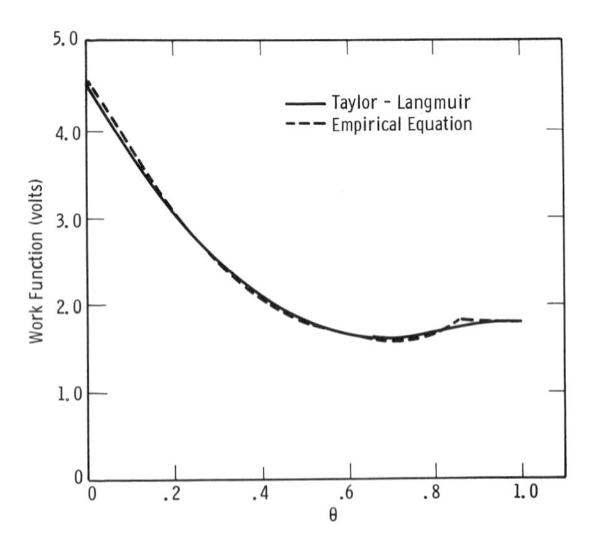

Fig. 17. Work function of tungsten as a function of fractional coverage by cesium.

* In principle, it is possible for a maximum to occur in each mode. However, the range of parameters for which this can happen is small. In several thousand cases calculated, this situation has never occurred.

found. Limited experimental data indicates that this assumption is not too inaccurate. Finally, the possibility of an arc being established is ignored. Although it is definitely known that an arcing mode can be set up, this possibility does not invalidate the above calculations. What it means, however, is that if the solution for the operating point found by the described procedure falls in a region of operation which will support an arc, it is not a valid solution for the maximum efficiency. A treatment of the operation of a diode in the presence of an arc is a difficult, unsolved problem. By comparing the optimum operating points calculated by the formulas given here to the limited experimental data available on arcs, it is found that, in general, these computed optimum points should not sustain arcs.

D. EMPIRICAL EVALUATION OF PARAMETERS

The relationship of effective work function to cesium pressure and surface temperature has not been well determined. The pioneering work of Taylor and Langmuir (9) provided the basis for the present empirical equation. While no theoretical justification of the equation can be given, the results are in close agreement with the original data and with the extrapolations of other investigators (1).

The effective work function of cesium-coated tungsten was assumed to be a function of only the fractional coverage, θ. The data of Langmuir and Taylor (9) are plotted in Figure 17. The empirical equation

$$\begin{aligned} \Phi &= 4.52 - 8.59\,\theta + 6.3\,\theta^2, \quad \theta \leq 0.86 \\ \Phi &= 1.81 \qquad\qquad\qquad\quad\;\;, \quad \theta \geq 0.86 \end{aligned} \tag{A38}$$

was fitted to this data. The results of this equation are also plotted in Figure 17.

The estimation of the fractional coverage as a function of cesium pressure and surface temperature was also made by the use of a set of empirical equations. Figure 18 shows the relation between surface temperature and coverage for two particular cesium temperatures (9). This can be well fitted by the equation

$$\theta = \left[1 + (T/T_0)^c\right]^{-1} \tag{A39}$$

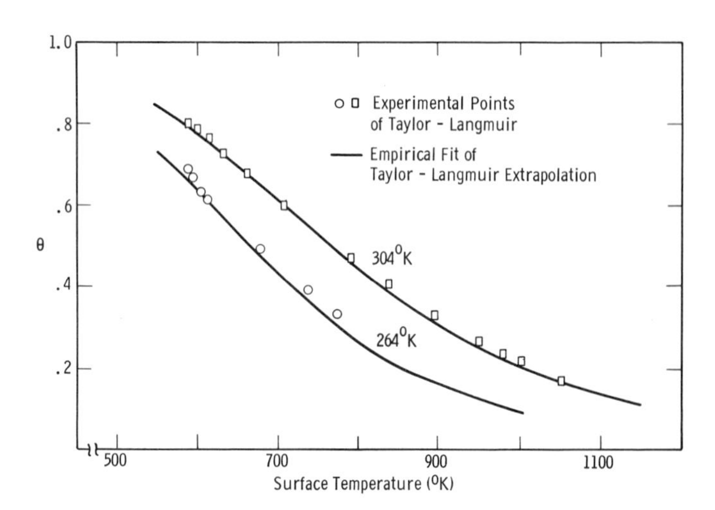

Fig. 18. Fractional coverage as a function of cesium temperature and surface temperature.

where T_o and c are constants, both of which are functions of the cesium pressure (or arrival rate).

If the relationship of cesium bath temperature, T_{Cs}, in °K, to cesium pressure, P, given by Nottingham (14) is used,

$$P = 2.45 \times 10^{-8} e^{-8910/T_{Cs}}, \text{ in mm Hg} \tag{A40}$$

The relationship of c and T_o to the temperature of the cesium bath T_{Cs} are found to be

$$\begin{aligned} c &= 2.73 + 720/T_{Cs} \\ 1/T_o &= 0.385\,(10^{-4} + 1/T_{Cs}) \end{aligned} \tag{A41}$$

Figure 18 shows the results of using these empirical equations for the two temperatures shown. Figure 5 shows the relationship of work function to cesium temperature and surface temperature. This is in fair agreement with the results of Schock (1), especially if the cesium temperature listed by Schock rather than the pressure is used to compare values. One particular difference is of interest. The measurements of Langmuir indicate that a minimum work function is obtained when θ is equal to 2/3. When a full monolayer is present, the work function is assumed to be that of cesium, as shown by Langmuir. This higher value of the work function at full coverage has a serious effect on the efficiency when low anode temperatures are used.

The thermal emissivity of tungsten is given by several references as a function of temperature. The data used here are those of Forsythe and Watson (8).

Values of plasma electrical resistance and thermal conductance are not available in the literature. Gottlieb and Zollweg (10) have recently measured the thermal conductance of cesium plasma.

Their data were fitted well by the empirical equation

$$\kappa_p = 1.5 \times 10^{-4} \left\{ 1 - \exp(-40\ Pd) \right\} \tag{A42}$$

where P was determined from eq. (40), d is the spacing in cm, and κ_p the thermal conductivity of the plasma in watt/(cm °K). The electrical resistance was estimated from J-V curves taken by Gottlieb and Zollweg. The results are quite preliminary in nature. For the calculations discussed hereafter, the plasma resistivity, ρ_p, was estimated by the equation

$$\rho_p = 2.67\,P, \text{ in ohm cm} \tag{A43}$$

E. RESULTS OF NUMERICAL CALCULATIONS

Because empirical data concerning the effect of cesium coverage on the thermionic properties of materials is well known only for tungsten, calculations were made only for a device in which both the anode and cathode were made of tungsten.

Calculations of the optimum operating point according to the method described above were carried out on a high speed digital computer for more than 5000 different combinations of spacing, and cathode, anode, and cesium bath temperatures. At the values of lead and load resistance which yielded the maximum efficiency for each of the given combination of parameters, some forty different properties, such as power delivered to load, external voltage, net current, etc., were tabulated in addition to the efficiency. As already shown, for given materials of construction, η optimized with respect to load and lead resistance is a function of only four variables: d, T_a, T_c, and P. Because of the number of independent variables, it is possible to plot lines of constant efficiency in a large number of ways. By keeping two of the parameters constant, efficiency contours can be plotted as a function of the other

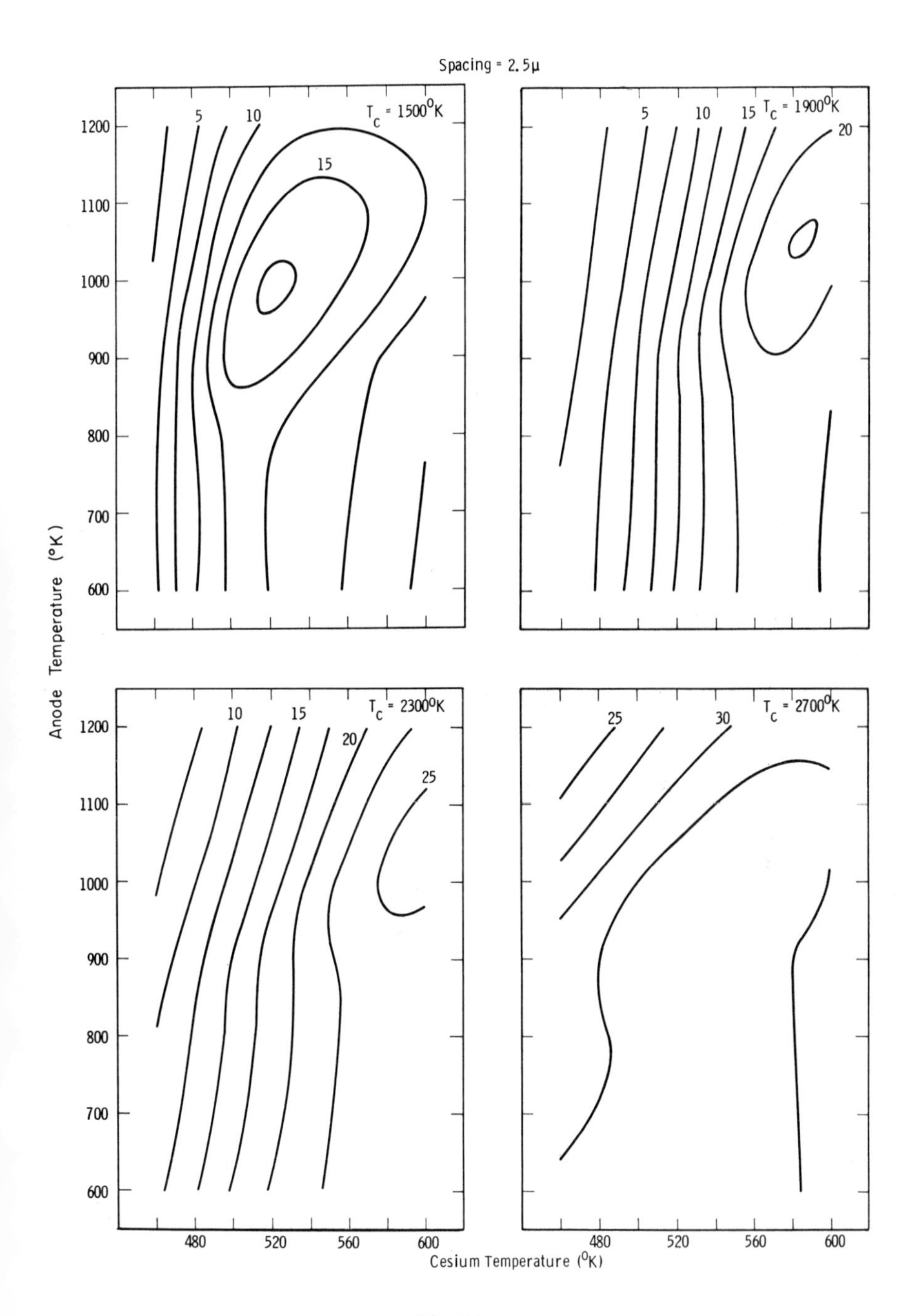
Spacing = 2.5μ
T_c = 1500°K
5
10
15
T_c = 1900°K
5
10
15
20
T_c = 2300°K
10
15
20
25
T_c = 2700°K
25
30
1200
1100
1000
900
800
700
600
Anode Temperature (°K)
480
520
560
600
Cesium Temperature (°K)

Fig. 19a.

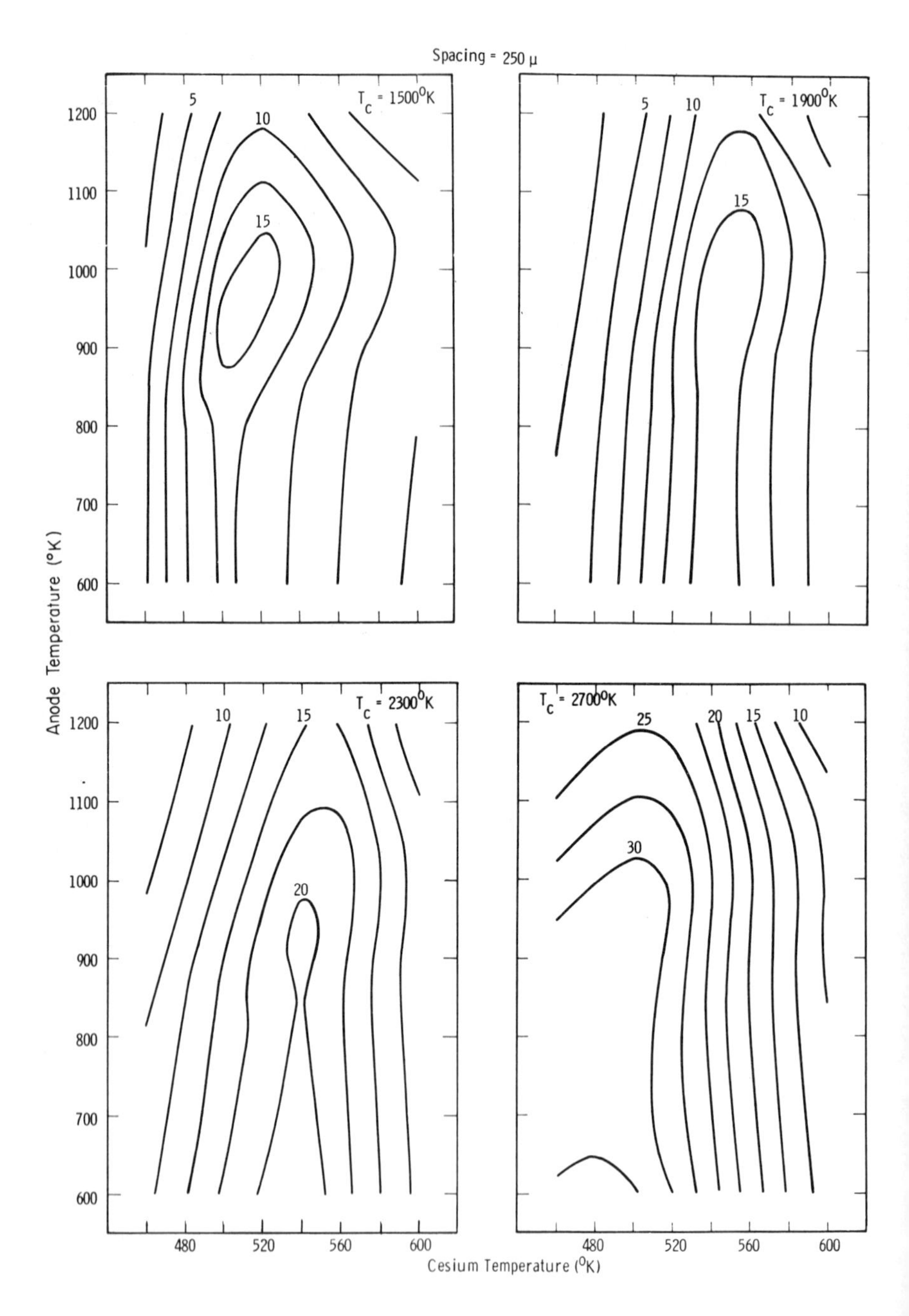
Spacing = 250 μ
T_c = 1500°K
T_c = 1900°K
T_c = 2300°K
T_c = 2700°K
5
10
15
20
25
30
1200
1100
1000
900
800
700
600
480
520
560
600
Anode Temperature (°K)
Cesium Temperature (°K)

Fig. 19b.

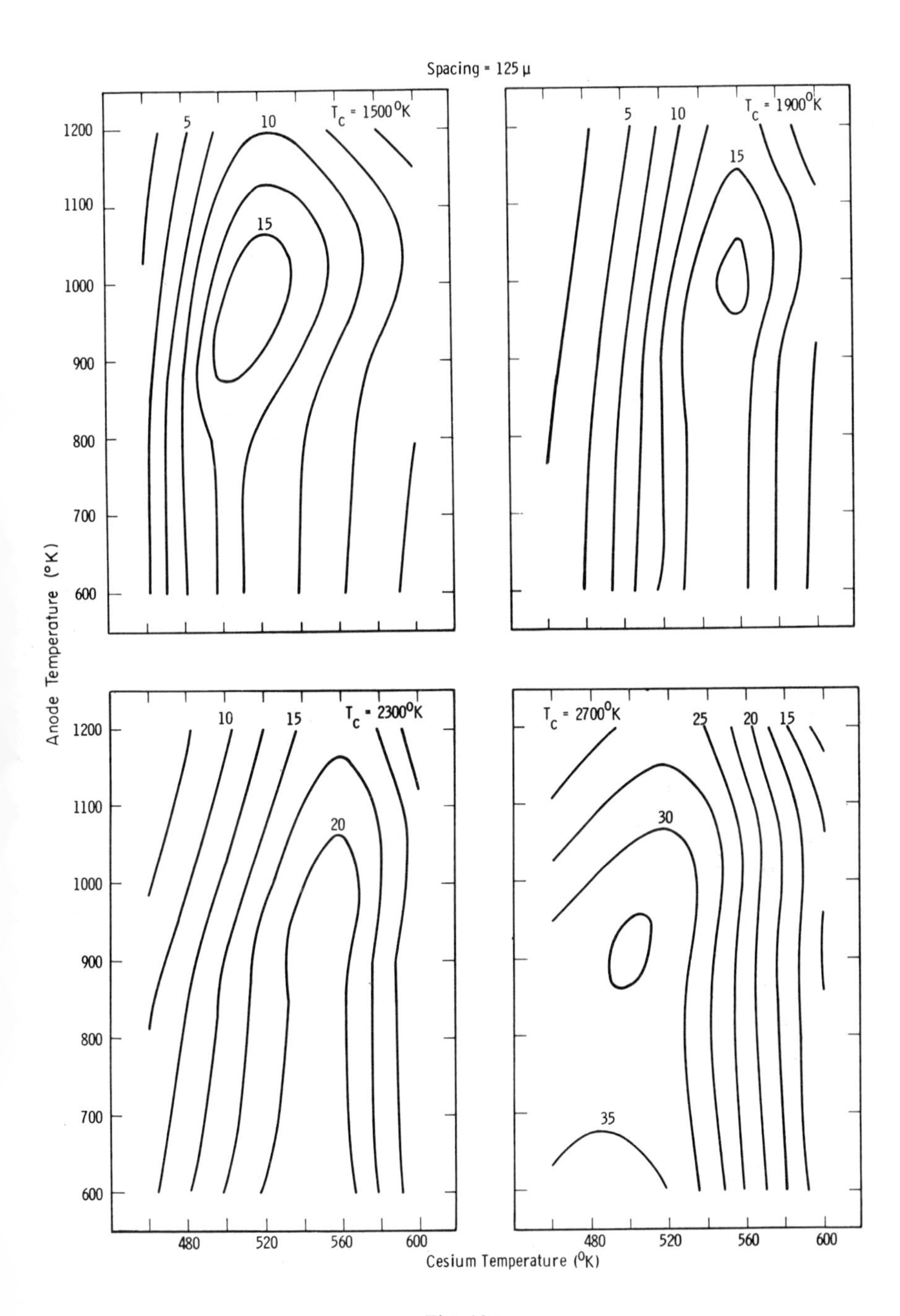
Spacing = 125 μ
T_C = 1500°K
5
10
15
T_C = 1900°K
5
10
15
T_C = 2300°K
10
15
20
T_C = 2700°K
25
20
15
30
35
1200
1100
1000
900
800
700
600
Anode Temperature (°K)
480
520
560
600
Cesium Temperature (°K)

Fig. 19c.

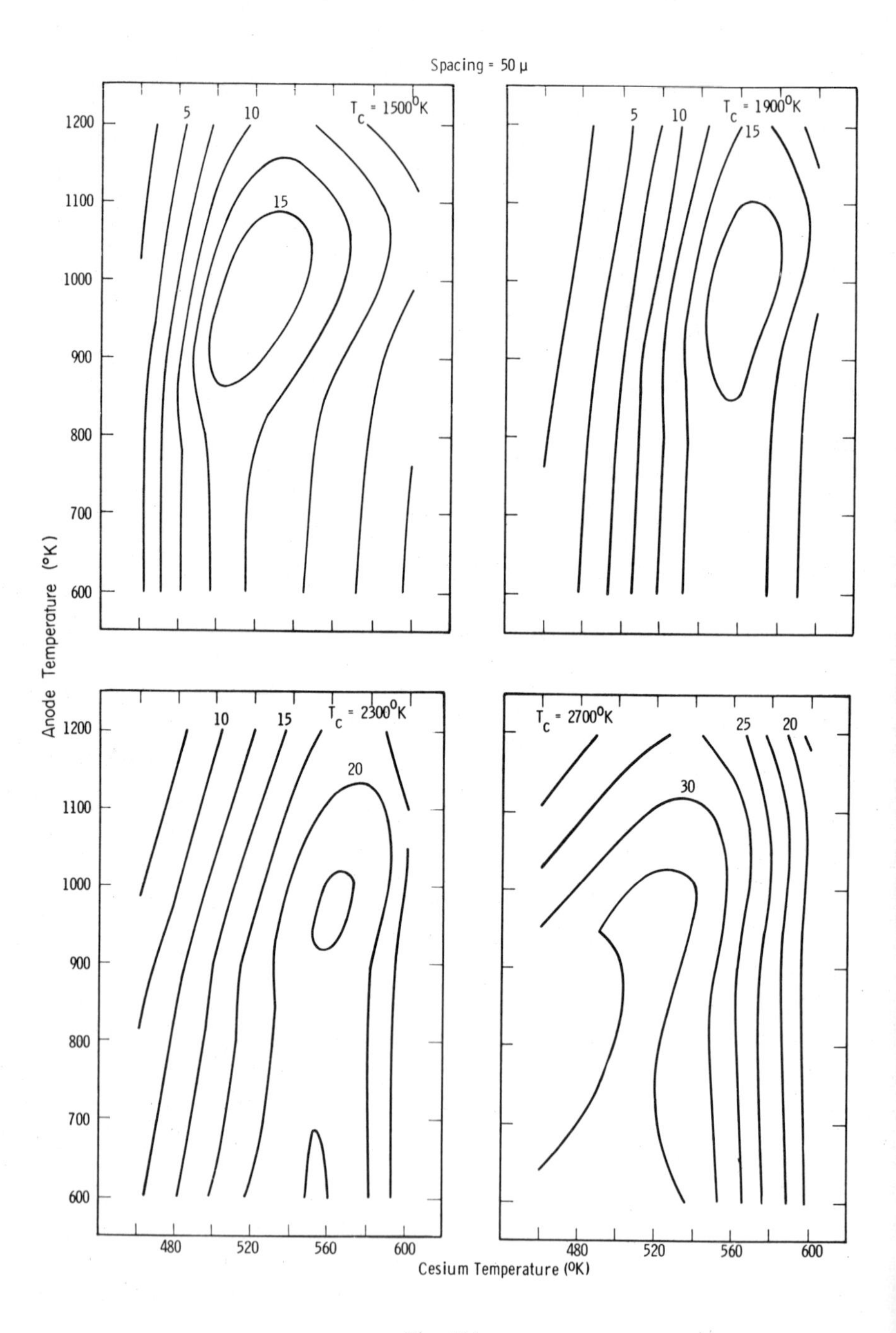
Spacing = 50 μ
T_c = 1500°K
5
10
15
T_c = 1900°K
5
10
15
T_c = 2300°K
10
15
20
T_c = 2700°K
25
20
30
1200
1100
1000
900
800
700
600
Anode Temperature (°K)
480
520
560
600
Cesium Temperature (°K)

Fig. 19d.

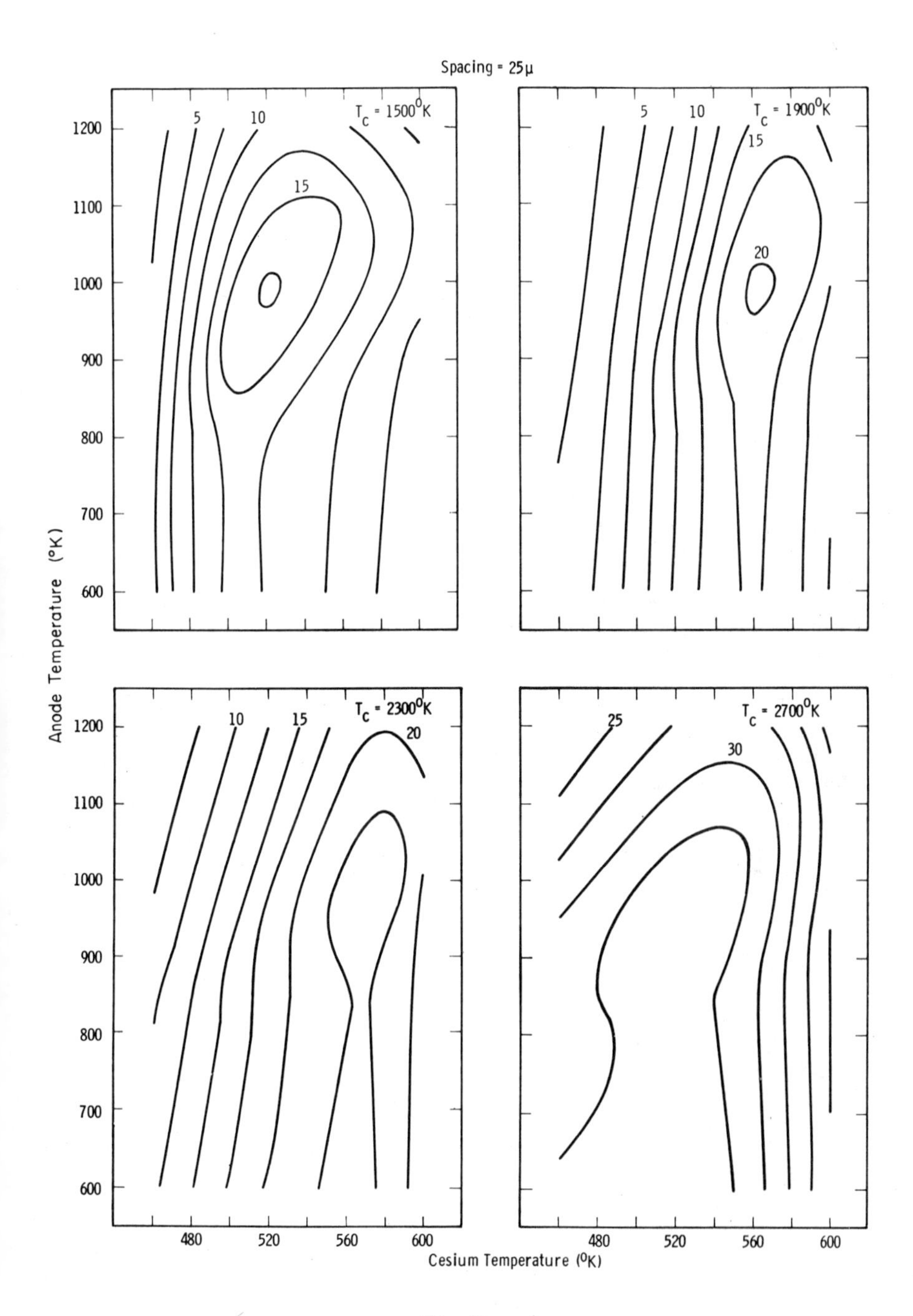
Spacing = 25μ
T_c = 1500°K
5
10
15
T_c = 1900°K
5
10
15
20
T_c = 2300°K
10
15
20
T_c = 2700°K
25
30
1200
1100
1000
900
800
700
600
Anode Temperature (°K)
480
520
560
600
Cesium Temperature (°K)

Fig. 19e.

two. Figures 7, 8, and 9 show, respectively, sample results of keeping fixed P and d, T_a and d, and P and T_a. It was found, however, that the most useful type of plot was obtained by fixing d and T_c and drawing curves of T_a and P with η as a parameter. A set of 20 such plots are shown in Figure 19a-e. A more complete set can be found in ref. 3. It should be noticed that in each case, the efficiency has a maximum occurring at a unique combination of cesium and anode temperatures. This point is discussed in the text.

References

1. Rasor, N. S., J. Appl. Phys., 31, 163 (1960); Houston, J. M., ibid. 30, 481 (1959); Ingold, J. H., ibid. 32, 769 (1961); Schock, A., ibid. 32, 1564 (1961); Hatsopolous, G. N., and J. Kaye, Proc. I.R.E. 46, p. 1574 (1958). Direct Conversion of Heat to Electricity, J. Kaye and J. A. Welsh, Ed., Wiley, New York, 1960.

2. U. S. Patent No. 2,980,819 issued to G. R. Feaster, April 18, 1961.

3. Richardson, L. S., and A. E. Fein, Interim Scientific Report No. 2, on AF33(616)-8292.

4. Edwards, J. W., H. L. Johnston, and P. E.Blackburn, J. Am. Chem. Soc., 74, 1539 (1952). Speiser, R., P. E. Blackburn, and H. L. Johnston, J. Electrochem. Soc., 106, 52 (1959). Edwards, J. W., H. L. Johnston, and P. E. Blackburn, J. Am. Chem. Soc., 73, 172 (1951). Kubaschewski, O., and E. L. Evans, Metallurgical Thermochemistry, Pergamon Press, New York, 1958, p. 335.

5. Dushman, S., Vacuum Technique, Wiley, New York, 1949, p. 17.

6. Dushman, S., Vacuum Technique, Wiley, New York, 1949, p. 74.

7. Maynard, C. W., Nuclear Sci. and Eng., 6, 174 (1959).

8. Forsythe, W. E., and E. M. Watson, J. Opt. Soc. Am., 24, 114 (1934).

9. Taylor, J. B., and I. Langmuir, Phys. Rev., 44, 423 (1933).

10. Gottlieb, M., and R. Zollweg, unpublished research.

11. Nottingham, W. B., "Thermionic Emission," Handbuch der Physik, Vol. 21, S. Flugge, Ed., Springer-Verlag, Berlin, 1956.

12. Webster, H. F., J. Appl. Phys., 30, 488 (1959).

13. (a) Jakob, M., Heat Transfer, Vol. II, ch. 31, New York, 1957. (b) Fein, A. E., Rev Sci. Inst., 31, 1008 (1960).

14. Nottingham, W. B., Tech. Rept. 373, Research Laboratory of Electronics, the Massachusetts Institute of Technology, 1960.

DISCUSSION

L. YANG (General Atomic Laboratories):In estimating the lifetime, what order of magnitude of the loss of material do you choose?

A. E. FEIN: It is one ten-thousandth of an inch. Figure 1 was ten-thousandth of an inch, one-tenth of one mil.

L. YANG: In Figure 4 which showed the effect of cesium on the rate of vaporization of tungsten the ordinate was of the order of 10, 20, and 30. Does that mean the vaporization rate is cut down by cesium by these factors?

A. E. FEIN: Yes.

L. YANG: Would your data indicate that scattering of the vaporized atoms is the main mechanism by which the rate of vaporization is reduced by cesium?

A. E. FEIN: The theory predicts that a plot of the ratio of vacuum evaporation to evaporation in the presence of cesium as a function of spacing should be a straight line. Further at zero spacing, the ratio should be equal to one. Therefore, each point of the experimental data determines a mean free path independently of the others.

Now, of the four points taken on these preliminary experiments, three of them fell very nicely on the same straight line with the point (0,1). The fourth fell far off the curve, but there is some reason to believe there was some warping of the filament in that location. Further, it appeared as if there was shorting at that point and that the evaporated material may have burnt off. Because of the wide variations in the mean free path one can assume that, for molybdenum through cesium there is no good choice of what the slope should have been. The fact that three out of the four points fell nicely on the same straight line with the point through which the curve must theoretically go indicates that the mechanism agrees well with the hypothesis of simply back scattering to the cathode.

We are performing more experiments of this type to verify this and to see if the mean free path follows the behavior it should for variations in cesium pressure. Results are still quite preliminary, but at any rate it can be said that there is considerable reduction in the evaporation rate.

W. L. TOWLE (Mallinckrodt Chemical Works): In your statement of conclusions, you said that in general longer lifetime corresponds to lower efficiency. Is this because of a difference in coverage of the electrode surface by the cesium?

L. S. RICHARDSON: This is based upon the increase in evaporation with increasing temperature; thus, the higher the temperature, the lower the lifetime.

W. L. TOWLE: But for two materials at a given temperature, then this conclusion does not hold?

L. S. RICHARDSON: That is correct.

Some Physicochemical Criteria for the Selection of Carbides As Cathodes in Cesium Thermionic Converters

LING YANG, F. D. CARPENTER, A. F. WEINBERG, and R. G. HUDSON

John Jay Hopkins Laboratory for Pure and Applied Science,
General Atomic Division of General Dynamics Corporation,
San Diego, California

Abstract

Despite their high melting points, the usefulness of carbides as cathodes in cesium thermionic converters depends on whether they can meet a number of material requirements. The purpose of this paper is to discuss these requirements on the basis of the mass- and energy-transport processes occurring at the interfaces between the cathode and its surroundings during the operation of a cesium thermionic converter. Carbide properties relevant to such application will be reviewed and the present status of carbide cathodes analyzed.

INTRODUCTION

In a cesium thermionic cell, the material problems center mostly around the cathode (1). This is because the cathode, operating in the temperature range from 1600 to 2000°C and separated from the anode by a very small distance (e.g., 20 mils), has to have a life of several thousand hours or more while delivering an electrical output of at least several watts per square centimeter if the cell is to be of any practical value. Metal carbides with melting points in the temperature range from 2500 to 3800°C, the carbides of hafnium, tantalum, niobium, zirconium, tungsten, and uranium are potential cathode materials; and carbide systems containing UC are of special interest as cathodes for the direct conversion of fission heat to electrical energy. To satisfy the requirements of longevity and useful performance, however, high melting point constitutes a necessary but not a sufficient condition; there are many other conditions to be met before a material can be used as a cathode in a cesium thermionic cell. It is the purpose of this paper to analyze the material problems associated with carbide cathodes and the criteria by which the usefulness of a carbide as a cathode can be judged.

The material problems of carbide cathodes can be delineated on the basis of the mass- and energy-transport processes occurring at the interfaces between the cathode and its surroundings during the operation of a cesium thermionic cell. Irrespective of the nature of the source of the thermal energy, the cathode can be treated as a system bound on one side by a partial vacuum containing cesium vapor and on the other side by a refractory-metal electrical lead. Figure 1 illustrates the mass- and energy-transport processes involved

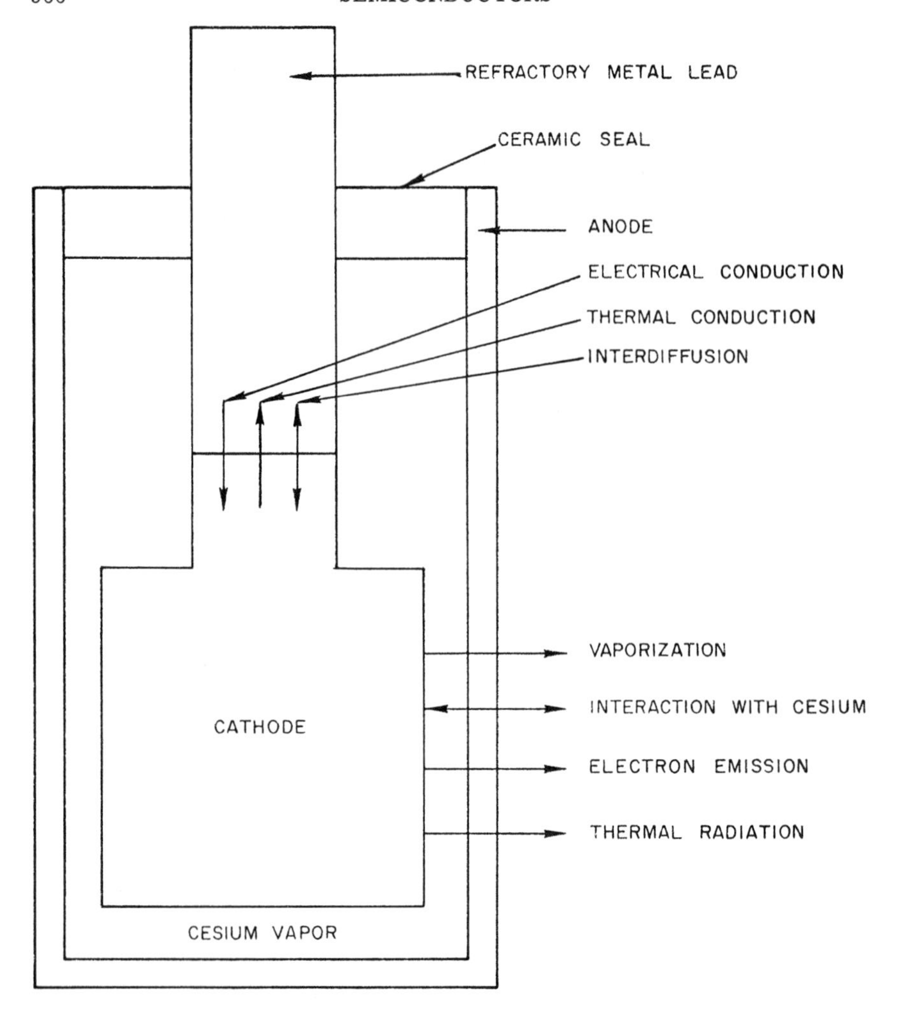

Fig. 1. Mass- and energy-transport processes occurring at the interfaces between the cathode and its surroundings in a cesium thermionic cell.

in such a system. It must be pointed out that these processes are not independent of each other. For instance, the change of the surface composition of the cathode by vaporization may seriously affect its thermionic properties and its interaction with cesium vapor; the condensation of the vaporized material on the anode surface may increase its thermal emissivity and thus the radiation thermal loss from the cathode; and the diffusion between the cathode and the refractory-metal lead may change the thermal and electrical conductivities at the interface. These processes jointly determine the life and the performance of the cell.

An ideal carbide cathode should meet the following requirements:

1. Low vaporization loss at temperatures where useful electron emission can be obtained.
2. Immunity to cesium vapor.
3. Compatibility with a refractory-metal lead.

4. High electrical conductivity to reduce Joule loss.
5. High thermal conductivity to facilitate heat transfer from source to cathode surface.
6. High electrical-to-thermal conductivity ratio so that the size of the part in connection with the refractory-metal lead can be reduced to minimize thermal conduction loss without leading to excessive Joule loss.
7. Low total thermal emissivity to reduce thermal radiation loss.
8. For fission heat conversion, negligible effect of radiation and fission products on the properties of the cathode.

These requirements will be discussed in more detail below.

VAPORIZATION LOSS VERSUS ELECTRON EMISSION

Both vaporization loss and electron emission increase with temperature. In the selection of carbide cathodes and their optimum operating temperatures, in order to obtain the life and the performance called for by the design, it is necessary to know their vaporization losses and their electron-emission properties at various temperatures in the presence of cesium vapor. As these data are rarely available, they are replaced by the vaporization losses and the saturation electron emission in vacuum as first approximations. Although the effect of cesium vapor on the vaporization losses of carbides is unknown and is a matter of great interest, the vacuum saturation emissions of ZrC and $(UC)_{20}$ $(ZrC)_{80}$ have been shown to be the same as their short-circuit currents in cesium thermionic cells (2).

Figure 2 summarizes the observed vaporization data and Figure 3 the vacuum saturation emissions of carbides of thermionic interest. In correlating these data, it is necessary to keep the following points in mind.

1. The vaporization-loss and electron-emission data are often obtained by different authors using samples prepared by different methods such as arc melting, hot pressing, carburization of metal, and electrophoretic deposition of carbide powder on tungsten supports. The similarities of the physical states and chemical compositions of these samples have not been established. Therefore, the results may not be comparable.
2. The vaporization loss and electron emission are measured in different temperature ranges. The data have to be extrapolated to the same temperature range before any correlation can be made. The justification of extrapolating these data for a few, to several, hundred degrees in temperature may be open to question.
3. The literature often specifies only the bulk composition of the carbide samples used for the vaporization and electron-emission studies. Unless the composition of the carbide sample is congruent or the rates of diffusion of the components of the sample are much faster than their rates of vaporization, the components will vaporize at different rates and the surface composition will change with time of heating. The observed rate of vaporization and electron-emission characteristics therefore do not correspond to those of the bulk composition. The relative roles of diffusion rate and rate of vaporization in determining the vaporization characteristics as a function of time of heating can be illustrated by the results obtained on the UC-ZrC solid solutions. For instance (3) for a UC-ZrC solid solution containing 15 mole-% of UC in the temperature range from 2120 to 2320°K, the rate of vaporization is low (see Fig. 2) and essentially independent of time of heating. The change of the uranium concentration near the surface is within the limit of experimental error for α-counting techniques, and x-ray diffraction studies indicate that the change of lattice constant with distance from the surface is negligible. It seems that the

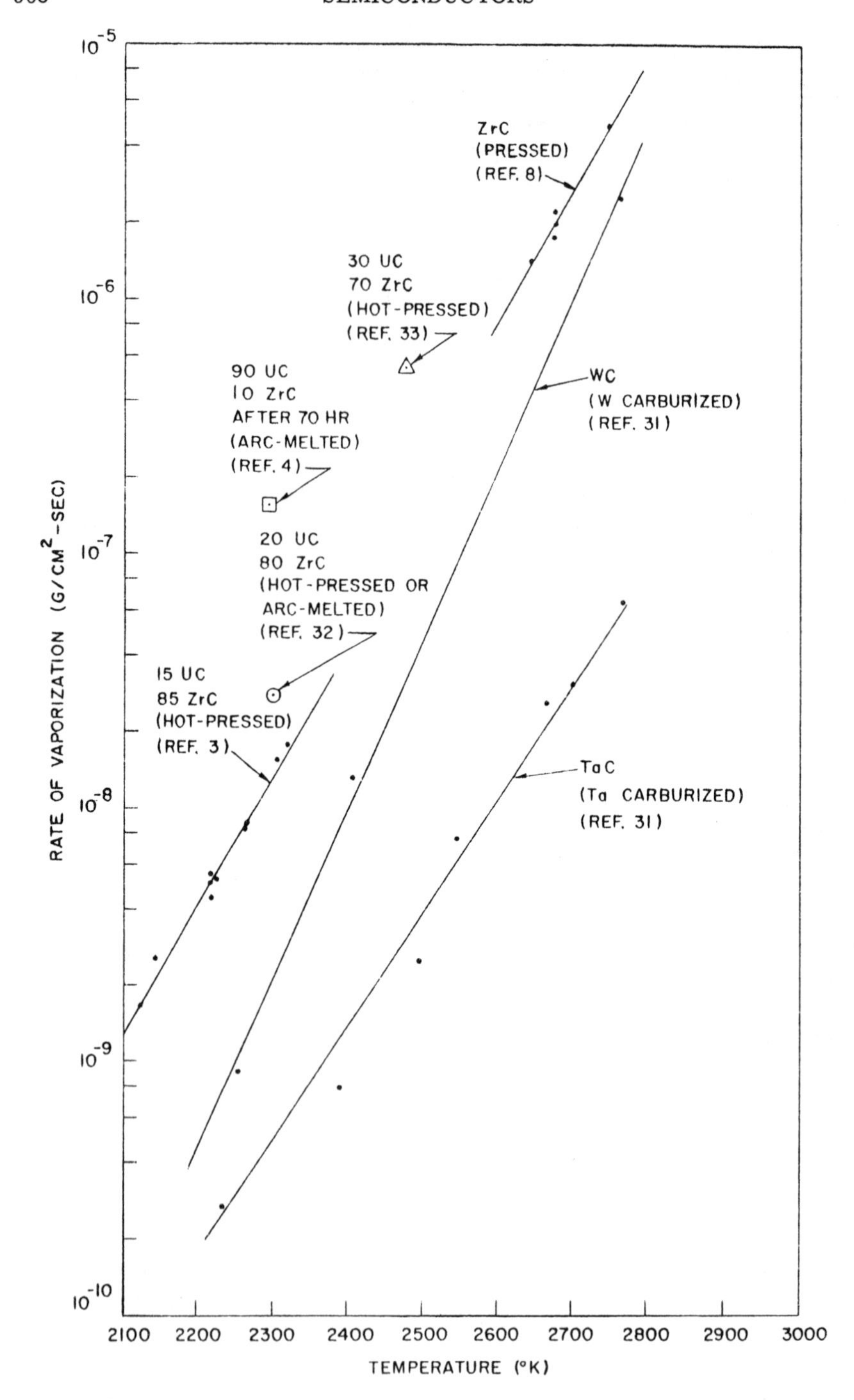

Fig. 2. Rate of vaporization in vacuum for various carbides; straight lines show the trend, not the mathematical relationship.

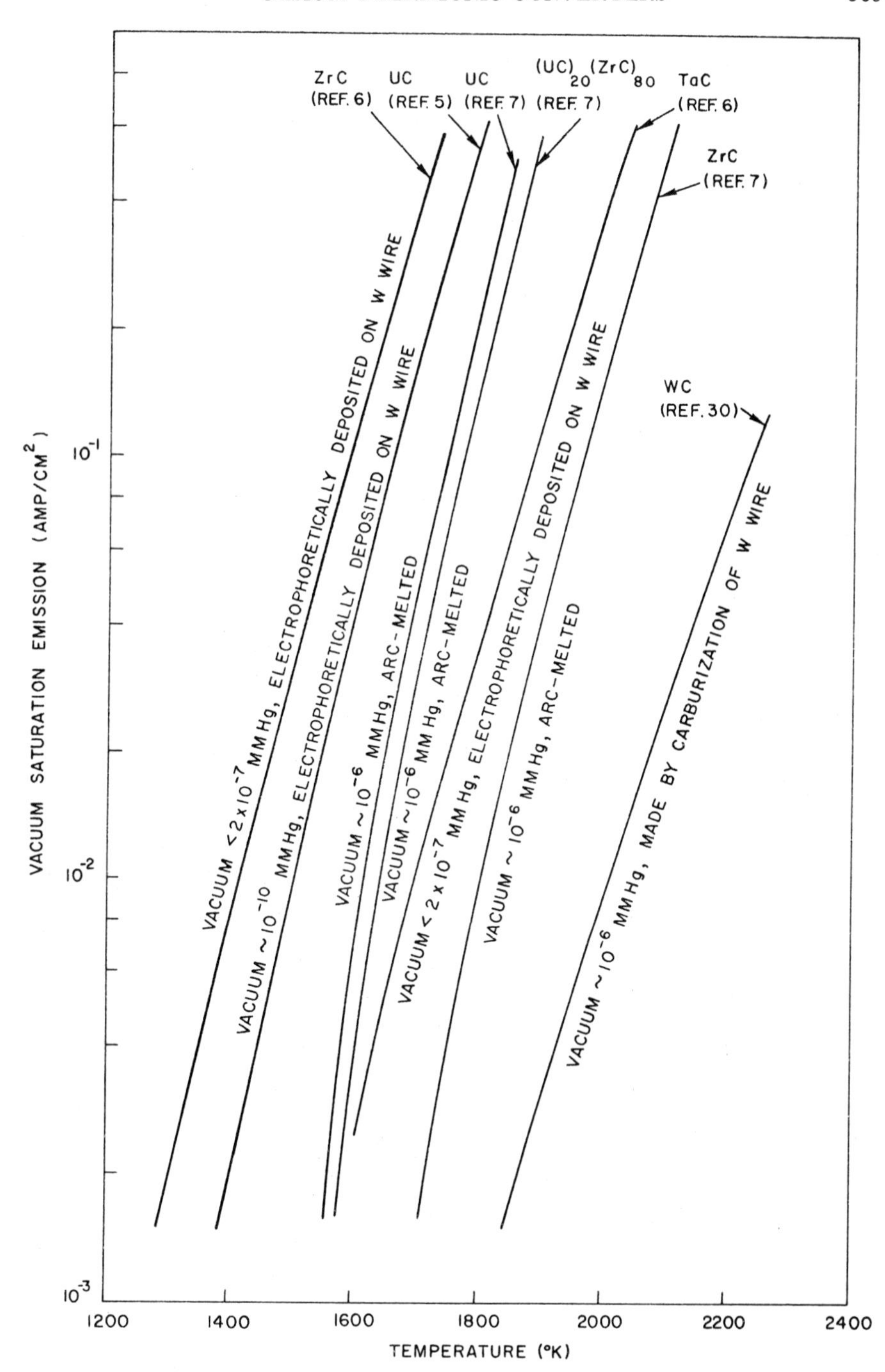

Fig. 3. Vacuum saturation emission for various carbides.

rates of diffusion of the components are fast enough to replenish the material lost by vaporization and that the concentration gradients in the sample, if any, are very small. On the other hand (4), for a UC-ZrC sample containing 90 mole-% of UC, the situation is different. At 2293°K, both the rate of vaporization and the uranium concentration near the surface decrease with time of heating (Figs. 4 and 5). A plateau region is finally reached after about 70 hr of heating where the rate of vaporization and the surface uranium concentration (down to about 35% of the initial value) vary only slightly with further heating.

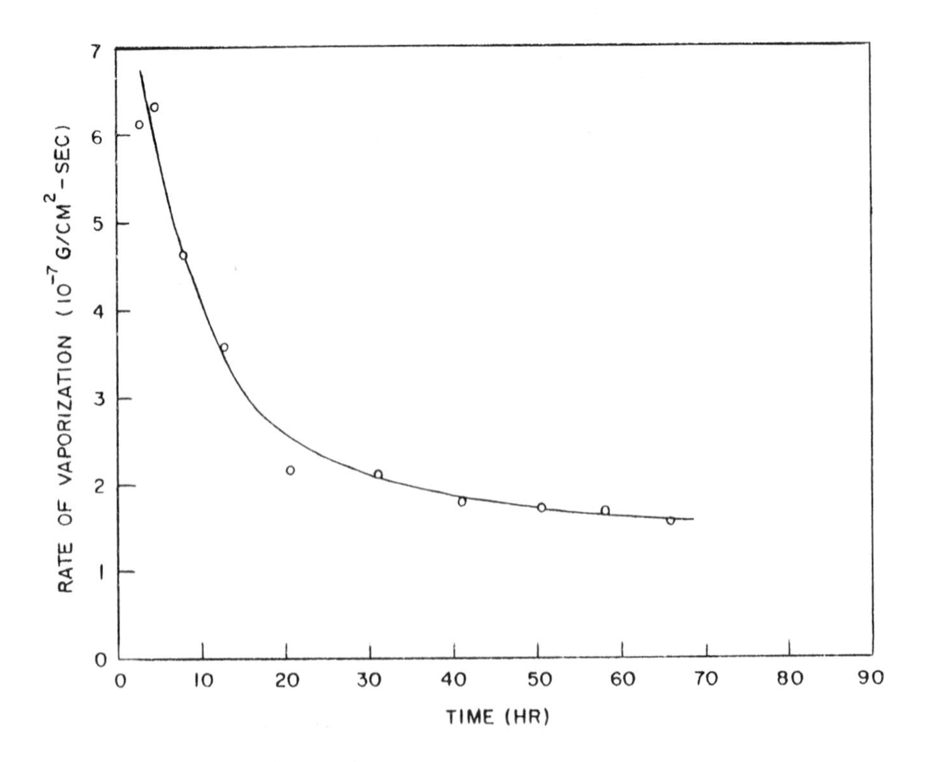

Fig. 4. Rate of vaporization versus time at 2293°K; initial composition, 90 mole-% UC and 10 mole-% ZrC.

It seems that the rates of vaporization are about balanced by the rates of diffusion only after the surface region is appreciably depleted of uranium concentration. It would therefore be misleading if the observed rate of vaporization were correlated with electron-emission data obtained with a $(UC)_{90}$ $(ZrC)_{10}$ sample which had not been "thermally aged" in such a way.

4. The electron-emission data of the same material vary widely with the method by which the sample is prepared. For instance, the electrophoretically deposited UC (5) and ZrC (6) samples on tungsten emit much better than do the arc-melted ones (7) (see Fig. 3). Whether this is due to the presence of metal-rich phases formed by reaction of the carbide with tungsten or to the better vacuum during the measurement of the emission is unknown. The lack of information on the physical and chemical conditions of the surface of the samples used in previous vaporization and electron-emission studies is a serious handicap in correlating these two properties.

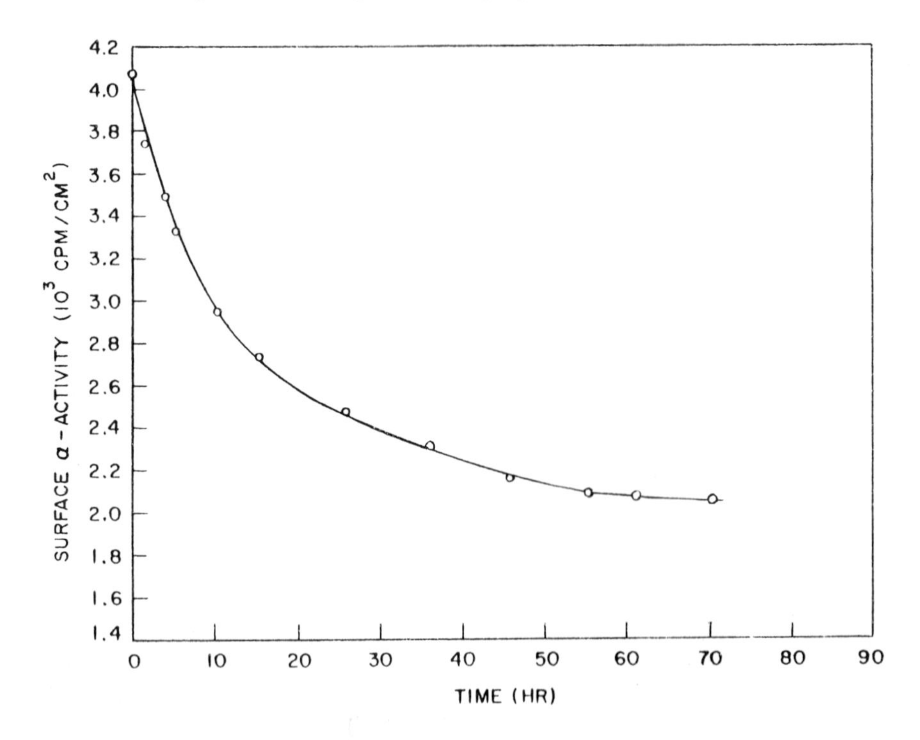

Fig. 5. Surface α-activity versus time at 2293°K; initial composition, 90 mole-% UC and 10 mole-% ZrC.

For the above-mentioned reasons, correlation of the rate of vaporization, J_0, and the vacuum saturation emission, J_-, at the same temperature is made only for ZrC and $(UC)_{15}(ZrC)_{85}$ in Figure 6. The Los Alamos J_- data (2) and extrapolated J_0 values from Pollock's vaporization data (8) were chosen for ZrC. For $(UC)_{15}(ZrC)_{85}$ the Los Alamos J_- data (7) for $(UC)_{20}(ZrC)_{80}$ and extrapolated J_0 values from the vaporization data obtained at our laboratory (3) have been used. The assumption that $(UC)_{15}(ZrC)_{85}$ emits as well as $(UC)_{20}(ZrC)_{80}$ is based on the observation made by Campbell and Hoffman at our laboratory that the vacuum saturation emissions of well-outgassed $(UC)_{10}$ $(ZrC)_{90}$ samples approach that of the Los Alamos $(UC)_{20}$ $(ZrC)_{80}$ data. It must

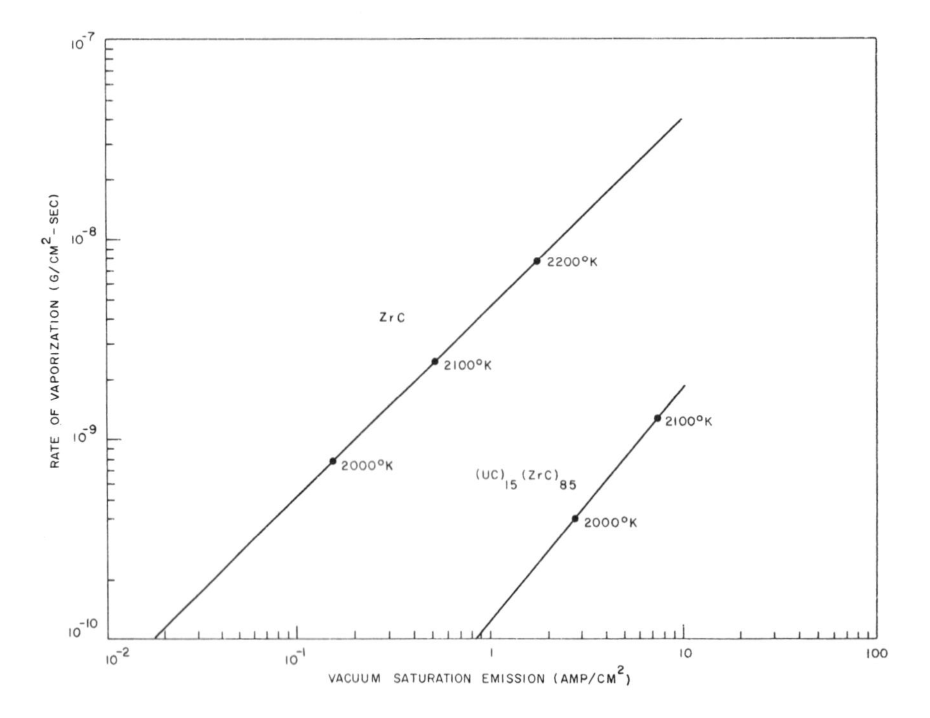

Fig. 6. Rate of vaporization versus vacuum saturation emission for ZrC and $(UC)_{15}(ZrC)_{85}$.

be emphasized that the correlations shown in Figure 6 are at best only approximations, since the vaporization data and the electron-emission results were not measured on the same samples. If it could be shown that by using the same sample and making the measurements in better vacuum (e.g., 10^{-9} mm of mercury), the J_0 and the corresponding J_- values are changed, then the positions of the curves in Figure 6 should be shifted accordingly. Efforts are being made along these lines in our laboratory.

If a value of 10^{-9} g/cm^2-sec is chosen as the maximum tolerable J_0, which is equivalent to a loss of 1 mil in thickness per year for a material of density equal to 10 g/cm^3, then $(UC)_{15}(ZrC)_{85}$ is capable of delivering 6 amp/cm^2. For the same output, ZrC would have to be operated at a temperature at which 50 mils in thickness would be lost in a year.

CHEMICAL INTERACTION BETWEEN CARBIDES AND CESIUM

Preliminary work carried out at our laboratory indicates that the chemical interaction between carbides and cesium is closely related to carbon content. If the carbon content is high enough to cause the separation of free graphite, the cesium will react with the graphite to form cesium-graphite lamellar compounds (e.g., CsC_8, CsC_{24}, CsC_{40}), in which the planes of carbon atoms alternate in a definite periodic sequence with planes of cesium (9,10). Thus, the

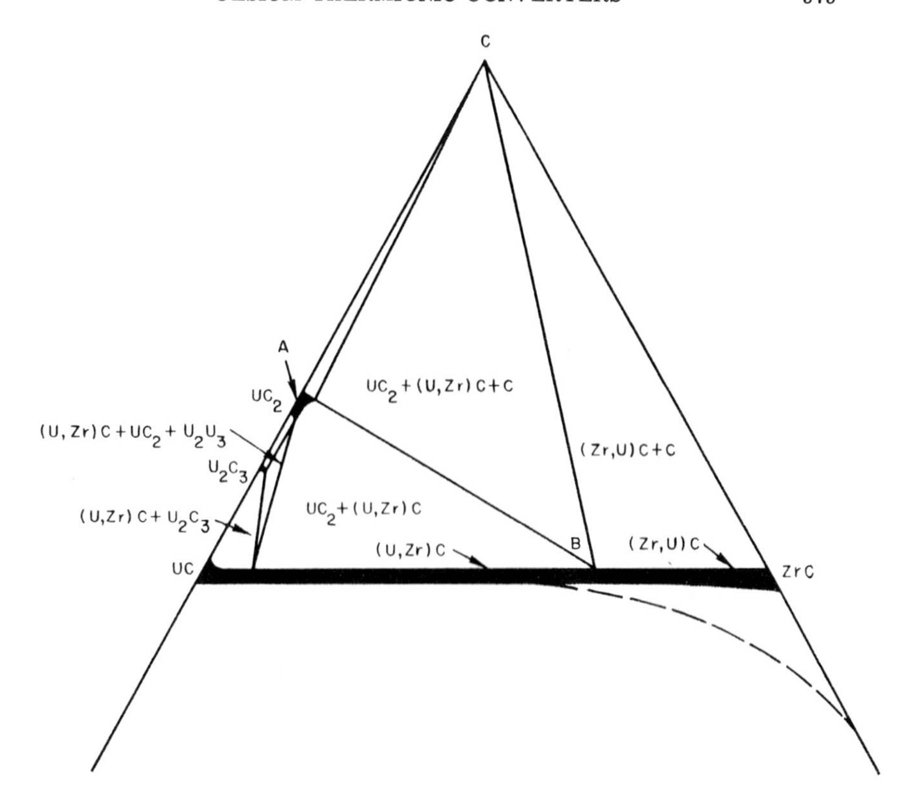

Fig. 7. The U-Zr-C ternary phase diagram at 1700°C (ref. 11).

interaction depends on the phase relationships among the components of the carbide as a function of temperature and composition. For binary systems such as U-C, Zr-C, Nb-C, Ta-C, and Hf-C, a carbon content higher than that of the higher carbide phases, UC_2, ZrC, NbC, TaC, and HfC is detrimental. For ternary systems such as U-Zr-C the situation is more complicated. Figure 7 shows the U-Zr-C ternary diagram at 1700 C given by Benesovsky and Rudy (11). It can be seen that for the pseudo-binary system UC-ZrC, with UC/ZrC $\leq 0.3/0.7$, any carbon content higher than that of the pseudo-binary will lead to the separation of graphite. On the other hand, for UC/ZrC > 0.43, precipitation of graphite occurs only when the carbon content reaches that given by points on the line AB in Figure 7. Figure 8 shows these critical carbon contents as a function of UC/ZrC ratio. At temperatures higher than 1700°C, point B in Figure 7 shifts slightly toward the UC side and the critical carbon content will be slightly lower than that shown in Figure 8.

Although corrosion will occur when graphite precipitates, the mode of corrosion will depend on how the precipitated graphite is distributed in the sample. Finely divided graphite precipitated at grain boundaries or in eutectic lamellae may cause the sample to crumble, while isolated graphite flakes in the grain may simply lead to pitting corrosion, which may be the case for NbC (12).

Figure 9 illustrates some examples of the corrosion of UC-ZrC by cesium. Figure 9a shows a stoichiometric UC-ZrC sample of UC/ZrC = 1.0 which has been in contact with liquid cesium at 250 C for two days and then kept in the capsule at room temperature for about one year. The sample is thoroughly

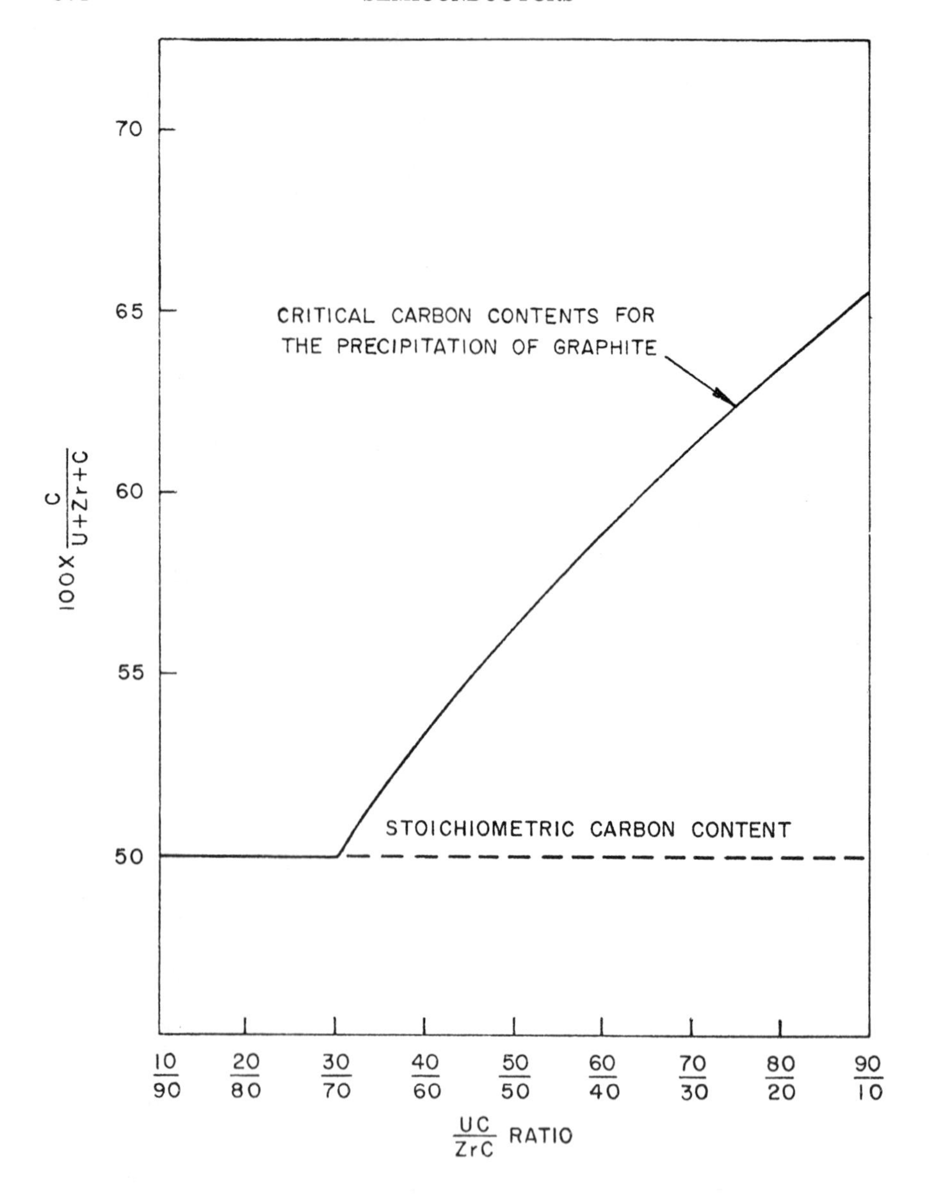

Fig. 8. Critical carbon contents for precipitation of graphite for various UC/ZrC ratios at 1700°C.

wetted by cesium but remains intact. Figure 9b shows another UC-ZrC sample of UC/ZrC = 1.0 but with C/(U + Zr + C) = 60 mole-%, which crumbled into powder after contact with liquid cesium at room temperature for only 5 min.

It must be pointed out that the alkali metal-graphite lamellar compounds are unstable at higher temperatures. For the potassium-graphite lamellar compounds, it has been shown (9) that for a potassium source temperature of 250°C, the graphite takes up practically no potassium if it is maintained at about 700°C. For the cesium-graphite lamellar compounds, only the decom-

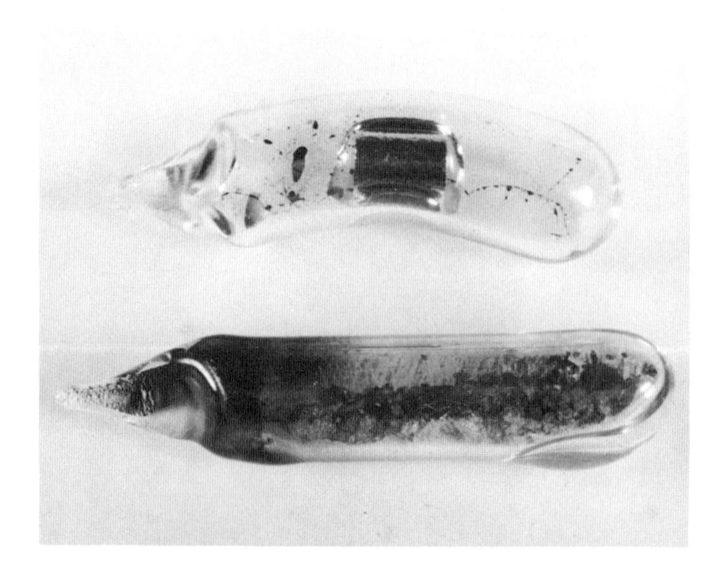

Fig. 9. Effect of carbon content on corrosion of UC-ZrC (UC/ZrC = 1.0) by cesium: top, 2 days at 250°C and 1 year at room temperature, C/(U+Zr+C) = 0.5; bottom, 5 min at room temperature, C/(U+Zr+C) = 0.6.

position pressures of the reaction $3CsC_8 \rightarrow CsC_{24} + 2Cs$ are available (9,10). Nevertheless, it is doubtful that the lamellar compounds could exist at the operating temperature of the carbide cathode (1600 to 2000°C). However, serious damage to the carbide cathode containing free graphite would occur when the cathode was thermal-cycled in the presence of cesium vapor. To avoid such difficulty, it is advisable to keep the carbon content below the critical value for graphite precipitation.

COMPATIBILITY BETWEEN CARBIDES AND REFRACTORY-METAL LEAD

Interdiffusion between the carbide cathode and the refractory-metal lead (e.g., tungsten, tantalum, molybdenum, and niobium) should not lead to the formation of voids or low melting phases. There have been very few systematic diffusion (13) and phase relationship (11) studies for the carbide-refractory metal systems. Creagh and Drell (14) carried out some 1/2 hr compatibility tests between UC and the refractory metals, the temperatures for liquid phases to appear being as follows: tungsten, 2425°C; tantalum, 2480°C; molybdenum, 1980°C; niobium, 2260°C. Results obtained at Los Alamos (15) for $(UC)_{30}(ZrC)_{70}$ are 2400°C for tungsten and 2100°C for tantalum. Figure 10 shows the appearance of a molybdenum crucible containing a $(UC)_{90}(ZrC)_{10}$ pellet, which has been heated in vacuum for 2 hr at 1900°C. Severe corrosion of the molybdenum and melting at the contact between the carbide and the crucible are evident, although the melting point of the carbide (~2700 C) is much higher than 1900°C.

It should be pointed out that the requirement of compatibility between the carbide cathode and the refractory-metal lead should be regarded as a condition for determining the operating temperature at the junction rather than for

Fig. 10. Corrosion of molybdenum crucible by $(UC)_{90}(ZrC)_{10}$ at 1900°C for 2 hr.

the selection of carbides for the cathode. By judicious engineering design, it should be possible to bring the junction temperature to the tolerable value.

JOULE AND THERMAL CONDUCTION LOSSES

The carbides of uranium and the refractory metals are fairly good electrical conductors. Their electrical resistivities at room temperature vary from 30 to 150 μohm-cm (16-18) as compared with 1.55 μohm-cm for copper. For the same carbide, the electrical resistivity changes with stoichiometry. For instance, the reported electrical resistivities of UC at 20°C vary between 33 and 99 μohm-cm (19), probably because of the difference in the carbon content of the samples. The electrical resistivity of the ZrC phase at room temperature changes from 300 μohm-cm at 3.50 wt-% carbon to 50 μohm-cm at 11.62 wt-% carbon (20). There have been only a few measurements at higher temperatures (21). The data indicated that in general the electrical resistivity at 2000°C is about three to four times higher than that at room temperature.

Most of the thermal conductivities of carbides have been measured at room temperatures. The data fall in the range 0.03 to 0.08 cal/cm-sec-°C (refs. 16, 17, 19, and 22-25), as compared with 0.9 cal/cm-sec-°C for copper. The thermal conductivity of UC has been studied up to 700°C, although its variation with temperature is a matter of controversy (23, 24).

The Joule and thermal conduction losses can be minimized by judicious engineering design. Owing to the lack of drastic difference among the electrical and thermal conductivities of the carbides of thermionic interest, the requirement is of secondary importance as a criterion in the selection of carbide cathodes.

THERMAL RADIATION LOSS

The thermal radiation loss from cathode to anode is determined by the cathode temperature and the emissivity factor, ϵ, of the cathode-anode system. Thus ϵ is related to both the total thermal emissivity of the cathode, ϵ_1, and the total thermal emissivity of the anode, ϵ_2. The spectral emissivity at 0.65 μ has been reported (7) to be 0.7 for ZrC, 0.9 for UC, and 0.8 for $(UC)_{20}$ $(ZrC)_{80}$, and the total emissivity was found to be 0.5 to 0.6 for $(UC)_{30}$ $(ZrC)_{70}$ in the temperature range from 1000 to 2800°K (26). Obviously the thermal emissivity varies not only with the chemical composition but also with the physical conditions of the cathode surface. Owing to the condensation of cathode materials on the anode surface, ϵ_2 is almost that of a blackbody, i.e., $\epsilon_2 \cong 1$. Like electrical and thermal conductivities, thermal emissivity is not a major determining factor in the selection of carbide cathodes.

RADIATION EFFECTS AND FISSION-PRODUCT RELEASE

For fission heat conversion, the carbide cathode is usually a UC-bearing polycarbide system which also serves as the reactor fuel. To select the suitable carbide system for such purposes, it is necessary to know whether radiation causes appreciable change in the dimensions of the cathode, how the fission products affect the performance of the cathode, and how fast each fission product diffuses out of the cathode at its operating temperatures into the cesium plasma. To date, irradiation tests have been carried out only on UC at relatively low temperatures (average surface temperature, 700 to 916°F; average core temperature, 1150 to 1376°F) (27). There is no information available on the effect of fission products on the performance of carbide cathodes; all the in-pile tests made so far have been for less than a few hundred hours and have failed because of condensation (28), cracking of the cathode, or other reasons which investigators have not been able to connect directly to fission-product buildup. The only fission-product release experiment (29) was conducted out-of-pile by heating an irradiated UC-ZrC cathode for 3 hr at 2400°C. Such information is badly needed for the selection of carbide cathodes for fission heat conversion and can only be obtained by making long-term in-pile tests.

CONCLUSION

In summary, out-of-pile selection of carbide cathodes should be made primarily on the basis of their vaporization properties and electron-emission characteristics in vacuum. This should be followed by studies of the life and the performance of these carbide cathodes in the presence of cesium and, for fission heat conversion, by in-pile tests. On the basis of the information available to date, the UC-ZrC system is still considered the most promising carbide cathode material for use in a cesium thermionic cell. Addition of small amounts of ZrC (~10 mole-%) to UC improves the mechanical properties and chemical stability and lowers the vaporization loss in vacuum but does not significantly alter the electron-emission property. The UC-ZrC cathodes of low UC content (~ 10 to 15 mole-% UC) should be able to operate for one year in vacuum at temperatures where useful electron emission (~6 amp/cm^2) can be obtained, with a loss of material equivalent to only a few mils in thickness.

The performance of these cathode materials in the presence of cesium and radiation must be determined by further tests. For UC-ZrC of high UC content (e.g., 90 mole-% UC), thermal aging in vacuum at high temperatures lowers its surface uranium concentration, and thus its vaporization loss in vacuum, while the high uranium concentration remains intact inside the sample. This is of special interest in cases where a high uranium concentration is needed in the bulk to provide a compact reactor core and a low uranium concentration is preferred on the surface to reduce the vaporization loss, such as in a thermionic reactor for space application. The data available are still insufficient for predicting the life and the performance of such high-UC-bearing cathodes. Studies of the vaporization properties and the electron-emission characteristics of UC-ZrC cathodes of various UC/ZrC ratios during and after thermal aging at different temperatures are being pursued in our laboratories for this purpose.

ACKNOWLEDGMENTS

This work supported in part by the Air Force Systems Command, U. S. Air Force under Contract AF 33(616)-7422; and in part by the National Aeronautics and Space Administration under Contract NAS 5-1253.

References

1. Yang, L., Ling, and F. D. Carpenter, J. Electrochem. Soc., 108, 1079 (1961).

2. Pidd, R. W., G. M. Grover, E. W. Salmi, D. J. Roehling, and G. F. Erickson, J. Appl. Phys., 30, 1861 (1959).

3. Carpenter, F. D., J. Dunlay, R. Howard, and L. Yang, "High Temperature, Vapor Filled Thermionic Converter," Annual Report for Contract AF 33(616) - 7422, August 31, 1961.

4. Weinberg, A. F., R. G. Hudson, B. Siefner, and L. Yang, "Investigations of Carbides as Cathodes for Thermionic Space Reactors," Quarterly Report to be submitted under NASA Contract NAS 5-1253.

5. Haas, G. A., and J. T. Jensen, Jr., J. Appl. Phys., 31, 1231 (1961).

6. Goldwater, D. L., and R. E. Haddad, J. Appl. Phys., 23, 70 (1952).

7. Pidd, R. W., G. M. Grover, D. J. Roehling, E. W. Salmi, J. D. Farr, N. H. Krikorian, and W. G. Witteman, J. Appl. Phys., 30, 1575 (1959).

8. Pollock, B. D., J. Phys. Chem., 65, 731 (1961).

9. Herold, A., Bull. soc. chim. France, 999 (1955).

10. Herold, A., Compt. rend., 232, 838, 1484 (1951).

11. Benesovsky, F., and E. Rudy, Planseeber. Pulvermet., 9, 65 (1961).

12. Wagner, P., and S. R. Coriell, Rev. Sci. Instr., 30, 937 (1959).

13. Elkins, P. E., Atomics International, Report NAA-SR-MEMO-5921.

14. Creagh, J. W. R., and I. L. Drell, in "Progress in Carbide Fuels," Notes from the Second AEC Uranium Carbide Meeting, Battelle Memorial Institute, Rept. TID-7589, April 20, 1960.

15. Glasstone, S. (comp.), "Quarterly Status Report of the LASL Plasma Thermocouple Development Program for Period Ending December 20, 1959," Los Alamos Scientific Laboratory Rept. LAMS-2396, Jan., 1960, p. 19.

16. Litz, L. M., "Asilomar Conference on High Temperature Technology," 1959, p. 134.

17. Schwartzkopf, P., and R. Kieffer, Refractory Hard Metals, MacMillan, New York, 1953.

18. Campbell, I. E., ed., High Temperature Technology, Wiley, New York, 1956.

19. Koenig, N. R., and B. A. Webb, "Properties of Ceramic and Cermet Fuels for Sodium Graphite Reactors," Atomics International Rept. NAA-SR-3880, June 30, 1960, p. 8.

20. Samsonov, G. V., and V. P. Latisheva, Fiz. Metal. i Metalloved, Akad. Nauk., S.S.S.R. Ural Filial, 2, 309(1956); translation, Rept. AEC-tr-3321.

21. Kolomoets, N. V., V. S. Neshpor, G. V. Samsonov, and S. A. Semenkovich, Zhur. Tekh. Fiz., 28, 2382 (1958).

22. Smith, C. A., and F. A. Rough, Nuclear Sci. and Eng., 6, 391 (1959).

23. Secrest, A. C., Jr., E. L. Foster, and R. F. Dickerson, "Preparation and Properties of Uranium Monocarbide Castings, Battelle Memorial Institute, Rept. BMI-1309, Jan. 2, 1959.

24. Boettcher, A., and G. Schneider, "Some Properties of Uranium Monocarbide," Proc. Intern. Conf. Peaceful Uses tomic Energy, Geneva, 1958, 6, 561 (1958).

25. Smith, C. A., and F. A. Rough, "Properties of Uranium Monocarbide," Atomics International Rept. NAA-SR-3625, June 1, 1959.

26. "Quarterly Status Report of the LASL Plasma Thermocouple Development Program for Period Ending December 20, 1960," Los Alamos Scientific Laboratory Rept. LAMS-2504, Jan. 31, 1961, p. 13.

27. Rough, F. A., and W. Chubb, eds., "An Evaluation of Data on Nuclear Carbides, Battelle Memorial Institute Rept. BMI-1441, May 31, 1960.

28. Howard, R. C., L. Yang, H. L. Garvin, and F. D. Carpenter, American Rocket Society Conference on Space Power Systems, Preprint 1287-60, Sept., 1960.

29. Glasstone, S. (comp.), "Quarterly Status Report of LASL Plasma Thermocouple Development Program for Period Ending December 20, 1959," Los Alamos Scientific Laboratory Rept. LAMS-2396, Jan. 1960, p. 9.

30. "Quarterly Status Report of the LASL Plasma Thermocouple Development Program for Period Ending March 20, 1961," Los Alamos Scientific Laboratory Rept. LAMS-2544, p. 10.

31. Hoch, M., P. E. Blackburn, D. P. Dingledy, and H. L. Johnston, J. Phys. Chem., 59, 97 (1955).

32. Bowman, M. G., "Chemistry of Fuel Element Cathode Materials," presented at American Rocket Society Space Power Systems Conference, Sept. 1960.

33. Glasstone, S. (comp.), "Quarterly Status Report of the LASL Plasma Thermocouple Development Program for Period Ending September 20, 1959," Los Alamos Scientific Laboratory Rept. LAMS-2364, Oct. 30, 1959, p. 14.

DISCUSSION

A. J. ROSENBERG (Tyco, Inc.): What happens to the work function as the composition changes with weight loss in the uranium-carbide rich phase?

L. YANG: We do not have data yet to answer your question in a quantitative way. However, we do know that the electron emission properties of the UC-ZrC system are insensitive to UC content if the latter is above a certain value. For instance it has been reported that $(UC)_{1.5}$ $(ZrC)_{98.5}$ emits almost like ZrC (ref. 29 of original paper), and $(UC)_{20}$ $(ZrC)_{80}$ emits almost like UC (ref. 7). Therefore it seems likely that the electron emission of the "thermally treated" samples will not be seriously affected until their surface uranium concentrations have been depleted to very low values by vaporization. This is being carried out in our laboratory.

K. A. SENSE (Astropower, Inc.): Do you know anything about the constant sublimation composition? Also, is anything known about the type of species that vaporizes from the surfaces?

L. YANG: There is no information available on the constant sublimation compositions of the UC-ZrC solid solutions. There are indications that for the zirconium carbide phase (ref. 8) and the uranium carbide phase (ref. 32), the constant sublimation compositions are close to ZrC and $UC_{1.1}$, respectively. We do know, however, that for the $(UC)_{15}$ $(ZrC)_{85}$ and the $(UC)_{90}$ $(ZrC)_{10}$ samples which we studied, the condensates are considerably richer in uranium than the samples themselves. There are no high-temperature mass spectrometer data on the vaporization species of the UC-ZrC solid solutions. For aluminum, beryllium, gallium, lanthanum, and titanium carbides in the temperature range 1600 to 2600°K [J. Phys. Chem., 62, 611(1958)] it has been shown that the vapor contains only traces of carbide molecule. For ZrC (ref. 8), experimental results also indicate the presence of mainly zirconium and carbon atoms in the vapor.

W. A. TILLER (Westinghouse Electric Corp.): Is the rate of evaporation found to be a function of the material, method of preparation, or the state of annealing?

L. YANG: The rate of vaporization of a new UC-ZrC sample is always much higher during the first 10 hr of heating. We believe that this is mainly due to the removal of impurities in the sample or the initial high uranium content at the surface. So if there is any annealing effect on the rate of vaporization at this stage, it is probably masked by other effects. After this initial outgassing stage, the rate of vaporization for the $(UC)_{15}$ $(ZrC)_{85}$ sample is essentially constant. For the $(UC)_{90}$ $(ZrC)_{10}$ sample, the continuous change of the rate of vaporization with time of heating is attributed to the change of the surface uranium concentration as indicated by counting. Our samples are prepared either by arc melting or by hot pressing and have densities better than 98% of their theoretical values. For less dense samples, it is highly possible that the replenishment of the constituents lost at the surface by evaporation could proceed by means of pore diffusion instead of volume diffusion and the rate of vaporization may vary with the density of the sample. As we have not studied the vaporization characteristics of low density samples, we are unable to offer experimental data to substantiate this.

W. A. TILLER: The point I am really getting at is that depending on your mode of deformation or annealing, the point defect concentration can be an order of magnitude higher than normal and will affect diffusion transport to the surface, which at short times presumably will be very rapid. At long times you return to the point defect concentration of the annealed material. I wonder if you are seeing this kind of effect.

L. YANG: No, there is no experimental evidence to relate the observed change in the rate of vaporization with the change in defect concentration with annealing.

J. F. ENGELBERGER (Consolidated Controls Corp.): Do you have any information on the effect of the cesium on the emission characteristics of these carbides.

L. YANG: On the basis of the early Los Alamos data (2), cesium does not seem to have any significant effect on the electron emission characteristics of these carbides.

Evaluation of Metal Emitters for Thermionic Converters

L. K. HANSEN and N. S. RASOR

Atomics International, Canoga Park, California

Abstract

Factors affecting the operation of emitters in thermionic energy converters are considered. The advantages of cesium coverage of a high work function emitter are discussed in some detail.

INTRODUCTION

The present interest in thermionic energy conversion is stimulating increased activity in the already active field of high temperature materials. Of the various requirements placed on materials in thermionic converters, the most difficult to satisfy are those placed on the converter emitter. There are two important criteria used to evaluate any prospective emitter material. First, it must allow conversion of heat to electricity with high efficiency. In general, the conversion efficiency is determined by the work function and thermal emissivity of the emitter material and increases with the temperature at which the emitter is operated. Secondly, the emitter material must allow a long converter lifetime. The lifetime is determined by such emitter properties as free energy of vaporization, chemical stability, and dimensional stability. In contrast to efficiency, the lifetime decreases with increasing emitter temperature. The interrelation of these parameters will be discussed to define the materials suitable for converter emitters.

EMITTER TYPES

Great interest has been shown in both the refractory high work function metals and the refractory low work function compounds, principally carbides. In both cases, these materials are used in an atmosphere of cesium vapor. This atmosphere can provide, space-charge neutralization for a diode converter since cesium atoms incident on the emitter surface are first adsorbed* and then re-emitted, some as ions. However, there is a fundamental difference in the operation of low work function and high work function emitters (see Fig. 1). With low work function materials (e.g., the carbides mentioned in Fig. 1) the lifetime of the adsorbed cesium is not sufficient for their surfaces to develop significant steady state cesium coverage. Therefore emitters of these materials are called "elementary" or "bare" emitters. With the high work function

*An extensive bibliography of surface adsorption phenomena is contained in W. B. Nottingham, Bibliography on Physical Electronics, Addison-Wesley, 1954. Two relatively recent reviews of this field are: J. A. Becker, Advances in Catalysis, 7, 135 (1955), and J. H. DeBoer, Advances in Catalysis, 8, 18 (1956).

EMITTER TYPES			TYPICAL MATERIALS
ELEMENTARY (BARE)		METAL	
		COMPOUND	ZrC, UC
ADSORBED FILM	REFLUXED	METAL	Cs on { Ta, Mo, W Re, or Ir
		COMPOUND	
	DISPENSER	PORE DIFFUSION	PHILIPS CATHODES (Ba -W)
		SOLID DIFFUSION	Th -W

Fig. 1. Emitter types and typical materials of current interest.

materials however (e.g., the refractory metals mentioned in Fig. 1), the lifetime of the adsorbed cesium can be sufficiently long to develop a comparatively large fractional coverage of cesium. Emitters of these materials operate with an adsorbed film which is maintained in equilibrium with the cesium vapor by a "refluxing" process. Since cesium tends to ionize while adsorbed on a high work function surface, the adsorbed ions appreciably lower the work function of the surface. The extent of the coverage and therefore the actual work function of such "adsorbed film" emitters depend not only upon the bare work function of the substrate material but also upon the emitter temperature and cesium pressure.

An adsorbed film can be maintained not only by vapor refluxing as described above but also by the dispensing of material from within the emitter. These "dispenser cathodes" (see Fig. 1), although not discussed in this volume have also been considered for thermionic converter emitters. They also have a low work function due to the adsorption of ions on a high work function substrate. In the case of Philips-type cathodes, barium for the adsorbed film is dispensed by diffusion through porous tungsten. With thoriated tungsten, on the other hand, thorium is dispensed by solid state diffusion. A cesium atmosphere for space-charge neutralization is also used with dispenser cathodes.

EVALUATION OF ELEMENTARY EMITTERS

The importance of an adsorbed film in obtaining a suitable converter emitter can be shown by a general evaluation of materials used without an adsorbed film. A figure of merit (1,2) which can be used to evaluate materials operated as elementary emitters, is the ratio of free energy of vaporization to work function, $\Delta F/\phi$. Since large electron currents and low emitter vaporization are required, it is desirable to have this figure of merit as large as possible. Table I shows the rate of emitter vaporization to be expected as a function of

TABLE I
$\Delta F/\phi$ and Vaporization Rates for Some More Promising Elementary Emitter Materials

Material	$\Delta F/\phi$	Vaporization rate, 10^{-3} in./yr
80-20 m/o ZrC-UC	1.4	0.2
	1.3	1.0
	1.2	6.0
UC	1.1	35
ZrC, Ta, W	1.0	200
	0.9	1100
Re	0.8	
	0.7	
Mo	0.6	

this figure of merit, for emitter operation at 10 amp/cm^2. Also listed in Table I is this figure of merit for several prospective emitter materials, evaluated at the temperature which would produce the above current density. From this we see that $\Delta F/\phi$ should be at least 1.2 and that none of the materials is entirely satisfactory. This is, the metals without adsorbed cesium are poor emitter materials and the carbides mentioned are only marginal.

Another figure of merit is the ratio of the heat of vaporization to work function, $\Delta H/\phi$. For a specific temperature and for materials which vaporize as a monomer, the difference between the heat of vaporization and free energy of vaporization is approximately constant. In this case, a value of $\Delta H/\phi$ greater than 2.1 is required for sufficient electron emission without excessive vaporization. The variation of $\Delta H/\phi$ with atomic number, Z, (3) is shown in Figure 2. Also included in the figure are evaluations for UC and ZrC. As can be seen, it is unlikely that any element can be operated satisfactorily as an elementary

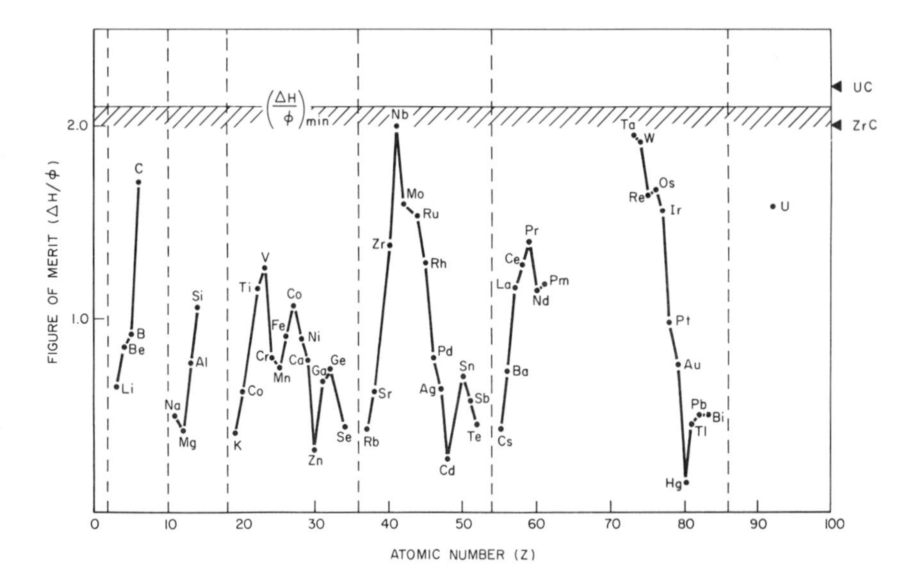

Fig. 2. Figure of merit ($\Delta H/\phi$) for the elements and two carbides.

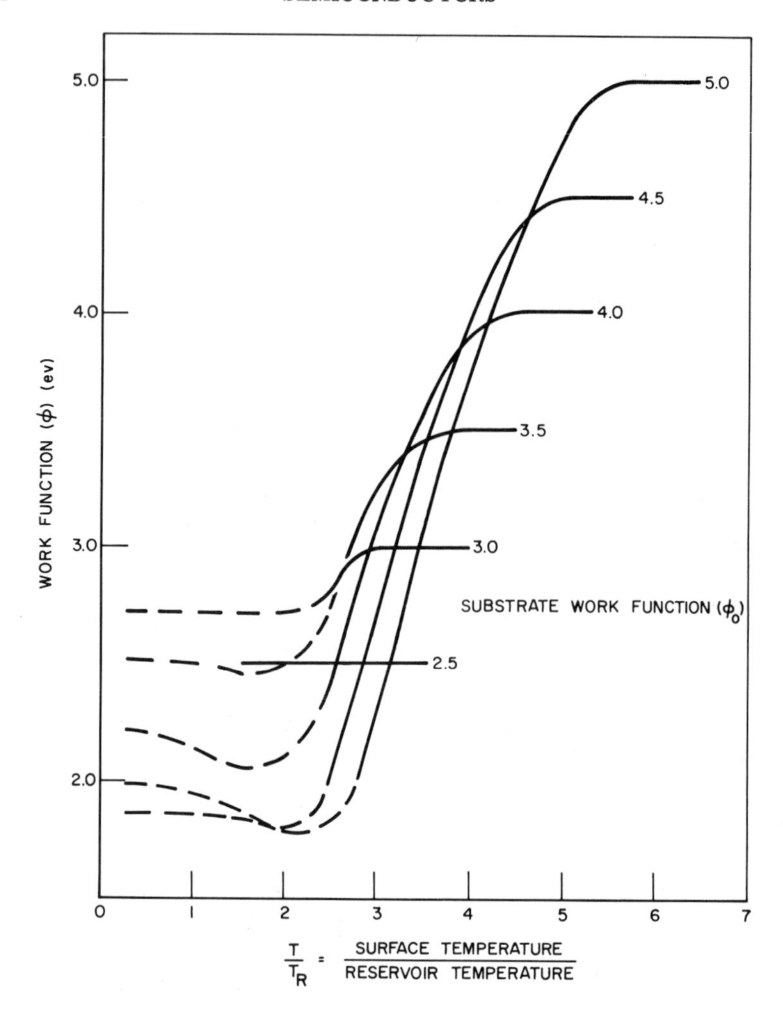

Fig. 3. Computed work function of surfaces with adsorbed cesium.

emitter. The prospects for finding suitable compounds, however, are more encouraging.

REGIONS OF CONVERTER PERFORMANCE

The effect of adsorbed cesium changes the above picture considerably. The actual operating work function of a converter emitter is then a function of the emitter temperature and cesium pressure (or equivalently, the cesium reservoir temperature*). However, knowing these parameters and the bare work function of the material, it is possible to determine the actual work function that results from the adsorbed cesium (4), (see Fig. 3). For a thermionic converter whose performance is limited only by the emitter electron and ther-

*The cesium reservoir temperature, T_{Cs} (°K), is related to the cesium vapor pressure by the following equation:

$$p = 2.45 \times 10^8 \, T_{Cs}^{-\frac{1}{2}} \exp \left\{ -8910/T_{Cs} \right\} \text{ mm of mercury}$$

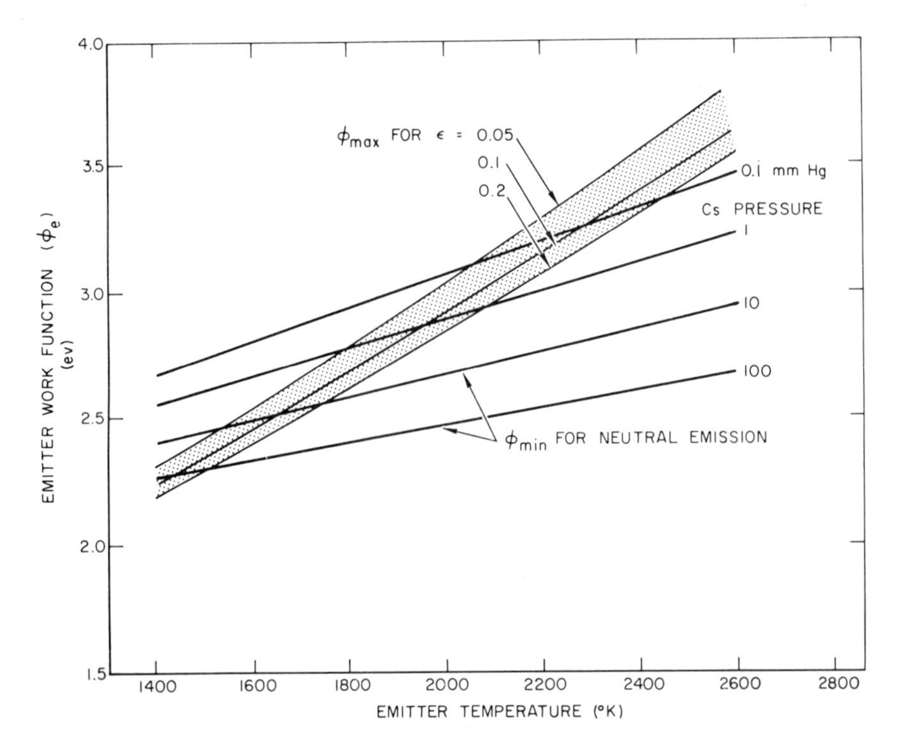

Fig. 4. Work function limits for neutralized, optimum, emitter currents.

mal emission characteristics, there is a maximum emitter work function which will allow maximum converter efficiency (2,5). This maximum efficiency depends upon the emitter temperature and upon the effective thermal emissivity of the electrode system.* The maximum work function for various emissivities is shown in Figure 4 within the shaded region. There is also a minimum work function allowing space-charge neutralization by ion emission from the emitter surface. This work function depends on the cesium pressure and emitter temperature. Figure 4 also shows this dependence.

Using the information given in Figures 3 and 4, the work function of the emitter in the presence of a cesium atmosphere can be eliminated as a variable. The regions of emission-limited converter performance for a given temperature can then be displayed as a function of the bare emitter work function and the cesium pressure. Such displays for several emitter temperatures and for an emissivity of 0.1 are shown in Figures 5a to 5d. In these figures, regions of adequate and inadequate ion and electron emission are defined by gradations in shading. In addition, the region of excessive pressure is indicated by cross-hatching. The boundary, particularly of the latter region, is not precisely defined. Contributing to this lack of definition is the present lack of understanding of the converter interelectrode processes. These processes become more important as the cesium pressure increases and will in general cause a degradation from the ultimate performance predicted for an

*Since part of the energy radiated by the emitter is reflected back by the collector, the effective emissivity of a plane parallel electrode system is given by $\epsilon = \frac{1}{(1/\epsilon_e) + (1/\epsilon_c) - 1}$ where ϵ_e is the emissivity of the emitter and ϵ_c is the appropriate emissivity of the collector.

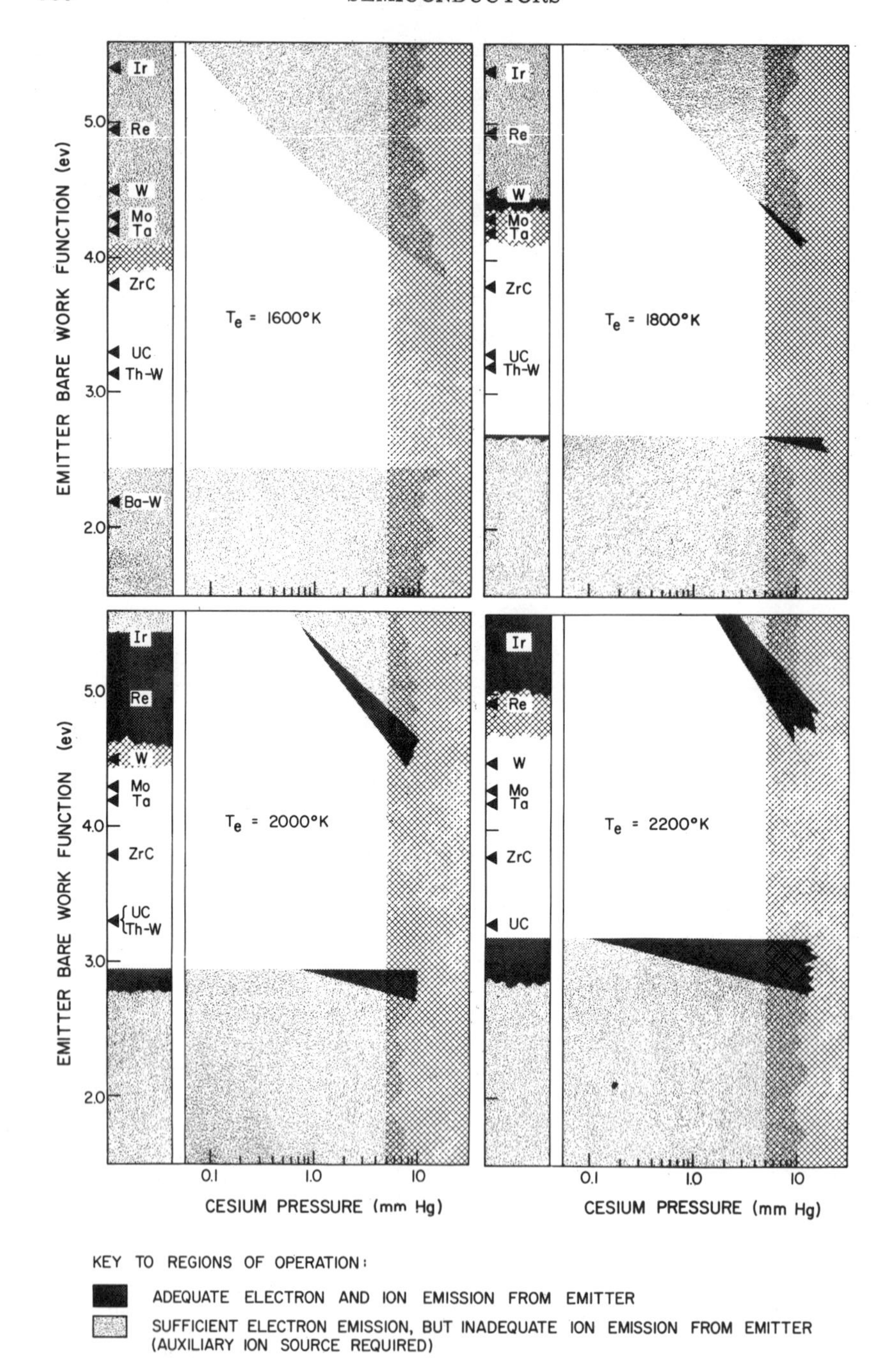

Fig. 5. Emission limitations on converter operations.

emission-limited converter. Another factor responsible for the lack of definition of the region of excessive pressure is the freedom for variation of the interelectrode spacing. To minimize the impedance losses accompanying high gas pressure, a narrow interelectrode spacing can be used. However, this approach for minimizing the effect is limited by the resulting increased heat conduction, the mechanical stability required of the cathode, and the extent to which volume ionization contributes to space-charge neutralization.

To the left of each figure is an index of some possible emitter materials with an indication, by shading, of the performance of these materials as inferred from the figures. In each figure, the upper region of acceptable performance represents operation with a reduced emitter work function due to cesium adsorption. The lower region of acceptable performance represents operation without changes in the emitter work function since significant cesium adsorption does not take place.

There are several important points that follow from Figure 5. Immediately evident is the fact that although many of the desired characteristics for emitter materials are quite specific, there are two choices as far as the work function is concerned. Either a material with a high bare work function or a material with a low bare work function can be used. Materials with bare work functions between 3 and 4 ev tend to be unsatisfactory for converter emitters operating at moderate temperatures (1600 to 2000°K) and pressures ($<$ 5 mm of mercury).

It also may be seen that, in the regions of acceptable performance, the appropriate bare work function is determined by the type of converter desired. In the upper region it is evident that without an auxiliary ion source, materials with bare work functions only slightly higher than 4 ev are required for operation at the lower temperatures. With an auxiliary ion source, however, materials with higher bare work functions ($>$5 ev) can operate at these lower temperatures with the advantage of low cesium pressure. It is only at higher temperatures that these high work function materials become useful for operation without an auxiliary source of ions. On the other hand, in the lower region, unless the work function is very low ($<$ 3 ev), efficient energy conversion is obtained only at excessively high temperatures (e.g., $>$2200°K in the case of uranium carbide).

GENERAL COMPARISON OF EMITTERS

The above delineation of the regions of converter performance provides a convenient framework for a general comparison of emitter materials. To maintain coherence in this comparison, those points which have already been discussed and some of those to follow are summarized in Figure 6. In this figure, areas of difficulty are shaded and cross-hatching is used where a qualified statement seems appropriate.

The figure of merit (column 1 of Fig. 6) for the elementary metal and compound emitters has already been discussed. This figure of merit could be generalized to include the adsorbed film emitters by taking into account the effect of the adsorbed film on the two parameters in the figure of merit ratio. From this point of view, the effect of the adsorbed film on the emitter work function improves the figure of merit considerably. Although the effect of the adsorbed cesium on electrode vaporization has not yet been determined, no difficulty is expected.

In the cross-hatched areas of columns 2 and 3 of Figure 6, attention is drawn to the correlation that exists between the free energy of vaporization and the work function (see Fig. 2.). No element has been found with both a very high heat of vaporization and a very low work function. Judging from

EMITTERS \ CRITERIA	1	2	3	4	5
	$\frac{\Delta F}{\phi}$	VAPORIZATION (LIFETIME)	EFFICIENCY OPTIMIZATION		
			ϕ_e	ϕ_c	ϵ
METAL (ELEMENTARY)	POOR	OPTIMUM PROPERTIES INCOMPATIBLE	NO CONTROL	DEPOSITS ON COLLECTOR	LOW
COMPOUND (ELEMENTARY)	MARGINAL	OPTIMUM PROPERTIES INCOMPATIBLE PREFERENTIAL VAPORIZATION	NO CONTROL	DEPOSITS ON COLLECTOR	HIGH
METAL (Cs REFLUXED)	GOOD	HIGH ΔF FOR SUBSTRATE EQUILIBRIUM FILM	ϕ_e GOOD (HIGH PRESSURE) CONTROL	GOOD	LOW
DISPENSER	GOOD	HIGH ΔF FOR SUBSTRATE DISPENSATE RUNS OUT	ϕ_e GOOD NO CONTROL	DEPOSITS ON COLLECTOR	LOW

Fig. 6. Performance comparison of some emitter types.

uranium carbide, this restriction is relaxed somewhat in the compounds. However, even if this particular case is representative of what can be expected from the compounds, the problem remains that for temperatures high enough to allow efficient energy conversion, significant vaporization will still occur. This correlation, therefore, is a fundamental obstacle in the search for satisfactory elementary emitters.

In the adsorbed film (refluxed and dispenser) emitters, however, a high bare work function for the substrate is desirable. This characteristic is compatible with the need for a high heat of vaporization. The high heat of vaporization refers only to the substrate, however. The adsorbed material vaporizes readily, but this is not a problem for refluxed emitters since the absorbed film is in equilibrium with the vapor. However, the dispenser cathodes have a lifetime limitation due to the eventual depletion of the dispensate. This same difficulty limits the use of Philips cathodes to relatively low temperatures.

In the vaporization (lifetime) column (column 2 of Fig. 6) attention is also drawn to the problem encountered with the compounds of preferential vaporization of particular species. This results in a change with time of the emission characteristics of the material. The actual operating conditions of such a material would depend not only upon the vaporization process but upon diffusion processes as well.

In columns 3 to 5 of Figure 6 a comparison of emitters in terms of efficiency is summarized. The significant parameters are the emitter work function, ϕ_e, and thermal emissivity, ϵ, and the collector work function, ϕ_c, (to the extent that it is affected by vaporized emitter material). As has already been discussed, a desirable work function is difficult to obtain with an elementary material without some compromise with the heat of vaporization. The adsorbed film emitters do not have this limitation. However, the refluxed metals can cause difficulty with impedance losses due to high cesium pressures. On the other hand, the refluxed metal emitter is the only emitter which has the advantage of easy variation of work function to allow constant optimization of performance as other parameters vary.

A low work function collector is essential to high efficiency. A consideration for all but the refluxed metal emitter is the change of the collector work function due to deposition of emitter materials on the collector. At cesium pressures and collector temperatures normal to converter operation, a high work function surface develops one of the lowest, stable work functions known. Therefore, the high work function materials used for refluxed metal emitters are compatible with optimum collector conditions.

Radiation loss is an important factor in limiting the efficiency of a converter, particularly at high temperatures. An emitter material with low thermal emissivity is therefore advantageous (e.g., see ref. 2 or 5). Generally, the thermal emissivities of the metals are low compared with the compounds. At typical emitter temperatures, clean metal surfaces usually have emissivities less than 0.4, whereas the emissivities of compounds may be as high as 0.8 or 0.9.

As may be seen in Figure 6, refluxed metal emitters appear to have fewer significant limitations than other types of emitters. In addition, metals are very attractive materials in some respects not considered in Figure 6. Generally the refractory metals are strong and have low creep rates at high temperatures. They are relatively ductile and easy to fabricate, and are readily available. The technology for their use in thermionic devices is already well understood. The importance of these and other factors for comparing emitter materials are strongly dependent on the particular design and application.

CONCLUSION

On the basis of the factors considered here, it appears that refractory metals employing adsorbed cesium are very attractive as thermionic emitters. The other types of emitters considered have serious limitations for normal converter operation. However, by choosing appropriately from among the known metals, most of the emitter requirements for the various regions of converter operation can be satisfied.

References

1. Rasor, N. S., "Review of Thermionic Conversion Research at Atomics International," NAA-SR-MEMO-5063 (March 1960).

2. Rasor, N. S., et al., "First Summary Report of Basic Research in Thermionic Energy Conversion Processes for Nonr-3192(00),) Paper B1, AI-6799 (Nov. 1961).

3. Feaster, G. R., et al., "An Investigation of Gas-Filled Thermionic Converters for Transforming Fission Heat Into Electricity," AD-251524 (Oct. 1960), Appendix F.

4. Rasor, N. S. et al., "First Summary Report of Basic Research in Thermionic Energy Conversion Processes for Nonr-3192(00)," Paper B2, AI-6799 (Nov. 1961); to be published.

5. Rasor, N. S., "Parametric Optimization of the Emission-Limited Thermionic Converter," in Progress in Astronautics and Rocketry, Vol. 3, Academic Press, New York, 1961.

DISCUSSION

D. E. KNAPP (Douglas Aircraft Co.): You emphasize the point that a high work function metal will give you greater cesium coverage for a given temperature and pressure. How did you calculate, then, the effect of the high cesium coverage on higher work function metals, or did you simply assume that this is the same as it would be for a lower work function metal?

L. K. HANSEN: The effect in reduction of work function for a given cesium coverage has a marked dependence on the bare work function of the substrate material (see Ref. 4 in text for details of the theory).

A. J. KENNEDY (Martin Co.): Is there experimental agreement for this prediction of work function as a function of emitter temperature and cesium pressure?

L. K. HANSEN: Until recently there has not been very much experimental data that could be compared with such a theory. However, now we have the data of John Houston of the General Electric Research Laboratory and of Lee Aamodt of the Los Alamos Scientific Laboratory. There is good agreement between the theory and the data which they have obtained.

Structural and Emitter Materials for Nuclear Thermionic Converters

L. N. GROSSMAN, A. I. KAZNOFF, and P. STEPHAS

General Electric Company, Vallecitos Atomic Laboratory
Pleasanton, California

Abstract

Material problems associated with a cesium-type nuclear-fueled thermionic converter are described, and several promising approaches to their solution discussed. Fuel-cladding compatibility is closely examined and likely combinations described.

INTRODUCTION

In developing a useful thermionic converter, several difficult problems concerning high temperature materials must be solved. Prior to designating the problem areas, it is desirable to describe the nuclear thermionic converter which gives rise to these problems.

Figure 1 illustrates the basic design features of a fuel rod thermionic converter of the cesium type. When installed in an operating power reactor, this converter rejects heat at a temperature suitable for other conversion schemes.

Typical operating temperatures of the nuclear thermionic converter are: cathode surface 2100°K, anode 1300°K. Localized transient temperature excursions, several hundred degrees higher and of short duration, will occur in the reactor. Materials capable of withstanding such excursions will be required. In addition, temperature and power distribution within the reactor core will require a design compatible with operation throughout an appreciable temperature range.

The main problems discussed in this paper are: (a) choice of fuel, (b) clad fuel versus self-cathodes (i.e., where the fuel is the cathode), and (c) materials compatibility.

Oxide-Fueled Cathodes

The high temperatures at which a nuclear thermionic cathode must operate will limit the choice of fuel-bearing compounds to the oxides, sulfides, and some intermetallic compounds of uranium. Of the possible choices, the best known fuel is uranium dioxide. Sufficient information on the thermodynamics of uranium dioxide is available to permit the formation of conceptual designs for nuclear thermionic fuel elements based on this material.

Two major problems associated with uranium oxide at high temperatures are: (1) UO_2 has a high decomposition pressure above 1700 C, and (2) UO_2 has low thermal and electrical conductivities.

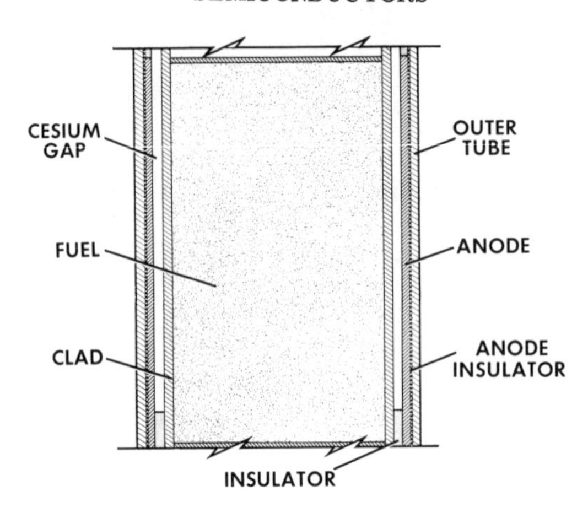

Fig. 1. Section of thermionic fuel element.

An oxide-bearing cathode therefore, must be clad in a material which will prevent decomposition of the oxide, and provide an electrical conduction path.

With the necessity for cladding, additional problems arise. They are: (3) UO_2 must be chemically compatible with its cladding at the operating temperature (2100°K); (4) at operating temperature, the cladding must be structurally and chemically stable; and (5) the clad becomes the emitter and therefore must have acceptable emission properties with the cesium vapor in order to maximize the cathode temperature.

In the following discussion, compatibility arguments are based mainly on available thermodynamic data. The conclusions should not be considered quantitative except where supported by experimental data. Thermodynamic data were extracted from refs. 1 through 11.

Vapor pressure and fabricability considerations limit the list of refractory metal to candidates to tungsten, molybdenum, tantalum, and niobium. At temperatures below 1000°C, uranium dioxide is stable with respect to reduction by any of these four metals. However, all four of them, as well as UO_2, form rather stable volatile oxides (4). Therefore, as the environmental temperature is increased, the entropy term of the free energy equation will favor reduction of the dioxide and the production of gaseous metal oxides.

In the case of niobium and tantalum, solid suboxides have been observed to form at temperatures near 1400°C. Although the ternary phase diagrams for tantalum-oxygen-uranium and niobium-oxygen-uranium are not available, a low-melting eutectic phase seems to exist in the system causing reaction to occur at temperatures too low for nuclear thermionic use (12). No similar behavior has been observed for either tungsten or molybdenum, and both retain appreciable creep resistance and strength above 1700°C (13).

The problem of reduction of the UO_2 fuel by tungsten or molybdenum requires further attention. Since the oxides formed are volatile, it is possible to limit the extent of this reaction by operating in a closed system. If the system (UO_2 + W or Mo) is vented (that is, an open system), then the reaction will be limited only by the kinetics associated with formation and volatilization of the gaseous oxides. The decomposition pressure of UO_2 makes the vented cathode impractical unless some other limiting step (such as diffusion of volatile products through a solid) can be introduced. Since no data are available on the diffusion of gaseous oxides through solids, the vented cathode concept should be discarded as a starting design for oxide-fueled cathodes.

Through the above process of elimination, the only nuclear thermionic fuel element, incorporating an oxide fuel, which appears immediately feasible is UO_2 completely sealed in molybdenum or tungsten. Tungsten is preferable to molybdenum in all respects except for ease of fabrication and capture cross section in the thermal range; for a fast reactor molybdenum and tungsten have comparable cross sections.

Although the unvented clad cathode is the only choice for an oxide-fueled cathode on the basis of the above arguments, the following potential problems are associated with it: (a) fission product gas buildup may rupture the cladding, and (b) volatile oxide products produced by the fuel-cladding reaction may be diffused through the cladding.

The problem of fission-gas release is the biggest unknown associated with the long-life operation of an oxide-fueled cathode. The capability of noble gases to permeate tungsten or molybdenum at elevated temperatures is unknown. The concentration gradient necessary to diffuse substantial quantities of gaseous fission products through the cladding requires gas solubilities on the order of 1 mole-% in the cladding metal. This high figure indicates that no appreciable quantity of gas will be diffused.

Carbide-Fueled Cathodes

Carbide fuels have desirable properties for thermionic applications and are being investigated. The high thermal and electrical conductivities of uranium carbide are very desirable properties in this application. Uranium carbide alloys (UC-ZrC) have a demonstrated limited capability in thermionic converters (14). Failure in cathodes based on UC solid solutions has apparently been due to decomposition of the fuel at elevated temperatures. Whether or not this is an intrinsic quality of a UC-fueled self-cathode (unclad) can be determined from an estimate of the partial pressure of uranium in equilibrium with the carbide at elevated temperatures.

Uranium Pressure over Uranium Carbide

The theoretical uranium pressure in equilibrium with UC can be estimated on the basis of available thermodynamic data. Some pertinent reactions, together with the free energy change associated with each reaction, are presented in Table I. The free energy change has been made linear for the temperature range 1600—2200°K.

The vaporization of most metal carbides takes place by dissociation to the elements (18). In the temperature range of interest, the $C_1(g)$ molecule domi-

TABLE I

Free Energy Change for U-C System

Reaction	Linearized free energy change, cal./mole	Ref. No.
$UC(s) = UC(g)$	$\Delta F_T^\circ = 123{,}700 - 32.1T$	15
$UC(s) = U(s) + C(s)$	$\Delta F_T^\circ = 21{,}600 - 3.7T$	16
$C(s) = C_1(g)$	$\Delta F_T^\circ = 170{,}340 - 37.3T$	17
$C(s) = C_2(g)$	$\Delta F_T^\circ = 193{,}700 - 44.4T$	17
$C(s) = C_3(g)$	$\Delta F_T^\circ = 184{,}000 - 45T$	17
$UC_2(s) = U(s) + 2C(s)$	$\Delta F_T^\circ = 36{,}000 - 3T$	16

nates the carbon species (17). If UC is assumed to dissociate by the reaction.

$$UC(s) = U(g) + C_1(g) \quad (1)$$

the resultant carbon pressure exceeds that which would be in equilibrium with solid graphite. The reaction

$$(1 + x)\ UC(s) = xU(g) + UC_{1+x}(s) \quad (2)$$

is consistent with the vapor pressure of carbon and with the tentative U-C phase diagrams available in the literature (16,19). It is therefore reasonable to assume that the decomposition of UC takes place as indicated by reaction (2).

An upper limit on the uranium pressure above UC is set by the vapor pressure of uranium. A more exact upper limit may be obtained from the phase diagram by taking into account the carbon dissolved in the liquid at any temperature. The curve labled "C-saturated U→U(g)" in Figure 2 should represent this upper limit fairly well.

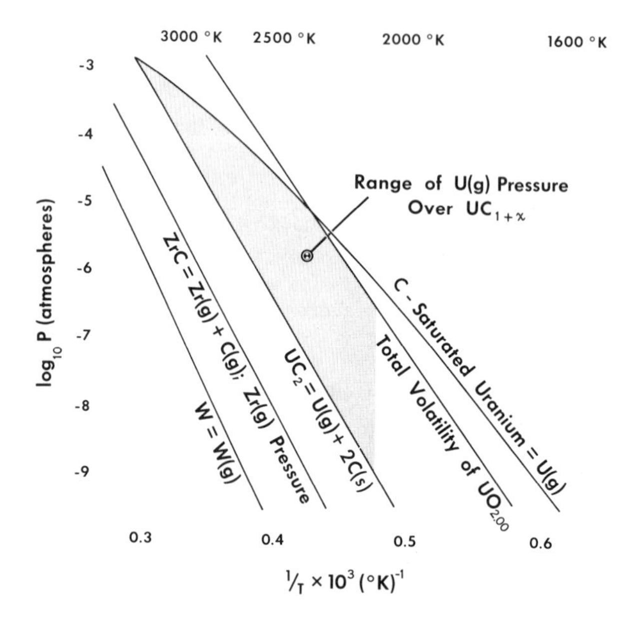

Fig. 2. Decomposition pressures of UC_{1+x}, UO_2, ZrC, and tungsten.

The lower limit on the uranium pressure above UC is given by the reaction

$$UC_2 = U(g) + 2C(s) \quad (3)$$

The lower limit may be further refined by assuming that UC and UC_2 form a complete solid solution above 2200°K even though the extent of the solid solution range is still in doubt. Under these conditions the pressure will vary more or less linearly with x, in the equation UC_{1+x}, between the carbon-saturated uranium value (assumed to be UC for simplicity) and the UC_2 value.

The resulting range of values for the partial pressure of U(g) over UC_{1+x} is shaded in Figure 2 together with the total volatility of UO_2 as measured by DeMaria et al. (20) and with the vapor pressure of tungsten. The accuracy of the data in Table I is probably poor for the carbide reactions. Hence the decomposition curve for UC_2 in Figure 1 may be incorrect by about $\pm 10\%$ uncertainty in log P.

To operate a converter for greater than a year, the decomposition pressure of any of its components must be kept below approximately 10^{-9} atm. Under this criterion, the upper temperature limit on a stoichiometric UC cathode is about 1900°K. A similar limitation would apply to a clad UC (or UO_2) cathode which is vented to release fission gases. For fuel temperatures above approximately 1900°K, mass transfer of the fuel would limit the life of the converter.

The decomposition pressure of the carbide fuel can be decidedly lowered by two approaches: (1) use a carbon-rich fuel such as UC_2 or (2) use a solid solution carbide such as (U, Zr)C which is rich in the more stable carbide.

Clad Carbide Cathodes

In addition to the fuel decompostion problem which arises with self-cathodes, compatibility problems exist associated with bonding the carbide to a metal substrate. In general the requirements on the cladding are the same as those described in connection with oxide fuels.

The discussion that follows is centered around UC and (U, Zr)C type fuels in metal clads. The information presented can be readily applied to bonding problems in unclad cathodes. Two cladding possibilities will be discussed: (1) metal cladding on a carbide fuel alloy and (2) metal cladding on a carbide fuel alloy with an interfacial diffusion barrier.

Before discussing the thermodynamic stability of the metal-carbide systems, it is essential to bracket the physical regime under consideration. The general requirements for a metal clad for a carbide fuel are identical to those described above for oxide fuels. The conditions are somewhat arbitrary, especially with respect to temperature; it is acknowledged that the lower temperatures will decrease the severity of many problems.

General Considerations for Compatibility

In the case of metal cladding the carbide fuel, the most desirable situation would be one where no reaction occurred. If the reaction does occur, it is desirable that the MC phase formed would be of a different structure than UC or have a much different lattice parameter so that intersolubility would be limited and diffusion rates low. It is also possible that ternary carbides may form at the reaction interface. Such compounds have been observed in several instances, but their stability is unknown. Further, the eutectic temperature for the carbides should be at least 2400°K to minimize the chances for local melting.

In the case of diffusion barriers, similar conditions must be met. In this paper the attention is limited to carbide barriers. The barrier must satisfy the above conditions on both its fuel and clad sides.

These requirements are guide lines and one would hardly expect any real material to satisfy all the requirements completely. In fact, the paucity of data related to lifetimes at high temperature precludes firm recommendations in almost all cases.

Even where compatibility data are available, it behooves the investigator to be very cautious. The experimental results of compatibility investigations are frequently difficult to interpret. They require precise metallographic and x-ray analysis, preferably at temperature. The experimental system has to be well defined: a poor vacuum with UC can be worse than an inert atmosphere. Pickup of materials from furnace heating elements and walls may also mislead the investigator. Reaction rates can be seriously influenced by type of surface contact.

Thermodynamics of Binary Carbide Systems

There are several sources for the thermodynamic data on binary carbides. Brewer (21), Krikorian (22), Latimer (23), Kubaschewski and Evans (24), and Rough and Chubb (25) have presented data on the free energies and entropies of formulation. Brewer's data are largely out of date and are superseded by Krikorian's revision. Latimer's review (23) relies largely on National Bureau of Standards (NBS) values (26) and is generally a noncritical compilation for condensed phases. Uncertainties in the thermodynamic values are not indicated. Kubaschewski and Evans have relied on NBS values and the evaluation of original data to tabulate values with indicated uncertainties. The

TABLE II
Thermodynamic Data for Metal Carbides[a]

	MP °K[b]	ΔF_{298}, kcal[b]	ΔH_{298}, kcal[b]	ΔS_{298}, eu	S_{298}, eu
Be_2C	2670	-12.4 ± 5	-13 ± 5	-2.0 ± 1.0	
B_4C	2720	-13.7 ± 0.7	-13.8 ± 2.7	-0.49 ± 0.12	
1/3 Al_4C_3	3100	-11.4 ± 3	-12 ± 3		
			15.6 ± 3		10.4 ± 0.8
TiC	3520	-42.98 ± 0.4	-43.85 ± 0.4	-2.90 ± 0.07	
			-43.9 ± 1.5		5.8 ± 0.2
ZrC	3810	-43.6 ± 1.1	-44.4 ± 1.1	-2.70 ± 1.0	
			(-48.2)		8.5 ± 1.5
VC	3100	(-14.5 ± 8)	(-15 ± 8)	-1.60 ± 0.11	
			(-12.5)		6.77 ± 0.1
V_2C	(2700)	-18.4 ± 6	-19 ± 6	-2.0 ± 1.0	
NbC	3770	-33.3 ± 0.7	-33.7 ± 0.6	-1.3 ± 1.0	
			-33.7 ± 1.5		8.9 ± 0.7
Nb_2C	3540	(-37.7 ± 8)	(-40 ± 8)	-1.0 ± 1.5	
TaC	4150	-38.1 ± 0.6	-38.5 ± 0.6	-1.19 ± 0.09	
			-38.5 ± 2.5		10.1 ± 0.2
Ta_2C	3670	-44.9 ± 6	-45 ± 6	-0.4 ± 1.0	
MoC	2970	- 2.0 ± 0.7	- 2.0 ± 0.7	-2.0 ± 0.7	0.0 ± 1.0
Mo_2C	2960	- 5.1 ± 1.5	- 4.5 ± 1.5	2.1 ± 1.0	
			-42 ± 5		(19.8)
WC	2870	- 8.6 ± 0.4	- 8.9 ± 0.2	0.6 ± 1.0	
			- 9.1 ± (2.5)	0.4	8.5 ± 1.5
W_2C	3023	(-11.7 ± 4)	(-11 ± 4)	2.5 ± 1.5	
1/2 UC_2	2773	-13.9 ± 4.0	-13.5 ± 4.0	1.4 ± 1.5	
			-19.4		
1/3 U_2C_3	2673	-23.2 ± 3.5	-23.3 ± 3.5	-0.3 ± 1.0	
UC	2663	(-27.6 ± 6)	(-28 ± 6)	-1.5 ± 1.5	

[a] Data taken from refs. 22 and 24.
[b] Parentheses indicate undetermined uncertainties.

values reported by Kubaschewski are more up to date and realistic than the other compilations with the exception of the Battelle Memorial Institute (BMI) (25) report on uranium carbides.

The pertinent thermodynamic values are given in Table II. Some of the values are plotted as a function of temperature in Figure 3. Because of the considerable uncertainties in the enthalpy, and particularly the entropy of formation of Mo_2C, the data from various sources are plotted in Figure 4. The temperature behavior of ΔF°_T for MoC, UC, and UC_2 are shown on the same plot.

The following is a brief explanation of the curves shown in Figure 4: Curve No. 1 represents the best data as chosen by Kubaschewski. Curve No. 2 utilizes ΔH°_T of Krikorian and ΔS°_T of the U. S. Bureau of Standards. Curve No. 3 and 4 represent Krikorian's values reflecting the spread due to uncertainties in ΔS°_T. Curve No. 5 is the BMI data on UC. Curve No. 6 is the BMI data on UC_2. Curve No. 7 represents the best data that are consistent with recent experiments. Curve No. 8 is Krikorian's values on MoC.

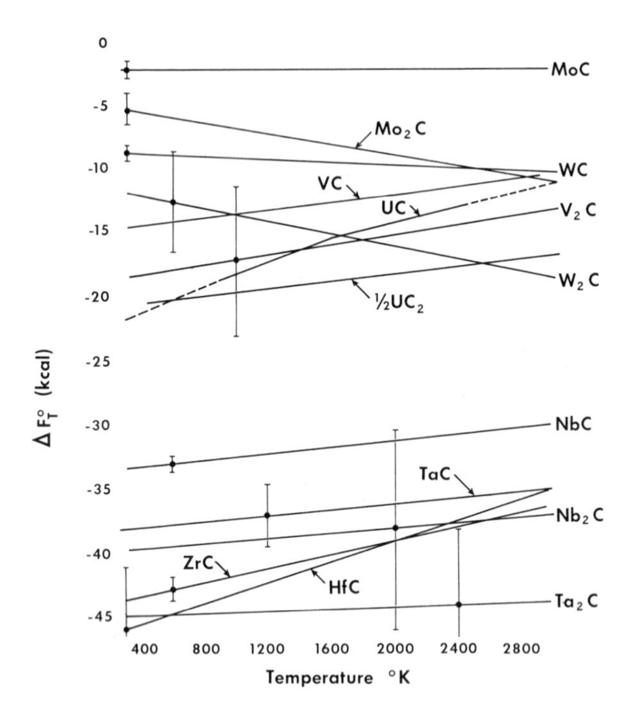

Fig. 3. Standard free energies of formation for carbides.

The shaded area reflects the lower uncertainty range due to uncertainties in ΔH°_T for Mo_2C.

Similar uncertainties exist in the data for W_2C (Fig. 5). The following is the explanation for the curves shown: Curve No. 1 represents Krikorian's data, which is deemed to be the most reliable for W_2C. Curve No. 2 and 3 reflect the uncertainty in the entropy values; the shaded area, represents the uncertainty in the values of ΔH°_T. Curve No. 4 is the BMI data on UC. Curve No. 5 is Kubaschewski's data on WC. Curve No. 6 is Kubaschewski's data representing the lower limit of stability due to the uncertainties in the entropy values.

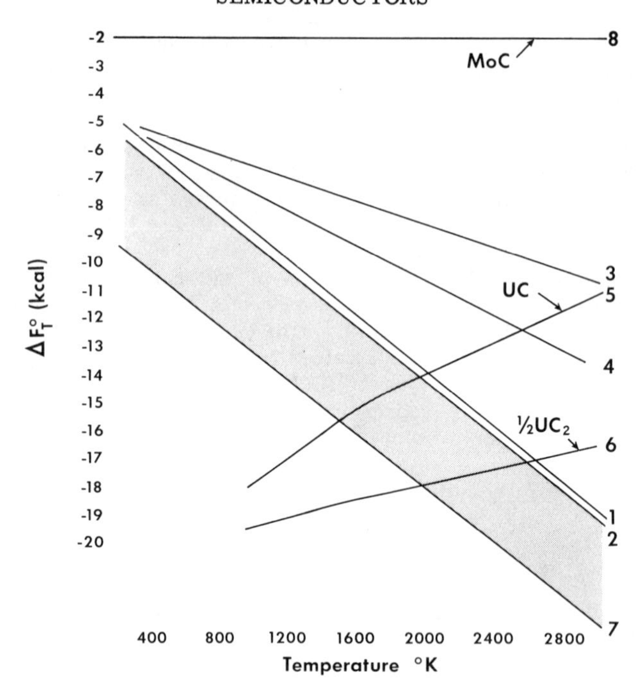

Fig. 4. Uncertainty in standard free energy of formation for Mo_2C.

Figures 4 and 5 are highly significant because they indicate that at low temperatures molybdenum and tungsten will not displace U from UC, but that at high temperatures such displacement will occur with the formation of Mo_2C and W_2C. Table II and Figure 3 indicate that the metals of subgroup 4 and 5 form the most stable carbides. Stability decreases going from left to right in the transition metals. Carbides on the right are metastable. In any particular subgroup, the carbides of elements with the higher atomic weights are the more stable. The lower carbides are generally more stable than the higher carbides of a given metal.

Feasibility of Metal Cladding of Uranium Monocarbide

Table II and Figure 3 indicate no possibility of a listed metal being compatible with UC at high temperatures. Niobium cannot be used since carburization is expected to be rapid; the use of molybdenum and tungsten depends on their rates of carburization. The use of niobium is particularly hopeless because the activation energy for diffusion of carbon in niobium is less than half that of tungsten. The same would hold for tantalum if that metal was considered.

In both cases the di-metal carbide is expected to form at the interface: Mo_2C, W_2C, Nb_2C, and Ta_2C. Some indication of the diffusion rates can be gleaned from the activation energy for carbon diffusion in molybdenum and tungsten which are in the range 1.45 to 1.95 ev and 1.72 to 2.2 ev (32), respectively; the pre-exponential factor is of the order of 10^3 - 10^4. Contrary to previously held notions (21), carburization of tungsten and molybdenum can be quite rapid at 1900°C for tungsten and at about 1600°C for molybdenum. Experimental carburization studies have shown that one may expect about a mil of carbide phase to form in less than 2 hr when in contact with pure

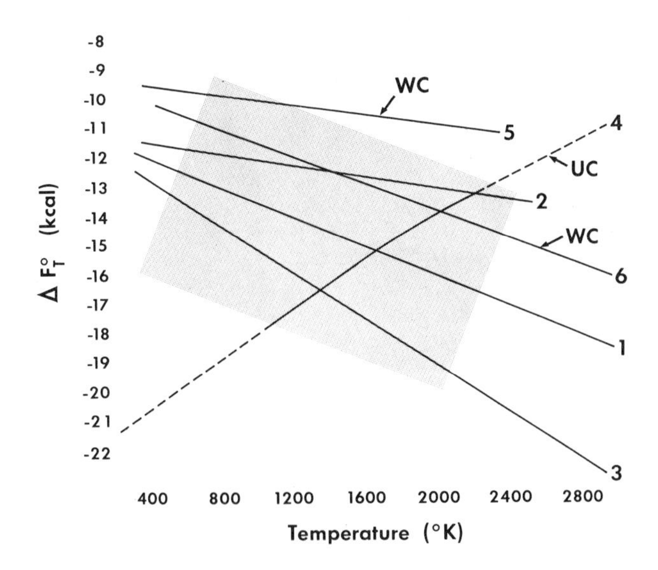

Fig. 5. Uncertainty in standard free energy of formation for W_2C.

carbon at the quoted temperatures. Battelle Memorial Institute investigators observed (27) that 1 mil of Mo_2C was formed on molybdenum when in contact with UC at 1200°C for 24 hr.

The above discussion shows that tungsten is the only real possibility as a clad for pure UC. Molybdenum has a remote chance at much lower temperatures and for shorter lifetimes.

Mo-U-C and W-U-C Phase Diagrams

The problems associated with Mo-U-C and W-U-C are discussed in somewhat greater detail in the following paragraphs. Rough and Chubb have issued a report describing the phase diagrams for the above systems (25) for temperatures of 1000°C. Further reports (27) were made on the compatibility of UC with various metals. The results obtained are confusing because of the failure to note that the free energies of UC cross those of Mo_2 C and W_2C. It is interesting to note that this crossover does not occur for the Nb-U-C nor the Ta-U-C systems.

The BMI ternary phase diagrams for Mo-U-C and W-U-C are of limited use, largely because the free energy crossover was not appreciated. One has to look to other sources for a more detailed picture. A standard reference source is the text of Schwarzkopf and Kieffer (28); other sources are the papers of Nowotny et al. (29, 30) and Westbrook and Stover (31). Studies of the psuedo-binary systems Mo_2C-UC and WC-UC (29) have indicated the the existence of ternary compounds such as WUC_2 lying between the indicated composition. In the case of WC-UC it is possible that more than one ternary phase exists at high temperatures. The close correspondence in behavior of molybdenum and tungsten suggest that the W_2C-UC system and the MoC-UC system also have ternary phases. Figures 6 and 7 represent the tentative composition triangles for the Mo-U-C and the W-U-C system above approximately 2100°K. The important differences between this representation and that of BMI is the existence of MoC, the disappearance of U_2C_3, and the existence of the ternary carbide phases.

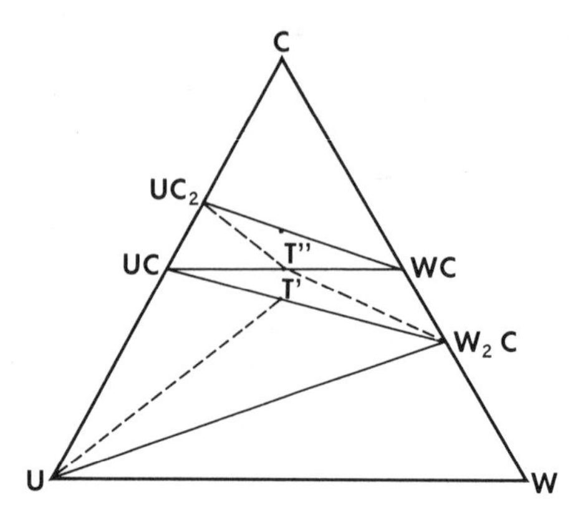

Fig. 6. Tentative W–U–C diagram for T > 2100°K: T' —ternary phase tentative; T'' —ternary phase established.

The existence of these ternary phases has a very important bearing on the lifetime of the clad. It can be expected that terminal solubilities will be low and that the presence of a ternary phase may have important consequence in the rates of diffusion and reaction. It has been observed that the solubility of UC in Mo_2C is less than 5% (29) and the solubility of Mo_2C in UC is somewhat larger at 2000°C.

It is clear that considerable elucidation on solid solubilities, ternary phases, eutectics, and liquidus lines are needed, along with basic diffusion studies, before designs utilizing UC with molybdenum and tungsten can be considered for long-term high-temperature service.

Feasibility of Using Diffusion Barriers between UC and the Metal Cladding

The following systems will be considered:

	Diffusion barrier	Clad
1	W_2C or WC	Nb
2	Mo_2C (MoC stable below 1300°C)	Nb
3	Mo_2C	Mo
4	W_2C	W
5	TiC	W, Mo

The diffusion barrier should not be expected to stop reactions but merely slow them down and therefore increase the lifetime of the clad.

In cases 1 and 2, one may expect solubility of the barrier carbides in UC at the high temperatures, so that the compatibility of the carbide barrier with the fuel is not satisfied. Furthermore, since niobium forms some of the most stable carbides, niobium will displace molybdenum and tungsten from the barrier carbides. Therefore, there is no compatibility between the barrier material and the clad.

Cases 3 and 4 are merely an extension of the cladding problem, i.e., if pure UC contacts molybdenum or tungsten, the dimetal carbide will form. Placing a layer of the reaction-produced carbide between the metal and the fuel may insure a more uniform carburization, which should be less catastrophic. It may also insure a slower rate of attack. Precarburization can increase the reliability of clad integrity and increase the time during which a

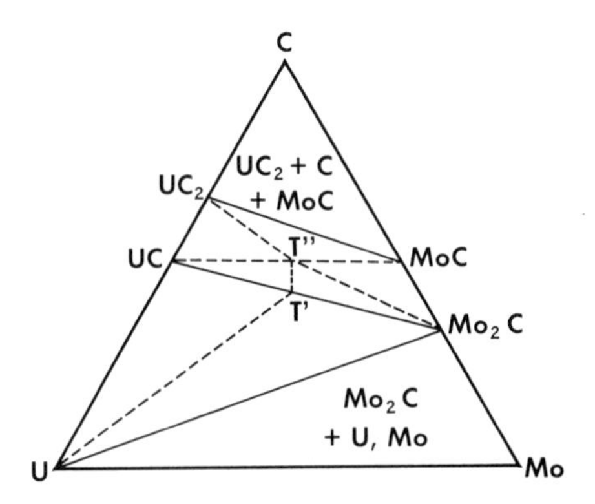

Fig. 7. Tentative **Mo–U–C diagram for T > 2100°K: T' —ternary phase established;** T'' —ternary phase tentative.

given amount of uranium is displaced. Displacement of uranium may be very undesirable, vaporization loss of the fuel may become severe, and a liquid phase could appear. Case 5 is similar to 3 or 4. At 2000°C the solubility of TiC in UC, and UC in TiC, is around 10%. At a high TiC content a UC_2 phase is reported and it is difficult to interpret unless excess carbon has been present (29). The system UC-UC_2-TiC, just as the systems UC-Mo_2C and UC-W_2C, needs considerable clarification, especially with respect to solid solubility limits, eutectics, and liquidus curves. In cases 3, 4, and 5 it appears desirable to have a sufficient amount of barrier materials to cause the occurrence of a two-phase system, regardless of whether or not there was a reaction at the metal interface. This would insure the presence of a diffusion barrier even if the interdiffusion, between the carbides, happens to be very rapid at these high temperatures.

There is another alternative for diffusion barriers. This would be use of metallic barrier materials that would be inert with respect to UC. One could consider ruthenium and osmium which do not form any stable carbides. These materials can be vapor deposited in thin layers on molybdenum or tungsten or on the fuel which is then inserted into the cladding. The interdiffusion of osmium (or ruthenium) into molybdenum (or tungsten) may produce surfaces which will not be attacked by the carbide fuel. In effect one would be producing a stabilized alloy. This approach may not be too expensive as the fabrication costs and the small amounts of these expensive metals would be low. Hansen (33) has presented the limited information on the molybdenum-osmium and other analogous systems, however, detailed information is not available for these alloy systems.

The use of osmium and ruthenium appears to be more interesting for molybdenum than for tungsten because of the faster rates of reaction in the UC-Mo system. For low temperature application molydenum is quite suitable; the ease of fabrication of molybdenum as compared with tungsten is a very important consideration for the choice of molybdenum.

Metal Cladding of a UC Alloy (UC-ZrC Type)

The reactivity of UC with metals and its comparatively high decomposition pressure (Fig. 2), suggest that one way of decreasing or halting the interface reaction is to use alloys of UC where the "activity of UC" would be lowered. The most popular alloy system is (U,Zr)C. Molybdenum and tungsten are the only metals that will probably be compatible with (U,Zr)C alloys with reasonably high UC content.

If one assumes that UC forms an ideal solid solution with ZrC, then the compatibility of the solid solution with molybdenum and tungsten may be determined from the thermodynamic data summarized in Table II. Whether the solid solution carbide will be reduced by the metal is a function of temperature and of the composition of the carbide. The UC-ZrC system is treated as a pseudobinary ideal solution with ZrC considered as an inert dilutent. In this way, Table III has been constructed; it lists the minimum temperature

TABLE III
Minimum Reaction Temperatures between (U, Zr) C and Tungsten or Molybdenum

UC, mole-%	W, temp. °K	Mo, temp. °K
1.0	1875 ± 450	2050 ± 500
0.9	2025 ± 500	2100 ± 500
0.7	2075 ± 525	2150 ± 525
0.5	2275 ± 550	2325 ± 525
0.3	2650 ± 550	2700 ± 550
0.1	>3000	>3000

at which a reaction may occur between the solid solution and the metals tungsten and molybdenum. The large uncertainity in minimum reaction temperature reflects the uncertainties in the thermodynamic data on W_2C and Mo_2C (Figs. 4 and 5).

Feasibility of Using a Diffusion Barrier between UC-ZrC and the Metal Cladding

The general trends discussed for pure UC apply in this case. The fact that TiC is miscible with ZrC indicates that TiC will not be an effective barrier.

The limited solubilities of Mo_2C in ZrC and UC (and vice versa) indicates that schemes utilizing Mo_2C and W_2C barriers may prove to be more effective than that using TiC.

As in the case of pure UC, osmium and ruthenium surface barriers may prove to be effective; the problems will certainly be less severe for UC-ZrC.

EXPERIMENTAL

The bulk compatibility of UC and (U,Zr)C with various refractory materials is being determined. Compatibility couples are formed by placing together lapped and polished, optically flat materials in a vacuum, tantalum element, tube furnace. Reaction products are studied by x-ray and metallographic techniques.

High density specimens of carbides are fabricated by vacuum hot pressing (34) in a graphite die at temperatures in excess of 2000°C, and by arc melting. For enriched fuel bodies, reactive hot pressing (36) is used to form dense (U,Zr)C from enriched UO_2, graphite powder, and ZrC.

The solid solutions $U_{0.3}Zr_{0.7}C$ and $U_{0.7}Zr_{0.3}C$ were observed to have reacted catastrophically with niobium metal after 1 hr at 1800°C in vacuum. The products were Nb_2C and metal-rich (U,Zr)C. The reaction had not gone to completion. During the same test, $U_{0.7}Zr_{0.3}C$ was observed to have reacted with tungsten; however, the reaction was not as severe as that with niobium. The chemical similarity between niobium and tantalum can be used to infer that neither metal can be used in contact with (U,Zr)C (with greater than about 30% UC) at elevated temperatures. This result is consistent with the thermodynamic data of Figure 3.

No major reaction occurred between $U_{0.3}Zr_{0.7}C$ and tungsten when heated to 2000°C for 100 hr. However, a slight surface discoloration, occurred at temperatures as low as 1800°C for 1 hr. The reaction products in the thin interfacial film of the 2000°C run were examined by x-ray diffraction. The presence of W_2C was determined. There were unidentified x-ray diffraction peaks. The solid solution was deficient in UC at the interface.

Arc-melted UC (stoichiometric) was heated for 5 hr at 1770°C against tungsten. A definite reaction occurred, indicated by the adhesion of the UC to the tungsten. There was a minor reaction between UC-UC_2 (about 80% UC_2) and tungsten as low as 1400°C after 1 hr of heating. These reactions qualitatively confirm the crossover position on Figure 3 between $\Delta F°$ for W_2C and for UC.

Arc-melted UC-UC_2 was heated in contact with tungsten. During heating, the pressure in the furnace was observed to rise sharply at 1880°C. After 1 hr at this temperature, the carbide had several pock marks and the tungsten had several corresponding protrusions, all a few mil in diameter. Heating 100 hr at 1880°C increased the number of pockmarks and protrusions. X-ray diffraction of the tungsten surface, after removal of the UC_2, showed W_2C present. These data qualitatively confirm the crossover position of $\Delta F°$ for W_2C and 1/2 UC_2 in Figure 3.

Preliminary studies indicate that a thin coating (5 mils) of W_2C uniformly carburized onto tungsten greatly decreases the rate of reaction between UC and tungsten at 2000°C. Further experimental work with this system and with UC-Mo_2C-Mo are in progress now.

CONCLUSIONS

Consideration of the decomposition pressure of fuel materials has led to the choice of cladding the fuel of a nuclear thermionic converter with a metallic emitter. Thermodynamic data and preliminary compatibility experiments indicate that likely fuel-cladding combinations are: UO_2-W; UO_2-Mo; UC-W; (U,Zr)C-W; (U,Zr)C-Mo; and UC_{1+x}-W. All of the uranium-rich carbide fuels will react with their claddings at operating temperatures (about 2100°K); in their case, carbide diffusion barriers are effective in reducing the rate of reaction.

References

1. Ackerman, R. J., P. W. Gilles, and R. J. Thorn, J. Chem. Phys, 25, 1089 (1956).

2. DeMaria, G., J. Drowart, and M. G. Inghram, J. Chem. Phys., 30, 318 (1959).

3. Chupka, W. A., J. Berkowitz, and C. F. Giese, J. Chem. Phys., 30, 827 (1959).

4. DeMaria, G., R. P. Burns, J. Drowart, and M. G. Inghram, J. Chem. Phys., 32, 1373 (1960).

5. Drowart, J., G. DeMaria, R. P. Burns, and M. G. Inghram, J. Chem. Phys., 32, 1366 (1960).

6. Inghram, M. G., W. A. Chupka, and J. Berkowitz, J. Chem. Phys., 27, 569 (1957).

7. King, E. G., W. W. Weller, and A. V. Christensen, "Thermodynamics of Some Oxides of Molybedenum and Tungsten," BM-RI-5664, 1960.

8. Ackermann, R. J., and R. J. Thorn, 2nd U. N. Conf. Peaceful Uses of Atomic Energy, Geneva, 28, 180, 15/P/715 (1958).

9. Degazarian, T. E., N. J. Dumont, L. A. duPlessis, W. E. Hatton, S. Levine, R. A. McDonald, F. L. Oetting, H. Prophet, G. C. Sinke, D. R. Stull, C. J. Thompson, Janaf International Thermochemical Tables, Vol. 2, Interim edition, prepared at Dow Chemical Corp., Midland, Mich. (1960).

10. Coughlin, J. P., U. S. Bur. Mines Bull., 542 (1954).

11. "Selected Values for the Thermodynamic Properties of Metals and Alloys," Mineral Res. Lab., Univ. of California, to be published.

12. Kingery, W. D., Symposium on High Temperature Technology, Asilomar, Calif., 1959.

13. Hall, R. W., and P. F. Sikora, "Tensile Properties of Molybdenum and Tungsten from 2700° to 3700°F," NASA Memo 3-9-59E, 1959.

14. LASL Plasma Thermocouple Development Program quarterly status report for period ending March 20, 1961 LAMS-2544.

15. Quarterly Progress Report for period ending Sept., 1960, "40-MW(E) Prototype High-Temperature Gas-Cooled Reactor Research and Development Program," GA-1774, 42, 1960.

16. Rough, F. A., and W. Chubb, eds., "An Evaluation of Data on Nuclear Carbides," BMI-1441, 62, 1960.

17. Thorn, R. J., and G. H. Winslow, J. Chem. Phys., 26, 1P6 (1957).

18. Ackermann, R. J., and R. J. Thorn, Proc. Second Intern. Conf. Peaceful Uses of Atomic Energy, Geneva, 28, 192, 15/P/715 (1958).

19. Wilson, W. B., J. Am. Ceram. Soc., 43, 77 (1960).

20. DeMaria, G., R. P. Burns, J. Drowart, and M. G. Inghram, J. Chem. Phys., 32, 1373 (1960).

21. Brewer, L., in Chemistry and Metallurgy of Miscellaneous Materials: Thermodynamics, McGraw-Hill, New York, 1950.

22. Krikorian, O. H., "High Temperature Studies," UCRL-2888, 1955.

23. Latimer, W. M., Oxidation Potentials, Prentice-Hall, Englewood Cliffs, N. J., 1952.

24. Kubaschewski, O., and E. L. Evans, Metallurgical Thermochemistry, Wiley, New York, 1956.

25. Rough and Chubb, "An Evaluation of Data on Nuclear Carbides," BMI-1441, 1960.

26. Rossini, F. P., P. P. Wagman, W. H. Evans, S. Levine, and I. Jaffee, U.S. Bureau of Standards, "Selected Values of Chemical Thermodynamic Properties," Circular 500 (1961).

27. Rough, F. A., and W. Chubb, eds., "Progress on the Development of UC-Type Fuels," BMI-1488, 1960.

28. Schwarzkopf, P., and R. Kieffer, Refractory Hard Metals, Macmillan, New York, 1953.

29. Nowotny, H., R. Kieffer, and F. Benesovsky, Rev. Met., 55, 453 (1958).

30. Nowotny, H., R. Kieffer, F. Benesovsky, and E. Laube, Acta Chim. Acad. Sci Hung., 18, 35 (1959).

31. Westbrook, and E. R. Stover, "Carbides for High Temperature Applications," G.E.R.L. No. 60-RL2565M, Nov., 1960.

32. Samsonov, G. V., Acad. Sci. Ukr. SSR, Kiev (1960).

33. Hansen, M., Constitution of Binary Alloys, McGraw-Hill, New York, 1958.

34. Hoyt, E. W., "Induction Heated Vacuum Hot Press," GEAP-3331, 1960.

DISCUSSION

L. YANG (General Atomic): Other papers in this volume have compared the merit of two types of emitter materials, namely, the refractory metal emitter versus the carbide emitter. In making such comparison, perhaps we should not neglect the following two important aspects.

First, the special field of application to which each type is better suited. It seems that for a fission heat source where a high uranium concentration is needed for a very compact reactor core, a carbide emitter may be more useful. On the other hand, for solar or chemical heat sources, refractory metal emitters may be more useful. Although the rate of vaporization of carbide emitters is much higher then that of the refractory metal emitter at their respective temperatures of operation, we believe that the vaporization characteristics of carbide emitters have not been studied thoroughly and that the possibility of finding methods to suppress the vaporization loss of carbides cannot be totally ruled out.

Second, for fission heat conversion using refractory metal emitters, the effect of the interaction between these metals and the uranium-containing fuels (carbides or oxides) on the life and the performance of these emitters has to be taken into consideration. Quantitative kinetic data on such matters are almost totally lacking in the temperature range where these emitters are operated.

Our present understanding of the emitter material problem is still far too limited. Experimental data are urgently needed for the judicious selection of emitter materials.

I would like to ask more details about the reaction between niobium, tantalum, and uranium dioxide.

L. N. GROSSMAN: I do not have quantitative data. I believe it is a liquid phase formation. The way that we have observed it, it is a melt down—a ceramic-metal type of reaction. A similar one is alumina-tantalum.

L. YANG: Your observation does not seem to agree with results reported by Gangler of NASA Lewis Research Center (NASA TND-262) and by Byerley at Chalk River (AECL-1126) on the compatibility between niobium, tantalum, and uranium dioxide.

L. N. GROSSMAN: Was that for long-time tests?

L. YANG: These are short-time studies (10 min to 1 hr). How long was the time of your experiments and under what conditions were these experiments carried out?

L. N. GROSSMAN: I do not know the details. I did not do this work. For a nuclear reactor, as far as I know, a 10% UC, 90% ZrC solid solution is not desirable. For instance, you have to get up to around 70 to 75% UC to have a uranium density as high as UO_2; and when you get that high you might as well use UC, and if you get below that you might as well use UO_2. From the data which I gave on higher UC content solid solutions, the vapor pressure would have to be reduced by many orders of magnitude to give a year's lifetime.

L. YANG: The vapor pressure of UO_2 is high and it has low electrical and thermal conductivities. It is useful only as the nuclear fuel to heat the refractory metal emitters. If we want materials to act as both the nuclear fuel and the emitter, uranium-containing carbides are by far still the best choice. Although the rates of vaporization of UC-ZrC solid solutions containing high UC concentrations are too high to give a year's life, it is possible that on heating at high temperatures the uranium concentration at the surface may be depleted because of the slow replenishment by diffusion from the bulk, as demonstrated in the case of $(UC)_{90}(ZrC)_{10}$. Thus the surface of the emitter may have a uranium concentration low enough to bring the vaporization loss down to a tolerable level and yet high enough to provide the electron emission needed; while the bulk of the emitter may still possess the high uranium concentration pertinent to a compact reactor core. The selection of optimum concentrations and operating conditions needs more experimental information. We are currently studying how the rate of vaporization and the surface uranium concentration vary with the time and the temperature of heating and the initial UC to ZrC ratio.

L. N. GROSSMAN: Would you reiterate the temperature employed?

L. YANG: The study of the vaporization characteristics of the $(UC)_{90}(ZrC)_{10}$ sample was carried out at 2030°C.

L. N. GROSSMAN: The center of the fuel will be quite a bit hotter than the surface. I do not know what final detailed effect this will have on diffusion, but there will be a temperature gradient throughout the sample.

R. A. CHAPMAN (Texas Instruments,): With its low thermal conductivity, UO_2 would possibly be a good fuel for a thermoelectric direct conversion element, where the surface temperature has to be 500 to 600°C. But with a thermionic element you would not be able to obtain the high surface temperature needed and at the same time a sufficient thermal heat flux at the surface for a reasonable diameter UO_2 pellet. Some "back-of-the-envelope" calculations will illustrate my point. For a 3/8-in. diam UO_2 pellet with a 50 w/cm^2 surface heat flux, the maximum surface temperature would be limited near 1500°C to keep the center line below the UO_2 melting temperature.

L. N. GROSSMAN: We thought of this, and the question that comes to my mind is, what is the matter with a molten center? Even if the center were molten, the vapor pressure inside the fuel and the structural shape of the fuel would be determined by the colder outside surfaces.

R. A. CHAPMAN: Hollow pellets could be used, but still the annular thickness of the pellet could result in a considerable temperature drop.

L. N. GROSSMAN: This exists right now in the present power reactors. The center of the UO_2 fuel is at such a high temperature that the vapor transport is extremely rapid. The UO_2 will be very near its melting point, if not over.

R. A. CHAPMAN: My point is that UO_2 is inferior to (U,Zr)C which would require a much smaller temperature drop to get even 100 w/cm^2 heat flux on the surface for the same annular thickness.

G. A. KEMENY (Westinghouse Research Labs.): One difficulty about themionic devices in general is that they operate at very high heat fluxes, and the introduction of another surface to which heat must be transferred invariably results in new problems. For example, a reasonable range for heat fluxes is 20 or 30 to maybe 100 w/cm^2. The question is, by what method is intimate contact maintained? If this contact is not maintained between the fuel and the cladding, the heat transfer from the fuel to the cladding would be by radiation, which would invariably cause overheating of the fuel, and therefore much faster deterioration and evolution of gases. Therefore, what method is used to assure that we will have conductive heat transfer between the fuel and the cladding?

L. N. GROSSMAN: I would think that the high-pressure, high-temperature fuel, which means high vapor pressure, and plating out on cold surfaces would tend to keep intimate contact between the fuel and the cladding material.

J. J. CONNELLY (Office of Naval Research): From our studies it appears that we can transfer 95% of the heat by radiation rather than by conduction at the temperature we are after. It might be easier to loosely compact your nuclear fuels inside and minimize some of the thermal expansion and other effects that enter during the start-up and shut-down periods of reactor systems. This is being investigated, and I think we can maintain the temperatures, even using radiation as a method of heat transfer inside the element.

F. WILLS (American Metal Climax, Inc.): I have seen only one reference to rhenium in your figures. Is there any reason why this material has no value in this application? Is the work function low, or is it the availability?

L. N. GROSSMAN: Commander Connelly mentioned the high nuclear cross section for rhenium. That would veto it pretty quickly in a reactor. I do not have its vapor pressure, but I think it is high also.

ANON.: Correction, the vapor pressure of rhenium is quite low.

J. J. CONNELLY: Rhenium does look very attractive for thermionic applications: mostly in space applications in solar heating. One does not have to go to very thin deposited surfaces with rhenium in order to keep the reactor considerations down to a reasonable size. It does have a high nuclear cross section.

R. CHANG (Atomics International): Concerning the radiation damage problem, my experience with pure UC at such temperatures is that it is very soft. You may have fuel swelling problems with your clad which will require strengthening the fuel by alloying with zirconium carbide.

L. N. GROSSMAN: That is another good reason for a clad, of course. I think you brought up the problem of fission-gas release; that is, all of the fission products would be released by the fuel, and then you have the problem of pressure building up inside the clad. For tungsten and molybdenum we do not expect the fission products to diffuse through. They require very high concentration gradients, e.g., the concentration gradients required to get these fission gases through a 10- to 20-mil clad demand gas solubilities on the order of 1 mole-% in the metal which I do not think you can get even at very high temperatures (at least while the clad is still structurally stable). I do not think you could get such solubilities in tungsten.

What we anticipate here is that we will use a fairly high vapor pressure fuel, but low enough so that we can leave a hole in the clad and not lose appreciable amounts of fuel while bleeding off fission products and gases.

Subject Index